P9-DUB-672

The Colorado Plateau II

BIOPHYSICAL, SOCIOECONOMIC, AND CULTURAL RESEARCH

The Colorado Plateau II

BIOPHYSICAL, SOCIOECONOMIC, AND CULTURAL RESEARCH

Edited by Charles van Riper III and David J. Mattson

The University of Arizona Press

Tucson

This volume is based on research presented at the Seventh Biennial Conference of Research on the Colorado Plateau. This conference was held at Northern Arizona University, Flagstaff, Arizona, and organized by the U.S. Geological Survey Southwest Biological Science Center, Colorado Plateau Research Station and the Center for Sustainable Environments at Northern Arizona University. The primary sponsors of this conference were the U.S. Geological Survey, U.S. National Park Service, U.S. Bureau of Land Management, and Northern Arizona University.

The University of Arizona Press

This book is printed on acid-free, archival-quality paper.
Manufactured in the United States of America

10 09 08 07 06 05 6 5 4 3 2 1

Library of Congress Cataloging-in-Publication Data
Conference of Research on the Colorado Plateau
(2001–) (7th : 2003 : Northern Arizona University)
The Colorado Plateau II : biophysical, socioeconomic, and cultural research / edited by Charles van Riper III and David J. Mattson.
p. cm.
Includes bibliographical references and index.
ISBN-13: 978-0-8165-2526-3 (hardcover : alk. paper)
ISBN-10: 0-8165-2526-9 (hardcover : alk. paper)
1. Natural history—Research—Colorado Plateau—Congresses. 2. Natural resources—Research—Colorado Plateau—Congresses. 3. Natural resources—Colorado Plateau—Management—Congresses. 4. Colorado Plateau—Research—Congresses. I. Van Riper, Charles. II. Mattson, David J. (David John) III. Title.
QH104.5.C58C66 2003
333.73′16′097913–dc22

2005013439

This book was published from camera-ready copy that was edited and typeset by the volume editors.

CONTENTS

FOREWORD

This is the seventh proceedings publication produced from papers presented at the series of biennial conferences over the past 14 years, highlighting research on the Colorado Plateau. I have had the pleasure of providing a preface for five of these proceedings. The 26 papers in this seventh volume are contributions from federal, state, and private sector researchers, who come together at Northern Arizona University every other year to share scientific information with land managers on the Colorado Plateau. This Biennial Conference series focuses on providing information to USGS partners, particularly land managers on the Colorado Plateau. The chapters of this book contribute to the ever-growing pool of scientific data that provides baseline information pertaining to physical, cultural, and biological resources of the Colorado Plateau. Support for many of these studies has come from a spectrum of federal, state, and private partners concerned about the well-being of the plateau's resources. I applaud the effort of the contributors. With modest funding and a broad base of public and institutional support, these authors have pursued important lines of work in the four states (Arizona, Utah, New Mexico, and Colorado) that comprise the Colorado Plateau biogeographic region.

The Colorado Plateau remains one of the richest eco-regions in North America in terms of its high rates of plant endemism and species richness of invertebrates and vertebrates. It is also a region that has experienced unprecedented levels of human use, in terms of the diversion of its water resources and dramatic increases in eco-tourist activities. Combine this natural richness with new stresses on plant and animal communities, and you have a justified need for more investment in science-based management of species and their habitats. The Colorado Plateau Biennial Conference has become the premiere forum where scientists and land managers come together for public discussion about current research and resource management issues on the Colorado Plateau.

As a people, we face the prospect of extensive local land and global environmental changes that continue to perturb the physical, cultural, and biological resources on lands of the Colorado Plateau. As the research branch for the Department of the Interior, we in the USGS are committed to identify, in a sound scientific manner, information that can be used by land managers to protect our resources from detrimental change due to modern human influences. We must develop the information necessary to alert our managers, elected officials, and the public to the importance of their natural surroundings as elements of the basic resources that sustain us, inspire us, and represent our natural biological and environmental heritage. There remains much to be done and our task has just begun.

DENNIS B. FENN

U.S. Geological Survey
Biological Resources Discipline
Center Director
Southwest Biological Science Center

DEDICATION

It is with great pleasure that we dedicate this book to Drs. "Chip" Groat and "Doug" Buffington. Both of these men have provided outstanding leadership for the USGS, and their individual efforts have been an integral part of the growth and success of the USGS Southwest Biological Science Center's Colorado Plateau and Sonoran Desert Research Stations. Chip has provided exemplary foresight in setting a long-range course for the U.S. Geological Survey. His efforts at integrating the four USGS disciplines have helped those of us in Biology to better integrate into the agency. Moreover, this effort will provide untold benefits for interdisciplinary research within the USGS for decades to come. Doug has provided leadership for the Western Region and was also a keystone person during the transition of the National Biological Survey into the U.S. Geological Survey. Over the years Doug has continued to provide the opportunity for us to work within the Southwest Strategy framework. In addition, he has personally had the foresight to provide financial and personal support for our efforts on the lower Colorado River. Without the leadership and guidance both of these men have provided, the Southwest Biological Science Center would not have been a reality.

Charles G. "Chip" Groat, born in Westfield, New York, completed his undergraduate degree in 1962 at the University of Rochester (BS in Geology), then earned a Master of Science in Geology (1967) from the University of Massachusetts and a Ph.D. in Geology

Charles G. "Chip" Groat

Doug Buffington

(1970) from the University of Texas at Austin. He is a distinguished professional in the science community with more than 25 years of direct involvement in geological studies, energy and minerals resource assessment, ground-water occurrence and protection, geomorphic processes and landform evolution in desert areas, and coastal studies. From 1978 to 1990, Chip held positions at Louisiana State University and the Louisiana Department of Natural Resources as director and state geologist. He also served as associate professor (1976–78) at the University of Texas at Austin, in the Department of Geological Sciences, and as associate director and acting director of the Bureau of Economic Geology. Prior to joining the University of Texas, Chip served as executive director (1992–1995) at the Center for Coastal, Energy, and Environmental Resources at Louisiana State University. From 1990 to 1992 he was executive director for the American Geological Institute. Then from 1983 to 1988, Chip served as assistant to the secretary of the Louisiana Department of Natural Resources, where he administered the Coastal Zone Management Program and the Coastal Protection Program. From May to November of 1998 he served as associate vice president for Research and Sponsored Projects at the University of Texas at El Paso, following three years as director of the Center for Environmental Resource Management. On 13 November 1998, Chip became the 13th director of the U.S. Geological Survey, U.S. Department of the Interior, and continues in that leadership role today.

Among his many professional affiliations, Chip is a member of the Geological Society of America, the American Association for the Advancement of Science, the American Geophysical Union, and the American Association of Petroleum Geologists. He has also served on more than a dozen earth science boards and committees and has authored and contributed to numerous publications and articles on major issues involving earth resources and the environment.

Born and raised in Jersey City, Doug Buffington spent most of his free childhood time roaming the rural areas of New Jersey. This set the stage for his lifelong interest in science. Doug received his BS in Biology from St. Peter's College in 1963, then his Master of Science (1965) and Ph.D. (1967) from the University of Illinois in Animal Ecology. He took a break from science during 1967–1969, serving in the U.S. Army as 1st Lt. and Captain at Fort Detrick, Maryland. In 1969 he returned to science and until 1972 was assistant professor of Biological Sciences at Illinois State University. Doug then joined the Argonne National Laboratory (1972–1977) and served in several capacities, including assistant director for the Environmental Impact Studies Division, thus setting the stage for his role in science administration. Doug served as staff and later as Senior Staff Member for Environmental Data and Monitoring on the President's Council on Environmental Quality (CEQ) 1977–1980. He joined the U.S. Fish and Wildlife Service (FWS) as chief of the Office of Biological Services in 1980. From 1982 until 1989 Doug was deputy regional director for research and development, and from 1989 until 1993 was the Fish and Wildlife Service's regional director for research and development. With the creation of the National Biological Survey (NBS) in 1993, he served as acting regional director until he became the director of the Alaska Science Center in 1995, where he remained for two years. In 1997 when the NBS was merged with the USGS, Doug was selected as the regional chief biologist at the U.S. Geological Survey in Seattle. In January of 2000 he became the regional director of the USGS, Western Region. He has had several national offices in the Ecological Society of America, received the U.S. Department of Interior's Meritorious Service Award, and served as a principal negotiator for the International Convention on Biological Diversity.

It is because of their commitment to quality science and their continued support of the USGS Southwest Biological Science Center that we dedicate the Proceedings of the Seventh Biennial Conference of Research on the Colorado Plateau to Dr. John "Doug" Buffington and Dr. Charles "Chip" Groat.

Introduction

INTRODUCTION: BIOPHYSICAL, SOCIOECONOMIC, AND CULTURAL RESEARCH ON THE COLORADO PLATEAU

Charles van Riper III and David J. Mattson

This is the seventh volume, and the second published by the University of Arizona Press, in a series of books that focus on research and resource management issues on the Colorado Plateau. These books highlight the integration of research into resource management efforts, as related to cultural, natural, and physical resources within the biogeographic province of the Colorado Plateau. This volume integrates aspects of biophysical and socioeconomic research through a mix of chapters that address management issues on the Colorado Plateau.

The scientific works published in this Biennial Conference Series contribute significantly to presenting peer-reviewed results of collaborative efforts between scientists and land managers. Many of the protocols and management techniques currently in use in land management units over the Colorado Plateau are a result of these collaborative works, including those of the U.S. Geological Survey Southwest Biological Science Center staff with university and other partner agency scientists, and with land managers from a variety of state and federal agencies. It has been clearly demonstrated that, because of similarities across the Colorado Plateau, techniques that work in one management unit are applicable to many other areas. This is due primarily to the similarity of habitat and climatological conditions over the Colorado Plateau.

The 29 chapters in this book were selected from the 116 research papers presented at the Seventh Biennial Conference of Research on the Colorado Plateau (5–8 November 2003 in Flagstaff, Arizona). The conference was hosted by the U.S. Geological Survey Southwest Biological Science Center's Colorado Plateau Research Station (CPRS) and the Center for Sustainable Environments at Northern Arizona University. The meeting theme revolved around research, inventory, and monitoring of lands over the Colorado Plateau, with a focus on aspects of biophysical and socioeconomic research.

Every paper that was selected for publication represents original research that has been peer reviewed by at least two scientists from that research discipline. Material presented in this book is information that has not been previously published in any other location. The 29 contributed studies each constitute a separate chapter. The book is divided into four sections: Socioeconomic, Biological, Cultural, and Biophysical.

The book begins with socioeconomic considerations. The opening chapter by Hecox and Holmes, from Colorado College, sets the stage for a theme that is carried throughout many of the chapters in this book, that of tying social and economic criteria into the management of resources over the Colorado Plateau. The chapter covers aspects of shifting socioeconomic patterns and resulting disparities over the Colorado Plateau. The authors track changes over the past three decades, examining data on population trends and types of employment from 31 counties in four states. One benchmark that they think has created the present disparity over the Colorado Plateau has been the shift in employment type. Using location quotient and mix-share analyses, they demonstrate that employment over the Colorado Plateau has shifted into resource-based jobs, manufacturing and the service industry. Concomitantly, there has been an employment shift away from the former heavy dependence on natural resource extraction. They conclude the chapter by recommending that future economic development on the plateau should focus on attracting higher-end producer services along with the self-employed and small-scale businesses. They believe that the spectacular natural and cultural resources of the Colorado Plateau will continue to draw people to the region, thus ultimately influencing all facets of the economy and political datums from which the present modern biogeographical pattern is based.

Chapter 2 presents survey results on perceptions of socioeconomic impacts of forest restoration policy over the Colorado Plateau. Ostergren and Ruther surveyed a random sample of residents in the northern Arizona region regarding their acceptance of various physical forest conditions. The authors also determined whether the surveyed residents were in agreement with different forest restoration and decision-making strategies. Basic demographic data gathered from respondents allowed judgments regarding sample bias. Although the authors found that respondents supported some restriction on appeals of administrative decisions, support was even stronger for continued and increased public involvement in official decision making. Ostergren and Ruther's other key findings include a preference for information from the U.S. Forest Service and newspapers compared to other sources, and support for forest restoration goals other than commercial timber production, but with an acceptance of permanent dirt roads and logging operations. They close the chapter by urging ongoing appraisal of public perspectives with an emphasis on reaching poorer and minority residents who were underrepresented in their initial survey.

Chapter 3 by Ruther and Ostergren provides results of another survey of northern Arizona residents, this one focused on perceptions and knowledge of mountain lions, which are currently a focus of controversy throughout the western United States. The authors found that almost all respondents thought it was important for mountain lions to be in Arizona, and that mountain lions played an important ecological role. However, older rural residents were significantly less accepting of mountain lions than were younger urban residents. Ruther and Ostergren point out that the majority of respondents preferred non-lethal devices for personal protection from lions and that anxiety about encountering mountain lions was actually greater for those who had never seen one, compared to those who had. The majority of respondents also preferred non-lethal resolution of conflicts between humans and mountain lions and limiting human development in prime lion habitat, which suggests considerable public support for conservation at a time of increasing human impacts.

The socioeconomic section of this book concludes with Chapter 4 by Hampton et al., who provide a demonstration and test of a spatial decision support system (SDSS) for forest restoration planning. The stated goal of this research was to test the system's efficacy and to elicit feedback for further development. The six participants in this test represented a spectrum of stakeholders regarding restoration of ponderosa pine forests throughout northern Arizona, including representatives from academe, the U.S. Forest Service, a community group, and environmental organizations. Participants developed various restoration scenarios that detailed evaluation criteria, spatial constraints, and priorities for implementing restoration treatments. Each scenario was appraised for overall impact on collectively accepted evaluation criteria and to determine spatial overlaps with other scenarios in preferred locations of treatments. The SDSS not only allowed participants to judge the potential effects of their preferences, played out as spatially explicit scenarios, but also allowed them to identify potential common ground in terms of both spatial overlaps of restoration treatments and shared acceptance of criteria for judging "success." Participants of this initial effort urged the authors to actively pursue engagement with decision makers to ensure that the newly developed SDSS was used to maximum positive effect.

The next chapter serves as a transition into the biological section of this book. Prather et al. incorporate protocols from the SDSS and Forest Ecosystem Restoration Analysis (ForestERA) to create Geographic Information System (GIS) models of the potential affects of forest restoration treatments on sensitive wildlife taxa. The information in Chapter 5 was collected throughout the southern transition area of the Colorado Plateau, along the Mogollon Rim from the Kaibab Plateau to the White Mountains in eastern Arizona. The authors use seven foundation GIS layers, ranging from dominant overstory vegetation, through slope and aspect, to tree density and basal area. From a survey of 40 academic institutions, they provide a list of sensitive taxa that can be incorporated into their model. The authors provide an in-depth example of how their habitat model might be used, taking the tassel-eared squirrel as an example. They demonstrate the usefulness of their habitat models as additional tools for managers throughout the Colorado Plateau.

Falzarano et al. in Chapter 6 continue the theme of habitat modeling, using decision tree methods on GAP program data to better classify and analyze land cover data from the northeast corner of Arizona. Their approach first develops Geographic Information Systems (GIS) data themes for land cover, vertebrate species distributions, and land stewardship. These GIS layers are seamless across all land management boundaries. The authors then utilize a decision tree to better evaluate habitats of concern and provide conservation of vegetation and vertebrate species. They believe that these newly created GIS products will allow land managers, policy makers, and planners to make better-informed land-use decisions, especially throughout northeastern Arizona.

Vegetation studies are introduced into the biological section of the book with Chapter 7, where Nabhan et al. deal with land-use effects on understory plant community composition in southwestern pine forests. Through the comparison of three disparate areas with long cultural histories—the Jemez Mountains (including Bandelier National Monument in New Mexico), Mesa Verde, including Mesa Verde National Park in Colorado, and Wupatki and Sunset Crater National Monuments in Arizona—the authors explore the effects of prehistoric human land-use practices and physical location on forest restoration plans. Using both the literature and personal interviews, the authors develop a hypothesis that human land use practices greatly influence plant species composition and that the degree of land-use differs considerably over the Colorado Plateau. They found that post-1850 fire histories among the three areas were much more varied than assumed due to widely varying cultural land-use factors. They suggest that restoration plans in pine forests throughout the Southwest should be guided by area-specific factors that are developed from cultural, biological, and physical histories of the landscape.

The next four chapters describe studies that examine vegetation changes in Grand Canyon National Park. Chapter 8 is the first of three chapters presenting the results of an analysis capitalizing on a recently discovered dataset from 1935 that documented forest conditions in Grand Canyon National Park. The authors compared forest conditions during 1935 with contemporary forest conditions documented by three recent studies within higher-elevation mixed conifer forests. They found that the total stand basal area has declined since 1935, primarily owing to declines in densities of small and mid-diameter conifers, primarily white fir, Douglas-fir, and spruce. The authors speculate that these declines were caused by self-thinning and outbreaks of western spruce budworm, and followed substantial increases in densities of small trees between the late 1800s and 1935 that had occurred in response to reduced fire frequencies. The authors close this chapter by recommending objectives and strategies for restoring mixed conifer forests within Grand Canyon National Park to pre-European settlement conditions.

Chapter 9 by Crocker-Bedford et al. extends the analysis introduced in Chapter 8 to the extensive zone within Grand Canyon National Park that transitions between mixed conifer and near-pure ponderosa pine forests. The authors begin this chapter by defining transition forests and observing that this vegetation type is not well described, primarily because of its relative scarceness in other parts of the Southwest. Unlike in the mixed conifer forests that are described in Chapter 8, densities of small trees have increased since 1935, primarily because of increases in white fir. By contrast, densities of quaking aspen and very large diameter trees (primarily ponderosa pine) have declined. The authors attribute increases in small white fir, and to some extent losses of aspen, directly to reduced fire frequencies, whereas they attribute losses of large pines to increased competition with small trees for moisture. As in Chapter 8, the authors close by recommending objectives and strategies for restoration of transition forests to pre-European ("Euroamerican") settlement conditions.

Chapter 10 concludes the trio of chapters analyzing change in forest conditions between 1935 and the present within Grand Canyon National Park, capitalizing on a rare dataset collected by the National Park Service Bureau of Forestry during the mid 1930s. In this chapter the authors examine changes in near-pure ponderosa pine forests, which since 1935 have exhibited substantial increases in small-diameter pines and substantial decreases among the largest classes of trees. As in transition forests (Chapter 9), the authors attribute the increase in small trees to substantially reduced fire frequencies since the late 1800s. They believe that declines in large-diameter trees were due to competition for limited moisture with burgeoning smaller trees. The authors conclude by recommending that Grand Canyon National Park the reduce numbers of small trees, maintain the numbers of mid- to large-diameter pines, and restore a regime of frequent low-intensity surface fires.

The last chapter to focus on Grand Canyon National Park vegetation presents an argument for the limitations of plant associations as a frame for mapping and management as compared to the advantages of addressing vegetation distribution and change on the basis of individual species. According to the authors, vegetation communities are ephemeral assemblages of species that often have no archaeological analogs and few prospects of persistence in the face of global climate change. Plant association types are often convenient human constructs that can obfuscate as much as edify regarding the distributions of individual plant species and that can lead researchers and managers to not fully appreciate, and thus not preserve or make available, plot data used to define and map association types. Cole and Cannella illustrate the potential benefits of plot data collected for mapping purposes with Utah agave in Grand Canyon National Park. The distribution of Utah agave is not closely linked to any mapped vegetation types, yet varies with land form, elevation, and geology such that plot data can be used to construct predictive models with relevance to anticipating the effects of global climate change. The authors urge those involved in vegetation mapping projects to save and make easily accessible their plot data, so that predictive models can be created for individual plant species.

The next four chapters cover management aspects of avian resources on the Colorado Plateau. Chapters 12 and 13 by Mark Sogge and his collaborators deal with birds in Grand Canyon National Park and Glen Canyon National Recreation Area. This work is the result of many years of monitoring activities along the Colorado River. The authors provide in Chapter 12 a quantitative model of avian community structure along the Colorado River in the two parks, and habitat structures associated with differing avian guild assemblages. Their various models predicted more than 75 percent of bird occurrence along the Colorado River. Of the major plants along the river corridor, only tamarisk and mesquite vegetation types functioned as good predictors of bird community parameters. One major finding was that as vegetation patch size increased, so did bird species numbers and abundance, thus leading them to support the continued management of large vegetation patches along the river corridor.

In Chapter 13 Felly and Sogge provide an annotated listing of birds that they observed along the Colorado River between Lees Ferry and Diamond Creek. This chapter was not meant to be an exhaustive list of all birds ever recorded along the river corridor, but this information does add substantially to previously published information and further refines the status, distribution, and seasonality of many bird species. The authors first provide the reader with a brief historical overview of the area, then some background on winter birds, migrating birds, and breeding birds. This work concludes with an exhaustive annotated species account of all birds that they observed between 1993 and 1995. This chapter should provide the reader with good insight as to why the Colorado River corridor is important to migrating and breeding birds in Arizona.

In Chapter 14 Wakeling and Lewis examine the effects of age and gender on mortality and reproduction in Merriam's turkey. They further explore reproduction in turkey demographics based on population modeling. The authors provide an exquisite example of stochastic population modeling using Monte Carlo simulations, with data from a wide array of turkey studies. They describe turkey population responses to varying levels of fecundity and mortality, and found that the most effective way to increase turkey populations throughout Arizona would be to enhance yearling female reproduction success. On the other hand, gender ratios skewed toward males have the potential to negatively impact turkey populations. The authors suggest in this chapter that, at current harvest rates, turkey populations in Arizona are not being overexploited by hunting.

In a second chapter on wild turkeys, Dubay et al. examine differences in the morphological characteristics of transplanted wild turkey populations in Arizona. The general theme in Chapter 15 is a comparison of Gould's and Merriam's turkey subspecies in an effort to determine which subspecies presently occurs in the southern Arizona Fort Huachuca flock. The authors present a series of measurements from birds captured at Fort Huachuca, in Sonora Mexico, and from northern Arizona. In all measurements the Gould's turkeys in southern Arizona and Mexico were more similar than the Merriam's turkey from northern Arizona. The authors conclude this chapter by suggesting that in the Fort Huachuca area, turkeys are more closely related to birds from Sonora, Mexico, and that perhaps this larger subspecies is better adapted to ranges in southern Arizona.

The next two chapters focus on resources and their management just on the southern edge of the Colorado Plateau, at National Park Service areas in the Verde Valley of Arizona. In Chapter 16 Charles Drost provides a summary of the present status and historic changes of all major vertebrate species found at Montezuma Castle National Monument. This chapter provides a much-needed summary of numerous years of a biological inventory that has been undertaken at this National Park Service site. Although the monument was created primarily to protect its remarkable deposits of cultural resources, Drost points out that it also contains diverse vertebrate species.

Chapter 17 describes inventory research done on western diamond-backed rattlesnakes at Tuzigoot National Monument. Erika Nowak presents the results of her long-term work, focusing on the life history and movement patterns of radio-implanted rattlesnakes at this monument. From information gathered on her numerous snake relocations, she provides the reader with an elegant series of western diamond-backed rattlesnake home-range maps. The findings of this study clearly show that rattlesnakes have annual activity ranges of less that one kilometer, but because of the monument's small size, more than half the sightings were outside the park boundary. Erika concludes this chapter with a discussion on the proper management of "nuisance" animals.

Chapter 18 is a unique application of inventory techniques, documenting invertebrate species assemblages along a vehicle disturbance gradient at Canyonlands National Park in southern Utah. Pech et al. examine the potential impact of off-road vehicles in Salt Creek Canyon under three usage regimes: no road and use since 1964, closed road since 1998, and open road where vehicle use still continues. Salt Creek is one of the few perennial riparian environments in this national park—thus the importance of this study. The authors found a wide variety of beetles, with some limited to specific locations while other species occur at all three locations. They demonstrate, through the use of principal components analysis, that beetle communities vary along the vehicle disturbance gradient. They conclude with the recommendation for continued monitoring in order to provide managers with better insights into the potential impacts of off-road vehicles on invertebrate species within Canyonlands.

Chapters 19 and 20 address the management of deer on the Colorado Plateau. Cunningham et al. examine habitat selection by female mule deer in the Mount Trumbell Resource Conservation Area (RCA) in northeastern Arizona. Deer were captured, radio-collared, and followed on the ground, with locations recorded while bedding and feeding. Using GIS vegetation data layers, the authors tested whether the deer sightings were distributed randomly over three treatment areas. They found that deer preferred to eat in the more open treatment area, but had a preference for bedding in more secure locations associated with denser vegetation patches. Throughout Chapter 19 the authors stress that oak and New Mexico locust patches should be left when removing vegetation during treatments. In Chapter 20 Munig and Wakeling examine mule deer harvest estimates from the Kaibab Plateau (Arizona Game Management Unit 12A). The authors compare results of hunter responses from a voluntary mail-in questionnaire with the hunter check station information from Jacob Lake. Biases were found in both datasets, but never exceeded 10 percent. Although the authors do not make any recommendations on which survey is best, they point out that the final decision should be based on data needs, fiscal considerations, and public acceptance.

The final two chapters focused on biology deal with management of carnviores on the Colorado Plateau. Chapter 21 by deVos and McKinney provides readers with background on why mountain lion numbers have increased throughout the western United States over the past several decades. The authors put forth a compelling argument that recent increases in mountain lion numbers are a direct result of decreased exploitive interference from bears and wolves, coupled with changes in vegetative cover that have improved hunting efficiency and caused increases in prey abundance. They attribute prey increases largely to human activities such as ranching and rural animal husbandry, coupled with historical increases in deer numbers in certain locations.

In Chapter 22 Reed and Leslie describe the efficacy of different non-invasive techniques for detecting carnivores, and they present preliminary results regarding the co-occurrence of carnivore species on the North Rim of Grand Canyon National Park. The authors randomly located, without overlap, 20 plots, and at each plot they established sign transects and placed hair snares and a single remote camera. Sign transects produced the most carnivore detections, cameras were effective for detecting coyotes and rare species, and hair snares were of little efficacy. The authors compared patterns of species detections with expectations based on a null model of random association and found that coyotes and bobcats were detected together less often than expected by chance. Reed and Leslie close by describing future directions for their work, which include a greater number of plots on both the North and South Rims of Grand Canyon

National Park, de-emphasizing the use of hair snares, and introducing track plates.

The cultural resources section of the book opens with a chapter by Smiley and Robins, focusing on factors that might help provide increased information from looted rock shelters over the Colorado Plateau. At present most archaeologists avoid working in looted rock shelters because the disturbed environment makes solid archaeological inference difficult. After working in previously looted rock shelters from the Comb Ridge of southern Utah, the authors discuss methods for investigating looted sites and provide techniques that will turn site disturbance to a scientific and cultural advantage.

Chapter 24 illustrates another cultural-social challenge on the Colorado Plateau—how to preserve cultural resources under extreme fire conditions, especially during the recent 10-year drought on the plateau. Hough et al. present survey information of 32 field house sites that were recently burned over by a 2002 lightning-caused fire at Wupatki National Monument. From this information, the authors have developed a fuel load assessment on 450 archaeological sites from the Flagstaff national park areas and have ranked those sites according to their risk potential for fire impacts. Their survey-sampling scheme permitted them to make comparisons across topographic and vegetative divisions at the three Flagstaff national monuments, finding among all site types that cliff dwellings were at disproportionately higher risk for damage from fire due to fuel loading and resource uniqueness and vulnerability. They close this chapter by recommending that on-site hazard fuel reduction should become a routine element of ruins preservation treatment and that archaeological sites should be included (not avoided) during large-block mechanical fuels treatment and prescribed fire projects.

Chapter 25 provides readers with an insight into the extent of cultural resources on the newly acquired lands at Walnut Canyon National Monument. The newly created area of the national monument includes 1420 acres and two new Anasazi "forts." The authors employed a standardized survey method that allowed them to quickly collect site attribute information on 210 sites. The age of these new archaeological sites ranged from early Archaic to the mid-twentieth century. The survey included 20 sites from the poorly understood Archaic period, many from the Sinagua culture, and several important historic structures. The most important historic structure was the Santa Fe Dam, built in Walnut Canyon to hold water for steam engines that ran between Winslow and Flagstaff in the early twentieth century.

Chapter 26 documents the challenges of including cultural resources in the park planning efforts at Grand Canyon National Park. Balsam et al. discuss and outline their attempts at using cultural resource information as part of the planning process, rather than facing unavoidable compliance at the end of the development process. They identify a series of efforts at Grand Canyon National Park that provide the reader with excellent examples of how cultural resources can be included in the national park management planning process. Considerable time is spent outlining the role that park cultural resources have played in shaping the Glen Canyon Dam Adaptive Management Program. They conclude their chapter by pointing out that the politics associated with planning efforts may at first appear intractable, but through providing cultural resource information early in the planning process, integration may be less of an obstacle than most people perceive.

The final three chapters of the book cover biophysical facets of the Colorado Plateau. Chapter 27 provides a transition from the cultural section of this book, examining the impacts of fires on springs in the White Mountain Apache Reservation in eastern Arizona. Long et al. examine the impact of floods on spring-fed wetlands following frequent fires on the western part of the reservation. After the Rodeo-Chediski wildfire, the authors took site information from 56 spring-fed wetland areas that were affected by the fire. They found that 14 percent of the sites had experienced rapid downcutting, and they discuss the mitigation measures that were employed to restore these wetlands. Within a proposed multi-objective decision-making framework, the authors present a preferred spring improvement technique that best meets all concerns on the reservation. The main message of this chapter pertains to a broader series of tribal wetland conservation goals.

Chapter 28 by Stevens et al. provides a new and novel protocol for the rapid assessment of southwestern stream-riparian ecosystems (RSRA). The authors recognize that stream-riparian ecosystems are among the most productive yet most threatened habitats in the Southwest, and that current monitoring methods rely on many subjective techniques. Throughout the chapter the authors build a case for the standardized RSRA, providing managers with a rapid assessment tool that provides information on the functional condition of riparian and associated aquatic habitats. They utilize 4–8 variables in the assessment protocol that range from water quality, through aquatic wildlife, to geomorphology. The protocol outlined in this chapter is designed for small- to medium-sized rivers in the Southwest, but with slight modification it would be applicable to temperate region riparian corridors.

The concluding chapter of the book (Chapter 29) examines the benefits of calculating large watershed areas with a global positioning system as compared to employing digital elevation models or traditional mapping methods. Poff et al. could find no difference in the final accuracy of the three methods, but they point out that their new GPS technique is much quicker for managers to use, allowing additional time for ground-truthing to ensure that areas are not omitted from the model. The end result of this technique would be an accurate assessment of surface runoff and contributions of watersheds to discharge from basins on the Colorado Plateau.

Any scientific work is never a single effort, but a direct result of assistance by many individuals. This book is no exception. We especially thank the following scientific peer reviewers: C. Allen, S. Arundel, T. Arundel, B. Barrett, P. Beier, B. Blackshear, M. Bogan, F. Brandt, M. Brooks, B. Brown, N. Cobb, K. Cole, C. Conway, K. Davis, M. Eaton, W. D. Edge, H. Fairley, D. Falk, P. Ffolliott, L. Floyd-Hanna, B. Gebow, S. Gloss, M. Goode, N. Gotelli, T. Graham, R. Guttierez, H. Hampton, K. Hays-Gilpin, T. Heinlein, J. Hilty, J. Holmes, C. Homer, A. Honoman, B. Howe, R. D. Johnson, T. Jones, P. Krausman, M. Kunzmann, C. Leib, F. Lindzey, K. Logan, J. Mast, M. McGinnis, K. Mock, M. Moore, P. Morgan, S. Nielsen, P. O'Brien, C. Olson, B. Powell, J. Prather, B. Ralston, R. Rasker, S. Rosenstock, C. Schwalbe, H. Shaw, C. Sieg, T. Sisk, M. Sogge, J. Spence, A. Springer, L. Stevens, R. Stoffle, D. Swann, A. Taylor, R. Tausch, T. Teel, T. Theimer, R. Thompson, R. Toupal, J. Unsworth, G. van Riper, D. Wilcox, and H. Zinn; they all unselfishly devoted their time and effort to improving each chapter that they reviewed.

Mark K. Sogge, Dennis B. Fenn, Anne Kinsinger, and Douglas Buffington all provided encouragement and/or financial assistance for this publication. Louella Holter helped in many ways with editorial details, and without her attention to detail this book would never have been a reality, or finished on time. We particularly thank Kimberly Anne van Riper who contributed the line drawings that are found throughout this book. The dedicated USGS Colorado Plateau Research Station staff (T. Arundel, C. Drost, K. Ecton, J. Hart, J. Holmes, S. Jacobs, M. Johnson, E. Nowak, C. O'Brien , E. Paxton, M. Saul, R. Stevens, M. Sogge, and K. Thomas) provided much-needed assistance during the 7th Biennial Conference. Finally, we express deep appreciation to our wives (Sandra Guest van Riper and Susan Bischoff) and to our children for their support and understanding during the time that this book was in production.

This work, like other research compilations that are centered on a particular theme, should help to focus attention on research currently being conducted over lands of the Colorado Plateau. In particular, we hope that the state land stewards of Arizona, Utah, Colorado, and New Mexico, and managers of the National Park system, U.S.

Forest Service, Fish and Wildlife Service, Bureau of Reclamation, tribal lands, and the many new BLM National Monuments will be able to utilize the ideas and concepts presented within each chapter, to launch efforts toward enhanced management and stewardship of their lands in the Southwest. Finally, if the material in this volume can act as a stimulus for future research support regarding the management of cultural, natural, and physical resources over the Colorado Plateau, the organizational and editorial work of the past 2 years will have been a very worthwhile and productive endeavor.

Socioeconomic
Resources
KIM VAN RIPER 05

THE COLORADO PLATEAU ECONOMY: SHIFTING PATTERNS AND REGIONAL DISPARITIES

Walter E. Hecox and F. Patrick Holmes

For the peoples and places of the American West, community economic and population stability tends to be of elemental concern. Many who have a "home on the range" fear that government will take actions in defense of a species or a wildland at the expense of whole communities and their economic vitality (Rasker and Roush 1996). These fears have originated from historical conceptions of the community's economic base and periodic exposure to cyclical boom-bust economies in the rural non-metropolitan West.

Researchers have achieved much in the past two decades to dispel these fears by demonstrating that a new economic base is evolving for these communities. They cite a declining resource-extraction based employment structure and an emerging relationship between public lands and community stability that emphasizes environmental quality and desirable lifestyles as the salient features of the structural transformation of the West's rural economy (Rasker 1993; Rasker and Roush, 1996; Power and Barrett 2001).

The preservation of natural amenity values like scenic beauty, water and air quality, and recreational opportunities has been demonstrated by some to help create new jobs by providing "attractive places to live, work, and do business" (Power 1996). Many have argued that this amenity-based "natural capital" is the foundation of future economic development in the rural West by citing long-term declines in the resource-extracting sectors and long-term growth trends in the service sector. Additional research demonstrating a link between business location decisions and areas with high-quality natural amenities has largely disproved the longstanding belief that natural resource extraction is the primary driver of economic prosperity (Johnson and Rasker 1995). In fact, a study by the U.S. Department of Agriculture entitled "Natural Amenities Drive Rural Population Change" found the relationship between the traditional economic base and economic and population growth to be largely nonexistent (McGranahan 1999).

Past studies of regional natural amenity–driven growth have focused mainly on the Greater Yellowstone region of Montana, Idaho, and Wyoming (Rasker 1991; Rasker et al. 1992). The Colorado Plateau's socioeconomic characteristics were first documented in *Charting the Colorado Plateau* (Hecox and Ack 1996). New data now available for 1970–2000 trace the economic characteristics that define the plateau and identify the dramatic changes facing the region. Additional analysis of the 31 counties incorporating the natural boundaries of the Colorado Plateau focuses here on employment as a measurement of structure and change.

METHODS

Through its Regional Economic Information System, the U.S. Department of Commerce has collected employment data from 1970 to 2000—the longest possible time series for which data are available—for all 31 counties of the Colorado Plateau. This county-level employment information was augmented with data from the Decennial Census of Population and Housing to provide a more refined view of service-based employment trends. Two types of regional employment composition analysis—location quotient analysis and mix-share analysis (Bendavid-Val 1991)—were then applied to the aggregated data.

For this study, location quotients (see below) were calculated by decade over the 30-year study period so as to compare changing regional employment specialization for the entire 31-county Colorado Plateau region to the data for the combined Four Corners area, the U.S. Census Bureau's Mountain Division, and the United States as a whole. Deviations from the expected results of these calculations suggest areas and industries where employment composition warrants further analysis. Second, mix-share analyses for 2000 and 1970 were compared for each employment sector. This additional analysis "disaggregated" net changes experienced by sub-regions of the Colorado Plateau and the overall Colorado Plateau economic region, thereby illuminating the causes of shifting sectoral competitiveness over 30 years.

Location Quotient Analysis

Location quotient analysis measures employment in selected industry categories or sectors in the region of interest relative to a benchmark region for a given year. A comparison of location quotients over 3 decades helps identify which employment sectors are becoming relatively larger and thus more competitive for the region of interest—the Colorado Plateau. The location quotient (LQ) is thus a relatively simple measure of the "proportionality" of employment, providing a rough measure of changes in a region's specialization and competitiveness. LQ = 1 signifies that a region's sectoral employment share is equivalent to that of the larger reference region, LQ < 1 signifies that it is below that of the reference region, and LQ > 1 signifies that it is above that of the reference region.

Mix-Share Analysis

Mix-share analysis is a measure designed to help regions describe and analyze labor market conditions and changes over a period of time. Regions or localities generally experience changes in their employment mix that are either more or less concentrated in certain employment sectors compared to the larger reference region. A locality might have some rapidly growing employment sectors while other sectors are experiencing slow growth or even absolute decline. In examining the region's labor market it is not sufficient merely to show that employment changes have occurred over the time period. More information can be obtained by "disaggregating" these employment changes into various structural effects that shine light on the reasons employment by sector is changing.

Mix-share analysis divides up changes in a region's employment over a period of time by sector into three different (and sometimes offsetting) effects: changes in the larger reference region's total employment, changes in the employment sector's share, and changes in the region's share of employment due to growth and slowdown in employment sectors. Mix-share analysis decomposes employment changes into mutually exclusive factors or "effects" on employment. It paints a picture of how well the region's current employment sectors are performing by systematically examining the national, local, and industrial components of employment change. It provides a dynamic account of total regional employment growth that is attributable to growth of the national economy (expressed here as the reference region growth effect), a mix of faster or slower than average growing industries (the industry mix effect), and the

$$\text{Location Quotient} = \frac{\dfrac{\text{subregional employment in industry } I \text{ in year } T}{\text{total subregional employment in year } T}}{\dfrac{\text{reference region employment in industry } I \text{ in year } T}{\text{total reference region employment in year } T}}$$

competitive nature of the local industrial sectors employing workers (the regional shares effect).

The reference region growth effect measures how much regional employment would have grown in each of its industries and in total IF they had grown at the same rate as total employment grew in the reference region. The industry mix effect measures how much regional employment would have changed in each of its industries and in total IF each regional sector had grown or declined to reflect changes in each sector's share of national employment over time. The regional shares effect measures how much regional employment would have changed in each of its industries and in total AFTER removing the national growth and industry mix effects. This amount of employment represents CHANGING regional shares in each sector beyond that explained by the first two effects. This can also be viewed as a proxy for increases and decreases in a region's competitiveness by employment sector.

Mix-share analysis can also be stated as

$$R = N + M + S$$

where R = total change in regional employment, N = national growth effect, M = industry mix effect, and S = regional shares effect.

RESULTS

Total employment on the Colorado Plateau increased by 225 percent, adding 419,318 new jobs, during the period 1970–2000. This was roughly 15 percent higher growth than in the Four Corners area, 25 percent more than in the U.S. Census Bureau's Mountain Division, and nearly 140 percent more than in the nation during this same period (Table 1 and Figure 1).

The structure of the Colorado Plateau economy and how it has changed over 30 years is illustrated in Table 2, which shows the percent shares of various employment sectors for 1970 and 2000. The Colorado Plateau continues to have the highest proportion of resource-based employment when compared to the benchmark regions of the whole United States, the U.S. Census Bureau's Mountain Division, and the Four Corners area. However, from 1970 to 2000 the Colorado Plateau saw the most drastic decline in this type of work, dropping from 14.8 to 6.0 percent of total employment.

The plateau increased its share of service-based employment from 78.1 percent in 1970 to 89.5 percent in 2000, leaving the region with the highest proportion of such employment in the three regions analyzed. This increase was substantially larger than the increases in the proportion of service-based employment in the Four Corners area and the Mountain Division, but was less than the increase the nation experienced as a whole from 1970 to 2000.

Location Quotient Analysis

Location quotient analysis enables us to paint a more complete picture of the changing employment structure on the Colorado Plateau relative to the Four Corner states, the Mountain Division, and the nation as a whole (Table 3). The location quotient for resource-based employment dropped from 2.59 to 1.66 relative to the United States, further indicating the decline in resource-based employment on the plateau. Location quotients for the plateau also declined relative to the Four Corners and the Mountain Division. However, the location quotient for mining employment and total farm employment, which includes ranching, increased relative to the Mountain Division and the Four Corners area from 1970 to 2000.

The location quotient for service-based employment dropped from 1.07 to 1.05 in 2000 relative to the United States, but increased from 0.97 to 1.01 and from 0.96 to 1.01 relative to the Mountain Division and the Four Corners area respectively. This relatively equal specialization in service-based employment between the regions is bolstered largely by high specialization in government and retail trade employment. The LQs for the service and financial industries remained below 1 in both 1970 and 2000, indicating that the plateau is less specialized in these sectors compared to all

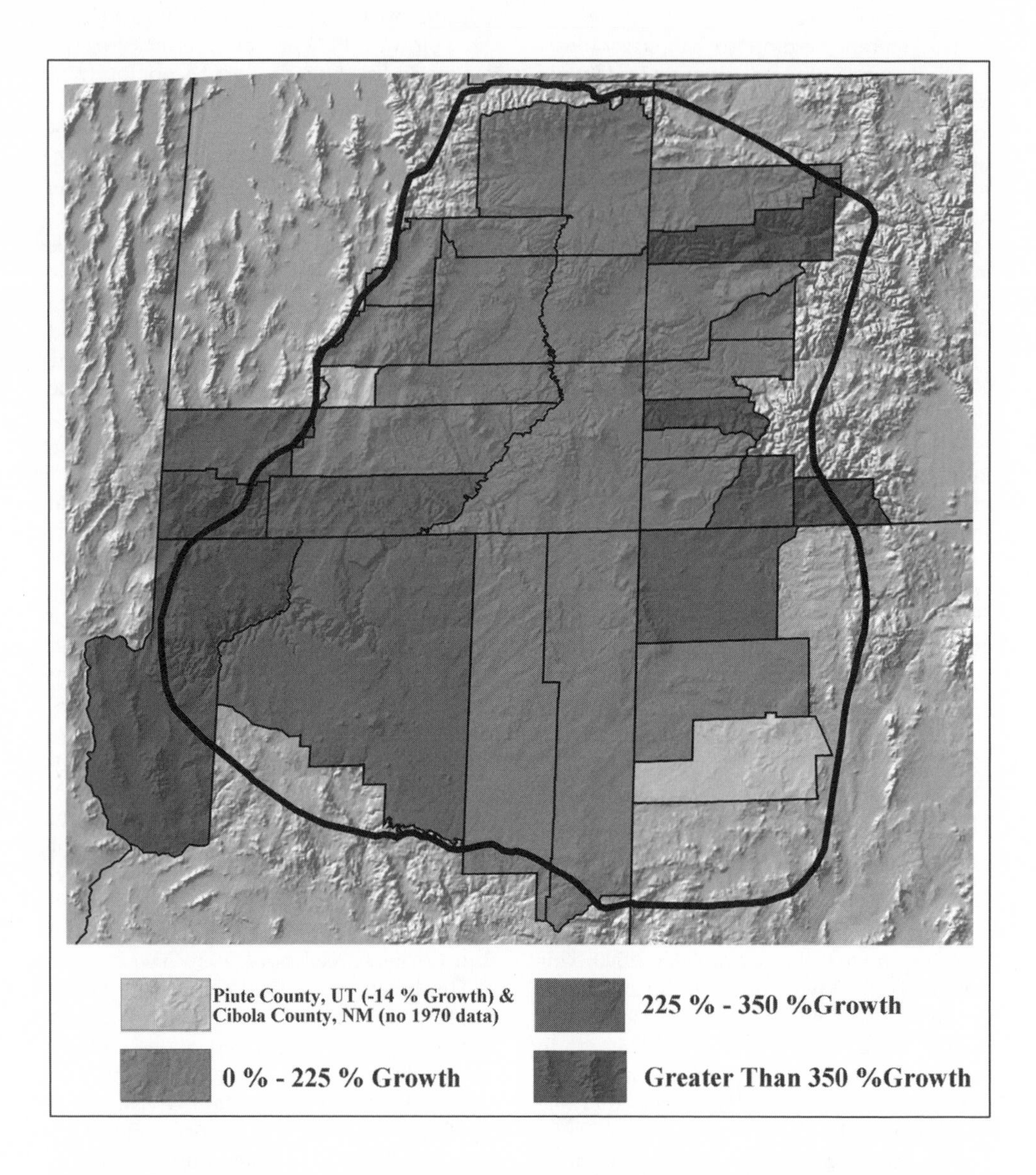

Figure 1. Employment growth from 1970 to 2000 for the United States, the Census Bureau's Mountain Division, the Four Corners, and the Colorado Plateau study region. Disparate rates of employment growth occur within counties of the plateau, with some counties growing faster and some slower than the regional average.

Table 1. Employment growth from 1970 to 2000 (see Figure 1).

Region	1970 Employment	2000 Employment	No. of New Jobs	Growth in New Jobs
United States	91,281,600	167,465,300	76,183,700	84.5%
Mountain Division	3,672,619	11,102,295	7,429,676	202.3%
Four-Corner states	2,631,894	8,156,359	5,524,465	209.9%
Colorado Plateau	185,910	605,228	419,318	225.5%

benchmark regions. Relative specializations in manufacturing for the plateau also remained well below 1 when compared to all benchmark regions, indicating that manufacturing is not a noteworthy specialty for the region.

A closer look at occupational concentrations in the Colorado Plateau is provided in Table 4. The service industry is composed of a wide variety of professions ranging from software design and manufacturing to food preparation and cleaning occupations. In general, the Colorado Plateau has higher LQ in the lower-paying service industries like food preparation, buildings and grounds cleaning and maintenance, education, and social services, but lower concentrations of higher-paying management, business, legal, financial, and computer-related occupations.

Thus, even this initial evidence suggests that the Colorado Plateau has begun to "mature" away from a heavy reliance on traditional resource-extraction forms of employment. Yet this shift from the "Old West" way of living to a "New West" with a modern, diversified service economy is not exemplified in these findings. Rather, evidence that the plateau has had a particularly pronounced swing in its employment base may be indicative of some of the "growing pains" occurring in the region, or more specifically, the inability of the regional economy to make a full transition from low-wage, low-skill industries to high-wage, high-skill industries.

Mix-Share Analysis

Mix-share analysis is used to construct an understanding of how well the industries of the Colorado Plateau are performing. The analysis shows how Colorado Plateau employment growth can be attributed to growth in the economy of a larger benchmark region, a mix of faster or slower growing industries, and the competitive nature of local industries. This analysis is shown in Figure 2.

Particular attention should be focused on the regional shares effect, which signals sectors whose competitiveness has grown over time, often enough to overcome a loss of employment due to the growth effect and the industry mix effect. In sectors where the shares effect is positive compared to the whole United States, and continues to be positive in benchmark regions closer to the plateau, a comparative advantage is indicated for the plateau from 1970 to 2000 in either attracting or retaining employment in that industry or sector. Conversely, in some sectors the shares effect declined as the benchmark region was refined, indicating a disadvantage in attracting or retaining employment in that industry or sector.

Total new jobs from 1970 to 2000 in resource-based employment on the plateau numbered only 8728, a mere 2 percent of total new jobs, but retention of jobs in mining and total farm employment yielded a strong advantage for the region over the broader areas of the Four Corners and Mountain states, where the decline in these industries was markedly more substantial. Similarly, the plateau experienced a positive shares effect in manufacturing employment compared to the Four Corners and Mountain Division states despite noteworthy industry declines. There were 14,038 new jobs

Table 2. Economic structure and change; the percent of total employment by industry classification for 1970 and 2000 for the United States, the Mountain Division, the Four Corner states, and the Colorado Plateau.

	Colo. Plat. 1970	Colo. Plat. 2000	4 Corners 1970	4 Corners 2000	Mtn. Div. 1970	Mtn. Div. 2000	United St. 1970	United St. 2000
A. Resource-based Employment								
Total farm employment	8.2	2.9	4.3	1.3	5.8	1.8	4.3	1.9
Agricultural services employment	0.8	1.0	0.7	1.4	0.7	1.4	0.6	1.3
Mining employment	5.8	2.1	2.8	0.8	2.8	1.0	0.8	0.5
Total resource-based employment	14.8	6.0	7.7	3.5	9.3	4.2	5.7	3.6
B. Manufacturing								
Manufacturing employment	7.1	4.5	11.1	7.7	10.2	7.2	21.6	11.4
Total manufacturing employment	7.1	4.5	11.1	7.7	10.2	7.2	21.6	11.4
C. Service-based Employment								
Construction employment	5.7	8.1	5.4	7.2	5.4	7.3	4.8	5.7
Transportation/utility employment	5.9	4.3	5.1	4.9	5.3	4.9	5.3	4.9
Wholesale trade employment	2.1	2.9	4.2	4.1	4.0	4.0	4.6	4.5
Retail trade employment	17.9	20.1	16.3	17.0	16.2	17.0	15.0	16.3
Financial sector employment	5.2	7.4	7.6	9.7	7.1	9.2	6.7	8.1
Services employment	18.1	29.8	19.1	32.0	20.1	32.5	18.7	31.8
Total government employment	23.2	16.9	23.4	14.0	22.5	13.8	17.6	13.6
Total services-based employment	78.1	89.5	81.2	88.8	80.5	88.6	72.7	85.0
D. Other Employment NES								
Other employment NES	5.2	7.4	7.6	9.7	7.1	9.2	6.7	8.1
Total other employment NES	5.2	7.4	7.6	9.7	7.1	9.2	6.7	8.1
E. Total Employment	100.0	100.0	100.0	100.0	100.0	100.0	100.0	100.0

Table 3. Location quotients for the Colorado Plateau versus the United States, the Mountain Division, and the Four Corners states. An LQ greater than 1 signifies higher specialization in that employment industry category than the benchmark region, an LQ equal to 1 signifies an equal degree of specialization, and an LQ less than 1 signifies a lower degree of specialization than the benchmark region.

Type of Employment	Location Quotient vs. the U.S. States 1970	Location Quotient vs. the U.S. States 2000	Location Quotient vs. Mtn. Division States 1970	Location Quotient vs. Mt. Division States 2000	Location Quotient vs. 4 Corners States 1970	Location Quotient vs. 4 Corners States 2000
A. Resource-based Employment						
Total farm employment	1.88	1.58	1.41	1.64	1.91	2.25
Agricultural services employment	1.43	0.78	1.12	0.70	1.17	0.73
Mining employment	7.17	4.33	2.12	2.15	2.12	2.62
Total resource-based employment	2.59	1.66	1.60	1.43	1.92	1.73
B. Manufacturing Employment						
Manufacturing employment	0.33	0.40	0.70	0.62	0.64	0.58
Total manufacturing employment	0.33	0.40	0.70	0.62	0.64	0.58
C. Service-based Employment						
Construction employment	1.17	1.41	1.05	1.11	1.04	1.13
Transportation/utility	1.10	0.87	1.10	0.89	1.14	0.88
Wholesale trade employment	0.46	0.64	0.54	0.73	0.50	0.71
Retail trade employment	1.19	1.23	1.10	1.18	1.10	1.18
Financial sector employment	0.78	0.91	0.74	0.80	0.69	0.76
Services employment	0.97	0.94	0.90	0.92	0.95	0.93
Total government employment	1.32	1.24	1.03	1.22	0.99	1.21
Total service-based employment	1.07	1.05	0.97	1.01	0.96	1.01
D. Other Employment NES						
Other employment NES	0.78	0.91	0.74	0.80	0.69	0.76
Total other employment NES	0.78	0.91	0.74	0.80	0.69	0.76
E. Total Employment (A + B + C + D)	1.00	1.00	1.00	1.00	1.00	1.00

Table 4. Location quotients for the Colorado Plateau versus the Mountain Division and the Four Corner states. An LQ greater than 1 signifies higher specialization in that employment industry category than the benchmark region, an LQ equal to 1 signifies an equal degree of specialization, and an LQ less than 1 signifies a lower degree of specialization (source Census 2000).

	Location Quotient vs. the Mtn Div. 2000	Location Quotient vs. the 4 Corners 2000
Service Occupations		
Management occupations, except farmers and farm managers	0.84	0.81
Business operations specialists	0.57	0.53
Financial specialists	0.67	0.63
Computer and mathematical occupations	0.26	0.22
Architects, surveyors, cartographers, engineers	0.47	0.42
Drafters, engineering, and mapping technicians	0.79	0.75
Life, physical, and social science occupations	0.82	0.81
Community and social services occupations	1.25	1.22
Legal occupations	0.67	0.65
Education, training, and library occupations	1.30	1.27
Arts, design, entertainment, sports, media	0.70	0.67
Health diagnose and treating and tech occupations	0.99	0.96
Health technologists and technicians	1.01	1.00
Healthcare support occupations	1.19	0.21
Fire fighting, prevention, law enforcement workers	1.18	1.15
Other protective services workers, including supervisors	0.93	0.99
Food preparation and serving-related occupations	1.19	1.32
Building and grounds cleaning and maintenance	1.20	1.27
Personal care and service occupations	0.98	1.13
Sales and related occupations	0.99	0.98
Office and administrative support occupations	0.89	0.98
Non Service Occupations		
Supervisors, construction and extraction work	1.53	1.53
Construction trades workers	1.30	1.31
Farming, fishing, and forestry	1.18	1.56
Farmers and farm management	1.26	2.01
Extraction workers	2.91	3.97
Installation, maintenance, and repair occupations	1.12	1.15
Production occupations	0.99	0.97
Supervisors: transportation and material moving	1.09	1.10
Aircraft and traffic control occupations	0.78	0.73
Motor vehicle operators	1.28	1.35
Rail, water, and other transportation occupations	1.49	1.73
Material moving workers	1.10	1.16
Total employed civilian population 16+	1.00	1.00

The Plateau Against the United States	N = Growth Effect	M = Industry Mix Effect	S = Regional Shares Effect
A. Resource-Based Employ.			
Total Farm Empl	-3,283	-15,932	21,762
Ag Serv Empl	4,797	3,516	-3,728
Mining Empl	752	-8,315	9,160
Total Resource-Based Employ.	**2,266**	**-20,731**	**27,194**
B. Manufacturing			
Manuf Empl	-391	-11,447	25,876
Total Manufacturing Employ.	-391	-11,447	25,876
C. Service-Based Employ.			
Construct Empl	12,448	3,669	22,379
Trans/Util Empl	7,592	-1,525	9,055
Whlsl Trade Empl	3,229	-67	10,449
Rtl Trade Empl	33,063	5,360	50,185
Fin Sector Empl	11,719	3,590	19,455
Services Empl	71,441	43,428	32,081
Total Gov Empl	17,929	-18,140	59,210
Total Service-Based Employ.	**157,421**	**36,314**	**202,814**
D. Other Employment NES			
Other Employment NES	11,719	3,590	19,455
Total Other Employment NES	11,719	3,590	19,455
E. Total Employment			
Total Employment	155,161	18,459	245,698
(A + B + C + D Above)			

The Plateau Against the Mountain States	N = Growth Effect	M = Industry Mix Effect	S = Regional Shares Effect
A. Resource-Based Employ.			
Total Farm Empl	-999	-31,660	35,206
Ag Serv Empl	7,525	4,420	-7,362
Mining Empl	548	-21,430	22,479
Total Resource-Based Employ.	**7,074**	**-48,670**	**50,323**
B. Manufacturing			
Manuf Empl	15,003	-11,797	10,832
Total Manufacturing Employ.	15,003	-11,797	10,832
C. Service-Based Employ.			
Construct Empl	32,583	11,303	-5,389
Trans/Util Empl	19,114	-2,985	-1,006
Whlsl Trade Empl	8,036	47	5,529
Rtl Trade Empl	72,417	5,268	10,923
Fin Sector Empl	28,472	8,768	-2,477
Services Empl	130,545	62,643	-46,238
Total Gov Empl	36,950	-50,478	72,527
Total Service-Based Employ.	**328,117**	**34,566**	**33,869**
D. Other Employment NES			
Other Employment NES	28,472	8,768	-2,477
Total Other Employment NES	28,472	8,768	-2,477
E. Total Employment			
Total Employment	376,094	-4,549	47,773
(A + B + C + D Above)			

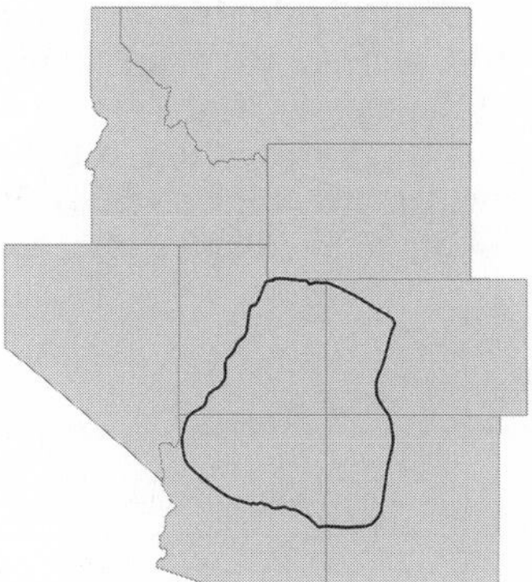

The Plateau Against the 4 Corner States	N = Growth Effect	M = Industry Mix Effect	S = Regional Shares Effect
A. Resource-Based Employ.			
Total Farm Empl	-850	-32,663	36,060
Ag Serv Empl	7,758	4,536	-7,711
Mining Empl	-1,271	-24,075	26,942
Total Resource-Based Employ.	**5,637**	**-52,202**	**55,291**
B. Manufacturing			
Manuf Empl	15,294	-12,514	11,259
Total Manufacturing Employ.	15,294	-12,514	11,259
C. Service-Based Employ.			
Construct Empl	32,736	10,656	-4,896
Trans/Util Empl	21,221	-1,709	-4,390
Whlsl Trade Empl	7,950	-339	6,000
Rtl Trade Empl	74,126	4,453	10,030
Fin Sector Empl	28,747	8,302	-2,286
Services Empl	140,773	70,319	-64,142
Total Gov Empl	36,709	-54,005	76,296
Total Service-Based Employ.	**342,262**	**37,677**	**16,612**
D. Other Employment NES			
Other Employment NES	28,747	8,302	-2,286
Total Other Employment NES	28,747	8,302	-2,286
E. Total Employment			
Total Employment	390,234	-8,947	38,032
(A + B + C + D Above)			

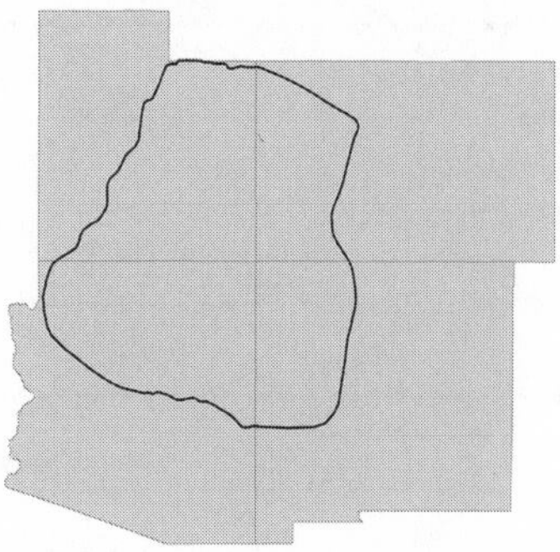

Colorado Plateau:
Total New Service Jobs - 369,552
Manufacturing Jobs - 14,038
Resource-Based Jobs - 8,728
Total New Jobs - 419,318

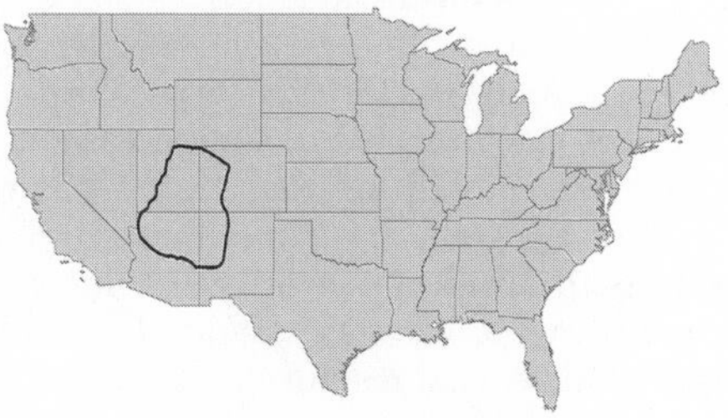

Figure 2. Mix-share analysis for the Colorado Plateau against the United States, the Mountain Division, and the Four Corners area.

in manufacturing from 1970 to 2000, about 3 percent of new job growth in the region.

The service-based industries accounted for 396,552 new jobs on the plateau, or about 95 percent of all new jobs. Of those, major contributions occurred in government employment (15%), in retail trade employment (22%), and in the services category, which comprised 37 percent of all service-based growth, or 146,950 new jobs. The "service" category covers a wide variety of industries, including the higher-paying business, engineering, and management services. It also includes lower-paying consumer services like gas station attendants, hotel workers, and restaurant employees, as well as social service work such as health care workers and police officers. Although the service industry produced the most new jobs for the plateau, the shares effect indicates that the plateau is at a strong relative disadvantage in attracting this type of employment into the region when compared to the rest of the mountain states, and even more so when compared to the Four Corners area.

The particularly pronounced structural shift in the plateau economy from resource-based to service-based employment may help to explain the region's most persistent problem over the past 30 years—the problem of falling earnings per job. Across the rural American West, average earnings per job (adjusted for inflation) have been falling for decades. This is in contrast to increasing earnings per job in the region's cities and in the nation as a whole (Rasker and Alexander 2003). Figure 3, which depicts this trend for the Colorado Plateau, the Four Corners area, and the entire nation, shows that recent upswings in average wages experienced by the Four Corner states and the United States failed to reach the counties of the Colorado Plateau. Explanations for this include the loss of relatively high wage occupations in mining, the influx of willing workers to the amenity-rich region which has caused the labor supply to outpace job growth, limited

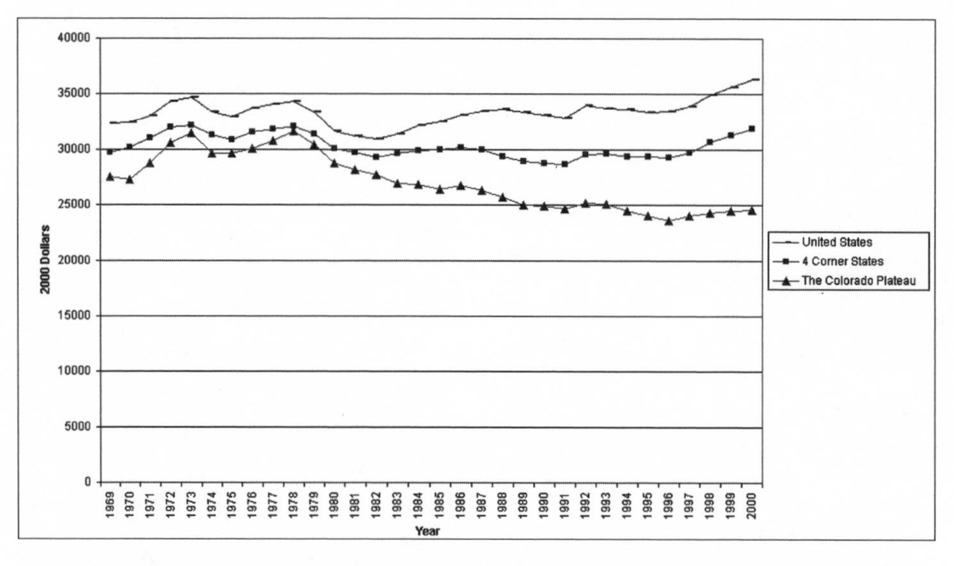

Figure 3. Recent upswings in average wages in the region and the nation have failed to reach the Colorado Plateau.

access to larger markets due to the relative isolation of the region, and a smaller proportion of the eligible work-force working full-time jobs.

DISCUSSION AND CONCLUSIONS

The structure and economic base of the Colorado Plateau has undergone dramatic changes in the period from 1970 to 2000. Because growth dynamics in this region have transformed sectoral shares of employment so drastically when compared with the Four Corners area and the rest of the Rocky Mountain West, the challenges of change for the plateau are in some ways both unique and heightened. Possibly more than any other region in the West, the Colorado Plateau has witnessed a rapid transition to "non-traditional" service-based industries as 15 percent of its employment base in resource extractive industries declined to just 6 percent by the turn of the century. Global market forces, including increased competition from abroad and improved technology requiring less labor and more capital, are likely the reasons that the traditional Western mainstays of forestry, agriculture, mining, and oil and gas extraction are in decline. Another contributing force has been the political struggle over land and resource management in an area with 85 percent public lands. These inter-regional dynamics indicate that competing as a low-cost producer of food, minerals, and timber can no longer be considered a long-term comparative advantage for the people and communities who call the Colorado Plateau home.

Mix-share analysis has shown that the Four Corners area and the Rocky Mountain region have both fared better than the Colorado Plateau in attracting a new influx of service industries. Nevertheless, nearly 95 percent of all new jobs on the plateau occurred in the service-based industries, indicating that the economy of the region will be structured differently for the next 30 years. Future economic development, to be consistent with the new "natural amenity" foundation for the region, should be geared toward attracting higher-end professional, management, and engineering service industries to the region. Development initiatives must recognize the newly evolving definition of the region's assets, principally the plateau's high levels of natural amenities and scenic beauty as well as cultural riches. Regional economic development promotion should focus on the area's need to gain access to larger markets through commuter air-travel destinations and high-speed Internet infrastructure. Such efforts must place a premium on generating and retaining a highly educated workforce. All told, the Colorado Plateau as a region has been far too reactive to its uniquely difficult economic transition. As we begin to chart a path for the prosperity of the region for the next 30 years, insightful proactive measures will be necessary to give the region's peoples and places the means to take advantage of and build upon the spectacular natural and cultural resources that define the Colorado Plateau.

LITERATURE CITED

Bendavid-Val, A. 1991. Regional and Local Economic Analysis for Practitioners. 4th ed. Praeger, New York.

Hecox, W. E., and B. L. Ack. 1996. Charting the Colorado Plateau: An economic and demographic exploration. Grand Canyon Trust, Flagstaff AZ.

Johnson, J. D., and R. Rasker. 1995. The role of economic and quality of life values in rural business location. Journal of Rural Studies 11 (4): 405–416.

McGranahan, D. A. 1999. Natural amenities drive rural population change. Economic Research Service, USDA. Agricultural Economic Report No. 781.

Power, T. M. 1996. Soul of the wilderness: Wilderness economics must look through the windshield, not the rearview mirror. International Journal of Wilderness 2 (1).

Power, T. M., and R. Barrett. 2001. Post Cowboy Economics: Pay and Prosperity in the New American West. Island Press, Washington D.C.

Rasker, R. 1991. Dynamic economy versus static policy in the Greater Yellowstone ecosystem. In Proceedings to the Conference on the Economic Value of Wilderness, Jackson WY, May 8–11. Southeastern Forest Experiment Station, U.S. Forest Service, Asheville NC.

Rasker, R. 1993. Rural development, conservation, and public policy in the Greater Yellowstone ecosystem. Society and Natural Resources 6: 111.

Rasker, R., and B. Alexander. 2003. Working Around the White Clouds. Sonoran Institute, Bozeman MT.
Rasker, R., and J. Roush. 1996. The economic role of environmental quality in Western public lands. A Wolf in the Garden: The Land Rights Movement and the New Environmental Debate, pp. 185–205. Rowman and Littlefield, Landham MD.
Rasker, R., N. Tirrell, and D. Kloepfer. 1992. The Wealth of Nature: New Economic Realities in the Yellowstone Region. The Wilderness Society, Washington D.C.

PUBLIC KNOWLEDGE, OPINION, AND SUPPORT OF FOREST RESTORATION: A SURVEY OF RESIDENTS IN NORTHERN ARIZONA

David Ostergren and Elizabeth J. Ruther

After several catastrophic fires since 1990, concern runs high in Arizona and the American West for the health of forest ecosystems. Forest managers, policy makers, researchers, timber industry experts, environmental NGOs and, most importantly, the general public are engaged in debate on future management. All interest groups appear to recognize that forest conditions have changed for the worse and that society needs to reconsider past practices and redirect future actions. However, as with many public land issues, solutions have been proposed and most have proved politically contentious. As management agencies design future strategies, assessing the opinions and perspectives of the public will be crucial to balancing a wide range of expectations for forest resources.

Public opinion is constantly changing and managers need data on social perceptions and expectations to determine desired future conditions. For instance, the perceived role of fire in forest ecosystems has changed dramatically since the 1970s. The public accepts and seems to recognize that prescribed fire is an effective management tool (Gardner et al. 1985; Shelby and Speaker 1990; and others). Land management agencies have used public relations and education programs to change public attitude toward fire, and related research shows the promise of such strategies (Jacobson et al. 2001; Loomis et al. 2001). In northern Arizona, ponderosa pine ecosystems are a focus of concern for city, county, state, and federal agencies, the general public, and researchers. Thus, as new information on ponderosa pine ecosystems is disseminated, all interested parties should be aware of how that information changes public opinion.

FOREST RESTORATION AND FIRE

Ponderosa pine ecosystems currently have higher stem densities and fuel loads compared to more open and park-like conditions prior to Euro-American settlement (Moore et al. 1999). Forest managers are considering strategies to restore those park-like conditions within a wide range of jurisdictions (Cooper 1960; Covington 1994; Dahms and Geils 1997; Fulé et al. 1997). Many approaches have been proposed, with solutions ranging from thinning as part of commercial production to thinning only around communities while the interior forest "restores" itself. Wholesale forest restoration via a mix of mechanical or human-powered treatments and the establishment of a pre-Columbian fire regime has also been advocated. Debate rages over both ends and means. Where should agencies employ non-mechanical or mechanical methods? Should preemptive action be taken in the interior forests or only to protect communities while waiting centuries for nature to restore itself? Further embedded in the debate are questions such as what role does fire have in forest health, whether timber from restoration should generate funds for federal agencies, and what potential impacts there might be on game, endangered species, or grazing activities (DeMillion 1999). As Herron stated (2001, p. 206), "the issue before the Forest

Service [and other land management agencies] is how to relate to fire within an evolving culture that is concerned with nature." For this study we asked questions about both "forest restoration" and "fire," though the public often associates one very closely with the other. The results shed light on what the public understands and expects from their forest.

This investigation focused on public perceptions of forest restoration, public participation, and possible future forest conditions. We assumed that everyone does not view restoration in the same manner, nor does everyone fully support ecosystem restoration (Brunson et al. 1997; Woolley and McGinnis 2000). For instance, many obvious objections would arise to restoring all farmland to prairies and desert, or removing all dams to re-create free-flowing rivers. However, even when there is general consensus on the need for small to mid scale restoration projects, people often otherwise disagree. In California, there is broad support to reintroduce native oaks, but there are opponents to cutting down century-old (but non-native) eucalyptus trees (Woolley and McGinnis 2000). Other hazards of restoration strategies are unintended effects. For instance reestablishing native fish populations by removing non-native trout species below Glen Canyon Dam would not only affect the recreational fishing industry, but might also impact a recently established bald eagle population (van Riper and Sogge 2003). Broad prescription of a particular management strategy will often be controversial and ought to be carefully considered and well communicated to the public before its application. In northern Arizona, managers and policy makers need to acknowledge that forest restoration will probably have unintended consequences, and act accordingly (Pyne 2001).

Forest restoration entails many issues and one of the most salient for the public is fire—prescribed or wild. Fire regimes have changed as Euro-Americans have settled the western United States. Since the early twentieth century fire frequency and intensity have been suppressed and nowhere is this more evident than on the wildland-urban interface (WUI). However, fire suppression increases fuel load and creates conditions for large, catastrophic forest fires that threaten ecosystems as well as human communities. One solution includes prescribed fire with the subsequent smoke and flames that may prompt public concern, a concern born out of ignorance. Thus a paradox has been created between managing for a fully functioning ecosystem and managing for the human desires of safety, predictability, and a "fire-free" forest. As the WUI continues to expand due to the population increase and dispersion that accompanies economic growth, the problems associated with the WUI will increase (Main and Haines 1974; Bradshaw 1987; Bailey 1991; Lavin 1997; Cardille et al. 2001). Northern Arizona is an excellent example of the WUI dynamic, as the proliferation of retirement communities and second homes accounts for 25 percent of total homes in Flagstaff.

FOREST MANAGEMENT AND PUBLIC PARTICIPATION

In general, land management agencies see restoration and prescribed fire as technical rather than social problems. This is due, in part, to an agency's grasp on ecological needs but lack of information on the social issues. National Forest managers have identified "environmental regulations" such as the Clean Air Act as the biggest barrier to prescribed burning. Other identified barriers are lack of funding, limited burn windows, and planning costs (Cleaves et al. 2000). Each barrier may have a technical solution, but social conditions may be more important in overcoming these barriers. For example how much smoke is society willing to accept to restore forest ecosystems and reduce wildfire risk? Managers may believe that the public is only concerned about safety and want someone to do something—anything—about fire danger. On the other hand, the public may not be informed about fire risk, potential solutions, or the cost of those solutions.

We suggest that elements of the public do care about root causes of fire hazards and that although public involvement can be complex and time consuming, educating the public about root causes will likely affect support for management strategies. For instance, are large wildfires the result of poor forest health, the collapse of the timber industry, poor urban planning, or unusual drought conditions? Theoretically, if the engaged public has a common understanding of "the problem," they may affect zoning regulations or help reduce planning costs by not appealing management decisions. Supportive communities can also pressure legislators to increase funding for agency initiatives. For instance, public concern after the California fires in the fall of 2003 supported passage of Public Law No. 108-148, the "Healthy Forest Initiative."

We assume that restoration practices cannot succeed without aid and support by the engaged public, especially those living in the WUI (Cortner 1991; Martin 1997). To successfully elicit public support, and to clearly understand their own roles, managers and policy makers need to understand the values, preferences, and attitudes of stakeholders. Research suggests that an involved public facilitates implementation of effective fire protection plans (Cortner et al. 1984; Chambers et al. 2001; Burns et al. 2002); very likely the same holds true for forest restoration. Public support is essential to managing forest ecosystems, and the social challenges of managing forests are at least as complex as the ecological and technical challenges (Findley et al. 2001).

Forest managers and policymakers clearly need to elicit, understand, and respond to public opinion and perceptions beyond what an Environmental Impact Statement may require. Our investigation contributes to the knowledge of public perception of forest restoration, focusing on opinions of northern Arizona residents. Our goal is to facilitate the integration of public opinions and perceptions into forest management decisions. This information will aid forest managers in communicating the goals of restoration, the benefits of prescribed fires, the potential dangers of reintroducing a fire regime, and the need for fire proofing as part of living in the WUI. We also assume that managers desire information on issues that pertain to values such as whether homes are more important than forests, how the communities share information, how communities might contribute to restoration or fire prevention/suppression, or how best to garner community support for a forest restoration plan. The results presented here are the first product of a larger study that will assess and link the values of residents to perceptions and knowledge. As we continue to disseminate information we hope to ease and facilitate future management decisions.

METHODS

We conducted a survey of Northern Arizona residents during the summer of 2003. The self-administered mail questionnaire was directed toward five major subject areas surrounding the forest restoration issue: (1) human behavior with regard to forest fire protection as fire is reintroduced into the ecosystem, (2) public expectations of forest restoration management, (3) perceptions of forest restoration, (4) knowledge about forest restoration, and (5) values toward nature. The questionnaire was designed by borrowing from other restoration questionnaires and in consultation with the Southwest Fire Social Science Collaborative, but the final form was the responsibility of the authors. We tested the survey on a small group of town residents and graduate students in Flagstaff and estimated 20–25 minutes to complete the survey. Length of residency and home location preceded specific questions regarding forest restoration. Demographic questions, such as age, gender, formal education, and occupation were asked at the end of the questionnaire to assist in explaining values toward nature. Respondents could select from five response options (strongly agree, agree, neither, disagree, or strongly disagree) for each value and opinion question. Results were compiled for each statement as percentages of responses to each category. A complete copy of the questionnaire is available upon request from the authors.

The sample spatial frame was determined

by laying a map of Arizona postal zip code regions over a map of the ponderosa pine ecosystem supplied by the Forest Ecological Restoration Analysis Project. The 35 zip codes in the sample area ran southeast from the Kaibab Forest on the North Rim of the Grand Canyon, to the Arizona–New Mexico border, and east of Show Low including Prescott, Sedona, Payson, and Flagstaff. Our original goal was a confidence interval of ± 3 percent and we estimated a 60 percent return rate. Thus a random sample of 1729 residents, stratified by postal zip codes, was purchased from Genesys Sampling in Fort Washington, Pennsylvania. The U.S. Postal Service approved 1644 addresses as current and in use. Our target population was adult (> 18 yrs old), seasonal or permanent residents who had lived in the region one year or longer. We were interested in potential voters and whether there was a difference between seasonal and permanent residents. We also assumed that one year of residence increases the possibility of knowledge about forest restoration and/or fire hazards.

Four mailings were sent, consisting of a pre-notice letter, a questionnaire with cover letter, a reminder postcard, and a replacement questionnaire, consistent with Dillman's (2000) Tailored Design Survey Method. The pre-notice letter was sent on June 23, 2003. One week later the 131-question survey was mailed with an explanatory cover letter, a self-addressed, stamped return envelope, and an incentive of two postage stamps. All questionnaires and correspondence were mailed from and returned to Northern Arizona University.

Twenty-seven surveys were undeliverable of the 1644 sent, resulting in an actual sample of 1617 residents. Forty-six percent (n = 750) of the surveys were returned. Fifty-seven of the surveys, 0.04 percent, were returned blank or were determined unusable. The completed surveys (n = 693) translate into a 43 percent return rate which is consistent with meta-analyses of decreasing response rates in natural resource mail surveys (Connelly et al. 2003). We attribute some of our low numbers to (1) conducting the survey in summer rather than the more successful months of January–March, (2) no major fires in northern Arizona during 2003, and (3) the length of the survey. To assess non-response bias, we contacted a sample of non-respondents in the winter of 2003–2004. Non-respondents were asked a sample of questions from each section of the questionnaire for comparison to the respondent sample. These results will be included in later publications.

RESULTS

We present our results as percentages of residents responding to one of five categories (strongly agree, agree, neither, disagree, or strongly disagree). Please note that the results presented here are descriptive and will be the basis for future explanatory analysis (e.g. regression analysis to determine if independent variables such as gender, ethnicity, age, education, income, or urban/rural living location influence opinions, values, and perception). In several cases these results indicate very strong trends in public opinion and potentially provide a reliable basis for future management. All data reported have a confidence interval of 95 percent and percentages are accurate ± 5 percent.

Demographics

Overall, the participating respondents were mostly white males, of middle age (45–65 yrs), and comparatively well educated. Mean respondent age was 57.8 years, with ages overall ranging from 19 to 93 (Figure 1). This age distribution was not consistent with the age distribution for residents of the region (U.S. Bureau of the Census 2003) but did resemble the age distribution of the voting population (Jamieson et al. 2002). Further analysis may correct for ages; however when forest managers make decisions, public participation will likely originate from the voting population. Respondents lived in northern Arizona from 1 to 82 years but most from 5 to 7 years and for an average length of 17.2 years. Ninety-one percent of the respondents described themselves as permanent residents. Respondents reported an average 15.5 years of formal education

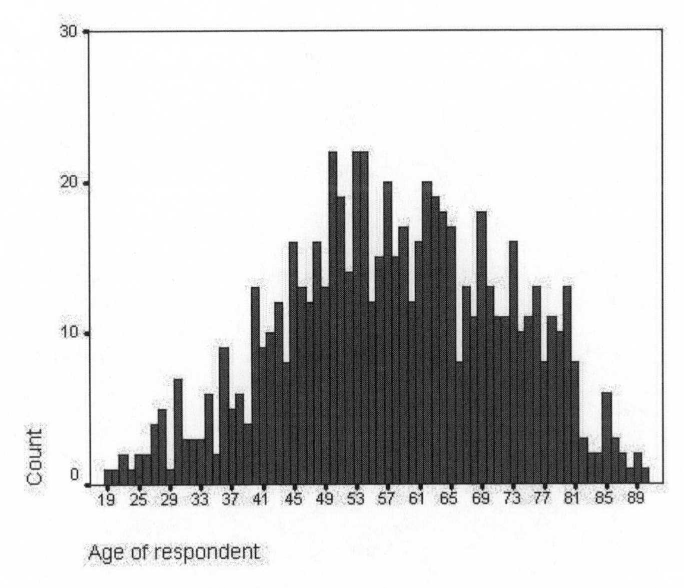

Figure 1. Age distribution of respondents.

which is consistent with the U.S. 2000 Census (U.S. Bureau of the Census 2003). Sixty-eight percent of the respondents were male and 32 percent were female, which is skewed from a more equal gender ratio reported by the U.S. Bureau of the Census 2000 for Flagstaff (49.6% male, 50.4% female). We used Flagstaff rather than counties as a basis for comparison because all relevant counties contained substantial areas not in ponderosa pine forest whereas Flagstaff was well within our spatial frame. Regional race distribution was also not reflected in our survey, with Hispanics and Native Americans substantially under-represented (Table 1). We will analyze the effect of race as indicated in subsequent reports so any conclusions drawn from this report are contingent on further analysis.

Public Involvement

Our first set of seven statements pertained to the environmental review process, public involvement, and communication. In light of the prevalent political rhetoric that blames litigation and administrative appeals for unhealthy forests, we were interested in assessing how residents view the rights and privileges of participation. Our respondents generally favored restrictions on the right to sue the government and the right to appeal forest management decisions (Table 2) but at the same time very strongly supported (90% ± 5%) community involvement in forest restoration and improved communication between locals and federal agencies (Table 3).

Of the 11 specific or generic information sources, newspapers and the U.S. Forest Service were by far the most preferred (Table 4).

Support for Restoration

Another series of nine statements queried levels of support for forest restoration. Respondents strongly supported (i.e. strongly agree or agree) forest restoration to promote working ecosystems (51.6% ± 5%) and to reduce risk of fire even if the forest looked different (56.2% ± 5%; Table 5). Respondents also offered moderate support for removing large trees in forest restoration projects and

Table 1. Comparison of our survey results with those of the 2000 U.S. Census for Flagstaff regarding race.

	White, other than Hispanic	Hispanic	Native American	Black, other than Hispanic	Other
Our survey results	93%	4%	1.2%	.5%	1.2%
2000 Census for Flagstaff	77.9%	10.9%	10%	1.8%	N/A

Table 2. Percentage distribution of responses to statements regarding limitations on how people or interest groups participate in environmental review (± 5%).

	Strongly Agree	Agree	Neither	Disagree	Strongly Disagree
Environmental review requirements should be bypassed to restore forests (n = 680)	24.71	25.29	13.97	18.24	17.79
The right to appeal forestry projects should be limited (n = 676)	26.18	30.33	15.83	13.46	14.2
Right to sue government about proposed forestry projects should be temporarily banned (n = 675)	36.89	20.44	16.59	13.48	12.59

Table 3. Percentage distribution of responses to statements regarding community involvement in forest management decisions (± 5%).

	Strongly Agree	Agree	Neither	Disagree	Strongly Disagree
Communication between federal agencies and locals needs to improve (n = 688)	57.12	32.41	8.28	1.31	0.15
Local residents should be informed before restoration projects begin in their communities (n = 686)	59.18	31.78	3.79	4.08	1.17
Local residents should be involved in community forest restoration projects (n = 686)	47.38	41.11	8.16	3.06	0.29

Table 4. Percentage of sources selected by respondents as preferred for information on forestry projects.

Source	Percent
Newspapers	33.3
United States Forest Service	27.9
TV	13.7
Radio	8.6
Local government agencies	6.1
Internet	5.4
Non-profit organizations	1.8
Family / friends	1.0
Other government agencies	1.0
Magazines	0.6
Other	0.6

taking action to help plant and animal species "recover." There was also some support for the U.S. Forest Service to generate funds from harvest, but without support for timber as the sole source of financial support for Forest Service restoration projects.

Responses were much less unified on other ostensible purposes or motivation for forest restoration. Surprisingly, protecting humans from fire was not a pervasive motivation among respondents (Table 6). Nor was there a unified belief that forest restoration would control nature or improve how the forest looks. Respondents differed substantially on whether forest restoration means removing "most" trees from a site.

Our survey included three statements that pertained to how respondents envisioned a "restored forest." Fundamentally, was it a wilderness with a natural fire regime, or was it a restored forest subject to human use, commercial activities, and access? The majority of respondents viewed restored forests as accessible to humans and subject to utilization (Table 7). Although somewhat ambivalent about removing timber every 20–30 years, respondents clearly perceived a restored forest as one not left "to go wild."

DISCUSSION AND CONCLUSION

We documented several strong trends in the opinions and perspectives of residents across northern Arizona who live in the ponderosa pine forest. These trends can be used by forest managers and policy makers as they determine the future of forest resources. However, our data may be biased. Though we can control for age and gender in future analysis, underrepresentation of Native Americans and Hispanics is a real concern. Our survey probably discriminated against lower-income and less-educated groups because it was long and in English. Our follow-up phone calls to "non-respondents" revealed that many numbers had been disconnected, which is more common for lower-income people. Public agencies are consistently challenged with adequately representing those who typically do not participate in public elections or attend information forums such as those mandated during the Environmental Assessment or Environmental Impact Statement process. Land management agencies may need to allocate funds and adjust priorities to facilitate participation, or conduct the extensive survey efforts that would be required to assess underrepresented groups. On the other hand, our data represent the segment of the general public who are essential to building political support for decisions on future forest conditions (i.e. the voting population).

Some of our strongest results concern public participation and communication. By and large, our respondents accepted temporary limits on rights to appeal or sue the government and/or streamlining of the environmental review process in forest projects (55% ± 5% strongly agreeing or agreeing, Table 2). However, our respondents almost universally thought that affected communities needed to be informed and involved in restoration projects (90% ± 5% strongly agree or agree, Table 3). Few survey results are this clear. The public wants to hear from federal land managers and they want to be included in planning. Furthermore, with such a high preference among our respondents for communication and "involvement" in forestry projects, this result also probably applies to underrepresented groups. The fact that our respondents also preferred to

Table 5. Percentage distribution of responses to statements regarding support for forest restoration (± 5%).

	Strongly Agree	Agree	Neither	Disagree	Strongly Disagree
The main purpose of forest restoration is to promote working ecosystems (n = 674)	51.6	33.2	7.1	3.8	1.7
Support efforts to reduce fire even if it changes the forest (n = 680)	56.2	35.6	3.8	2.8	2.1
Forest restoration should be used to help recover plant and animal species that are rare and endangered (n = 681)	30.3	42.6	11.8	9.5	4.0
National forest management should generate income (n = 671)	21.3	36.7	22.4	10.9	8.8
The Forest Service's budget should not depend on how much commercial timber is produced (n = 685)	46.2	32.9	11.5	5.5	2.5

Table 6. Percentage distribution of responses to statements regarding purposes of forest restoration (± 5%).

	Strongly Agree	Agree	Neither	Disagree	Strongly Disagree
Main purpose of forest restoration is to protect humans from fire (n = 684)	10.7	21.1	18.3	33.6	16.4
Restoration efforts will help us to control nature (n = 675)	16	33.3	22.4	14.1	14.2
Forest restoration should try to improve how the forest looks (n = 679)	19.1	37.1	23.7	12.4	5.8
I would support forest restoration even if most trees, large and small, are removed (n = 693)	11.4	26.7	12.8	26.8	20.8

Table 7. Percentage distribution of responses to statements regarding conditions of a restored forest (± 5%).

	Strongly Agree	Agree	Neither	Disagree	Strongly Disagree
It's okay to remove timber from a restored forest every 20 to 30 years (n = 685)	17.0	38.0	29.1	11.3	3.5
It is acceptable for a restored forest to have permanent dirt roads for future management needs (n = 687)	39.2	46.5	5.8	5.1	2.6
For restoration, a forest may be thinned out by logging but only once and then allowed to go wild (n = 686)	2.7	14.7	20.6	35.2	25.7

receive their information from the U.S. Forest Service should be encouraging to federal agencies and may indicate a rising level of trust in federal agencies. Similarly, given the trust placed in newspapers by our respondents, the U.S. Forest Service would likely benefit by making it clear when they are the source of information presented in a newspaper.

In our statement, "The main purpose of forest restoration is to promote working ecosystems," we deliberately chose the phrase "working ecosystems" to try to assess whether the public believes that the purpose of forest restoration is more than producing goods and services for humans. Eighty-four percent (± 5%) of respondents "strongly agreed" or "agreed" with that statement, 91 percent agreed that it was okay to change how the forest looks, and 73 percent agreed that the purpose of forest restoration is to help recover plants and animals. These results suggest that respondents are aware that there is a biological underpinning to forest restoration practices—that is, that the purpose of forest restoration is more than managing a crop or producing timber products. And though there was some support among our respondents for generating income from timber, there was clear opposition to timber as the sole source of income. Based on these results, the public supports active intervention on public forests, but not for purely commercial ends.

Our respondents did not view forest restoration primarily as a means of protecting humans from fire. However, previous literature shows broad acceptance of fire prevention, suggesting that the public distinguishes between forest restoration projects to reduce fire hazards. Many of our respondents also accepted the removal of large trees during forest restoration. This result may support those who oppose diameter caps on forest restoration projects, but, on the other hand, 47 percent of respondents were ambivalent about or agreed that large trees should never be removed. Similarly, our respondents exhibited a wide range of opinions about whether forest restoration ought to improve how a forest looks or if restoration will help to "control nature."

An important feature of forest restoration debates pertains to final conditions, or the role or purpose of a "restored" forest. For instance, is a ponderosa pine forest "restored" when the ecological system is complete with park-like conditions, a pre Euro-American fire regime, vigorous understory vegetation, and most of the native fauna? More importantly, once restored, is the forest open to human access and, if so, for what? This issue is important because a healthy ecosystem, and the model for a restored forest, has often been defined as pre-Columbian. Though Native Americans undoubtedly influenced ponderosa pine ecosystems, their impact pales in comparison to the impacts of thousands of miles of dirt roads that allow access for hunting, hiking, logging, and other human activity. Is a restored forest one that is essentially a "wilderness area" with minimal human impact? Or may a restored forest contain all the ecological components as well as bulldozers and road graders? Our respondents believed that timber operations may be and dirt roads definitely are part of a restored forest (85% ± 5% strongly agreeing or agreeing). Given that roads potentially have many adverse effects (Forman and Alexander 1998), forest managers will have to grapple with how many miles of dirt roads and how much timber, especially if they want to also sustain or restore organisms such as large carnivores.

Public land management is a contentious and difficult undertaking in the best of circumstances. Nonetheless, the public is very interested in participating in forest management decisions and the concept of forest restoration appears to be gaining ground and becoming better understood. These trends may or may not facilitate decision making, but forest managers and other public decision makers need to realize how important ongoing, two-way communication is for public support and success in management. This survey is only one source of information for forest managers and may only reflect perspectives prevalent during the summer of 2003. We recommend that regular surveys (every 2 years for instance) be conducted that emphasize the collection of information from culturally underrepresented groups. Surveys may be conducted by different investigators with different foci, but each should attempt to increase our insight into public perceptions. The type of data presented here should also be combined with information from focus groups and personal contacts to clarify what the public expects, perceives, and is willing to trade for future forest conditions. No solution can be applied in all situations nor is any solution final. As information and public opinion change, the challenge for land managers will be to assess public expectations and respond by adjusting long-term management strategies.

ACKNOWLEDGMENTS

This research was supported by a grant from the Ecological Restoration Institute at Northern Arizona University. The authors also thank the Southwest Fire Social Science Collaborative for their repeated input and review of the questionnaire.

LITERATURE CITED

Public Law No: 108-148, 117 Stat. 1887. 2003. Healthy Forests Restoration Act of 2003. Signed into law 12/3/03. U.S. Government Printing Office, http://thomas.loc.gov/.

Bailey, D. W. 1991. The wildland-urban interface: Social and political implications in the 1990's. Fire Management Notes 52: 11–18.

Bradshaw, T. D. 1987. The intrusion of human population into forest and range lands of California. Proceedings of the Symposium on Wildland Fire 2000. April 27–30, 1987, South Lake Tahoe, CA, General Technical Report PSW-101, USDA Forest Service, Pacific Southwest Forest and Range Experiment Station, Berkeley CA.

Brunson, M. W., B. Schindler, and B. Steel. 1997. Consensus and dissension among rural and urban publics concerning forest management in the Pacific Northwest. In Public Lands Management in the West: Citizens, Interest Groups, and Values, edited by Brent Steel. Greenwood Publishing, Westport CT.

Burns, S., C. Sperry, and R. Hodgson. 2002. People and fire in Western Colorado: Methods of engaging stakeholders. Conference presentation at Fire, Fuel Treatments, and Ecological Restoration: Proper Place, Appropriate Time. April 16–18. Fort Collins CO.

Cardille, J. A., S. Ventura, and M. G. Turner. 2001. Environmental and social factors influencing wildfires in the upper Midwest, United States. Ecological Applications 11: 111–127.

Cleaves, D. A., J. Martinez, and T. K. Haines 2000. Influences on prescribed burning activity and costs in the national forest system. Southern Research Station, GTR SRS-37. USDA Forest Service.

Connelly, N. A., T. J. Brown, and D. J. Decker 2003. Factors affecting response rates to natural resource- focused mail surveys: Empirical evidence of declining rates over time. Society and Natural Resources 16: 541–549.

Cook, S. 1997. Wildfire adapted ecosystems meet man's development. Australian Journal of Emergency Management 12: 24–31.

Cooper, C. F. 1960. Changes in vegetation structure, and growth of southwestern pine forests since white settlement. Ecological Monographs 30: 129–164.

Cortner, H. J. 1991. Interface policy offers opportunities and challenges: USDA Forest Service strategies and constraints. Journal of Forestry 89: 31–34.

Cortner, H. J., P. D. Gardner, J. G. Taylor, E. H. Carpenter, M. J. Zwolinski, T. C. Daniel, and K. J. Stenberg. 1984. Uses of public opinion surveys in resource planning. Environmental Professional 6: 265–275.

Covington, W. W. 1994. Postsettlement changes in natural fire regimes and forest structure. Journal of Sustainable Forestry 2: 153–181.

Dahms, C. W., and B. W. Geils, Technical Editors. 1997. An assessment of forest ecosystem health in the Southwest. USDA Forest Service, Rocky Mountain Forest and Range Experiment Station, General Technical Report RM-GTR-295.

DeMillion, M. A. 1999. Mount Logan Wilderness reference condition and social preferences for ecological restoration. Master's thesis, Northern Arizona University, Flagstaff.

Dillman, D. A. 2000. Mail and internet surveys: The tailored design method. 2nd ed. John Wiley and Sons, New York.

Findley, A. J., M. S. Carroll, and K. A. Blatner. 2001. Social complexity and the management of small diameter stands. Journal of Forestry 99: 18–27.

Forman, R. T. T., and L. E. Alexander. 1998. Roads and their major ecological effects. Annual Review of Ecology and Systematics 29: 207–231.

Fulé, P., W. Covington, and M. Moore. 1997. Determining reference conditions for ecosystem management of southwestern ponderosa pine forests. Ecological Applications 7: 895–908.

Gardner, P. D., H. J. Cortner, K. F. Widaman, and K. J. Stenberg. 1985. Forest-user attitudes toward alternative fire management policies. Environmental Management 9(4): 303–311.

Herron, J. 2001. "Where there's smoke": Wildfire policy and suppression in the American Southwest. In Forests under Fire, edited by C. J. Huggard and A. R. Gomez. University of Arizona Press, Tucson.

Jacobson, S. K., M. C. Monroe, and S. Marynowski. 2001. Fire at the wildland interface: The influence of experience and mass media on public knowledge, attitudes, and behavioral intentions. Wildlife Society 29: 929–937.

Jamieson, A., H. B. Shin, and J. Day. 2002. Current population reports: Voting and registration in the election of November 2000. U.S. Census Bureau, Washington D.C.

Lavin, M. J. 1997. Managing fire risk to people, structures, and the environment. Fire Management Notes 57: 4–6.

Loomis, J. B., L. S. Bair, and A. Gonzalez-Caban. 2001. Prescribed fire and public support: Knowledge gained, attitudes changed in Florida. Journal of Forestry 99: 18–23.

Main, W. A., and D. A. Haines. 1974. The causes of fire on northeastern national forests. USDA Forest Service Research Paper NC-102.

Martin, R. E. 1997. Prescribed fire as a social issue. Conference Proceedings: Environmental Regulation and Prescribed Fire: Legal and Social Challenges. Tampa Airport Hilton, Tampa, FL, March 1417, 1995. Center for Professional Development, Florida State University, Tallahassee.

Moore, M. M., W. W. Covington, and P. Z. Fulé. 1999. Reference conditions and ecological restoration: A Southwestern ponderosa pine perspective. Ecological Applications 9: 1266–1277.

Pyne, S. J. 2001. The perils of prescribed fire. Natural Resources Journal 41: 1–8.

Shelby, B., and R. W. Speaker. 1990. Public attitudes and perceptions about prescribed burning. In Natural and Prescribed Fire in Pacific Northwest Forests, edited by J. D. Walstad, S. R. Radosevich, and D. V. Sandberg, pp. 253–260. Oregon State University Press, Corvallis.

U.S. Bureau of the Census. 2003. http://www.census.gov/main/www/cen2000.html (Web site last revised January 2, 2004).

van Riper III, C., and M. K. Sogge. 2004. Bald eagle abundance and relationships to prey base and human activity along the Colorado River in Grand Canyon National Park. In The Colorado Plateau: Cultural, Biological, and Physical Research, edited by C. van Riper III and K. A. Cole, pp. 163–185. University of Arizona Press, Tucson.

Woolley, J. T., and M. V. McGinnis. 2000. The conflicting discourses of restoration. Society and Natural Resources13: 339–357.

ATTITUDES TOWARD AND PERCEPTIONS OF MOUNTAIN LIONS: A SURVEY OF NORTHERN ARIZONA RESIDENTS

Elizabeth J. Ruther and David M. Ostergren

In the last 20 years mountain lion–human conflict has increased dramatically throughout the western United States (Beier 1991; Green 1991; Foreman 1992). Recently, the public and media have been paying more attention to the issue (Reid 2003). Reasons for this trend include an expanding human population, rebounding mountain lion populations from historic lows in the early twentieth century, and rising prey populations surrounding residential developments. Often, the most accessible and productive landscapes that are developed for human use are also prime mountain lion habitat (Halfpenny et al. 1991; Sunquist and Sunquist 2001). These elements join to create interaction and conflict between human and lion.

Because of their large size and high trophic level, mountain lions require large home ranges with an adequate prey base to sustain a viable population (Logan and Sweanor 2001). Human development of mountain lion habitat results in habitat encroachment and fragmentation, both of which directly threaten the mountain lion's existence. Core undeveloped habitat becomes smaller and edges connected to human activity become more numerous. Many times developed areas act as impenetrable boundaries, reducing the cats' ability to find mates or disperse as juveniles (Beier and Barrett 1993; Beier 1995). Eventually, the ratio of intact mountain lion habitat to development reaches a critical threshold where the quality of habitat has diminished to a level in which remaining suitable habitat is virtually nonexistent (Sunquist and Sunquist 2001).

According to the U.S. Census Bureau, the national rate of human population growth from 1990 to 2000 was 13.2 percent. Arizona is the second fastest growing state, expanding at three times the national growth rate, or 39.6 percent (U.S. Census 2000). As the West continues to experience the most extreme population redistribution since the late 1800s (Riebsame 1997), human-lion conflict research has the potential to help mitigate future conflict. Human dimensions of carnivore management studies have not previously been investigated in northern Arizona, and mountain lion–human conflict has increased in recent years in this region. Therefore, results from this report have substantial implications for the future of mountain lion conservation in Arizona, posssibly including a reduction in the discrepancies between what the public actually desires and what managers think the pubic desires, which may help reduce both public-agency conflict and human–mountain lion conflict.

Interest in the human dimension of wildlife research is intensifying as the search for the root causes and solutions to predator-human conflict continues (Kellert 1985; Manfredo et al. 1998; Zinn et al. 1998; Riley and Decker 2000; Peine 2001). Looking at the foundation of behaviors and attitudes toward mountain lions may hold the key to reducing conflict. Values are stable beliefs that individuals use as standards for evaluating life; they are unlikely to change as rapidly as behaviors and attitudes (Fulton et al. 1996). Value orientations, attitudes and norms, behavioral intentions, and behaviors build upon the underlying values that individuals hold, which are difficult to change (Vaske and Donnelly 1999). Therefore, when we understand the public's value orientations toward wildlife and nature we will be able to better predict and influence reactions to management decisions and we will significantly improve managers' ability to meet wildlife management goals (Manfredo et al. 1998). Peine et al. (1999) asserted that "managers need to learn to expect, as a matter of routine, sociological analysis of contentious issues related to the interface of natural and social systems" (p. 79). This investigation seeks to contribute to the growing base of information on human dimensions research on mountain lion–human conflict and to contribute to future solutions.

We conducted a mail survey of northern Arizona residents during the summer of 2003. The results presented here are part of a larger investigation that seeks to understand residents' basic value orientations toward wildlife and nature and that aims to identify key factors that explain how such value orientations are formed.

METHODS

The mail survey was an endeavor to assess northern Arizona residents' value orientations and perceptions of mountain lions and also their opinions and knowledge of forest restoration issues. It was therefore important to reach residents who live in the ponderosa pine forest ecosystem. This population sample is also appropriate for mountain lion–human dimensions research because it contains the main societal components that contribute to lion-human conflicts throughout the West. Some of these components include, but are not limited to, the higher density of lions in northern Arizona compared to other parts of the state, the quantity and availability of outdoor recreational activities, increasing development, and the level of encounters or conflict many towns in northern Arizona have experienced. The sample was determined by laying a map of Arizona postal zip code regions over a vegetation map of the state. Thirty-five zip codes were recorded that overlaid the ponderosa pine forest ecosystem in northern Arizona. A random sample of 1729 residents, stratified by postal zip codes, was purchased from Genesys Sampling in Fort Washington, Pennsylvania. The U.S. Postal Service approved 1644 addresses as current and in use. Our target population was adult (> 18 yrs old), seasonal or permanent residents who have lived in the region one year or longer.

Four mailings, including a pre-notice letter, questionnaire, reminder postcard, and replacement questionnaire consistent with Dillman's (2000) mail survey methods, were sent to the selected respondents. The second mailing, which contained the 131-question questionnaire, an explanatory cover letter, a self-addressed, stamped return envelope, and an incentive of two postage stamps, was mailed approximately one week after the pre-notice letter, on 23 June 2003. A reminder postcard was sent 2 weeks later urging recipients to reply. On 18 July 2003, a replacement questionnaire was mailed as the final contact. All correspondence and questionnaires were mailed from and returned to Northern Arizona University.

Of the 1644 questionnaires sent, 27 were undeliverable, resulting in a sample of 1617 residents. Forty-six percent (n = 750) of the questionnaires were returned. Fifty-seven of the questionnaires, 0.04 percent, were returned blank or were determined unusable. The completed questionnaires (n = 693) translate into a 43 percent return rate, which is consistent with other natural resource focused mail survey return rates (Connelly et al. 2003).

To assess non-response bias, we contacted a sample of non-respondents during the winter of 2003/2004. Non-respondents were asked a sample of questions from each section of the questionnaire for comparison to the respondent sample. We attempted to call 174 non-respondents. Twenty-two people participated, 37 refused, 44 numbers were disconnected or wrong, 3 respondents were deceased, and 66 non-respondents could not be contacted after repeated calls.

Demographic characteristics were approximately equal to the sample population, with the exception that the mean length of residency was 23.1 years in contrast to 17.2 years. Most of the participants reported they did not remember why they did not send the survey back or stated they were too busy. The sample was too small to compare to the sample population statistically; however the characteristics of both populations were very similar and therefore non-respondent bias should not be a significant factor in this study.

We developed a self-administered mail questionnaire directed toward four major subject areas that affect attitudes toward wildlife and mountain lions: values toward wildlife and nature, knowledge about mountain lions, perceptions of mountain lions, and human–mountain lion conflict. The subject areas were adopted from Kellert's (1994) framework (Figure 1).

Each subject area was addressed using a series of statements rated on a scale ranging from strongly agree/support (1) to strongly disagree/oppose (5). Three was the neutral point on the response scale. Questions about location and length of residency, as well as frequency of recreational activities preceded questions regarding mountain lions. Demographic questions, such as age, gender, formal education, and occupation were asked at the end of the questionnaire to help explain the underlying correlates of values toward nature, as well as perception questions. A complete copy of the questionnaire is available upon request from the authors.

Frequencies and percentages were obtained through preliminary analyses using SPSS (version 11.5) software. Significant relationships between categorical variables were determined through chi-square tests.

RESULTS

Overall, the participating respondents were mostly white males with a mean age of 57.8 (minimum = 19, maximum = 93; Figure 2a). Respondents reported living in northern Arizona from 1 to 82 years. The average length of residency was 17.2 years; the most frequent responses were from 5 to 7 years (Figure 2b). Ninety-one percent of the respondents were permanent residents. Respondents have completed an average of 15.5 years of formal education (Figure 2c).

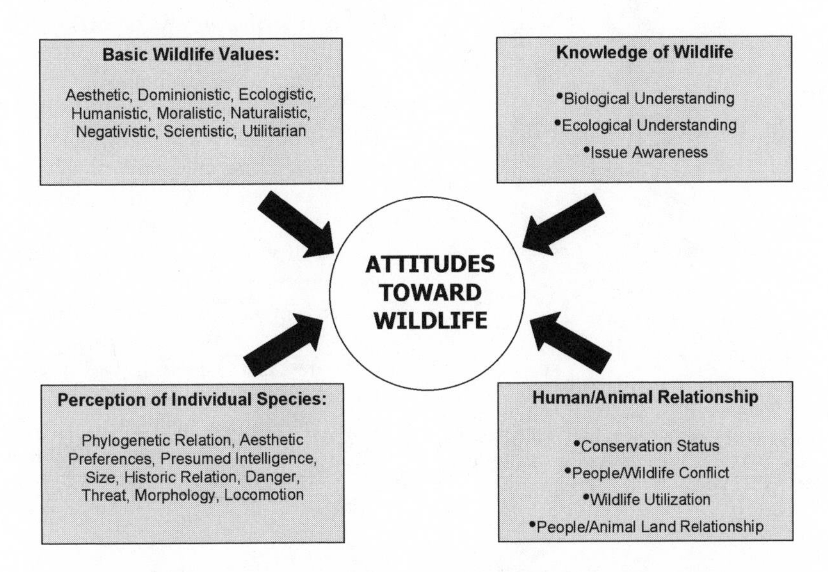

Figure 1. Factors affecting attitudes toward wildlife (from Kellert 1994).

(a)

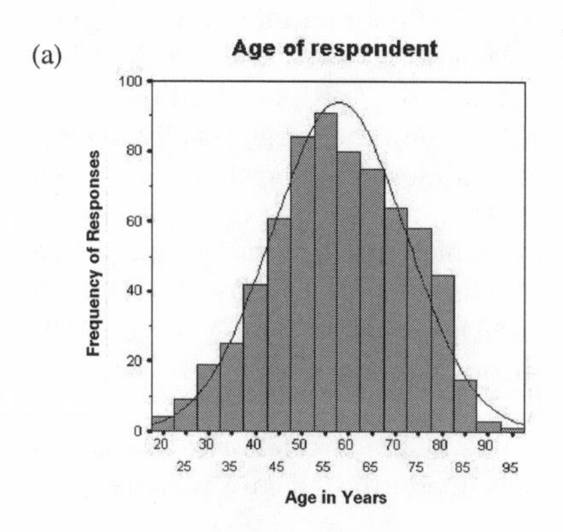

(b)

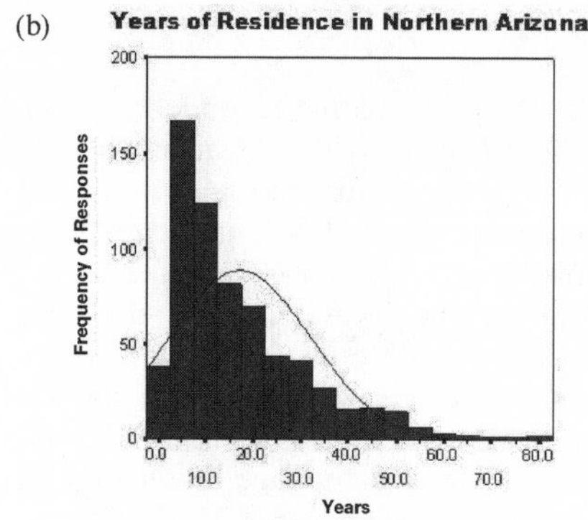

(c)

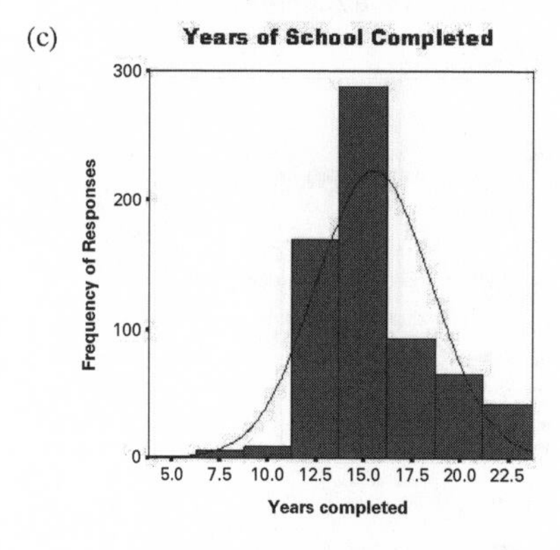

Figure 2. Respondent demographics. (a) Age distribution: mean = 57.8 years, n = 676, SD = 14.33; (b) years of residency: mean = 17.2 years, n = 661, SD = 14.83; (c) years of formal education: mean = 15.5 years, n = 673, SD = 3.01.

Opinions and Perceptions

Respondents were given a number of statements regarding the mountain lions' role and importance in nature, their existence value, and their worth to humans. When asked whether or not it was important to know that mountain lions existed in Arizona (even if they never saw one in the wild; Hook and Robinson 1982), more than 82 percent of the respondents (n = 682) agreed that it was important (Figure 3a). Responses to this positively framed question were mirrored by responses to a negatively framed question. Eighty-six percent of the respondents (n = 680) disagreed with the statement, "the mountain lion does not play an important role in nature" (Figure 3b).

Two factors were significantly related to how participants responded to the previous statement. There appears to be an inverse relationship between age and the importance that respondents placed on the mountain lion's role in nature ($p = <.0005$, $\chi^2 = 24.85$). The older the respondents, the more likely they agreed that the mountain lion did not have an important role in nature (Figure 4). Whether respondents lived in an urban (defined as 50,000+, i.e. Flagstaff, Arizona) or rural setting affected their answers; urban dwellers disagreed with the negative statement more consistently ($p = .036$, $\chi^2 = 11.06$) than rural residents. Gender, hunting as a hobby, and whether respondents thought that they had seen a mountain lion in the wild did not have a significant relationship to the aforementioned statement.

In December of 2001, two mountain lions were shot by Arizona Game and Fish because of negative interactions with pets and humans on a popular hiking trail near Flagstaff (AZ Daily Sun 2001b). Many concerned citizens protested the shootings and heated criticism followed in the weeks after the lions were destroyed (AZ Daily Sun 2001a). Therefore, we asked respondents whether or not they worry about contact with mountain lions. There appears to be an

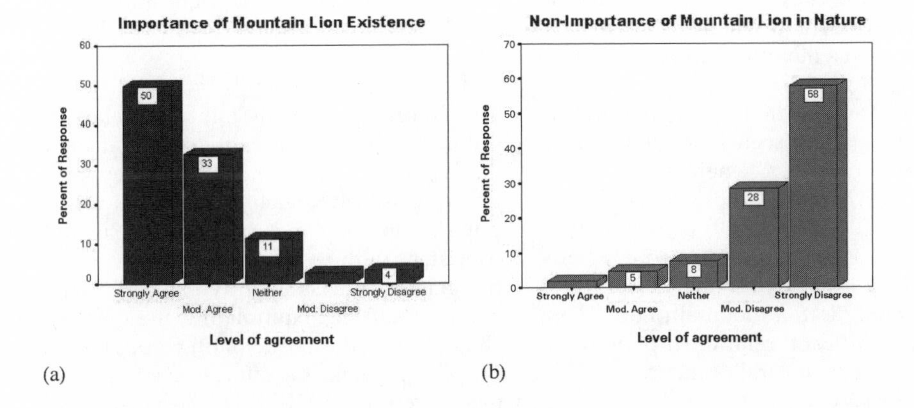

Figure 3. Percentages of people who strongly agreed to strongly disagreed with the statements (a) I may never see a mountain lion in the wild, but it is important to me to know they exist in Arizona (n = 682), and (b) the mountain lion does not play an important role in nature (n = 680).

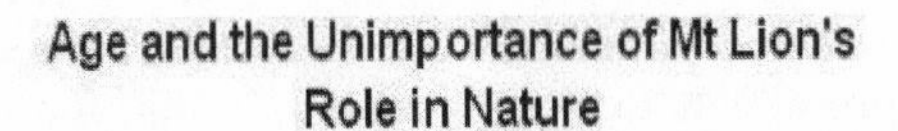

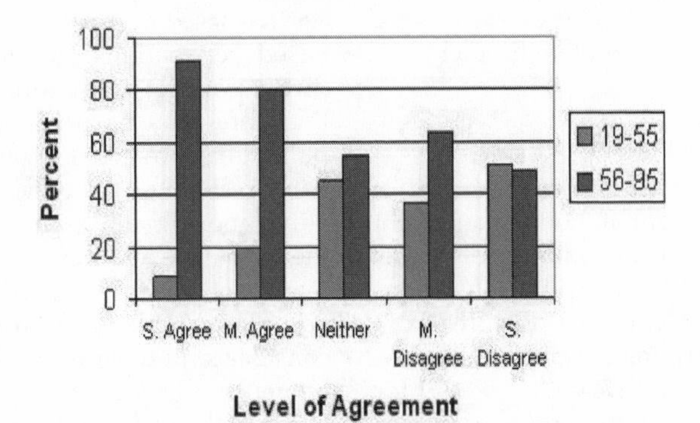

Figure 4. Relationship between age and the unimportance of the mountain lion's role in nature ($p = <.0005$, $\chi^2 = 24.85$).

inverse relationship between how much respondents worry and whether they believe they have seen a mountain lion at least once in the wild (Figure 5).

It is interesting that respondents who think that they have seen a mountain lion at least once worry less, whereas those who have not, worry more. Respondents who are older (p = .001, $\chi^2 = 17.99$) or female (p = .008, $\chi^2 = 13.84$) tend to worry more, whereas hunters (p = <.0005, $\chi^2 = 32.17$) worry less about running into a mountain lion. There was no significant relationship between worry and urban or rural residence.

Personal Safety Devices

Participants were also asked which device they would like to use to protect themselves while in mountain lion country. More than half (55.6%) of the respondents (n = 672) preferred non-lethal protection devices such as pepper spray or air horns, whereas 30.7 percent preferred to carry a gun. Approximately 14 percent of the respondents indicated that they would rather not carry, or use, any protection device. Although more than 85 percent of the respondents would like to carry a protection device, lethal or non-lethal, only 13.2 percent of the respondents reported actively worrying about running into a mountain lion while in the woods.

Mountain Lion Management Preferences

Respondents were also asked a series of questions about mountain lion management actions to determine the amount of public support that may be expected if the actions were actually implemented. Results from ratings of three actions are presented here. There is high support (77.3%, n = 686) for the live trapping and relocation of "problem" mountain lions (Figure 6a). There is also relatively high support (61%, n = 683) for limiting development of lands in mountain lion habitat (Figure 6b). The respondents (n = 683) had mixed opinions about implementation of a lottery system to regulate mountain lion hunting. More than half (52.2%) of the respondents supported the implementation of such a lottery system, approximately 22 percent were undecided, and 25 percent opposed it (Figure 6c).

DISCUSSION

This information sheds light on trends in public perceptions of carnivores and their management. The demographics reported do not precisely match our original target population. The age distribution is not consistent with residents of the region and the gender ratio is slightly skewed. However, instead of manipulating the data we chose to use the participating population. People who make the effort to participate in natural resource surveys are most concerned about these issues and engage most frequently in activities they feel sympathetic about (Theodori and Luloff 2002). Therefore, these respondents are probably the most vocal about mountain lion management policy and the most likely to interact with game and wildlife managers.

It is interesting that the most frequently reported length of residency in northern Arizona was 5–7 years, although the mean was 17.2 years. An influx of people from other parts of the country may affect overall mountain lion management policy preferences. Newcomers, with increasingly service-based occupations rather than land-based occupations, might experience less direct conflict with mountain lions and therefore may be more likely to support policy that protects the lions. On the other hand, residents new to the area may be more naive about living with mountain lions on the landscape and may be surprised or may overreact when encounters or conflicts occur. Agencies and city officials could consider supplying tourists and new homeowners with educational materials about living with mountain lions on the landscape to increase awareness and reduce the possibility of a negative encounter.

Our data indicate that respondents value the existence of mountain lions in the wild and that they believe mountain lions are important to ecosystems. There appears to be considerable support for conserving mountain lion populations. Furthermore, management actions such as limiting

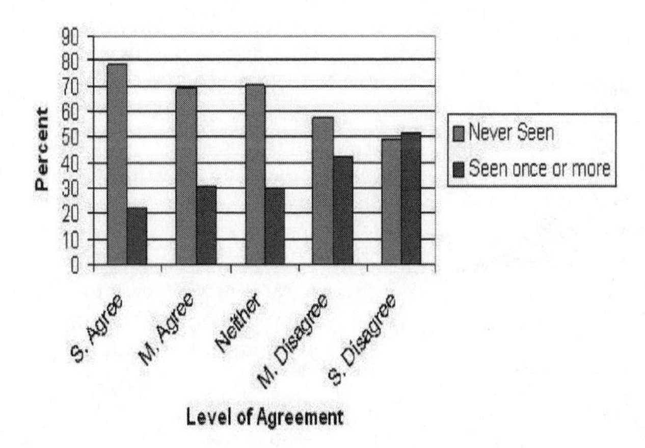

Figure 5. Relationship between whether people think that they have seen a mountain lion in the wild and active worry about them ($p < .0005$, $\chi^2 = 23.695$).

development in mountain lion habitat may have more support by the local community than expected. Broadly stated, actions such as limiting development or implementing a lottery system for hunting tags may be more complex issues in reality, but this information provides a starting place for discussion of alternative mountain lion management strategies.

This preliminary report and other studies (Bright et al. 2000; Chase et al. 2000; Miller and McGee 2001; Peine 2001) suggest that management of a species that ranges across several states and ecosystems is difficult and may be most successful at regional and local levels. Public perceptions, values, and opinions vary regionally and locally because of cultural and geographic differences. A local or statewide plan is needed to address these differences. After each state has a plan, a national or multi-state effort may be more feasible to address travel corridor and habitat fragmentation issues. Recognizing the scientific and political complexity of large carnivore management is vital to the overall success of future conservation strategies (Kellert 1994). The knowledge of how these forces interact on a community level increases the usefulness and practical application of pubic input for land and wildlife managers.

Our respondents from northern Arizona have placed value on mountain lions and support protecting them from population decline even if it limits human activity in some areas. They believe that mountain lions are integral to ecosystem health and agree that there are benefits to maintaining top predators in nature. Residents believe that a healthy number of cats should remain in the forest and they will support measures to maintain such populations. Thus, we suggest that management agencies should conduct an extensive, accurate assessment of mountain lion populations. Any resulting hunting policy should assure the general public that mountain lions will be a part of northern Arizona in the foreseeable future.

Conducting research and taking preventive measures may circumvent the heated debate that occurred in 2001 surrounding the Mt. Elden lion management decision. For example, when confronted with the question regarding protection devices, a majority (85%) of respondents chose to carry a safety device even if they responded that

(a)

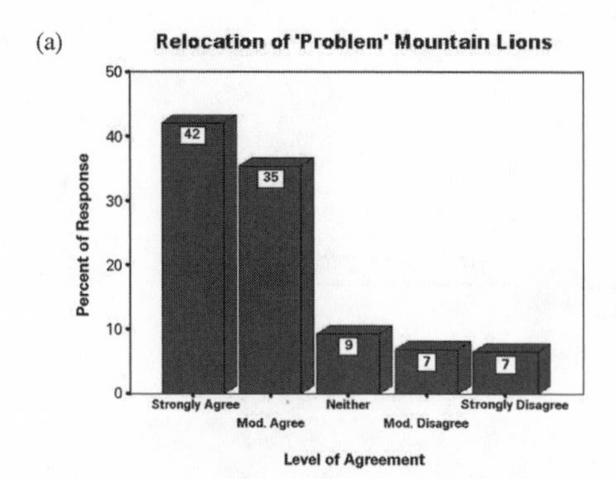

(b)

Lottery System Implementation

Percent of Response

40 30 20 10 0

29 23 22 7 18

Strongly Agree, Mod. Agree, Neither, Mod. Disagree, Strongly Disagree

Level of Agreement

(c)

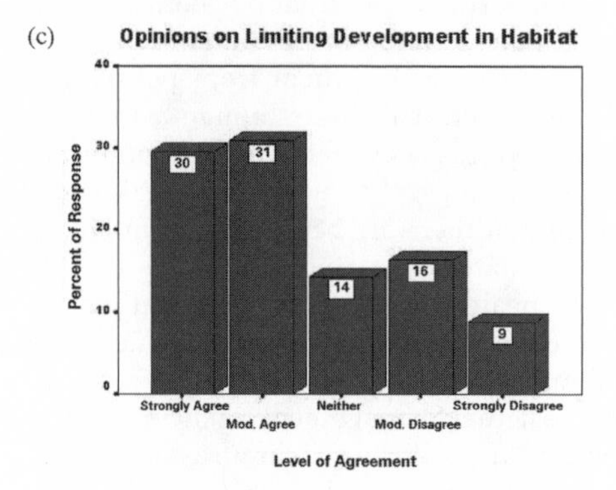

Figure 6. Percentages of people who strongly agreed to strongly disagreed with (a) relocating "problem" mountain lions (n = 686), (b) implementing a lottery system for permits like they do for elk (n = 683), and (c) limiting development of lands in mountain lion habitat (n = 683).

they do not actively worry about running into mountain lions while in the woods. Hence, widespread fear concerning mountain lion encounters does not exist. However, in areas where managers warn recreationists that there is a higher than average likelihood of encountering a mountain lion, suggesting additional safety strategies, such as air horns or pepper spray, may add to the public's peace of mind.

Further research on mountain lion encounters will provide managers and decision makers with a more complete picture of human-lion conflict. Money allocated and set aside to conduct a brief phone survey directly after a well-publicized mountain lion encounter would provide invaluable information regarding the threats and risks perceived by those living in mountain lion habitat. For instance managers have broader management options if they find that the public is tolerant of mountain lion sightings and/or more frequent encounters with pets. In addition, the Arizona Game and Fish Department could assess the support for various actions, including lethal and nonlethal options, and also assess whether recent encounters influence those opinions.

Increased competition between humans and wildlife for space and resources makes the expression of public values a growing necessity (Kellert and Clark 1991). Human dimensions research combined with wildlife research fills an important knowledge gap for managers. Inviting social scientists to collaborate with natural scientists and managers creates an entirely new arena where innovative policies and better management plans can respond more quickly and effectively to reduce human-lion conflict and to relieve the mounting human pressures faced by mountain lions.

ACKNOWLEDGMENTS

This project was supported in part by a grant from the Ecological Restoration Institute. A special thanks to Lisa Taylor for assistance with data entry.

LITERATURE CITED

AZ Daily Sun. 2001a. Elden lion killings caused an uproar. Arizona Daily Sun, Flagstaff. 12/30/2001.

AZ Daily Sun. 2001b. Elden lion set to be killed. Arizona Daily Sun, Flagstaff. 12/01/2001.

Beier, P. 1991. Cougar attacks on humans in the United States and Canada. Wildlife Society Bulletin 19: 403–412.

Beier, P. 1995. Dispersal of juvenile cougars in fragmented habitat. Journal of Wildlife Management 59: 228–237.

Beier, P., and R. H. Barrett. 1993. The cougar in the Santa Ana Mountain Range, California. Final Report: Orange County Cooperative Mountain Lion Study. Department of Forestry and Resource Management, University of California, Berkeley. 105 pp.

Bright, A. D., M. J. Manfredo, and D. C. Fulton. 2000. Segmenting the public: An application of value orientations to wildlife planning in Colorado. Wildlife Society Bulletin 28(1): 218–226.

Chase, L. C., T. M. Schusler, and D. J. Decker. 2000. Innovations in stakeholder involvement: What's the next step? Wildlife Society Bulletin 28(1): 208–217.

Connelly, N. A., T. L. Brown, and D. J. Decker. 2003. Factors affecting response rates to natural resource-focused mail surveys: Empirical evidence of declining rates over time. Society and Natural Resources 16(6): 541–549.

Dillman, D. A. 2000. Mail and Telephone Surveys: The Total Design Method. 2nd ed. John Wiley and Sons, New York.

Foreman, G. E. 1992. Pumas and people. Cat News 16: 11–12.

Fulton, D. C., M. J. Manfredo, and J. Lipscomb. 1996. Wildlife value orientations: A conceptual and measurement approach. Human Dimensions of Wildlife 1: 24–47.

Green, K. A. 1991. Summary: Mountain lion–human interaction questionnaires. In Proceedings of the Mountain Lion–Human Interaction, Symposium and Workshop, edited by C. E. Braun, pp. 4–9. April 24–26, 1991, Denver. Colorado Division of Wildlife, Denver.

Halfpenny, J. C., M. R. Sanders, and K. A. McGrath. 1991. Human-lion interactions in Boulder County, Colorado: Past, present, and future. In Proceedings of the Mountain Lion–Human Interaction, Symposium and Workshop, edited by C. E. Braun, pp. 10–16. April 24–26, 1991, Denver. Colorado Division of Wildlife, Denver.

Hook, R. A., and W. L. Robinson. 1982. Attitudes of Michigan citizens toward predators. In Wolves of the World: Perspectives of Behavior, Ecology, and Conservation, edited by F. H. Harrington and P. C. Paquet, pp. 382–394. Noyes Publishing, Park Ridge NJ.

Kellert, S. R. 1985. Public perceptions of predators, particularly the wolf and coyote. Biological Conservation 21: 167–189.

Kellert, S. R. 1994. Public attitudes toward bears and their conservation. International Conference on Bear Research and Management 9: 43–50.

Kellert, S. R., and T. W. Clark. 1991. The theory and application of a wildlife policy framework. In Public Policy Issues in Wildlife Management, edited by W. R. Mangun, pp. 17–38. Greenwood Press, New York.

Logan, K. A., and L. L. Sweanor. 2001. Desert Puma: Evolutionary Ecology and Conservation of an Enduring Carnivore. Island Press, Washington.

Manfredo, M. J., H. C. Zinn, L. Sikorowski and J. Jones. 1998. Public acceptance of mountain lion management: A case study of Denver, Colorado, and nearby foothill areas. Wildlife Society Bulletin 26(4): 964–970.

Miller, K. K., and T. K. McGee. 2001. Toward incorporating human dimensions information into wildlife management decision-making. Human Dimensions of Wildlife 6: 205–221.

Peine, J. D. 2001. Nuisance bears in communities: Strategies to reduce conflict. Human Dimensions of Wildlife 6: 223–237.

Peine, J. D., R. E. Jones, M. R. English, and S. E. Wallace. 1999. Contributions of sociology to ecosystem management. In Integrating Social Science with Ecosystem Management: Human Dimensions in Assessment, Policy, and Management, edited by H. K. Cordell and J. C. Bergstrom. Sagamore Publishing, Champaign IL.

Reid, E. 2003. Stalker. Outside Magazine, p. 11.

Riebsame, W. E. 1997. Atlas of the New West: Portrait of a Changing Region. W.W. Norton, New York.

Riley, S. J., and D. J. Decker. 2000. Wildlife stakeholder acceptance capacity for cougars in Montana. Wildlife Society Bulletin 28(4): 931–939.

Sunquist, M. E., and F. Sunquist. 2001. Changing landscapes: Consequences for carnivores. In Carnivore Conservation, edited by J. L. Gittleman, pp. 399–418. Cambridge University Press, Cambridge.

Theodori, G. L., and A. E. Luloff. 2002. Position on environmental issues and engagement in proenvironmental behaviors. Society and Natural Resources 15: 471–482.

U.S. Census. 2000. Censusscope. Social Science Data Analysis Network, University of Michigan, www.ssdan.net. Available online at www.censusscope.org.

Vaske, J. J., and M. P. Donnelly. 1999. A value-attitude-behavior model predicting wildland preservation voting intentions. Society and Natural Resources 12(6): 523–537.

Zinn, H. C., M. J. Manfredo, J. J. Vaske, and K. Wittmann. 1998. Using normative beliefs to determine the acceptability of wildlife management actions. Society and Natural Resources 11(7): 649–663.

DEMONSTRATION AND TEST OF A SPATIAL DECISION SUPPORT SYSTEM FOR FOREST RESTORATION PLANNING

Haydee M. Hampton, Ethan N. Aumack, John W. Prather, Yaguang Xu, Brett G. Dickson, and Thomas D. Sisk

The well-documented increase in fuels and decline in health in southwestern ponderosa pine forests since the late nineteenth century have been attributed to livestock grazing, logging, fire suppression, and human development (Cooper 1960; Harrington and Sackett 1990; Swetnam 1990; Covington and Moore 1994; Allen et al. 2002). Historical high-frequency, low-intensity fire regimes have been dramatically altered, leading to the increased likelihood of catastrophic fire. Although restoration planning does occur at the project level, treatment implementations have been hindered by the lack of broader-scale analyses in which impacts of alternative plans are compared (Sisk et al., in press). Planning at scales commensurate with key ecosystem processes, such as fire, allows estimation of aggregate effects of management on biodiversity and fire hazard. Given the need for broadscale planning, we developed a spatial decision support system (SDSS) as part of the Forest Ecosystem Restoration Analysis (ForestERA) project (Hampton et al. 2003; Hampton et al., in press) for application to hundreds of thousands to millions of acres. This SDSS may be used to rank and compare alternative forest management plans according to a set of user preferences.

The aim of ForestERA is to provide land managers and the public with the data and tools needed to efficiently use increasing quantities of ecological information, so that forest restoration problems can be addressed at spatial scales broader than individual projects. We have developed a flexible framework for addressing multiple questions regarding management implementation, including forest restoration prescriptions, at the landscape scale. Our SDSS is linked to a collection of integrated models (Figure 1; please note that the original color figures are all available at www.forestera.nau.edu) that predict effects of treatments on forest structure (Xu et al., in press), wildlife distributions (Prather et al., this volume; Prather et al., in press), fire hazard, and other parameters relevant to fire and forest ecology. It can be used more widely than the study region described in this paper; we are currently adapting the underlying models to ecosystems in north-central New Mexico and eastern Arizona.

The SDSS helps stakeholders to define objectives and apply criteria for designing and prioritizing forest treatments and to explore the tradeoffs between alternative management strategies. The first step involves defining the problem to be addressed and objectives to be met by the forest management scenario (Figure 2). In the second step stakeholder(s) build a prioritized management action scenario using spatial data that represent values, risks, and other factors. In the third step they select and assess criteria to evaluate the degree to which the scenario achieves its objectives. Estimates of the change in forest structure following various treatments allow us to assess changes in evaluation criteria, such as fire behavior and wildlife habitat characteristics. Stakeholders

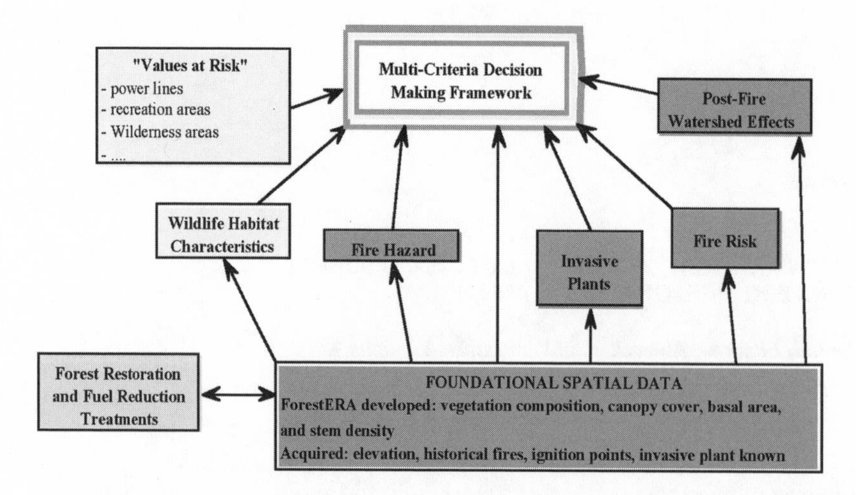

Figure 1. Flowchart showing the integration of foundational spatial data layers (green) which support many of the spatial models representing risks (pink) and values (yellow) across the study area. Stakeholders can select from the model outputs, other ForestERA-provided spatial data, and their own layers to develop alternative forest management scenarios in a multi-criteria decision making framework (or SDSS).

build alternative scenarios to explore the range of management options (step 4 in Figure 2) and then compare the alternatives spatially and in terms of their predicted effects (step 5). Finally, a sensitivity analysis could be performed to assess the relative impact of various decision criteria on scenario characteristics (step 6).

Through extensive outreach to diverse stakeholders over the last 3 years we have found that there is demand for the types of regional spatial data and tools in our SDSS to explore alternative forest management scenarios in the Southwest. The design of our SDSS and other aspects of our project have been guided by substantial input from stakeholders during initial scoping and throughout our efforts to finalize SDSS datasets and tools and test them in real-world planning forums. Because our SDSS is designed to model different combinations of stakeholder preferences in order to define a range of acceptable management scenarios, we have worked extensively with stakeholders in the region to characterize values and preferences. We designed our SDSS to support and foster a "civic science" in which, as explained by Cortner (2003), citizens participate directly in a discursive research process designed to promote democratic deliberation and collaborative learning. Harris (1995) has explained the need for scientists, resource managers, and the public to learn from each other to effectively bring science and policy together. This paper presents the results of a demonstration and test of our SDSS applied to a 2 million acre study area on the western Mogollon Plateau in northern Arizona (Figure 3).

METHODS

In order to demonstrate and test the capability of the ForestERA SDSS to represent stakeholder values and preferences in forest management, we asked six stakeholders from agencies, environmental groups, and academia to participate in a series of half-day workshops in September and October of 2003, in Flagstaff, Arizona. The goals of the exercise (referred to hereafter as the "SDSS test") were to test the capacity of the SDSS to support landscape-scale restoration analysis, solicit feedback on project deliverables from ForestERA stakeholders, demonstrate facets

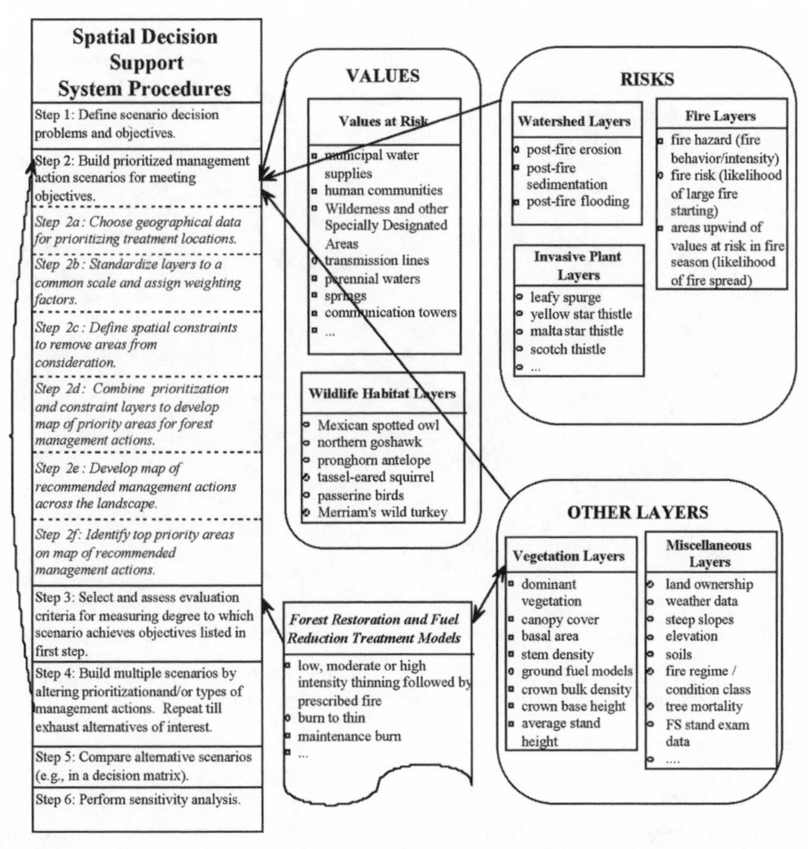

Spatial Decision Support System Procedures
Step 1: Define scenario decision problems and objectives.
Step 2: Build prioritized management action scenarios for meeting objectives.
Step 2a: Choose geographical data for prioritizing treatment locations.
Step 2b: Standardize layers to a common scale and assign weighting factors.
Step 2c: Define spatial constraints to remove areas from consideration.
Step 2d: Combine prioritization and constraint layers to develop map of priority areas for forest management actions.
Step 2e: Develop map of recommended management actions across the landscape.
Step 2f: Identify top priority areas on map of recommended management actions.
Step 3: Select and assess evaluation criteria for measuring degree to which scenario achieves objectives listed in first step.
Step 4: Build multiple scenarios by altering prioritizationand/or types of management actions. Repeat till exhaust alternatives of interest.
Step 5: Compare alternative scenarios (e.g., in a decision matrix).
Step 6: Perform sensitivity analysis.
VALUES
Values at Risk
municipal water supplies
human communities
Wilderness and other Specially Designated Areas
transmission lines
perennial waters
springs
communication towers
...
Wildlife Habitat Layers
Mexican spotted owl
northern goshawk
pronghorn antelope
tassel-eared squirrel
passerine birds
Merriam's wild turkey
RISKS
Watershed Layers
post-fire erosion
post-fire sedimentation
post-fire flooding
Fire Layers
fire hazard (fire behavior/intensity)
fire risk (likelihood of large fire starting)
areas upwind of values at risk in fire season (likelihood of fire spread)
Invasive Plant Layers
leafy spurge
yellow star thistle
malta star thistle
scotch thistle
...
OTHER LAYERS
Vegetation Layers
dominant vegetation
canopy cover
basal area
stem density
ground fuel models
crown bulk density
crown base height
average stand height
Miscellaneous Layers
land ownership
weather data
steep slopes
elevation
soils
fire regime / condition class
tree mortality
FS stand exam data
....
Forest Restoration and Fuel Reduction Treatment Models
low, moderate or high intensity thinning followed by prescribed fire
burn to thin
maintenance burn
...

Figure 2. Flowchart of ForestERA spatial decision support system showing relationships between system components and data layers. After objectives are identified, stakeholders select layers representing values and risks to identify priority areas (steps 2a–2d) and management actions (2e). Forest restoration models are used to adjust vegetation layers, which allow evaluation criteria, such as fire hazard and wildlife habitat characteristics, to be estimated.

of landscape-scale restoration analysis to a diverse audience, and identify points of (dis)agreement among stakeholders during the process of prioritizing treatments. To achieve our goal to test ForestERA capacities, we structured this SDSS test to explore the manner in which ForestERA tools would actually be used in a collaborative assessment process. We sought to find points of (dis)agreement among stakeholders as one small step toward finding mutually agreeable solutions to historically contentious issues. Fostering deeper understanding among stakeholders of various perspectives has been identified as a valuable outcome in and of itself (Oelschlaeger 2003).

Before the first workshop we distributed to the six participants introductory information that described the ForestERA SDSS as well as a questionnaire prompting them for decision objectives and criteria so that each of them could develop a landscape-scale forest treatment scenario for the western Mogollon Plateau region. We asked them to specify a title and brief description for their scenario, fire mitigation or other objectives, a list of weighted criteria for prioritizing treatment locations, and a characterization of the types of treatments that would be simulated within the prioritized locations,

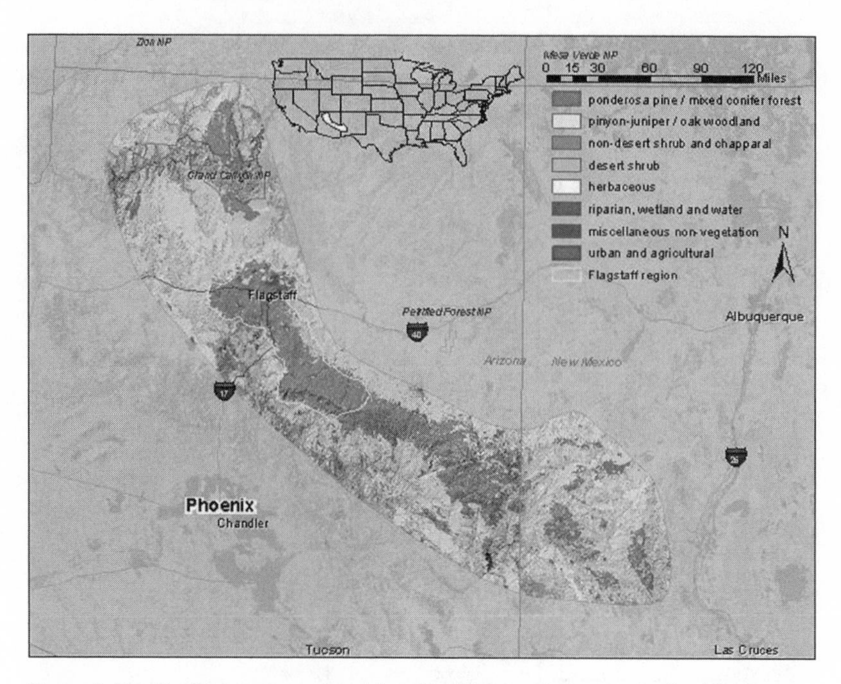

Figure 3. The 2 million acre western Mogollon Plateau study area (yellow) is shown within the major ponderosa pine dominated areas in northern Arizona.

including areas they thought should be excluded from treatment. We limited the area of simulated treatments to 150,000 of the 2 million acres in the study site as a rough estimate of the area that could be treated in 10–15 years. We also asked the participants to recommend additional criteria for developing prioritized treatment scenarios. Due to limited time for the SDSS test, we limited each stakeholder representative to developing and running one scenario. This limitation has been overcome since our SDSS test by fully automating our models; collaborators can now explore multiple what-if scenarios in an iterative fashion. We informed the participants that we would synthesize the results of our test and present them at the 7th Biennial Conference of Research on the Colorado Plateau in November of 2003.

We reviewed five scenarios developed from information provided by participants and identified techniques for comparing them at our first stakeholder meeting in late September of 2003. At the second meeting, in mid October, we compared scenarios and presented methods for creating "hybrid" scenarios to the group. The group roughly outlined three hybrid scenarios of interest aimed at protecting community values, ecosystem values, and specially designated lands such as Wilderness Areas. After the second meeting, our team refined the specifications for each hybrid and asked participants to fill out a second questionnaire to identify weighting factors for prioritization criteria we had selected for the three hybrid scenarios. At the third meeting, in late October, we reviewed the three hybrid scenarios, identified points of agreement and disagreement, and discussed participant concerns and suggestions about datasets, tools, and prioritization and comparison processes. We asked them to fill out a third and final questionnaire prioritizing future refinements of the ForestERA SDSS.

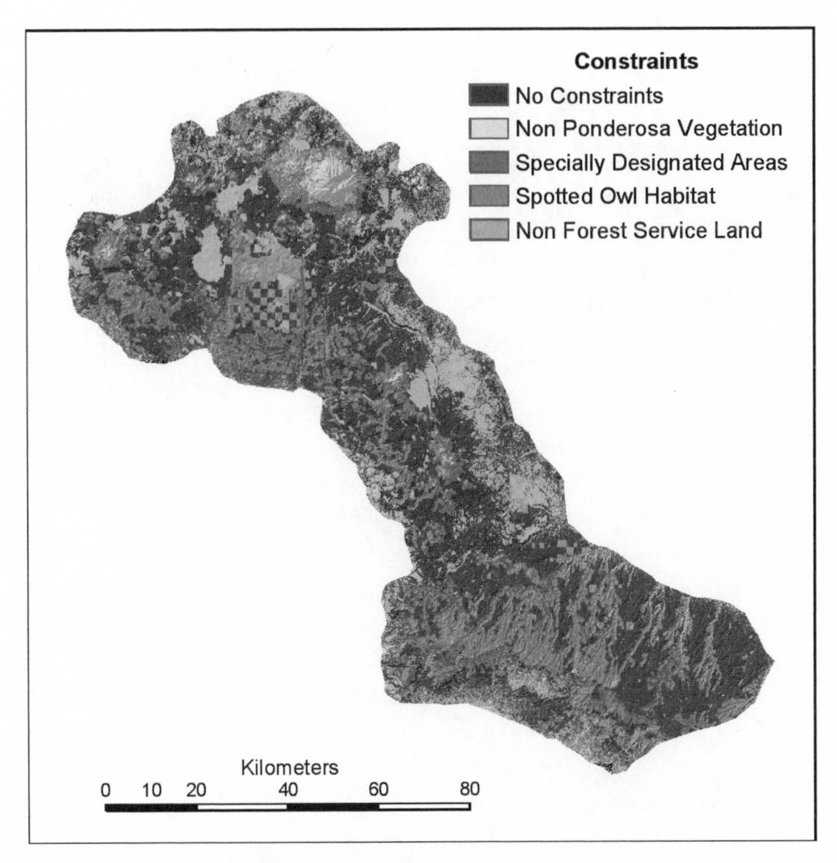

Figure 4. Areas restricted from treatment in the Wildland Fire Mitigation and Preparation scenario in the western Mogollon Plateau study area.

We strove to find participants for the SDSS test who were informed and active in forest restoration activities in the study region and whose combined experience provided overall diversity in terms of both expertise (e.g., forest policy, forest treatments, fire ecology, wildlife biology) and organizational affiliations (e.g., management, academia, environmental groups). However, the participants made it clear that they were not representing their organizations, but instead were using their individual expertise and perspectives to explore the extent to which the ForestERA SDSS could represent either their own values and preferences or those values that would form a better challenge or test of the SDSS. The participants were Bill Block, Project Leader, Rocky Mountain Research Station, USDA Forest Service; Pete Fulé, Assistant Professor, Forestry, Northern Arizona University; Steve Gatewood, Executive Director, Greater Flagstaff Forests Partnership; Heather Green, Community Forestry Liaison, Region 3, USDA Forest Service; Taylor McKinnon, Forest Conservation Program Manager, Grand Canyon Trust; and Todd Schulke, Forest Policy Director, Center for Biological Diversity.

The six stakeholders who participated in this test and demonstration represented a significant level of diversity in terms of the interests and areas of expertise of forest professionals and other stakeholders in

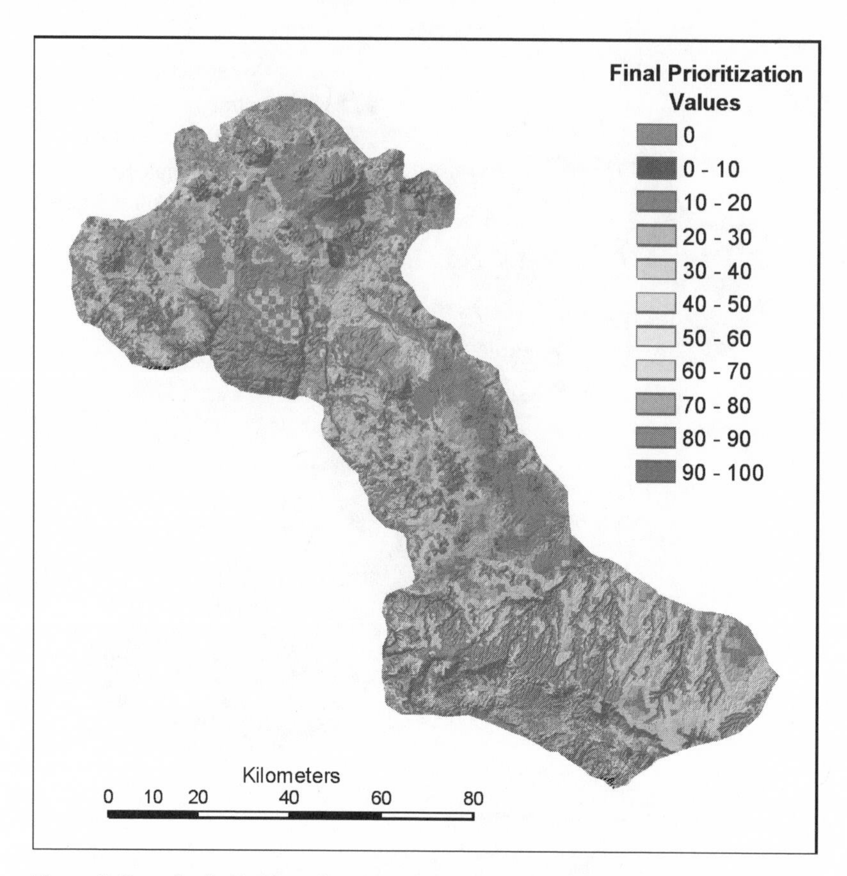

Figure 5. Overall prioritization of areas in the western Mogollon Plateau study area for application of treatments under the Wildland Fire Mitigation and Preparation scenario. Areas constrained from treatment are labeled as zero.

northern Arizona. However, due to the small number of participants compared to subsequent collaborative workshops, such as the Western Mogollon Plateau Adaptive Landscape Assessment (Sisk et al. 2004), it was not possible to capture the full spectrum of perspectives. For example, we did not have representation from commodity interest groups, watershed specialists, or political office holders.

RESULTS

Workshop participants gained familiarity with SDSS data and methods, saw and compared the results of prioritized treatment scenarios, and provided valuable feedback on how we could most efficiently deliver our data and tools to the public. Five of the six participants filled out the initial questionnaire and developed forest management scenarios using the data and methods presented to them. All six participated in two or more of the three workshops. Different scenarios prioritized or otherwise highlighted different values.

As an example of scenario development, we present assumptions and results pertaining to the "Wildland Fire Mitigation and Preparation" scenario. This scenario prioritizes three values: human communities, municipal watersheds, and at-risk focal species habitats. The first goal is to immediately mitigate threats of incoming crown fire to critical landscape values. The second goal is to create a landscape relatively safe from crown fires, thereby decreasing risks from subsequent applications of prescribed fire to achieve needed restoration goals in the matrix of wildland forests. The Wildland Fire Mitigation and Preparation scenario assumes that subsequent restoration will minimize manipulations of forest structure to reestablish a naturally variable fire regime over large areas.

For each scenario participants could specify jurisdictional, ecological, political, or other constraints for limiting the placement of modeled treatments across the landscape. The participant who created the Wildland Fire Mitigation and Preparation scenario decided to constrain treatments from areas with slopes greater than 40%, areas not dominated by ponderosa pine, non U.S. Forest Service land outside a half-mile distance from human developments (i.e., Wildland Urban Interface; WUI), predicted Mexican spotted owl (MSO) nesting and roosting habitat, and specially designated areas including Inventoried Roadless and Wilderness Areas and national monuments (Figure 4). Conversely, participants were asked to specify which criteria for placing treatments (and relative weights applied to each) would best address their specified goals and objectives (Table 1). We then prioritized areas for treatment based on spatially explicit application of weighted criteria (Figure 5).

Each participant specified which type of treatment to place within the highest priority 150,000 acres. In the Wildland Fire Mitigation and Preparation scenario, the participant specified that "full restoration" treatments (i.e., high-intensity thinning followed by prescribed burning) should occur within communities and a half-mile buffer surrounding them with the exception that all predicted goshawk nesting habitat would receive intermediate treatments (moderate thinning plus prescribed burning). In municipal watersheds, full restoration treatments were mandated in areas with higher fire hazard, except in predicted goshawk habitat and areas of high recruitment for tassel-eared squirrels, in which case treatments would only be applied where they would not drop habitat below critical thresholds. These specifications resulted in a mixture of high and medium intensity treatments (Figure 6) spread throughout the scenario's highest priority 150,000 acres for restoration (Figure 5).

Table 1. Problems, scenario objectives, prioritization criteria, and weighting factors used in the Wildland Fire Mitigation and Preparation scenario. Areas with higher combined values of the weighted criteria in this table are classified as higher priority areas for management.

Problems	Scenario Objectives	Prioritization Criteria	Weights
Forest ecosystems are inadequately protected from high-intensity crown fires.	Minimize fire threat (hazard) to overall ecosystem.	Fire hazard (measure of the amount of fuel available to a burning fire).	10
		Fire risk (relative chance of a fire >40 acres starting).	1
Human communities are inadequately protected from high-intensity crown fires.	Reduce fire threat (hazard) to human communities.	Areas upwind of values during fire season.	20
		Community proximity.	20
		Municipal watershed proximity.	10
Focal species habitats are inadequately protected from high-intensity crown fires.	Reduce fire threat (hazard) to focal species habitats.	MSO nesting and roosting habitat.	20
		Northern goshawk nesting habitat.	10
		Tasseled-eared squirrel high density.	5
		Turkey high-quality habitat.	2
		Passerine birds species richness.	2

MSO = Mexican spotted owl.

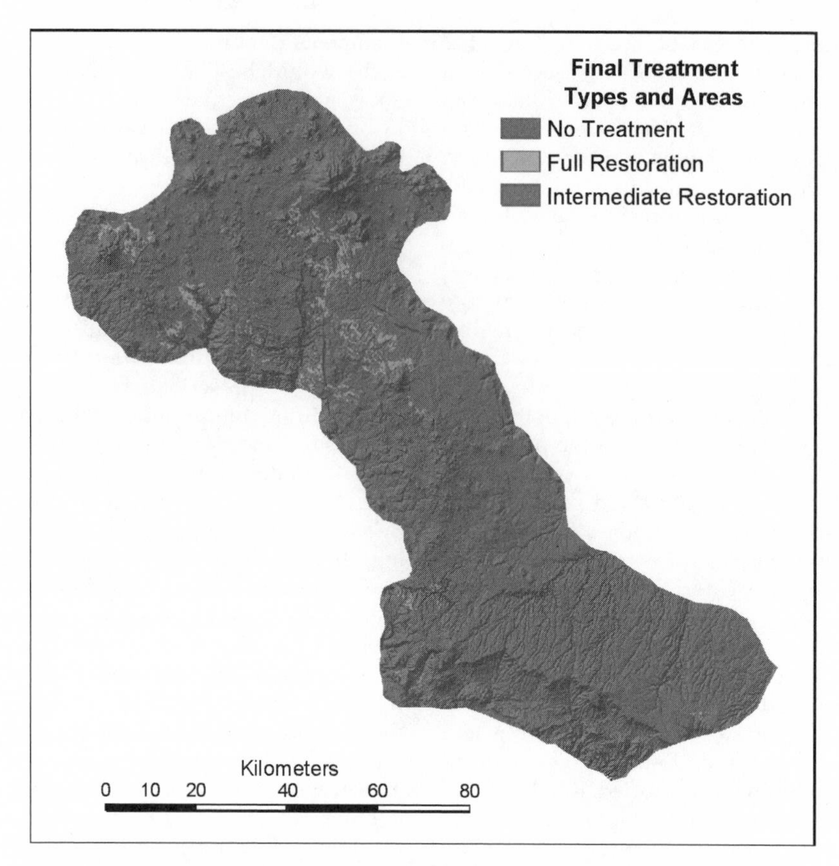

Figure 6. Areas within the western Mogollon Plateau study area targeted for different levels of treatment under the Wildland Fire Mitigation and Preparation scenario. Treatments are limited to the highest-priority 150,000 acres.

We addressed several issues identified during the review of initial scenarios, with additional information during subsequent meetings. The group recommended that we focus on making our data and tools more usable by land management agencies (e.g., switching from metric to English units), developing additional forest treatment characterizations, refining WUI definitions, and calculating the portion of non-ponderosa pine vegetation within the WUI. To make our data more useful to U.S. Forest Service (USFS) managers we demonstrated that ForestERA vegetation data could be made more similar to existing data used by managers by averaging the relatively fine scaled ForestERA vegetation data for individual USFS stands. In addition, we developed a new MSO-related data layer representing management designations in the MSO Recovery Plan. We explained our ongoing efforts to characterize pre and post treatment vegetation structure via literature reviews, research, and consultations with experts. We also demonstrated various definitions of the WUI, including various-sized buffers around communities, power lines, cellular phone towers, municipal watersheds, and areas upwind of communities. We calculated that non-ponderosa pine vegetation types encompassed approximately 26,000 acres or 9 percent of the area within half a mile of urban areas, power lines, and cell towers. This figure was of interest because we had only developed simulated forest treatment prescriptions for ponderosa pine–dominated forests, whereas prescriptions for other forest types may be needed.

Participants requested the following additional data layers for use in prioritizing locations and types of management actions: presence of mollisol soils as an indicator of historical grassland and savannah, USFS MSO protected areas, which include Protected Activity Centers (PACs) and restricted habitat, because the USFS is legally mandated to manage these lands in specific ways, biodiversity hotspots, wildlife corridors, riparian areas, archaeological sites, springs, power lines, communication towers, old-growth forest, and insect- and drought-related tree mortality on USFS lands. They also requested additional models and methods, such as the ability to place treatments according to Finney's herringbone strategy (Finney 2001) and application of weighting factors that varied with distance upwind of values to reflect varying degrees of risk associated with dominant wind directions during the fire season. We achieved this by linearly decrementing weights from highest at community boundaries to progressively lower approaching our study area boundary in a windward direction.

Some participants were uncomfortable selecting weighting factors for prioritization layers, as unpredictable consequences sometimes occurred in final prioritizations. However this concern diminished as the group gained familiarity with our models. For example, one member realized that it was more effective to use categorical restrictions rather than weights in order to exclude treatments from certain areas. Also, some of the unexpected model results offered insights such as uncovering locations where numerous values and risks overlapped to form high-priority areas. We also explored scale issues with the group, such as representing fine-scaled riparian areas at the comparatively coarse grain of 90 m used in the majority of data layers in this demonstration of our SDSS.

Participants felt constrained by only two simulated management actions: high and medium level thinning treatments followed by prescribed burning. We provided only two prescriptions at the time because our ability to model the effects of treatments on forest structure was still in development. However, we did add a light thinning/burning prescription for simulation because of the participants' strong interest. There was also interest in modeling variable-intensity thinning that would return any given location to its natural fire regime. This spurred a discussion on the assumptions and other uncertainties involved with use of fire behavior models.

After reviewing each scenario during the first workshop, we discussed ways to compare them. In the past we used graphs and tables of predicted changes in wildlife and fire hazard characteristics for the entire study area (Table 2) and for only areas upwind of communities, decision matrices (Table 3), and visual comparisons of maps showing changes in significant evaluation criteria. The group suggested that we map the overlap of prioritizations for the different scenarios (Figure 7), and compare scenario results for additional areas, such as within areas in which treatments were simulated, within the Nature Conservancy Ecoregional Conservation sites, and within various distances of communities (for example, Table 4). In addition, we explored patterns of weights among scenarios and prioritization criteria (Table 5 and Figure 6), mapped constraints to find commonalities among scenarios (Figure 8), and investigated the effect on total priorities of the distribution of weighted criteria values (Figure 9). To investigate weights, we determined the percentage of each prioritization criterion on the total priority score for the highest priority areas.

After comparing the group's individual scenarios, we discussed defining additional scenarios with the goal of further identifying points of (dis)agreement among stakeholders. We suggested they consider refining existing scenarios or define hybrids that

Table 2. Predicted effects of each of the five scenarios on evaluation criteria in the overall study area.

Evaluation Criteria	Wildland Fire and Mitigation	Sustainable Functioning Forest	Fix it and Don't Screw it Up	Critical Ecosystem Protection	Strategic Restoration
Fire hazard reduction	3.2%	6.1%	4.2%	4.0%	4.3%
MSO habitat	–1.0%	7.5%	–3.1%	–15.2%	–11.9%
Goshawk habitat	–9.2%	–15.7%	–5.1%	–7.8%	–12.0%
Pronghorn habitat quality	3.5%	4.7%	2.0%	3.1%	3.8%

MSO = Mexican spotted owl.

Table 3. Decision matrix based on equal weighting of each evaluation criteria for all five workshop scenarios showing predicted change in evaluation criteria upwind of communities. A decision matrix allows a total combined score to be determined for each scenario.

Evaluation Criteria	Wildland Fire and Mitigation	Sustainable Functioning Forest	Fix it and Don't Screw it Up	Critical Ecosystem Protection	Strategic Restoration
Fire hazard reduction	7.1%	7.8%	4.3%	5.2%	5.2%
MSO habitat	–2.5%	11.4%	–3.5%	–23.8%	–17.4%
Goshawk habitat	–21.24%	–19.4%	–6.8%	–13.3%	–16.4%
Canopy cover	4.7%	3.3%	1.7%	3.6%	4.2%
Tree density	10.1%	9.2%	3.7%	9.7%	6.4%
Total score	–1.86%	–10.6%	–0.6%	–18.6%	–18.0%

MSO = Mexican spotted owl.

Table 4. Performance of the five workshop scenarios with respect to six evaluation criteria in or within half a mile of communities. Note that "Critical Ecosystem Protection," which focused on protection of Wilderness and other specially-designated areas, had little predicted effect on evaluation criteria near communities compared to within the overall study area (see Table 2).

Evaluation Criteria	Wildland Fire and Mitigation	Sustainable Functioning Forest	Fix it and Don't Screw it Up	Critical Ecosystem Protection	Strategic Restoration
Fire hazard reduction	13.8%	8.8%	6.3%	0.0%	2.5%
MSO habitat	–8.1%	–22.3%	–8.7%	0.1%	–16.2%
Goshawk habitat	–27.1%	–19.9%	–13.3%	0.0%	–8.4%
Canopy cover	–10.2%	–2.8%	–1.7%	0.0%	–1.3%
Tree density	–27.8%	–14.7%	–11.0%	0.0%	–5.0%
% Area treated	30.0%	9.8%	10.8%	0.0%	4.1%

MSO = Mexican spotted owl.

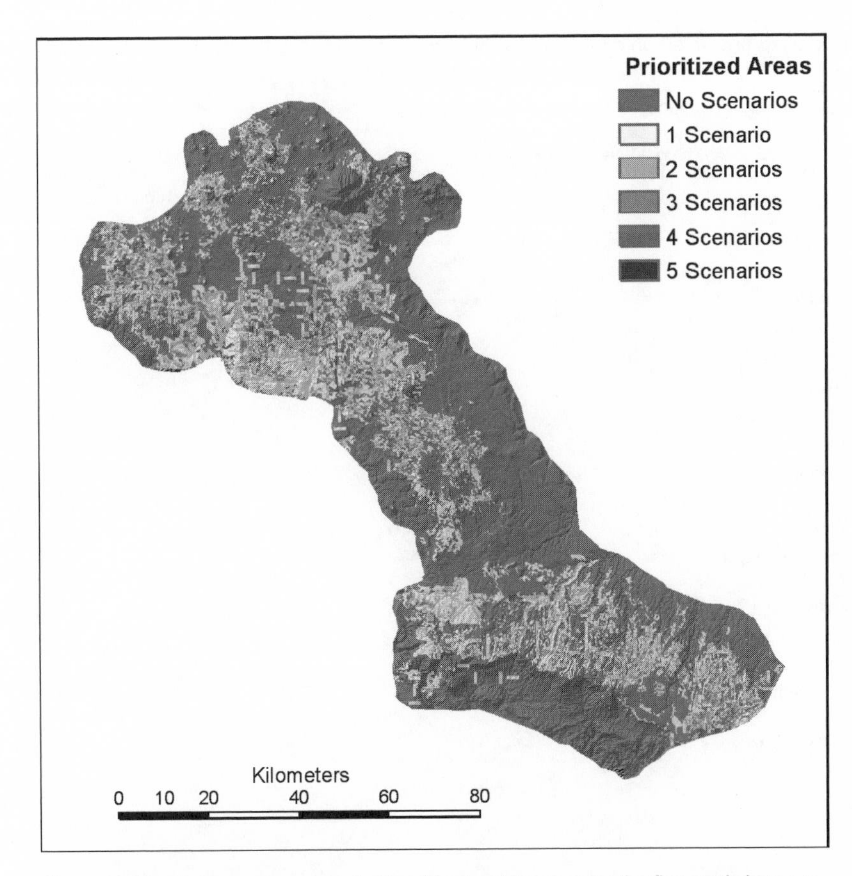

Figure 7. Overlap among areas prioritized for treatment under the five workshop scenarios. Approximately 530,000 acres (27%) of ponderosa pine–dominated lands were not identified under any scenarios; 400,000 acres under only one scenario; 105,000 acres under two; 22,000 acres under three; 3,000 acres under four; and only about 300 acres under all five.

better distinguished different values. The group chose to develop three hybrid scenarios for which our team specified objectives, constraints, prioritization criteria, and rules for placing treatments. Participants filled out a questionnaire to provide weighting factors for each criterion of these "hard-wired" scenarios, which we averaged for use in simulations (Table 6).

The three hybrid scenarios reflected different priorities likely to emerge in landscape-scale restoration planning. The objectives of these scenarios were, in turn, to (1) minimize threat of fire to communities and associated values (Community Protection), (2) minimize threat of fire to specially designated areas (e.g., Inventoried Roadless Areas), habitat of important species, and other areas of ecological importance (e.g., springs, riparian areas; Special Element Protection), and (3) reduce the threat of fire and slow its spread across the entire landscape (Ecosystem Protection). We chose to keep many characteristics constant among the hybrid scenarios, including total area treated (at 150,000 acres), most constraints (Figure 10), and most treatment rules (Table 7). We did this, first, because we tried to reflect similarities among the original five scenarios, and second because we thought it

Table 5. Selected prioritization criteria and weighting factors assigned to each by participants for each of the five workshop scenarios.

Prioritization Criteria	Scenarios					Weights	
	Wildland Fire and Mitigation	Sustainable Functioning Forest	Fix it and Don't Screw it Up	Critical Ecosystem Protection	Strategic Restoration	Total Wts	Avg Wt
Fire hazard	10	25	40	10	35	120	24
Fire risk	1	–	–	10	10	21	4.2
Fire spread	–	–	–	–	35	35	7
Community proximity	20	25	30	–	–	75	15
Municipal watersheds	10	–	–	–	–	10	2
Prevailing winds	20	–	–	10	15	45	9
Mexican spotted owl	20	–	10	5	5	40	8
Goshawk	10	–	10	5	–	25	5
TE squirrel density	5	–	–	–	–	5	1
Turkey	2	–	–	–	–	2	0.4
Passerines	2	–	10	5	–	17	3.4
Basal area > 100 sq ft per acre	–	25	–	–	–	25	5
Canopy cover > 45%	–	25	–	–	–	25	5
Wilderness areas	–	–	–	15	–	15	3
Roadless areas	–	–	–	10	–	10	2
Ecoregional conservation areas	–	–	–	15	–	15	3
Aquatic/riparian areas	–	–	–	15	–	15	3

Table 6. Weighting factors provided by participants for each criterion in each hybrid scenario. A value of zero was used to calculate averages if no weight was provided, except in the case of the threshold provided to define high tree density, which is an average of two participants' recommended thresholds.

Hybrid Scenarios	Weighting Factors					
	Member 1	Member 2	Member 3	Member 4	Member 5	Average
Community Protection						
Fire hazard	50	20	20	75	20	37
Fire risk	0	0	10	0	20	6
Half-mile buffer	0	40	30	15	20	21
Municipal watersheds	20	20	10	0	20	14
Upwind areas	30	20	30	10	20	22
Special Element Protection						
Fire hazard	50	34	30	75	25	43
Fire risk	0	0	10	0	25	7
Treatable MSO habitat	0	33	20	25	25	21
Half-mile buffer	50	33	40	0	25	30
Ecosystem Protection						
Fire hazard	50	25	40	80	25	44
Fire risk	0	25	10	0	25	12

MSO = Mexican spotted owl.

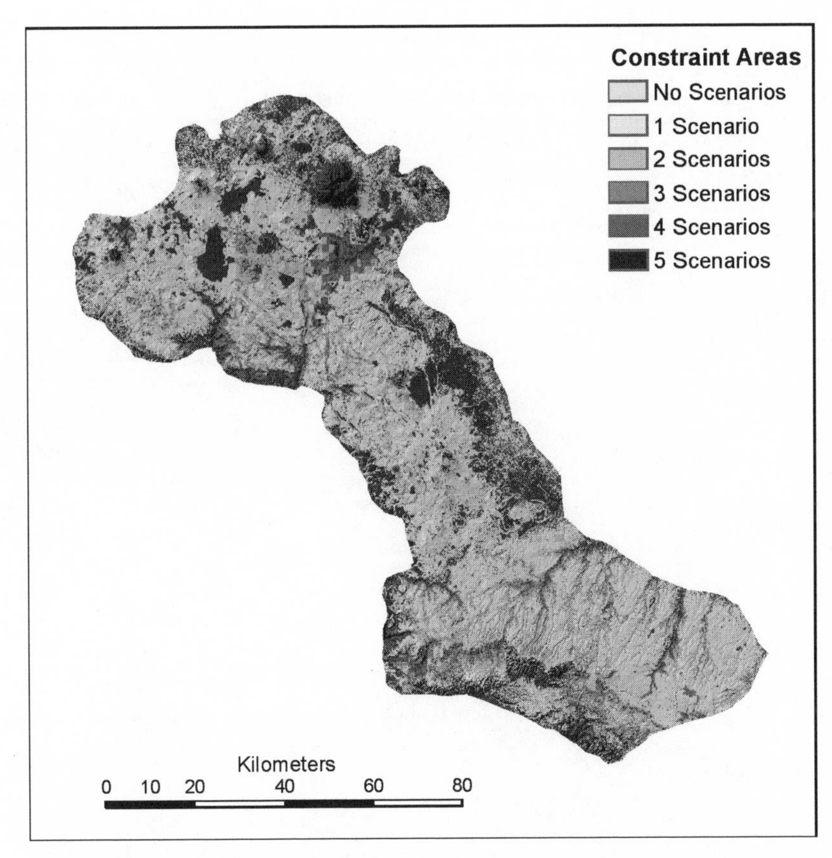

Figure 8. Overlap among areas constrained from treatment under the five workshop scenarios. Approximately 771,000 acres (37%) within the western Mogollon Plateau study area were not constrained from treatment under any scenario; 309,000 acres under only one scenario; 352,000 acres under two; 51,000 acres under three; 79,000 acres under four; and 504,000 acres under all five.

would be easier to understand differences in results of these scenarios if some characteristics were held constant. Resulting patterns of treatments among the hybrid scenarios were significantly different (Figure 11). As expected, treatments were clustered around and upwind of communities in the Community Protection scenario, but were distributed more evenly in the Ecosystem Protection scenario. The Ecosystem Protection scenario had the highest decision matrix score (Table 8), suggesting the preferred impact on evaluation criteria for the entire study area. It had the highest score primarily due to its high reduction in fire spread and low impact on MSO habitat compared to the other two scenarios. However, this could change given different evaluation criteria and weighting factors. For example, if only fire hazard and goshawk habitat were used to evaluate the scenarios, the Special Element Protection scenario would be preferred.

The third and final questionnaire distributed to the participants asked them to rank priorities for future work in landscape-scale restoration as not important, important, or critical (Table 9). Forest growth modeling, fire spread modeling, and estimating the uncertainty of model results received the

Table 7. Rules applied to each hybrid scenario for placement of simulated treatments. We used the fire hazard layer to place and specify treatments in the Wildland Urban Interface (WUI), defined here as half a mile from communities, whereas predicted MSO habitat and slope were used as well outside of the WUI. For example, we did not simulate any low-intensity treatments in the WUI, but placed them outside the WUI in (1) areas with slopes of 30–40%; (2) areas predicted to have surface or passive crown fire if ignited in 90th percentile dry conditions; and (3) predicted MSO habitat where higher-intensity treatments are likely to degrade the habitat quality below minimum thresholds.

Treatment Type	Placement Within the WUI	Placement Outside of the WUI
Low-intensity thinning followed by prescribed burning	None	In areas with 30–40% slope
		In areas of predicted ground or passive crown fire (torching)
		MSO habitat where medium treatment is not possible
Medium-intensity thinning followed by prescribed burning	In areas of predicted ground fire, passive crown fire, and active crown fire with lower heat output	In areas of predicted active crown fire with lower heat output
		MSO habitat where medium treatment is possible
High-intensity thinning followed by prescribed burning	In areas of predicted active crown fire and higher heat output	In areas of predicted active crown fire and higher heat output

MSO = Mexican spotted owl.

Table 8. Decision matrix showing the total score, as percent change from baseline summed over chosen evaluation criteria, for each hybrid scenario within our western Mogollon Plateau study area, giving each evaluation criterion equal weight.

	Hybrid Scenarios		
Evaluation Criteria	Community Protection	Special Element Protection	Ecosystem Protection
Fire hazard reduction	3.7%	3.2%	5.9%
Fire spread reduction	2.7%	2.6%	4.5%
Mexican spotted owl habitat	–3.3%	–4.4%	–1.4%
Goshawk habitat	–11.1%	–9.6%	–12.6%
Total	–7.9%	–8.3%	–3.6%

Table 9. Preferences among workshop participants for future analysis and model development related to prioritization and effects analysis of forest restoration treatments. Zero = not important, 1 = important, and 2 = critical.

Topics of Future Analysis and Model Development	Preference Scores				
	Mean	Std Dev	Min	Max	Range
Forest growth	1.5	0.6	1	2	1
Fire spread	1.5	0.6	1	2	1
Model uncertainty	1.5	0.6	1	2	1
Additional species	1.3	1.2	0	2	2
Wildlife corridors	1.0	0.0	1	1	0
Economics	1.0	0.8	0	2	2
Wildfire effects	1.0	0.8	0	2	2
Additional treatment	0.8	1.0	0	2	2
Social data	0.8	1.0	0	2	2
Reference conditions	0.8	1.0	0	2	2
Increase resolution	0.3	0.5	0	1	1
Carbon budget	0.3	0.5	0	1	1

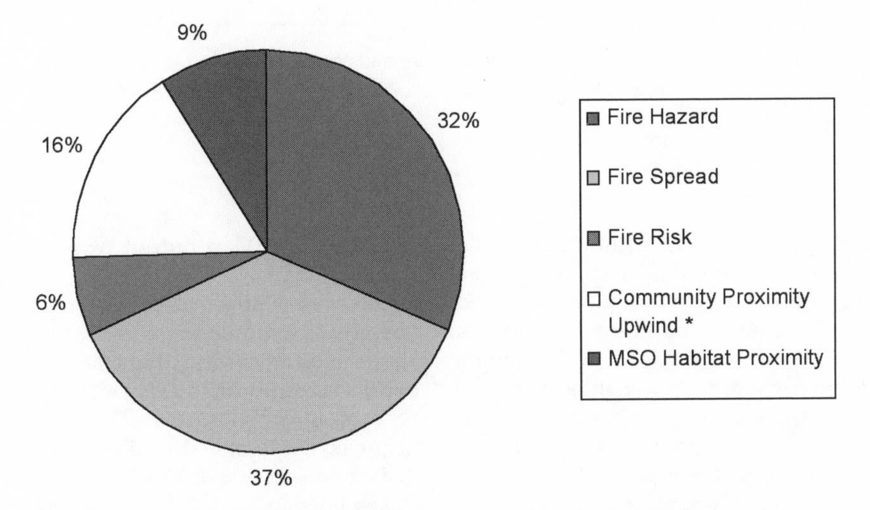

Figure 9. Percent contribution of each prioritization criterion in Strategic Restoration scenario to overall priorities on the 150,000 acres selected for treatment, incorporating effects of both criterion weights and spatial distribution of criteria values.

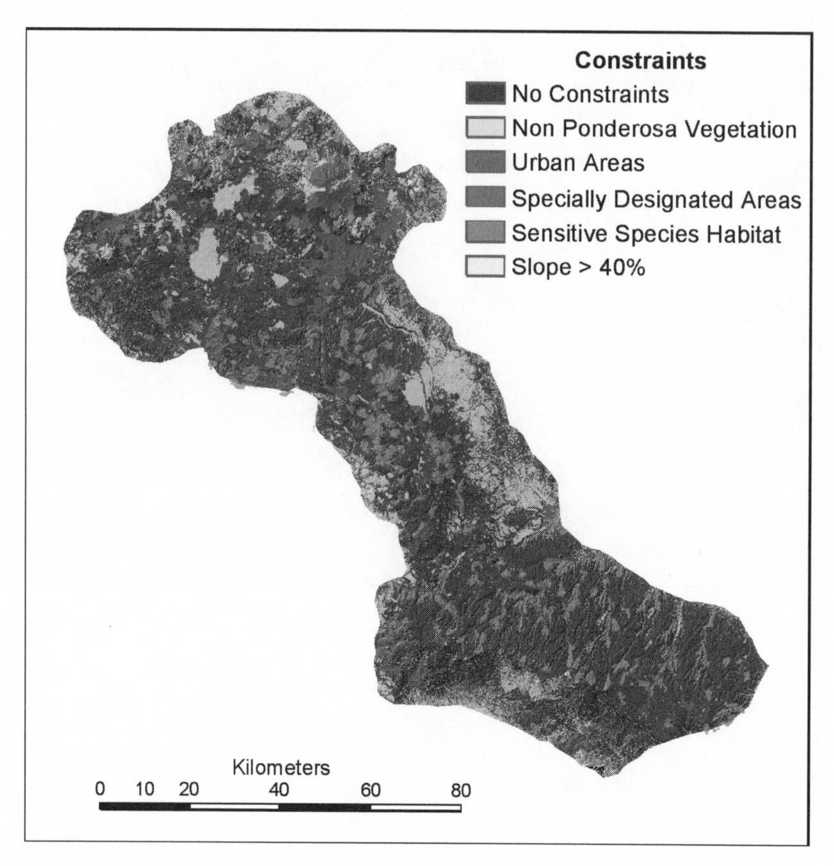

Figure 10. Areas within which treatments were and were not constrained for all three hybrid scenarios, amounting to 740,000 acres or 36 percent of our western Mogollon Plateau study area.

highest mean scores. Although adding models for additional species had a high average score, participant opinions ranged widely from not important to critical. In contrast, it was unanimous that developing a wildlife corridor map was important. The participants also rated economics and wildfire effects as important on average, although with less consensus.

DISCUSSION

The participants in our SDSS test introduced novel approaches that expanded both the types of data and the techniques that we used. For example, we had not previously used the FlamMap output of fire spread (Finney 2003) as a measure of treatment effects, nor had we used forest structure (directly as opposed to indirectly via fire behavior models) and soil type (i.e., mollisols) to prioritize areas for treatment. Gaining this type of information fulfilled one of our major goals: to solicit feedback on project deliverables from a diverse set of ForestERA stakeholders.

We were also able to demonstrate some facets of landscape-scale restoration analysis to a broad audience given the diversity of the group and their high interest level. When asked for their opinion on the process,

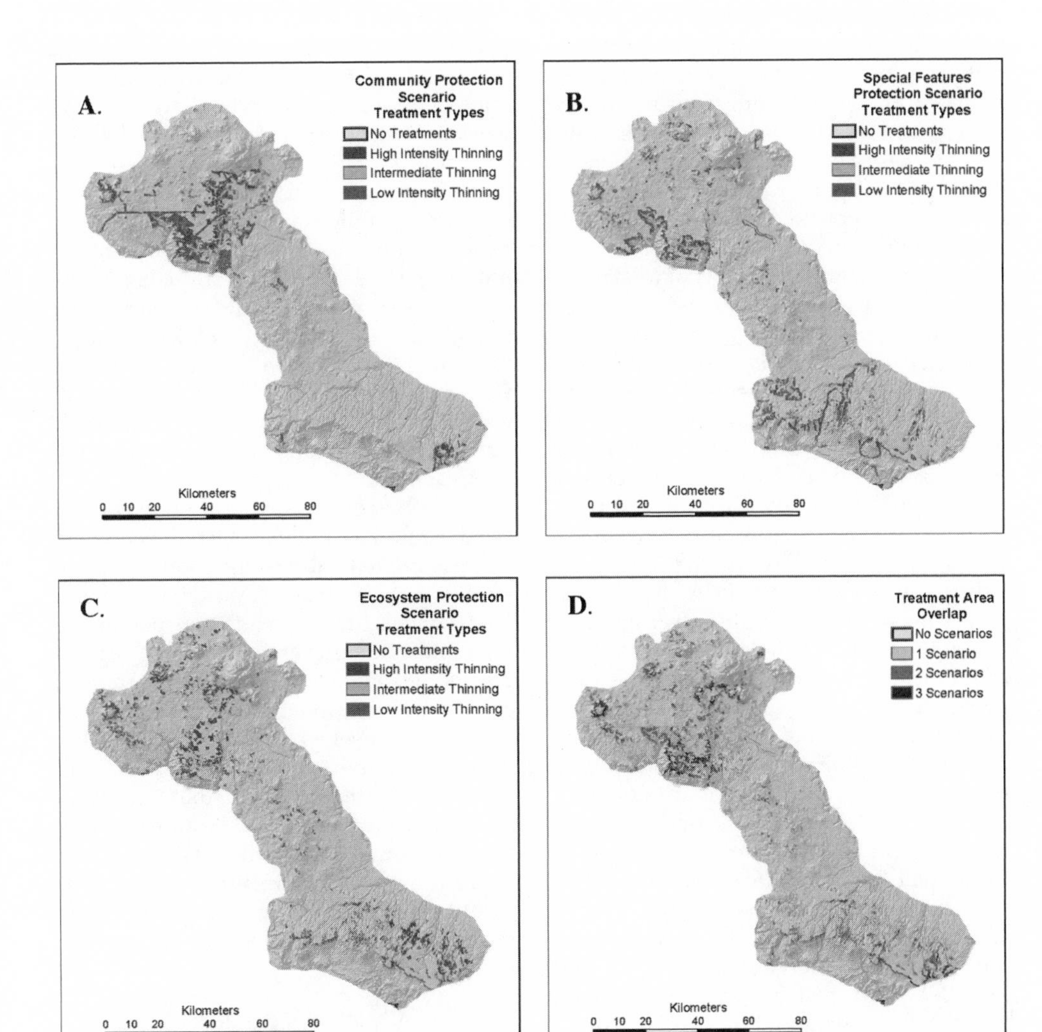

Figure 11. Distribution of prioritized treatments, by type, within our western Mogollon Plateau study area for the three hybrid scenarios (A, B, and C), and for areas of agreement among the three scenarios (D).

participants stressed that developing and distributing data and tools that facilitated forest restoration decision making was not sufficient for successfully transferring our technology, even if the project was designed with substantial stakeholder input. In order for our tools to be used, they recommended that we participate in planning forums in order to train potential users in appropriate applications.

As far as identifying points of (dis)agreement among stakeholders, we observed seemingly greater understanding of values and perspectives held by other participants, although we did not formally assess this outcome. Participants expressed common interests by more often selecting certain prioritization criteria such as fire hazard and MSO habitat in contrast to other criteria, such as habitat for individual passerine species. We are unsure to what extent the hybrid scenarios captured the diversity of values in our study area; however one participant opined that the scenarios captured about two-thirds of stakeholders' values and preferences.

Our SDSS contains many data layers and tools for combining them into meaningful products. This complexity is necessary to represent the "multi-criteria" nature of risks and values. It takes time for each individual to consider the choices available, even when provided with key information in written and oral forms. The group found that even with their considerable interest in and knowledge of restoration management, when they considered multiple criteria across large areas, new issues arose involving factor interactions that took time to fully understand. As researchers familiar with our own work, it is easy to lose sight of this important consideration. Time must be allowed for users of the SDSS to become familiar with the multiple criteria, understand the significance of spatial arrangements (e.g., interactions between overlapping and proximal elements), and account for the unavoidable scale issues inherent in landscape-scale analysis.

All landscape-level analyses must somehow account for fine to coarse scale processes, for example snags (fine scale) and precipitation (coarse). Many people involved in forest restoration are used to working with finer-scaled processes at the project level. However, practitioners work with coarser-scaled data over extents larger than the 2 million acre western Mogollon Plateau study area. To be robust, an analysis must recognize and consider the significance of processes that fall outside study area boundaries or operate at smaller or larger scales than the dominant spatial layers (e.g., 90 m for most ForestERA layers). The SDSS test heightened our awareness of the need to clarify what assumptions can be made at the landscape scale.

Our SDSS test and demonstration emphasized a few key points. Most stakeholders will need time to absorb the complexities of multi-criteria landscape-scale data and tools; model details must be made transparent; the SDSS must be sufficiently flexible to capture diverse stakeholder values and preferences; and some components will need to be refined and developed. The most interesting recommendation supported by nearly all workshop participants was that successful implementation of the SDSS depended on our team actively supporting its use in planning forums. They were sure that our tools would get little use in real-world planning without this assistance in technology transfer.

ACKNOWLEDGMENTS

We thank Bill Block, Pete Fulé, Steve Gatewood, Heather Green, Taylor McKinnon, and Todd Schulke for the considerable time and effort that they graciously volunteered to our test and demonstration of ForestERA data and tools. Their input has significantly influenced the design of our spatial decision support system. Funding for the ForestERA project was provided by the Ecological Research Institute at Northern Arizona University.

LITERATURE CITED

Allen, C. D., M. Savage, D. A. Falk, K. F. Suckling, T. W. Swetnam, T. Schulke, P. B. Stacey, P. Morgan, M. Hoffman, and J. T. Klingel. 2002. Ecological restoration of southwestern ponderosa pine ecosystems: A broad perspective. Ecological Applications 12: 1418–1433.

Cooper, C. F. 1960. Changes in vegetation, structure, and growth of southwestern pine forests since white settlement. Ecological Monographs 30: 129–164.

Cortner, H. J. 2003. The governance environment: Linking science, citizens, and politics. In Ecological Restoration of Southwestern Ponderosa Pine Forests, edited by P. Friederici, pp. 70–80. Island Press, Washington D.C.

Covington, W. W., and M. M. Moore. 1994. Southwestern ponderosa forest structure and resource conditions: Changes since Euro-American settlement. Journal of Forestry 92: 39–47.

Finney, M. A. 2001. Design of regular landscape fuel treatment patterns for modifying fire growth and behavior. Forest Science 47: 219–228.

Finney, M. A. 2003. FlamMap: FlamMap2 Beta Version 1. USDA Forest Service, Rocky Mountain Research Station, Missoula Fire Lab, Missoula MT.

Hampton, H. M., Y. Xu, J. W. Prather, E. N. Aumack, B. G. Dickson, M. M. Howe, and T. D. Sisk. 2003. Spatial tools for guiding forest restoration and fuel reduction efforts. Proceedings of the 23rd Annual Environmental Systems Research Institute (ESRI) International User Conference, San Deigo CA. Environmental Systems Research Institute, http://gis.esri.com/library/userconf/proc03/.

Hampton, H. M., E. N. Aumack, J. W. Prather, B. G. Dickson, Y. Xu, and T. D. Sisk. In press. Development and transfer of spatial tools based on landscape ecology principles: Supporting public participation in forest restoration planning in the southwestern U.S. In Forest Landscape Ecology: Transferring Knowledge to Practice, edited by A. Perera, L. Buse, and T. Crow. Springer, New York.

Harris, F. W. 1995. Policy and Partnership. BioScience (Supplement): S64–S65.

Harrington, M. G., and S. S. Sackett. 1990. Using fire as a management tool in southwestern ponderosa pine. In Proceedings of the Symposium on Effects of Fire Management of Southwestern Natural Resources, edited by J. S. Krammes, pp. 122–133. USDA Forest Service General Technical Report GTR-RM-191. Fort Collins CO.

Oelschlaeger, M. 2003. Ecological restoration as thinking like a forest. In Ecological Restoration of Southwestern Ponderosa Pine Forests, edited by P. Friederici, pp. 81–91. Island Press, Washington D.C.

Prather, J. W., N. L. Dodd, B. G. Dickson, H. M. Hampton, Y. Xu, E. N. Aumack, and T. D. Sisk. In press. Landscape models to predict the influence of forest structure on tassel-eared squirrel populations. Journal of Wildlife Management.

Sisk, T. D., H. M. Hampton, J. Prather, E. N. Aumack, Y. Xu, M. R. Loeser, T. Munoz-Erickson, B. Dickson, and J. Palumbo. 2004. Western Mogollon Plateau Adaptive Landscape Assessment (WMPALA) Report. In Forest Ecological Restoration Analysis (Forest ERA) Project Report, 2002–2004, vol. 2. Center for Environmental Sciences and Education, Northern Arizona University, Flagstaff.

Sisk, T. D., H. M. Hampton, J. W. Prather, Y. Xu, and E. N. Aumack. In press. Modeling fire and biodiversity to guide ecological restoration of pine forests in arid North America. Landscape and Urban Planning.

Swetnam, T. W. 1990. Fire climate and history in the southwestern United States. In Proceedings of the Symposium on Effects of Fire Management of Southwestern Natural Resources, edited by J. S. Krammes, pp. 6–17. USDA Forest Service General Technical Report GTR-RM-191. Fort Collins CO.

Xu, Y., J. W. Prather, H. M. Hampton, B. G. Dickson, E. N. Aumack, and T. D. Sisk. In press. Advanced exploratory data analysis for mapping regional canopy cover. Photogrammetric Engineering & Remote Sensing.

Biological Resources

MODELING THE EFFECTS OF FOREST RESTORATION TREATMENTS ON SENSITIVE WILDLIFE TAXA: A GIS-BASED APPROACH

John. W. Prather, Haydee M. Hampton, Yaguang Xu, Brett G. Dickson, Norris L. Dodd, Ethan N. Aumack, and Thomas D. Sisk

Over the past century, human activity has dramatically altered ponderosa pine (*Pinus ponderosa*) and associated forests in the southwestern United States (Covington and Moore 1994; Covington et al. 1994; Belsky and Blumenthal 1997). Humans have harvested large trees across millions of hectares, grazed their livestock extensively across the land, and actively suppressed wildfire—activities that have changed forest conditions to favor dense stands of small trees and interlocking canopies (Covington and Moore 1994; Mast et al. 1999). These forest conditions are capable of carrying crown fires over very large areas, where, during the previous thousand years at least, frequent ground fires typically burned across large areas, but seldom spread through forest canopies (Covington et al. 1997). Large crown fires, coupled with the expansion of human communities and infrastructure into ponderosa pine ecosystems, have increasingly caused the unprecedented destruction of property, creating the perception that wildfire poses unacceptable levels of risk to humans and the ecosystems that supply them with water, timber, recreation, and a host of other public values (Bosworth 2002). In addition, the competition for resources within dense stands of trees has resulted in reduced tree vigor, and could contribute to higher mortality during drought and larger and more frequent insect outbreaks (Zimmerman 2003). High-intensity crown fires and insect outbreaks can significantly degrade wildlife habitat, including critical areas for sensitive, threatened, and endangered species (Chambers and Germaine 2003). For these reasons, many ecologists, conservation biologists, and land managers have called for large-scale management that is designed to reduce the threat of stand-replacing wildfires by restoring degraded ecosystem conditions and ecosystem function, such as frequent ground fire (Covington and Moore 1994; Moore et al. 1999).

Although there is a general consensus on the need for forest restoration, stakeholders remain divided about how to do it (Covington 2000; Kloor 2000; Wagner et al. 2000). In particular, many stakeholders have focused on the potentially negative effects that fuels reduction treatments may have on biodiversity, leading some to urge caution and call for more study before treatments across large landscapes are implemented (e.g., Tiedemann et al. 2000; Dodd et al. 2003). Although some field studies have focused on the effects of restoration treatments on wildlife (Allen et al. 2002; Chambers and Germaine 2003), the possible impacts on most taxa remain largely unknown. In addition, data from small-scale restoration plots may not adequately predict effects from multiple treatments and/or large-scale restoration projects (Battin and Sisk 2003). Vegetation composition and structure data are rarely up to date and available in a comprehensive and consistent form across large landscapes. Thus, our knowledge about the distributions and abundance of many sensitive species remains limited, hindering our ability to design forest treatment plans to benefit those taxa.

To incorporate wildlife concerns into the prioritization and implementation of restoration treatments in the Southwest, managers need predictive models that can identify critical habitat for important wildlife taxa across large landscapes and predict the potential effects of treatments on those taxa. Such models, based on the best scientific data available, would be invaluable aids in prioritizing treatment areas, predicting the impacts of forest management on wildlife, and designing management plans that mitigate the possible negative impacts on certain species. In addition, predictive habitat models could form a baseline for additional hypothesis testing and for long-term monitoring as landscape-scale restoration treatments progress.

The Forest Ecosystem Restoration Analysis (ForestERA) project is creating a package of landscape-scale spatial data and modeling tools designed to support scenario analysis and planning in ponderosa pine forests and related vegetation types across northern Arizona. Currently the project is mapping forest structure and composition across 8 million acres of forested land on the southern Colorado Plateau. A primary goal of the project is to link forest structural characteristics to wildlife distributions and fire behavior. In this way, we hope to predict the effects of proposed forest treatments on those attributes and use this information to help prioritize the timing and location of management actions. We present here an overview of the process we used to build habitat models for wildlife in this region and details about some of those models. In addition, we present a simple example of how the ForestERA toolset can be used to assess impacts of treatments on a sensitive species, the tassel-eared squirrel (*Sciurus aberti*).

STUDY AREA

The ForestERA project is focused on a broad area of the southern Colorado Plateau, extending from the Kaibab Plateau in northern Arizona across the forest belt along the Mogollon Rim to the White Mountains of western Arizona and the Gila Mountains of eastern New Mexico (Figure 1; please note that the original color figures are all available at www.forestera.nau.edu). Within this region, we have identified four focal areas with relatively similar ecological and physiographic characteristics: the Kaibab Plateau, the western Mogollon Plateau, the eastern Mogollon Plateau and White Mountains, and the Gila Mountains region of New Mexico. To start, the project is focusing research efforts on the 2 million acres of the western Mogollon Plateau (see Figure 1).

DEVELOPMENT OF PREDICTIVE DATA LAYERS

We used seven primary data layers to model wildlife habitat: dominant overstory vegetation, slope, sine of aspect, cosine of aspect, canopy cover, basal area, and tree density. The slope and aspect layers were derived from a 30 m resolution Digital Elevation Model (DEM) obtained from the U.S. Geological Survey (USGS). The aspect layer was converted to radians, with the sine and cosine used in modeling. This converted aspect into an east-west component and a north-south component, rather than using it as a categorical variable (Beers et al. 1966). All of the vegetation layers were derived from remotely sensed imagery and were classified using data provided by a host of collaborative scientists and public agencies.

The dominant overstory vegetation layer was derived from Enhanced Thematic Mapper (ETM) imagery, using a classification tree methodology (Breiman et al. 1984; Hansen et al. 1996) and a machine-learning procedure known as "boosting" (Freund and Schapire 1995). Training data for the model came from more than 1100 ground plot locations. The resulting map had nine vegetation classes: open (grassland or shrubland), ponderosa pine, quaking aspen (*Populus tremuloides*), mixed conifer, pinyon-juniper (*Pinus edulis* and *Juniperus* spp.), juniper-dominated mix, ponderosa pine–quaking aspen mix, ponderosa pine–Gambel oak (*Quercus gambelii*) mix, and mixed conifer–quaking aspen mix (Figure 2a). A

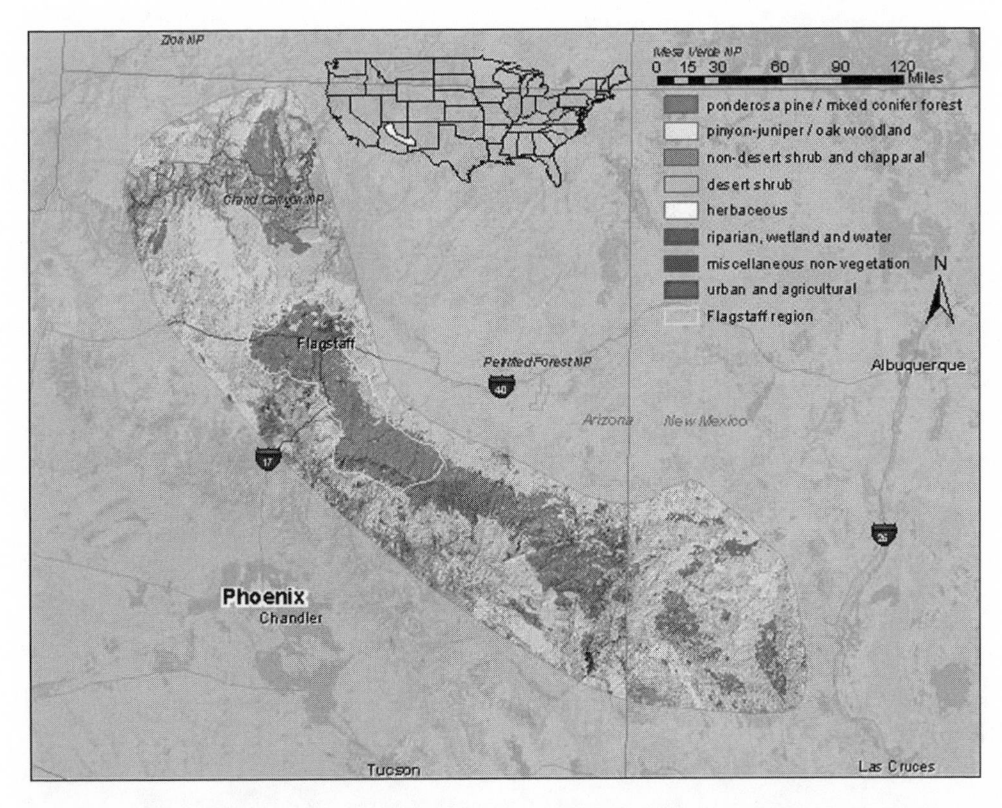

Figure 1. The ForestERA study area, encompassing 8 million acres of forested land across northern Arizona and western New Mexico. The 2 million acre western Mogollon Plateau is outlined in yellow.

ten-fold cross-validation assessment of the accuracy of this layer (misclassification error rate = 23.4%, Kappa = 0.57) indicated that it meets the suggested standards of accuracy for vegetation classification layers (Thomlinson et al. 1999).

The basal area (Figure 2b) and tree stem density (Figure 2c) layers were derived from ETM imagery using a regression tree methodology (Breiman et al. 1984) and training data from more than 800 ground plots. The accuracy of these layers was assessed using data from 567 ground plots taken in 63 locations across the study area. The results of this regression analysis indicated that a highly significant relationship exists between the values for these attributes measured on the ground plots and the predicted values in the spatial data layers (basal area, $r^2 = 0.508$, $P < 0.0001$; tree stem density, $r^2 = 0.584$, $P < 0.0001$).

The canopy cover layer was derived from digital orthophoto quads (DOQs) using an advanced exploratory data analysis method (Xu et al., in press) similar to the widely used fractal C-A methodology (Cheng et al. 1994). The accuracy of this layer was assessed using data from 200 ground plots taken by Dodd at 18 locations across the western Mogollon Plateau (Dodd et al. 1998). The results of this regression analysis indicated that a highly significant relationship exists between values for canopy cover measured on the ground and the predicted values for canopy cover in the spatial data layers ($r^2 = 0.545$, $P < 0.0001$). We believe that these spatial data layers were the most accurate available for this region at the time

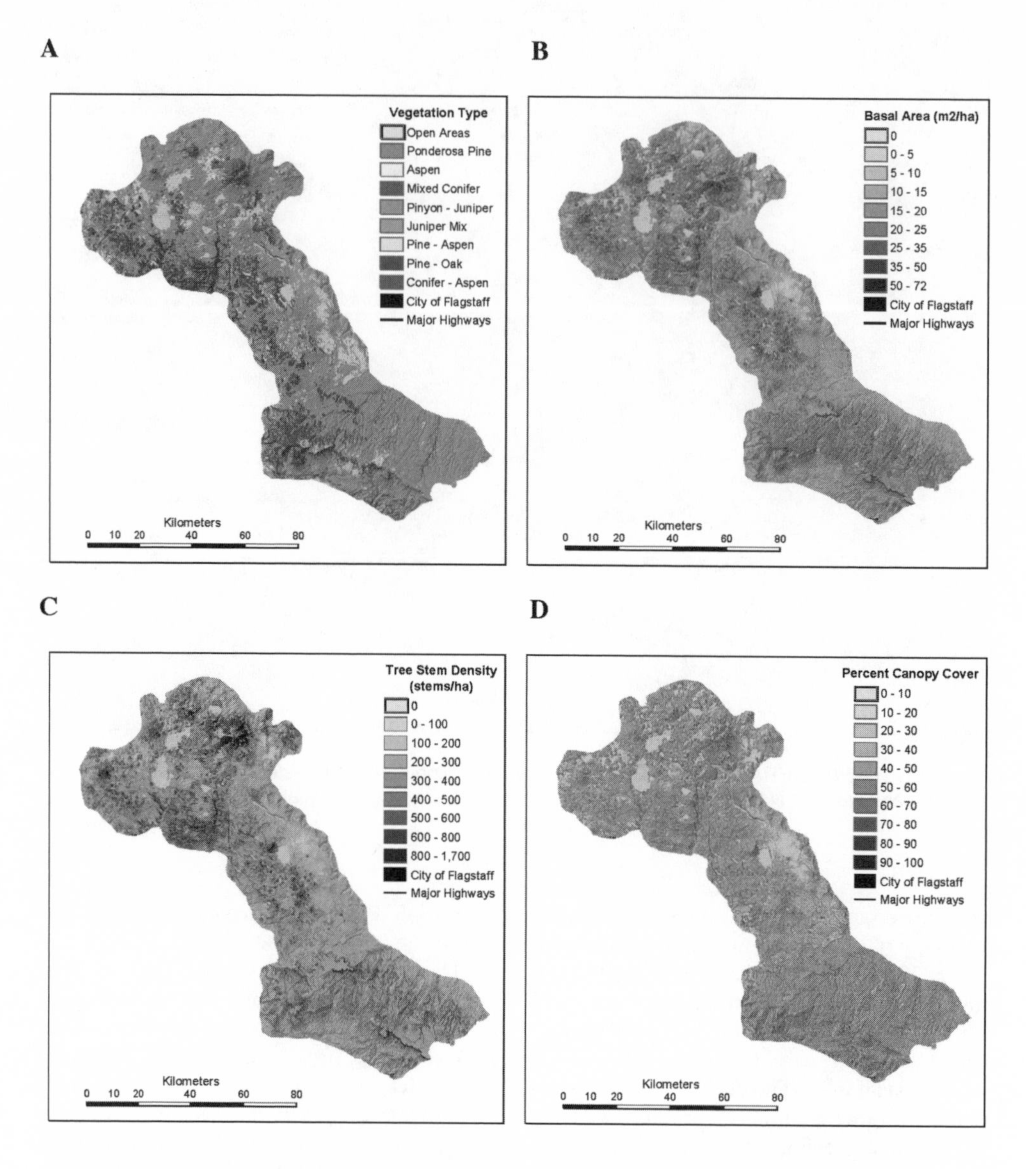

Figure 2. Examples of ForestERA spatial vegetation layers for the western Mogollon Plateau: (a) dominant overstory vegetation type, (b) basal area (m^2/ha), (c) tree stem density (stems/ha), and (d) canopy cover (%).

of our modeling efforts and were suitable for our modeling purposes. Further details of the mapping effort will be presented in future publications.

CHOOSING WILDLIFE TAXA

We surveyed wildlife experts from more than 40 academic institutions, government agencies, and environmental groups (hereafter stakeholders) to identify taxa important for inclusion in our habitat modeling efforts. In addition, we identified taxa that were specially designated as important indicator, threatened, or endangered species. Finally, we conducted a thorough literature search to determine which taxa had been well studied within our focal areas. Based on this information, we chose to model habitat for seven individual species and three taxonomic groups (Table 1).

The Mexican spotted owl (*Strix occidentalis lucida*) and northern goshawk (*Accipiter gentilis*) were the species for which habitat models were of highest interest to stakeholders. Both are listed as management indicator species by the U.S. Forest Service (USFS) and are considered highly sensitive to management actions (Beier and Maschinski 2003). In addition, the owl is listed as a federally threatened species (Friederici 2003). Fortunately, researchers have studied both goshawks (Bright-Smith and Mannan 1994; Beier and Drennan 1997) and owls (Ganey and Balda 1994; Reynolds et al. 1994; Peery et al. 1999) extensively; their habitat requirements are well documented and management guidelines are readily available (Reynolds et al. 1992; U.S. Fish and Wildlife Service 1995). We obtained georeferenced nest site location data for owls from the USFS and the Fish and Wildlife Service, and for goshawks from the USFS and the Arizona Game and Fish Department. Using these data we were able to develop and test nesting habitat models for both of these taxa.

Avian communities have been well studied in southwestern ponderosa pine forests (Rosenstock 1996; Block and Finch 1997; Griffis-Kyle and Beier 2003) and are thus likely to be good species for monitoring the effects of restoration treatments (Chambers and Germaine 2003). Several species, including the pygmy nuthatch (*Sitta pygmaea*) and hairy woodpecker (*Picoides villosus*), are considered management indicator species by the USFS. We were able to develop habitat models for 10 avian species, including 2 management indicators. We also developed a species richness model using 30 species of typical ponderosa pine forest birds. Data for the modeling effort were obtained from georeferenced point count locations in the region. The passerine bird habitat models were developed only in ponderosa pine and pine-oak habitats, due to very limited data availability in other vegetation types.

Tassel-eared squirrels are specialists in ponderosa pine, being dependent on pine seeds, terminal buds, and mycorrhizal truffles as food sources. They play a key role in dispersing spores from mycorrhizal fungi symbionts of ponderosa pine (States and Gaud 1997), are an important prey species for the goshawk (Reynolds et al. 1992), and are considered a management indicator species in national forests. Over the past 10 years, Norris Dodd of the Arizona Game and Fish Department (AGFD) has studied the habitat requirements of squirrels extensively (Dodd et al. 1998, 2003). A wealth of additional information on the species exists in the published literature (e.g., Patton et al. 1985; Pederson et al. 1987). We were thus able to develop models for squirrel density and recruitment.

The AGFD has conducted a number of studies on pronghorn (*Antilocapra americana*; Ockenfels et al. 1994). Together with ancillary data in the form of other publications (e.g., Lee et al. 1998), we were able to gather enough information to develop habitat models for this species. In addition, we obtained a sample of pronghorn radio-tracking locations in northern Arizona. Pronghorn are considered sensitive to management in Arizona (Friederici 2003) and are listed as a management indicator species by the Forest Service.

Merriam's wild turkeys (*Meleagris gallopavo merriami*) are considered a management indicator species by the USFS, and are an important game species in Arizona. As with

Table 1. Criteria used to prioritize and select wildlife taxa for modeling in the ForestERA project.

Taxon	Stakeholder Interest[1]	Literature / Data Availability	Special Designation[2]
Northern goshawk (*Accipiter gentilis*)	High	High	Sensitive
Mexican spotted owl (*Strix occidentalis lucida*)	High	High	Threatened
Tassel-eared squirrel (*Sciurus aberti*)	High	High	Management indicator
Pronghorn (*Antilocapra americana*)	High	Moderate	Sensitive
Merriam's wild turkey (*Meleagris gallipavo merriami*)	Moderate	Moderate	Management indicator
Black bear (*Ursus americanus*)	High	Low	Management indicator
American elk (*Cervus elaphus*)	Low	Moderate	Management indicator
Avian species[3]	High	High	Management indicators
Rodents[3]	Moderate	Moderate	Management indicators
Bats[3]	Moderate	Moderate	Threatened or sensitive

[1]As determined by surveys of stakeholders and meetings with collaborating scientists.
[2]Special designations include threatened or endangered status, sensitive to management, or management indicator species.
[3]Indicates that some, but not all members of the group have this special designation.

pronghorn, the AGFD has studied turkeys extensively (Wakeling 1991; Mollohan et al. 1995). Because turkey roost locations are good predictors of where they will be found at other times (Wakeling and Rodgers 1996), and roosting habitat has been particularly well studied (e.g., Rumble 1992; Wakeling and Rodgers 1996), we chose to model this aspect of turkey habitat use. We used a small number of georeferenced locations for turkey roosting, loafing, and feeding sites to test our turkey habitat model.

We considered several other taxa—black bear (*Ursus americanus*), American elk (*Cervus elaphus*), rodents, and bats—for modeling, but rejected them for various reasons. Some data were available on black bear habitat requirements in the region (e.g., LeCount and Yarchin 1990), but we could not create the necessary predictive layers to model bear habitat. Bears prefer areas with high amounts of horizontal cover (shrubs and/or saplings), a habitat characteristic that cannot be measured with the remote sensing imagery available to us. In addition, no spatial data were available to help test the predictive model. For American elk, data were available on habitat preferences and requirements (e.g., Brown 1991), but stakeholders had little interest in developing a habitat model for this species due to its abundance (Table 1). Small mammals and bats have both been studied in the region, and data were available from studies in the Flagstaff area on each of these taxonomic groups, but high variability in capture rates of bats and rodents makes results from these studies difficult to interpret. Patterns in the data related to forest structural characteristics may be overwhelmed by seasonal and annual variation in captures. In addition, we could not develop predictive data layers for downed woody debris, snags, and other critical habitat features for rodents and bats (Chambers and Germaine 2003).

CHOOSING MODELING PROCEDURES

After a list of wildlife taxa was finalized for the modeling effort (Table 2), we could determine which modeling procedures to use. Many techniques have been used to create predictive models of taxonomic distributions (Guisan and Zimmermann 2000), and major differences between these techniques are critical in deciding which procedure is most appropriate. Differences include the amount of data required, the types of predictive variables that can be used, the value of the model outputs (e.g., presence/absence, probability of occurrence or abundance of the organism), the statistical rigor of the models, the ease of interpretation of the models, and the amount of statistical and biological knowledge necessary to understand and apply the models. We took all these factors into consideration, placing high importance on using procedures that were easy to understand and apply, in determining which modeling procedures could be used given the available data.

For Mexican spotted owl and northern goshawk, we first created models of potential nesting habitat using simple rules. The owl has very specific habitat requirements; it is mainly found in areas of mixed-conifer and pine-oak habitat and prefers places with high basal area (U.S. Fish and Wildlife Service 1995). We were thus able to use our vegetation composition and basal area layers to identify owl nesting habitat. The goshawk is not restricted to any particular habitats, but it does prefer to breed in places with high basal area and high canopy cover (Reynolds et al. 1992), so we used these attributes to define potential nesting habitat. In both cases, we further refined the potential habitat models using the Mahalanobis distance statistic, along with physiographic (slope, sine and cosine of aspect) and vegetation structural characteristics (basal area, tree stem density, canopy cover) at nest sites. Mahalanobis distance is mathematically similar to certain types of ordination, but it functions as a multivariate environmental envelope model (Farber and Kadmon 2003). It has been successfully applied in a few wildlife habitat modeling efforts (Clark et al. 1993; Knick and Dyer 1997; Corsi et al. 1999), and is one of only a few multivariate statistics that can be used with presence-only data. Final results of the model subdivide the habitat into areas where one would expect to find a given proportion of the nests of each species.

Table 2. Final list of taxa chosen for wildlife habitat modeling, the type of data available for the modeling effort, and the modeling procedures used to create each model.

Taxon	Type of Data Available	Modeling Procedures
Northern goshawk (*Accipiter gentilis*)	Nest sites	Rules and Mahalanobis distance
Mexican spotted owl (*Strix occidentalis lucida*)	Nest and roost sites	Rules and Mahalanobis distance
Tassel-eared squirrel (*Sciurus aberti*)	Density and recruitment	Multiple linear regression
Pronghorn (*Antilocapra americana*)	Radio-tracked locations	Rules
Merriam's wild turkey (*Meleagris gallipavo merriami*)	Roost, feeding, and loafing sites	Rules
Avian species richness	Number of species	Regression tree
Passerine birds (individual species)	Presence/absence	Classification tree

We chose Classification and Regression Tree (CART) analysis (Breiman et al. 1984) as our modeling procedure for birds. CART models require large amounts of data, which were available only for avian taxa. We only created these models for areas with ponderosa pine and pine-oak vegetation because we lacked data for other vegetation types. Predictor variables in the models included slope, sine and cosine of aspect, basal area, tree stem density, and canopy cover. CART procedures have been shown to be very useful in ecological contexts because continuous and discrete predictive variables can be used in the models, the models are statistically rigorous, and outputs are easily understood (De'ath and Fabricius 2000). Because they are nonparametric and divide datasets into independent groups, CART models also have several additional advantages. Input data do not need to be normally distributed, it is not necessary for predictor variables to be independent, and relationships between predictor variables and observational data are modeled well when relationships between them are not linear. In published studies, CART models have performed well when compared to models created by regression (Anderson et al. 2000; De'ath and Fabricius 2000; Dettmers et al. 2002) and environmental envelope (Skidmore et al. 1996) techniques. Using presence/absence data, we generated classification tree models for 10 species: hairy woodpecker (*Picoides villosus*), western wood-pewee (*Contopus sordidulus*), plumbeous vireo (*Vireo plumbeous*), pygmy nuthatch (*Sitta pygmaea*), white-breasted nuthatch (*Sitta carolinensis*), brown creeper (*Certhia americana*), western bluebird (*Sialia mexicana*), Grace's warbler (*Dendroica graciae*), western tanager (*Piranga ludoviciana*), and chipping sparrow (*Spizella passerina*). We considered additional species for modeling, but some, such as the yellow-rumped warbler (*Dendroica coronata*) and the mountain chickadee (*Poecile gambeli*), proved too common to model effectively, and others, such as the red-faced warbler (*Cardellina rubifrons*), proved to be too rare. For abundant species, it was difficult to find variables that explain the few locations where birds were absent, and rare species occurred so infrequently that it was difficult to identify factors correlated with their presence. In addition to models for the individual species, we generated a species richness model using a regression tree methodology and presence/absence data for 30 species of typical ponderosa pine forest birds.

We created tassel-eared squirrel habitat models using standard generalized linear regression techniques (McCullagh and Nelder 1983; Nicholls 1989). Research has indicated very strong relationships between squirrel populations and forest structural characteristics, such as basal area and canopy cover (Dodd et al. 1998, 2003). We were able to relate squirrel density and recruitment estimates taken by Dodd on 18 study sites across the Coconino National Forest (Dodd et al. 2003) to these forest structural characteristics, which we estimated using ForestERA remote-sensing data on the same 18 sites (Prather et al., in press).

We used rule-based models to classify habitat suitability for pronghorn and turkeys. In both cases, there were insufficient data to produce statistical models of habitat suitability. However pronghorn are known to strongly avoid areas of high slope and high canopy cover (Lee et al. 1998; Ockenfels et al. 1994), so we used these attributes to model habitat suitability. In the case of turkeys, roost sites are usually associated with areas of high slope and high basal area (Rumble 1992; Wakeling and Rodgers 1996), so we used these relationships to model their habitat.

SOME EXAMPLES OF MODELS

The final step in our modeling efforts was the creation and validation of spatial models for all of the above-mentioned taxa. As examples, we present here the final models for Mexican spotted owl, tassel-eared squirrel density, avian species richness, and pronghorn habitat suitability.

In the Mexican spotted owl model (Figure 3a), nesting and roosting habitat is predicted to cover approximately 226,000 ha (28%) of the study area. We validated this model using georeferenced owl nest sites. Of the

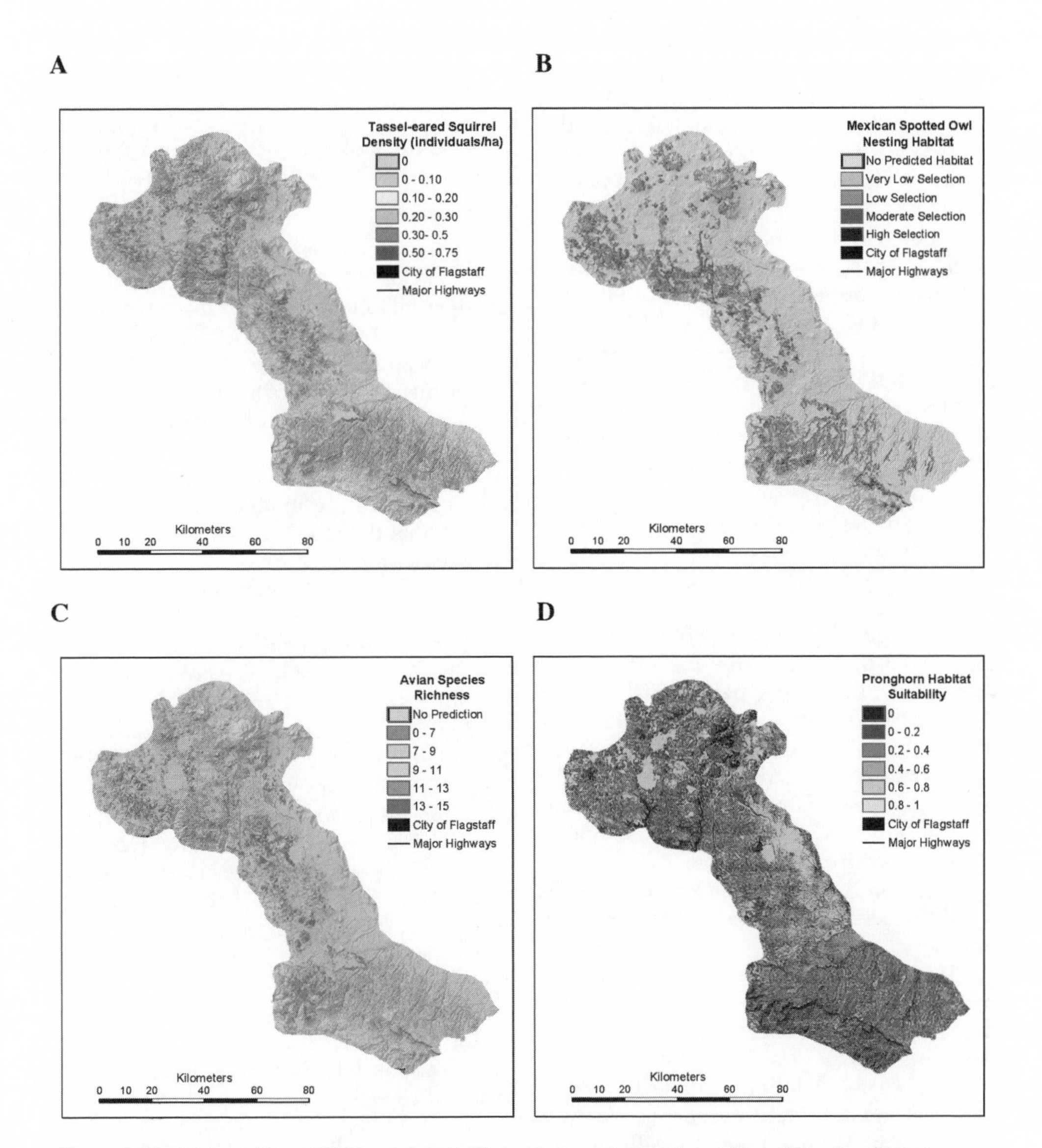

Figure 3. Examples of ForestERA spatial wildlife habitat models on the western Mogollon Plateau: (a) predicted Mexican spotted owl nesting habitat, (b) predicted tassel-eared squirrel density, (c) predicted avian species richness, and (d) predicted pronghorn habitat suitability.

132 sites with known locations, 111 fell within the extent of predicted nesting habitat. This was significantly more than would be expected by chance (chi-square test, v = 2, $\chi^2 = 92.2$, $P < 0.0001$). The Mahalanobis distance statistic was effective at identifying areas within the extent of predicted habitat in which 50, 75, 90, and 100 percent of nests could be found. Actual numbers of nests in each of these portions of the landscape followed expected patterns (chi-square test, v = 3, $\chi^2 = 1.02$, $P = 0.80$).

The best predictor of tassel-eared squirrel density was basal area (m^2/ha). We built the model (Figure 3b) using the relationship determined by using linear regression (density = –0.1815 + 0.0206 * basal area). To test this model we used simple linear regression to compare squirrel density estimates taken by Dodd on seven 60 ha field plots during 1996 and 1997 (Dodd et al. 1998) with estimates of density for the same locations from our model. The results of this analysis suggested that the model was effective at identifying the proper pattern of high and low relative densities across these plots ($n = 7$, $r^2 = 0.60$, slope of regression line = 0.84). We would not expect this model to predict exact densities, as many density-independent factors, such as weather, cause fluctuations in squirrel populations (Keith 1965). See Prather et al. (in press) for more details on the tassel-eared squirrel habitat modeling effort.

The avian species richness model predicts the number of species (out of 30 species typical of ponderosa pine habitats) that would be found in areas of pine or pine-oak habitat on the western Mogollon Plateau (Figure 3c). The model was built using richness values from 312 point count locations scattered across the plateau. It has six terminal nodes, and explains 44 percent of the variation in species richness at those count locations. The following rules occur in the model:

- In pure ponderosa pine, an average of 8.5 species are expected to occur in locations with basal area < 23.5 m^2/ha. In locations with basal area > 23.5 m^2/ha, an average of 11.5 species are predicted to occur when slope is less than 7.5 degrees, while an average of 6.8 species are predicted when slope exceeds 7.5 degrees.
- In pine-oak, an average of 10.1 species are predicted to occur in areas with slope above 8.5 degrees. In areas with slope below 8.5 degrees, an average of 10 species are predicted to occur when tree density is less than 187 stems/ha, and an average of 14.3 species are predicted when tree density exceeds 187 stems/ha.

We assessed the model using data from 56 point count locations reserved for use as a testing dataset. For each of the rules above, we averaged the species richness value from points that fell within portions of the landscape meeting those criteria. We used linear regression to assess the relationships between average species richness at those points and the predicted average species richness from the model. The results suggested that the model was doing a good job of predicting patterns of species richness ($n = 6$, $r^2 = 0.91$, slope of regression line = 1.04) across the landscape.

The pronghorn habitat suitability model (Figure 3d) predicts higher habitat suitability for pronghorn in areas with low slope and low canopy cover. The value for habitat suitability is 1 wherever slope is less than 5 degrees and canopy cover is less than 20 percent. As slope and/or canopy cover increases, the value for suitability decreases. If slope exceeds 20 degrees or canopy cover exceeds 60 percent, the habitat suitability value is zero. We assessed this model using 455 pronghorn locations that were identified using georeferenced signals from radio-tracking collars. The mean habitat suitability value at these locations was 0.85, and only 48 locations fell in areas with suitability values below 0.5.

DETERMINING TREATMENT EFFECTS ON WILDLIFE TAXA

We gathered information on potential forest treatments and their expected effects on forest structure through a literature search and consultation with experts in the fields of forestry and forest management. We used empirical data from experimental studies on treatments in ponderosa pine in Arizona (Fulé et al. 2001a, 2001b, 2002), Colorado

(Lynch et al. 2000), and Montana (Scott 1998) to assess treatment effects. Based on these data we modeled the effects of low, intermediate, and high intensity thinning, and low intensity broadcast burning, on several forest structural attributes (Table 3).

By linking forest structure to wildlife habitat models, we estimated how changes in forest structure, due to management, might affect those taxa (Table 4). Several species, including wild turkey, northern goshawk, Mexican spotted owl, and tassel-eared squirrel, are likely to be negatively affected by most restoration treatments. These effects were largely related to after-treatment reduction in basal area and canopy cover. For the same reasons, treatments were likely to have a positive effect on pronghorn habitat suitability. Passerine birds showed highly variable responses to treatments (Table 4). The models suggest that the majority of the passerine bird species are most likely to be present in areas of intermediate tree densities or canopy cover. If treatments reduce high tree densities to intermediate levels, they would have positive impacts, whereas treatments that reduce intermediate tree densities to low levels would have largely negative impacts. Thus, placement and intensity of treatments in reference to existing conditions will, in some cases, determine whether forest restoration has positive or negative effects.

DISCUSSION

Wildlife modeling for the ForestERA project was developed around the need to understand and guide the location and timing of forest restoration treatments across northern Arizona. From the beginning of the project, stakeholders advocated for improvements in the capacity to analyze the impacts of restoration on wildlife habitat and species at the landscape scale, though they recognized that populating these models would be challenging. Stakeholders stressed the need for improved analytical capacity to address various species of concern, including threatened, endangered, and sensitive species, management indicator species, ponderosa pine–dependent species, and common or abundant species within the study area. They also recognized a need for models that would allow stakeholders to compare forest treatment scenarios so that forest management decisions could be influenced by wildlife values as well as other concerns, such as fire.

After a thorough literature review, and a search for existing datasets that could be used in the modeling process, we identified a list of taxa for which model development was considered important (Table 1). This list was then reduced (Table 2) after we determined whether our spatial data layers and modeling procedures were useful for modeling each of those taxa. Collaborating with wildlife experts from a number of different agencies was critical to the success of this effort, as they provided the necessary data and expertise for success. These collaborators were also instrumental in guiding the process of data selection and the choice of modeling procedures to be used.

Table 3. Mean (and range) percent reduction of the different restoration treatment types on forest structural attributes used for our analysis. Values are based on literature review and expert opinion.

Treatment Type	Basal Area (%)	Canopy Cover (%)	Tree Density (%)
High-intensity thinning	60 (50–70)	40 (30–50)	80 (70–90)
Intermediate-intensity thinning	40 (30–50)	30 (20–40)	65 (55–75)
Low-intensity thinning	30 (20–40)	20 (10–30)	50 (40–60)
Low-intensity broadcast burning	5 (0–10)	5 (0–10)	5 (0–20)

Table 4. Predicted short-term responses of various wildlife taxa to restoration treatments in their habitat based on preliminary results from our modeling of restoration-type treatments across the western Mogollon Plateau.

Taxon	Response to Treatment[1]
Northern goshawk (*Accipiter gentilis*)	Negative - neutral
Mexican spotted owl (*Strix occidentalis lucida*)	Negative - neutral
Tassel-eared squirrel (*Sciurus aberti*)	Negative - neutral
Pronghorn (*Antilocapra americana*)	Positive - neutral
Merriam's wild turkey (*Meleagris gallipavo merriami*)	Negative - neutral
Avian species richness	Positive - negative
Individual avian species	Highly variable both within and between species

[1]Responses are assumed to be short term only. Species that are initially negatively impacted by treatments may show positive responses over time as forest health increases. Response will also vary depending on the location and intensity of treatments.

We spent a considerable amount of time determining which sorts of modeling procedures would be most appropriate. Critical factors included the types (presence/absence, density, etc.) and amounts of data that were available for a particular species, the types of predictor variables to be used in the models, the complexity of the procedure, and the ease with which outputs from the modeling procedure could be understood by stakeholders. The latter factor was very important to us, as even very good models are unlikely to be used by managers if they cannot be easily understood. It was also critical that our modeling effort cover large landscapes and that each model was linked to forest structure so that we could explore and compare the effects of proposed management actions on forest condition and wildlife populations.

Among the chosen modeling procedures, only the linear regression techniques used for the tassel-eared squirrel density and recruitment models can be considered "traditional" statistical measures. Both CART and Mahalanobis distance have not yet been widely used in wildlife habitat modeling, although both are becoming more popular. In addition, due to data limitations, we were forced to use non-statistical rules-based techniques to build some models. Stakeholders required information for certain taxa, even if data limitations precluded creation of statistical models. Flexibility in choosing modeling procedures was key to creating models for many species.

In total we were able to create habitat models for 15 individual species, as well as a model of passerine bird species richness. All of the taxa modeled were birds or mammals, in part due to lack of available data and information on other taxonomic groups, and in part due to the stated interest of stakeholders in this region. Nevertheless, the taxa we chose were a diverse group that exhibited a variety of predicted responses to treatments (Table 4). Because models were linked to forest structure, we predicted not only the response of the species to treatments, but the reasons for that response. For example, our models suggested that many passerine bird species were most likely to be present in areas of lower tree densities. However, within those areas they were more likely to be present in areas of higher basal area or canopy cover. Depending on the effect of the treatment on tree density and basal area, the probability of occurrence of a particular species might increase, decrease, or remain the same in a particular area.

Integration with management is a final important step in preparing the wildlife habitat models for use in prioritization planning across forests in the region. Discussions with land managers from the USFS indicated that it was important to apply Forest Service definitions to habitat models. For example, guidelines for the Mexican spotted owl (U.S. Fish and Wildlife Service 1995) call for different management practices in "protected" and "restricted" habitat. Protected habitat includes protected activity centers, habitat on slopes greater than 40 percent, and specially designated areas such as wilderness and roadless areas. Restricted habitat includes habitat in which basal area exceeds 32 m^2/ha, as well as areas of high-density oak within pine forests. In this case and others, our models can be combined with other GIS layers or reclassified to meet the needs of management agencies.

TASSEL-EARED SQUIRREL EXAMPLE

We present the effects of treatments on squirrel density as predicted using our linear regression model to show how the ForestERA toolset can be used to model the impacts of treatments on wildlife. Extensive previous research has suggested that tassel-eared squirrel population attributes can be related to forest structure (Patton et al. 1985; Pederson et al. 1987). Dodd also found that squirrel density and recruitment were significantly related to various aspects of ponderosa pine forest structure (Dodd et al. 1998, 2003). We were able to use these data to identify relationships between squirrel population attributes (density and recruitment) and the structural attributes (basal area and canopy cover) of ponderosa pine forests across the landscape. Figure 3a shows the predicted densities of squirrels across the western Mogollon Plateau under current conditions.

Because treatments reduce basal area, they were predicted to have negative effects on tassel-eared squirrel populations. We compared the effects of two treatment types—high-intensity thinning and low-intensity thinning—within 200,000 acres of treatments across the western Mogollon Plateau. The treatment areas were chosen as part of a prioritization process that is described elsewhere (see Hampton et al. 2003; Hampton et al., this volume). Figure 4a shows squirrel densities in the area southwest of the city of Flagstaff, and Figure 4b shows the treatment areas in that part of the landscape. High-intensity thinning over the treatment areas results in a predicted 10.1 percent decrease in tassel-eared squirrel density across the region and a predicted 70.6 percent decrease in treatment areas (Figure 4c). In contrast, low-intensity thinning over all treatment areas results in a predicted 4.2 percent decrease in squirrel density across the region and a predicted 35.5 percent decrease in squirrel density in treatment areas (Figure 4d). Users could apply this information when planning landscape-level treatment implementation.

The ForestERA toolset was designed to allow the user maximum flexibility in prioritizing and designing treatments. The user may choose to mitigate the impacts on wildlife taxa in a variety of ways. For example, treatments may be moved out of areas of critical habitat for a species (see Hampton et al. 2003), or other treatments that have lighter impacts may be chosen, as is represented in this example. In the case of tassel-eared squirrels, treatments will almost always have a negative effect, but lower intensity treatments will have less negative impacts. The ForestERA toolset thus allows users to compare multiple criteria, including impacts to wildlife, across alternative treatment scenarios, thus facilitating the planning and prioritization process for forest restoration.

ACKNOWLEDGMENTS

This research was undertaken as part of the Forest Ecosystem Restoration Analysis (ForestERA) project, a landscape-level assessment of forest conditions across northern

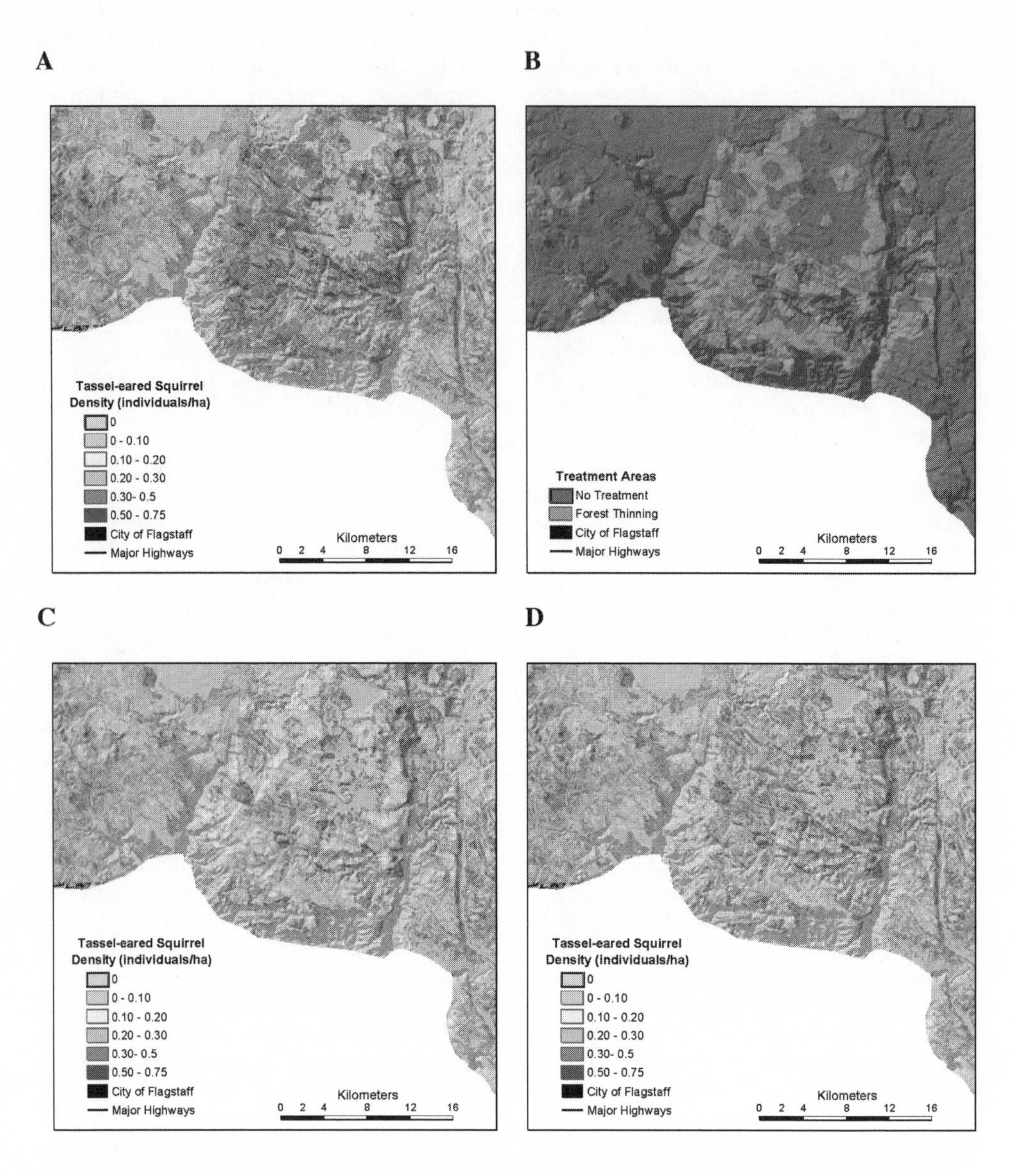

Figure 4. An example of the predicted effects of treatments on wildlife: (a) tassel-eared squirrel density southwest of Flagstaff under current conditions, (b) the locations in which treatments are placed in the example modeling scenario, (c) predicted squirrel densities after high-intensity thinning treatments are carried out in all treatment areas, and (d) predicted squirrel densities after low-intensity thinning treatments are carried out in all treatment areas.

Arizona. Funding was provided by the Ecological Restoration Institute (ERI) under a U.S. Department of the Interior emergency appropriations bill (2001). Collaborators and data providers for the habitat modeling efforts included Joseph Ganey (spotted owl), Bill Block (spotted owl, passerine birds), and Jeff Jenness (spotted owl) of the USDA Forest Service Rocky Mountain Research Station, Paul Beier (northern goshawk) and Carol Chambers (passerine birds) of Northern Arizona University, Micheal Ingraldi (northern goshawk), Norris Dodd (tassel-eared squirrel), Steve Rosenstock (passerine birds), Brian Wakeling (turkey), and Richard Ockenfels (pronghorn) of the Arizona Game and Fish Department, Kerry-Griffis Kyle (passerine birds) of Syracuse University, and Shaula Hedwall (spotted owl) of the U.S. Fish and Wildlife Service. We thank Bill Romme of Colorado State University and Pete Fulé and John Bailey of Northern Arizona University for their help in defining the effects of restoration treatments on forest structure. Without input from these collaborative scientists much of this research would not have been possible.

LITERATURE CITED

Allen, C. D., M. Savage, D. A. Falk, K. F. Suckling, T. W. Swetnam, T. Schulke, P. B. Stacey, P. Morgan, M. Hoffman, and J. T. Klingel. 2002. Ecological restoration of Southwestern ponderosa pine ecosystems: A broad perspective. Ecological Applications 12: 1418–1433.

Anderson, M. C., J. M. Watts, J. E. Freilich, S. R. Yool, G. I. Wakefield, J. F. McCauley, and P. B. Farhnestock. 2000. Regression-tree analysis of desert tortoise habitat in the central Mojave Desert. Ecological Applications 10: 890–900.

Battin, J., and T. D. Sisk. 2003. Assessing landscape-level influences of forest restoration on animal populations. In Ecological Restoration of Southwestern Ponderosa Pine Forests, edited by P. Friederici, pp. 175–190. Island Press, Washington D.C.

Beers, T. W., P. E. Dress, and L. C. Wensel. 1966. Aspect transformation in site productivity research. American Scientist 54: 691–692.

Beier, P., and J. E. Drennan. 1997. Forest structure and prey abundance in foraging areas of northern goshawks. Ecological Applications 7: 564–571.

Beier, P., and J. Maschinski. 2003. Threatened, endangered, and sensitive species. In Ecological Restoration of Southwestern Ponderosa Pine Forests, edited by P. Friederici, pp. 306–327. Island Press, Washington D.C.

Belsky, A. J., and D. M. Blumenthal. 1997. Effects of livestock grazing on stand dynamics and soils in upland forests of the interior West. Conservation Biology 11: 315–327.

Block, W. M., and D. M. Finch. 1997. Songbird ecology in Southwestern Ponderosa Pine forests: A literature review. USDA Forest Service General Technical Report RM-GTR-292.

Bosworth, D. 2002. Fires and forest health: Our future is at stake. Fire Management Today 63: 4–11.

Breiman, L., J. J. Friedman, R. A. Olshen, and C. J. Stone. 1984. Classification and Regression Trees. Wadsworth and Brooks, Monterey CA.

Bright-Smith, D. J., and R. W. Mannan. 1994. Habitat use by breeding male northern goshawks in northern Arizona. Studies in Avian Biology 16: 58–65.

Brown, R. L. 1991. Effects of timber management practices on elk. Arizona Game and Fish Department Technical Report No. 10. Phoenix.

Chambers, C. L., and S. S. Germaine. 2003. Vertebrates. In Ecological Restoration of Southwestern Ponderosa Pine Forests, edited by P. Friederici, pp. 268–285. Island Press, Washington D.C.

Cheng, Q., F. P. Peterberg, and B. B. Ballantynge. 1994. The separation of geochemical anomolies from background by fractal methods. Journal of Geochemical Exploration 51: 109–130.

Clark, J. D., J. E. Dunn, and K. G. Smith. 1993. A multivariate model of female black bear habitat use for a Geographic Information System. Journal of Wildlife Management 57: 519–526.

Corsi, F., E. Dupre, and L. Boitani. 1999. A large-scale model of wolf distribution in Italy for conservation planning. Conservation Biology 13: 150–159.

Covington, W. W. 2000. Helping western forests heal: The prognosis is poor for United States forest ecosystems. Nature 408: 135–136.

Covington, W.W., and M. M. Moore. 1994. Southwestern ponderosa forest structure: Changes since Euro-American settlement. Journal of Forestry 92: 39–47.

Covington, W. W., R. L. Everett, R. Steele, L. L. Irwin, T. A. Daer, and A. N. D. Auclair. 1994. Historical and anticipated changes in forest ecosystems of the Inland West of the United States. Journal of Sustainable Forestry 95: 13–63.

Covington, W.W., P. Z. Fulé, M. M. Moore, S. C. Hart, T. E. Kolb, J. N. Mast, S. S. Sackett, and M. R. Wagner. 1997. Restoring ecosystem health in ponderosa pine forests of the Southwest. Journal of Forestry 95: 23–29.

De'ath, G., and K. E. Fabricius. 2000. Classification and regression trees: A powerful, yet simple, technique for ecological data analysis. Ecology 81: 3178–3192.

Dettmers, R., D. A. Buehler, and J. B. Bartlett. 2002. A test and comparison of wildlife-habitat modeling techniques for predicting bird occurrence at a regional scale. In Predicting Species Occurrences: Issues of Accuracy and Scale, edited by J. M. Scott, P. J. Heglund, M. L. Morrison, J. B. Haufler, M. G. Rapheal, W. A. Wall, and F. B. Sampson, pp. 607–616. Island Press, Washington D.C.

Dodd, N. L., S. S. Rosenstock, C. R. Miller, and R. W. Schweinsburg. 1998. Tassel-eared squirrel population dynamics in Arizona: Index techniques and relationships to habitat conditions. Arizona Game and Fish Department Research Technical Report No. 27. Phoenix.

Dodd, N. L., J. S. States, and S. S. Rosenstock. 2003. Tassel-eared squirrel population, habitat condition, and dietary relationships in north-central Arizona. Journal of Wildlife Management 67: 622–633.

Farber, O., and R. Kadmon. 2003. Assessment of alternative approaches for bioclimatic modeling with special emphasis on the Mahalanobis distance. Ecological Modelling 160: 115–130.

Friederici, P. 2003. Ecological restoration of Southwestern ponderosa pine forests. Ecological Restoration Institute, Flagstaff AZ.

Freund, Y., and R. E. Schapire. 1995. A decision-theoretic generalization of on-line learning and application to boosting. In Proceedings of the Second European Conference on Computational Learning Theory, pp. 23–27. Spring-Verlag, London.

Fulé, P. Z., A. E. M. Waltz, W. W. Covington, and T. A. Heinlein. 2001a. Measuring forest restoration effectiveness in hazardous fuels reduction. Journal of Forestry 99: 24–29.

Fulé, P. Z., C. McHugh, T. A. Heinlein, and W. W. Covington. 2001b. Potential fire behavior is reduced following forest restoration treatments. In Ponderosa Pine Ecosystems Restoration and Conservation: Steps Toward Stewardship, compiled by G. K. Vance, C. B. Edminster, W. W. Covington, and J. A. Blake, pp. 28–35. USDA Forest Service Proceedings RMRS-P-22.

Fulé, P. Z., W. W. Covington, H. B. smith, J. D. Springer, T. A. Heinlein, K. D. Huisinga, and M. M. Moore. 2002. Comparing ecological restoration alternatives: Grand Canyon, Arizona. Forest Ecology and Management 170: 19–41.

Ganey, J. L., and R. P. Balda. 1994. Habitat selection by Mexican spotted owls in northern Arizona. Auk 17: 162–169.

Griffis-Kyle, K. L., and P. Beier. 2003. Small isolated aspen stands enrich bird communities in Southwestern ponderosa pine forests. Biological Conservation 110: 375–385.

Guisan, A., and N. E. Zimmermann. 2000. Predictive habitat distribution models in ecology. Ecological Modelling 135: 147–186.

Hampton, H. M., Y. Xu, J. W. Prather, E. N. Aumack, B. G. Dickson, M. M. Howe, and T. D. Sisk. 2003. Spatial tools for guiding forest restoration and fuel reduction efforts. Proceedings of the 23rd Annual Environmental Systems Research Institute (ESRI) International User Conference (CD-ROM).

Hansen, M., R. Dubayah, and R. Defries. 1996. Classification trees: An alternative to traditional land cover classifiers. International Journal of Remote Sensing 17: 1075–1081.

Keith, J. O. 1965. The Abert squirrel and its dependence on ponderosa pine. Ecology 46: 150–163.

Kloor, K. 2000. Returning America's forests to their "natural" roots. Science 287: 573–575.

Knick, S. T., and D. L. Dyer. 1997. Distribution of black-tailed jackrabbit habitat determined by GIS in southwestern Idaho. Journal of Wildlife Management 61: 75–85.

LeCount, A. L., and J. C. Yarchin. 1990. Black bear habitat use in east-central Arizona. Arizona Game and Fish Department Research Technical Report No. 4. Phoenix.

Lee, R. M., J. D. Yoakum, B. W. O'Gara, T. M. Pojar, and R. A. Ockenfels. 1998. Pronghorn Management Guides. Eighteenth Biennial Pronghorn Antelope Workshop. Arizona Antelope Foundation.

Lynch, D. L., W. H. Romme, and M. L. Floyd. 2000. Forest restoration in Southwestern ponderosa pine. Journal of Forestry 98:17–24.

Mast, J. N., P. Z. Fulé, M. M. Moore, W. W. Covington, and A. E. M. Waltz. 1999. Restoration of presettlement age structure of an Arizona ponderosa pine forest. Ecological Applications 9: 228–239.

McCullagh, P, and J. A. Nelder. 1983. Generalized Linear Models. Chapman and Hall, London.

Mollohan, C. M., D. R. Patton, and B. F. Wakeling. 1995. Habitat selection and use by Merriam's turkey in northcentral Arizona. Arizona Game and Fish Department Research Technical Report No. 9. Phoenix.

Moore, M. M., W. W. Covington, and P. Z. Fulé. 1999. Reference conditions and ecological restoration: A southwestern ponderosa pine perspective. Ecological Applications 9: 1266–1277.

Nicholls, A. O. 1989. How to make a biological survey go further with generalized linear models. Journal of the Royal Statistical Association A135: 370–384.

Ockenfels, R. A., A. Alexander, C. L. D. Ticer, and W. K. Carrel. 1994. Home ranges, movement patterns, and habitat selection of pronghorn in central Arizona. Arizona Game and Fish Department Research Technical Report No. 13. Phoenix.

Patton, D. R., R. L. Wadleigh, and H. G. Hudak. 1985. The effects of timber harvesting on the Kaibab squirrel. Journal of Wildlife Management 49: 14–19.

Pederson, J. C., R. C. Farentinos, and V. L. Littlefield. 1987. Effects of logging on habitat quality and feeding patterns of Abert squirrels. Great Basin Naturalist 47: 252–258.

Peery, M. Z., R. J. Gutierrez, and M. E. Seamans. 1999. Habitat composition and configuration around Mexican spotted owl nest and roost sites in the Tularosa Mountains, New Mexico. Journal of Wildlife Management 63: 36–43.

Prather, J. W., N. L. Dodd, B. G. Dickson, H. M. Hampton, Y. Xu, E. N. Aumack, and T. D. Sisk. In press. Landscape models to predict the influence of forest structure on tassel-eared squirrel populations. Journal of Wildlife Management.

Reynolds, R. T., R. T. Graham, M. H. Reiser, L. Bassett, P. L. Kennedy, D. A. Boyce Jr., G. Goodwin, R. Smith, and E. L. Fisher. 1992. Management Recommendations for the Northern Goshawk in the Southwestern United States. USDA Forest Service General Technical Report RM-GTR-217.

Reynolds, R. T., S. M. Joy, and D. G. Leslie. 1994. Nest productivity, fidelity, and spacing of northern goshawks in Arizona. Studies in Avian Biology 16: 106–113.

Rosenstock, S. S. 1996. Habitat relationships of breeding birds in northern Arizona ponderosa pine and pine-oak forests. Arizona Game and Fish Department Research Technical Report No. 23. Phoenix.

Rumble, M. A. 1992. Roosting habitat of Merriam's turkeys in the Black Hills, South Dakota. Journal of Wildlife Management 56: 750–759.

Scott, J. H. 1998. Fuel reduction in residential and scenic forests: A comparison of three treatments in a western Montana ponderosa pine stand. USDA Forest Service Research Paper RMRS-RP-5.

Skidmore, A. K., A. Gauld, and P. Walker. 1996. Classification of kangaroo habitat distribution using three GIS models. International Journal of GIS 10: 441–454.

States, J. S., and W. S. Gaud. 1997. Ecology of hypogeous fungi associated with ponderosa pine I. Patterns of distribution and sporocarp production in some Arizona forests. Mycologia 89: 712–721.

Thomlinson, J. R., P. V. Bolstad, and W. B. Cohen. 1999. Coordinating methodologies for scaling landcover classification from site-specific to global: Steps toward validating global map products. Remote Sensing of the Environment 70: 16–28.

Tiedemann, A. R., J. O. Klemmedson, and E. L. Bull. 2000. Solution of forest health problems with prescribed fire: Are forest productivity and wildlife at risk? Forest Ecology and Management 127: 1–18.

U.S. Fish and Wildlife Service. 1995. Recovery plan for the Mexican spotted owl (*Strix occidentalis lucida*). U.S. Department of the Interior Fish and Wildlife Service, Albuquerque NM.

Wagner, M. R., W. M. Block, B. W. Geils, and K. F. Wenger. 2000. Restoration ecology: A new forest management paradigm or another merit badge for foresters? Journal of Forestry 98: 22–27.

Wakeling, B. F. 1991. Population and nesting characteristics of Merriam's turkey along the Mogollon Rim, Arizona. Arizona Game and Fish Department Research Technical Report No. 7. Phoenix.

Wakeling, B. F., and T. D. Rodgers. 1996. Winter diet and habitat selection by Merriam's turkeys in north-central Arizona. Proceedings of the 7th National Wildlife Turkey Symposium, pp. 175-184.

Xu, Y., J. W. Prather, H. M. Hampton, E. N. Aumack, and T. D. Sisk. In press. Advanced exploratory data analysis for mapping regional canopy cover. Photogrammetric Engineering and Remote Sensing.

Zimmerman, G. T. 2003. Fuels and fire behavior. In Ecological Restoration of Southwestern Ponderosa Pine Forests, edited by P. Friederici, pp. 126–143. Island Press, Washington D.C.

USING DECISION TREE MODELING IN GAP ANALYSIS LAND COVER MAPPING: PRELIMINARY RESULTS FOR NORTHEASTERN ARIZONA

Sarah Falzarano, Kathryn Thomas, and John Lowry

The methods used to map land cover (vegetation community and land use) for large regions have developed from extrapolating field-based surveys to interpretation of aerial photography to classification of satellite imagery. These methods typically rely on field data, expert knowledge, and remotely sensed data, and have not incorporated the full potential of important data sources like elevation, geology, and climate. A new method using decision tree modeling (Friedl and Brodley 1997; Brown de Colstoun et al. 2003; Pal and Mather 2003) allows the incorporation of a wide variety of predictor variables to map land cover that include both categorical and continuous inputs.

Creating the land cover map for the second-generation gap analysis in Arizona has provided a great opportunity to use the decision tree approach. The state's topography ranges from low deserts to high mountains, and includes a diversity of vegetation communities. Decision trees allow the incorporation of copious field data and multiple satellite imagery scenes into the model, as well as other mapped data.

DECISION TREE MODELING

Decision trees are a predictive modeling methodology. In the literature, decision trees are also called classification trees, classification and regression trees, and CART; however, CART also refers to a specific software program (Salford Systems) which was not used in this study. Using decision trees in land cover mapping on a regional scale is a new approach to Gap Analysis Program (GAP) land cover mapping. GAP land cover mapping has previously relied on supervised or unsupervised classification of remotely sensed data, but more recently, decision trees have consistently been found to be more accurate than other classification methods (Friedl and Brodley 1997; Brown de Colstoun et al. 2003).

Decision trees recursively split training data into more and more homogeneous groups until a terminal node (representing a land cover class) is reached. In its graphical representation, a decision tree looks like an upside-down tree; the top of the graph is the trunk where all of the data are initially represented, the branches of the tree split the data into different groupings, and the bottom of the graph are the leaves (terminal nodes) which represent the land cover classes (Figure 1). The lengths of the branches correspond to the magnitude of deviance described by the split. In general, the branches at the top of the graph explain more of the deviance in the training data than the branches at the bottom. In its purest form, a decision tree fits the training data until all of the observations are assigned to "pure" terminal nodes where no deviance remains. However, this can lead to overfitting of the data when the results are extrapolated because the training data may not include the true heterogeneity of the population. A tree can be shrunk or pruned so that a certain deviance in the group is acceptable and its extrapolation potential is maximized. The graphical decision tree can also be interpreted through textual rules.

Decisions for splitting the data are derived from independent, or predictor,

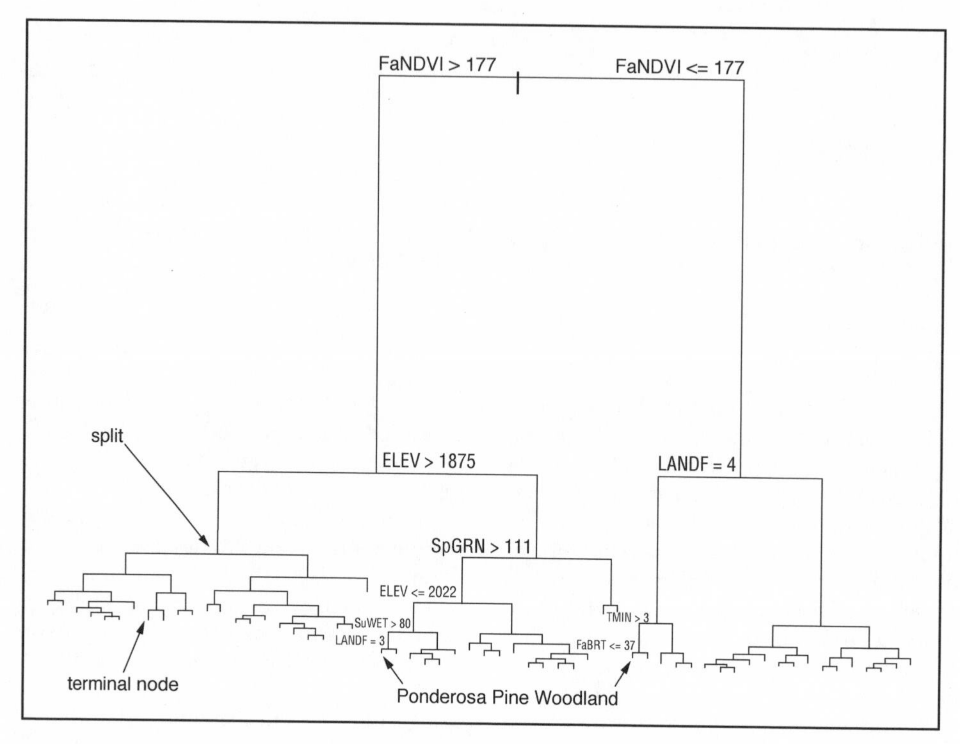

Figure 1. Classification tree. Sample rules show that two different paths can lead to the same land cover class.

variables. A decision tree is similar to a regression analysis, with one dependent variable (the vegetation community represented at each field plot) that is influenced by one or more predictor variables.

Decision trees offer many advantages. They allow the incorporation of both numeric (e.g., elevation) and categorical (e.g., geology) data in the classification process as predictor variables (Quinlan 1993). They also do not make any assumptions about normal distributions in the predictor variables (Friedl and Brodley 1997). Their ease of interpretation in graphical plots and textual rules facilitates an examination of each class prediction for obvious outliers or mistakes. They are also computationally efficient, which allows for standard computer processors to complete analyses in a reasonable amount of time (Friedl and Brodley 1997).

GAP ANALYSIS IN ARIZONA

The National Gap Analysis Program is coordinated by the U.S. Geological Survey. The program's ultimate goal is to keep common species common by looking for gaps in the existing preserve system (Scott et al. 1993). This is accomplished by producing maps of land cover, terrestrial vertebrate distributions, and land stewardship. These maps facilitate determination of the conservation status of biota using a GIS analysis. A GAP project has traditionally been conducted on a state-by-state basis.

The first generation, state-based GAP projects often produced the first detailed, statewide land cover maps (Homer and Crist 1999). The projects had to be innovative in methods and funding to map land cover for the entire state. However, the land cover maps are relatively coarse by today's standards, which limits the use of the maps.

The first-generation gap analysis in Arizona (AZ GAP) was one of the first GAP projects to be started. The project was based on Landsat 5 satellite imagery from 1990 (Halvorson et al. 2001). The imagery was classified into spectrally similar types and assigned a land cover class based on field work and ancillary data. Although this method is similar to an unsupervised classification and is fairly quantitative, there is enough subjectivity in the process to make it unique to Arizona. In addition, the vegetation classification was based on the Brown, Lowe, and Pase system (Brown et al. 1979), a system unique to the Southwest. First-generation, state-based GAP projects typically used classification systems that were unique to their state or region and were hard to cross-walk with neighboring states. Finally, the AZ GAP land cover map has a minimum mapping unit, or resolution, of one square kilometer. This means that the smallest feature the map identifies is at least one square kilometer in size, making the map usable and applicable only at coarse scales.

The National GAP has sought to use the lessons learned from the first-generation, state-based projects to "generate more consistent and efficient methods for land cover mapping" (Jennings 1997) and facilitate advances in technology and methodology that create more thematically and spatially detailed maps. The second-generation gap analysis in Arizona is being conducted as part of the Southwest Regional GAP Analysis Project (SWReGAP), which is a cooperative effort between five Southwest states—Arizona, Colorado, Nevada, New Mexico, and Utah—to produce seamless (with respect to internal state boundaries) land cover, vertebrate species, and land stewardship maps (Jacobs et al. 2001).

To accomplish the goal of a seamless land cover map, we are using the National Vegetation Classification (Grossman et al. 1998). The map classes represent ecological systems (Comer et al. 2003) that describe recurring groups of vegetation associations that are found in similar physical environments and that are influenced by similar ecological processes. NatureServe developed the ecological system concept of vegetation descriptions (http://www.natureserve.org/publications/usEcologicalsystems.jsp) as well as the map class definitions used for this project, in consultation with SWReGAP state project representatives.

Another way to achieve a seamless land cover map is to use mapping zones. The five-state Southwest region is divided into mapping zones (Figure 2) that allow each state project to focus and specialize on fewer, spectrally similar vegetation types at a time, which improves accuracy. There are regional standards for making the land cover map, but each state project has some leeway in adapting the methods to fit the expertise and knowledge of their team. The Arizona project is responsible for mapping the land cover in five of the mapping zones, labeled AZ-1 through AZ-5 in Figure 2. This chapter focuses on using decision tree modeling to produce preliminary second-generation land cover maps in AZ-2, the area roughly corresponding to northeastern Arizona.

METHODS

The Arizona project incorporates an iterative process using five steps in land cover mapping. First is field measurement of vegetation community (land cover) characteristics. Field work gives the most accurate information on the land cover type because a person is directly viewing the ground. However, a statewide map cannot be created using field work alone because of constraints on access, logistics, and money. The second step is the acquisition and preprocessing of remotely sensed imagery, which provides complete spectral coverage for the entire Southwest region. The third step is to use the field, ancillary environmental, and spectral data in the decision tree modeling process. The fourth step is to use GIS to bring everything together and create the land cover map, and finally, the land cover output is verified using withheld data and expert review and, as this whole process is iterative, parameters are modified to refine the map.

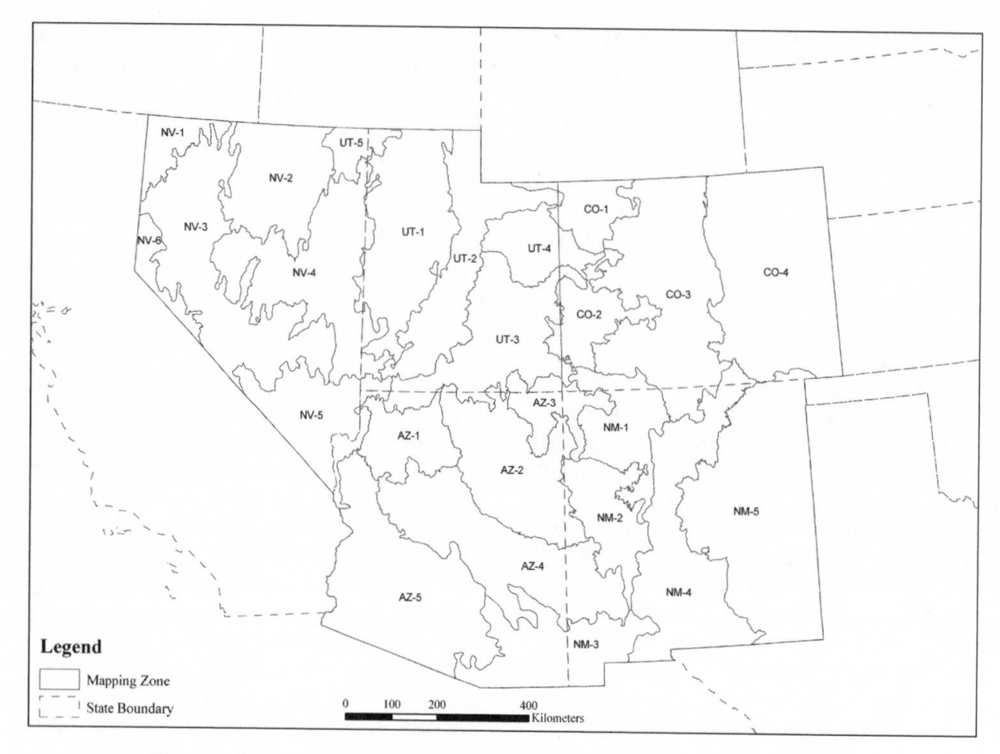

Figure 2. SWReGAP mapping zones; the focus here is on mapping in AZ-2.

Vegetation Community Sampling

We took a very opportunistic approach to field sampling in order to collect a plethora of vegetation data because decision tree modeling requires a lot of input data. In fact, the more field data collected, the more splits are possible in the decision tree which allows a more refined map. Although the non-random sample design may reduce the predictive power of the decision tree, the logistics of access to remote and private land areas and the need for lots of field data dictated the opportunistic approach. We used the existing road network in Arizona to access a wide geographic area and obtain as many observations as possible. In general, the field teams assessed the vegetation type every mile or whenever the vegetation changed. They surveyed a 1-hectare area, about 2.5 acres, because the initial target resolution of the second-generation land cover map was no less than 5 hectares.

Figure 3 shows the locations of the field data used in the first iteration map for AZ-2. In total 1710 observations were collected in the 2001 and 2002 field seasons (June through October), along with pre-existing data collected in 1997. The lack of observations for the first iteration in the middle of the mapping zone is the Hopi Reservation; we did not have permission from the Hopi Tribe to do field work until later in the project. We tried to get a good geographic spread of the area to capture the natural variability of vegetation communities in the landscape.

The field data were compiled and each observation was assigned an ecological system. We then examined the representation of each ecological system in the mapping zone. If there were not enough samples of a particular system, it was grouped with another similar system, if possible, or omitted from the data set. The land cover

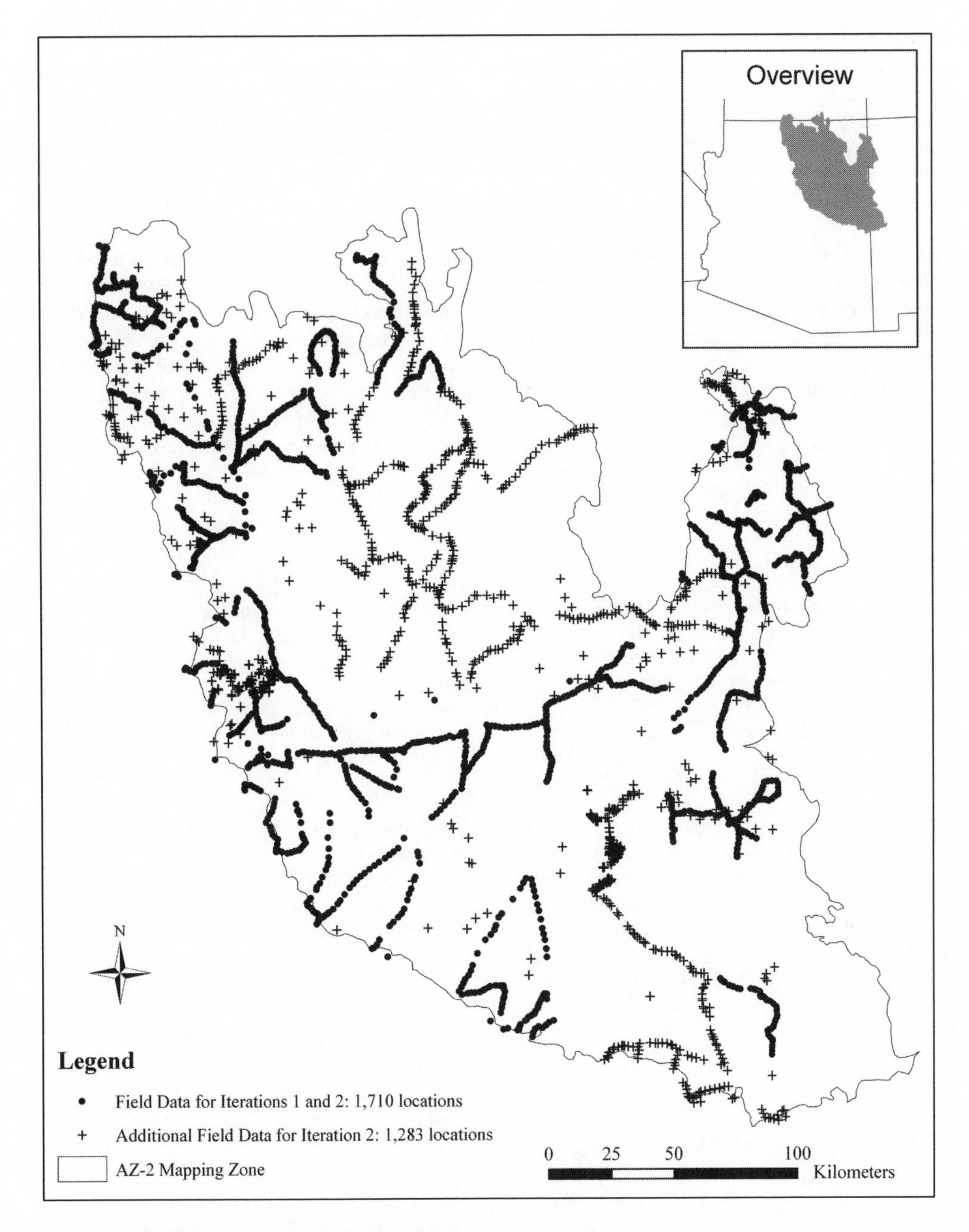

Figure 3. Field plots in AZ-2. The first iteration map used field data from 1710 locations. The second iteration map included an additional 1283 locations, for a total of 2993.

classes in the map consist of ecological systems, groups of ecological systems, altered (e.g., by fire, invasive species) ecological systems, and land use classes. For the AZ-2 mapping zone, the vegetation information from the 1710 field observations was collated and classified into 15 land cover classes (Table 1).

Remotely Sensed Imagery

For the remote sensing aspect of the project, we used data from the Landsat 7 Satellite Enhanced Thematic Mapper Plus sensor. The scenes are mostly from the year 2000, but some scenes are from 1999 and 2001. For each scene, spring, summer, and fall imagery are acquired to look at the differences in phenology of the vegetation among the seasons. The five-state group worked together to determine the standard processing steps for the imagery. First, the images were terrain and atmospherically corrected by the regional office at Utah State University. At the state level, multiple scenes were mosaicked and clipped to the mapping zone for each season. Value-added layers were derived by the state projects for each mosaic for each season. The Normalized Difference Vegetation Index (NDVI) was used to focus on the red and near-infrared bands of the satellite imagery (Kidwell 1990), and tasseled cap indices give information about soil brightness, vegetation greenness, and soil wetness (Crist and Cicone 1984). Finally, a

Table 1. Land cover classes for the first and second iteration maps.

Class	Code	Vegetation Type
First Iteration		
1	N30	Barren
2	S059	Colorado Plateau Blackbrush–Mormon Tea Shrubland
3	S010	Colorado Plateau Mixed Bedrock Canyon and Tableland
4	S039	Colorado Plateau Pinyon-Juniper Woodland
5	S999	Colorado Plateau Sand Shrubland
6	N82	Cultivated Crops
7	N22	Developed, Low Intensity and Open Space
8	N21	Developed, Medium and High Intensity
9	S054	Inter-Mountain Basins Big Sagebrush Shrubland
10	S075	Inter-Mountain Basins Juniper Savanna
11	S065	Inter-Mountain Basins Mixed Salt Desert Scrub
12	S090	Inter-Mountain Basins Semi-Desert Grassland
13	S079	Inter-Mountain Basins Semi-Desert Shrub Steppe
14	S032034	Rocky Mountain Montane Dry-Mesic Mixed Conifer Forest and Woodland
15	S036	Rocky Mountain Ponderosa Pine Woodland
Second Iteration		
16	S056	Colorado Plateau Mixed Low Sagebrush Shrubland
17	S045	Inter-Mountain Basins Mat Saltbush Shrubland
18	S012	Inter-Mountain Basins Active and Stabilized Dunes
19	S096	Inter-Mountain Basins Greasewood Flat
20	S014	Inter-Mountain Basins Greasewood Wash
21	S011	Inter-Mountain Basins Shale Badland
22	S013	Inter-Mountain Basins Volcanic Rock and Cinder Land
23	D08	Invasive Annual Grassland
24	D04	Invasive Southwest Riparian Woodland and Shrubland
25	S046	Rocky Mountain Gambel Oak–Mixed Montane Shrubland
26	S028	Rocky Mountain Subalpine Dry-Mesic Spruce-Fir Forest and Woodland
27	S023	Rocky Mountain Aspen Forest and Woodland

cluster map of each mosaic for each season was produced through unsupervised classification. The cluster map provides information on the natural groupings of spectral signatures of the imagery.

Decision Tree Modeling

In decision tree modeling, the dependent variable is the land cover class determined at each field plot. Twenty-five spatial layers were examined as predictor variables (Table 2). The satellite imagery and derived layers like NDVI and tasseled cap indices are important predictor variables. Digital Elevation Models (DEMs) and derived layers such as slope, aspect, soil moisture, and landform were included in the modeling. Finally, other mapped layers like geology, average annual precipitation, and average annual minimum and maximum temperature were incorporated into the modeling process.

All 25 predictor variables were spatially intersected in a GIS with the land cover class at each of the 1710 field data locations and incorporated into decision tree modeling. This spatial intersection of predictor variables with the land cover class at each field observation results in a table with 1710 rows and 26 columns (the 25 predictor variables and one dependent variable), the cells of which are the value of the predictor variable at each location. This table, which was produced with a CART module (developed for USGS, EROS Data Center by Earth Satellite Corp; copyright held by USGS) for ERDAS IMAGINE (Leica Geosystems), provides the input data for running decision trees.

See5 (Rulequest Research) is an easy to use software program that uses the information from the table produced with the

Table 2. Predictor variables examined for use in the decision tree.

Source	Value-Added Derivation	Independent Variable	Code
Imagery (Landsat 7 ETM+)	Normalized Difference Vegetation Index	Fall NDVI	fandvi
		Spring NDVI	spndvi
		Summer NDVI	sundvi
	Tasseled cap indices	Fall soil brightness	fabrt
		Fall vegetation greenness	fagrn
		Fall soil wetness	fawet
		Spring soil brightness	spbrt
		Spring vegetation greenness	spgrn
		Spring soil wetness	spwet
		Summer soil brightness	subrt
		Summer vegetation greenness	sugrn
		Summer soil wetness	suwet
	Cluster map with 40 classes	Fall cluster map	faclust40
		Spring cluster map	spclust40
		Summer cluster map	suclust40
DEM (30 m from USGS)		Elevation	elev
	Slope	Slope	slope
	Aspect	Aspect	aspect
	Topographic relative moisture index[1]	TRMI	trmim
	Landform[2]	Landform	landf
DAYMET[3]		Minimum temperature	tmin
		Maximum temperature	tmax
		Precipitation	precip
		Shortwave radiation	radiat
Geology		Geology[4]	geol

[1]Parker 1982; [2]Manis 2001; [3]Thornton et al. 1997; [4]Kirby 2003.

CART module to produce a decision tree. Since a tree can overfit the data (in effect, fit the field data too closely to account for heterogeneity in the landscape), it was shrunk by establishing a minimum number of samples for each terminal node. That is, a terminal node that defines a land cover class had to have 50 plots or else it was lumped into the next higher node.

The See5 boosting capacity produced higher classification accuracies by creating a series of decision trees in an iterative fashion, with each successive tree focusing on errors of the previous tree (Brown de Colstoun et al. 2003). The final tree is determined by voting among the different trees.

GIS

The See5 program outputs not only a graphical tree, but also textual rules predicting land cover classes, which are used in a GIS. For example, vegetation type A may consist of a rule where elevation is greater than 1000 m, aspect is southwest, and spring NDVI is greater than 0.75. Each of the predictor variables, in this case elevation, aspect, and spring NDVI, is used in implementation of the rules to get a spatial model of land cover classes. That is, wherever the above rule is met, vegetation type A is predicted to occur.

The rules from the decision trees were implemented in a GIS to produce the first iteration land cover map for AZ-2. Agriculture and urban classes were screen digitized with raw satellite imagery and existing data, and overlaid on top of the classified map.

The deviance at each terminal node of a decision tree can be mapped to produce a map of probabilities of correct classification. This probability map is used to target areas for further field data collection to improve the map. Field crews were thus sent to low-probability areas to gather more data, with the intent of including additional data for map improvement.

Refining the Map

The mapping process is iterative. The land cover map is verified using withheld data and expert review, and parameters are modified to refine the map. Additional field data, refining of the land cover class definitions, and selective input of the predictor variables can change and improve the output.

Validation of the map using withheld field data provides valuable insight into areas needing improvement. Checking the land cover class determined in the field of the withheld data against the land cover class of the map in those same locations produces an error matrix (Congalton and Green 1999) that shows accuracy by map class, along with overall accuracy. The matrix provides information on a class-by-class basis, as well as an overall accuracy statistic. Locations for improvement in mapping are identified by class and location.

Adding more field data allows decision trees to make better splits, or more splits if more map classes are desired. Decision trees require a lot of data because they need to split the data into smaller and smaller groups. Additional field data can also alter the map classes by encouraging different lumping or splitting of classes based on total numbers in each class.

Additional field data were found in several ways. The Hopi Tribe granted permission to collect field data on the Hopi Reservation in March of 2003. This allowed us to fill in the gap of field data in the middle of the mapping zone. We also incorporated existing data from other projects for Sunset Crater, Wupatki, and Petrified Forest National Park units. Finally, we used the probability map, error matrix, and expert review to determine areas to send field crews to collect data.

Another technique for refining the map is to limit the number of predictor variables in the decision trees; decision trees yield higher accuracy with a limited number of predictor variables (Pal and Mather 2003). With many predictor variables, it is likely that some variables are related, making it difficult for the classifier to perform well.

One method of limiting the number of predictor variables looks at the usage of each predictor variable in the decision tree to decide which variables to use. Presumably, the more times a variable is used in the tree, the more important it is in the classification. Using a script and the textual rules, it is possible to count the number of times a predictor variable is used for each land cover class, and the average percentage and/or maximum percentage can be used to determine which predictor variables are most important. The more important variables are used in a second decision tree modeling process.

Another method of limiting the number of predictor variables looks at the correlations between predictor variables to decide which variables to use. Using the table of spatially intersected field data with all of the predictor variables, a correlation analysis is conducted on the predictor variables. The predictor variables that are most correlated with each other may contribute the same information to the decision trees, and should be reduced to one variable in a subsequent decision tree model. This technique, combined with the above technique to determine the important variables, reduces the number of predictor variables used in decision trees and improves results.

All 25 predictor variables were spatially intersected with the land cover class at each field plot and incorporated into decision tree modeling. Before implementing the rules in a GIS, the predictor variables were evaluated for their importance.

RESULTS

Figure 1 shows the output from running a decision tree. A sequence can be followed all the way through the tree to arrive at a terminal node that defines the land cover class. The long length of the lines at the top of the graph indicates that the splits are important, but the splits at the bottom explain less and less of the deviance in the field data.

A sample textual rule from the decision tree is

> Ponderosa Pine Woodland = FaNDVI > 177 and ELEV > 1875 and SpGRN > 111 and ELEV ≤ 2022 and SuWET > 80 and LANDF = 3.

Notice that ELEV (elevation) is used twice in the rule, meaning that it was encountered in the tree at two different splits. The rule could be simplified to incorporate both splits by saying 1875 < ELEV ≤ 2022. In addition, there can be several paths to the same label. Another rule to the same ecological system appearing in the same decision tree is

> Ponderosa Pine Woodland = FaBRT ≤ 37 and LANDF = 4 and TMIN > 3.

This second rule is much simpler and does not use as many predictor variables as the first. These rules are implemented in a GIS to get a thematic map of land cover classes.

Figure 4 (left) shows the probability map of correct classification for the first iteration map of AZ-2. Overall accuracy determined from an error matrix comparing the map to withheld data is 99 percent.

Refining the Map

Although the overall accuracy determined from the error matrix is exceptional, it is really just a verification of the work, not a true, independent accuracy assessment because the field data used to make the map were collected in the same manner and at the same time as the data used for verification. The high accuracy also indicates that mapping more map classes is possible.

The additional field data from the Hopi Reservation and elsewhere almost doubled the amount of data going into decision tree modeling, from 1710 in the first iteration to 2993 observations in the second iteration (Figure 3). These data allowed classification of 27 land cover classes, a significant increase over the 15 classes in the first iteration (Table 1). The additional data also provided field observations in areas that showed low probability of correct classification in the first iteration map.

Table 3 shows the percentage that each predictor variable was used in the second iteration decision tree modeling using the 2993 plots. Predictor variables with high average and/or maximum percentages were considered important. In addition, Table 4 shows the correlation matrix of the predictor variables, which was used to determine which variables were most unique. These two tools were used to reduce the number of

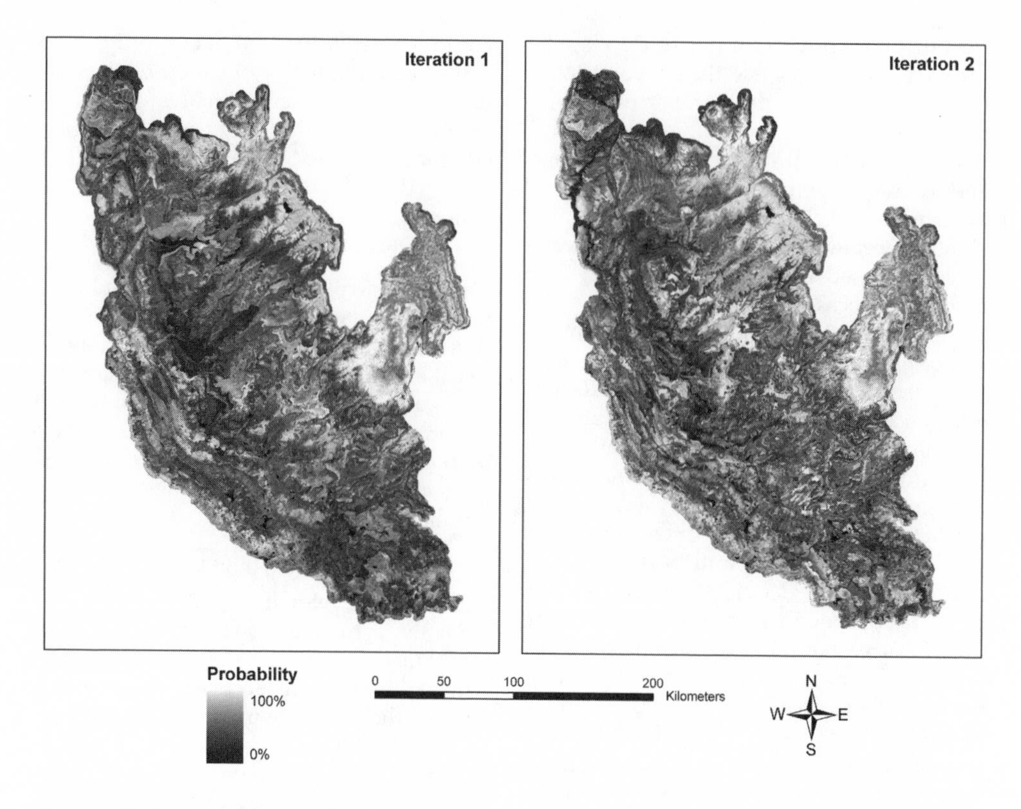

Figure 4. Probability of correct classification for the first iteration land cover map for AZ-2 (left) and the second iteration map (right).

predictor variables from all 25 to just 10: elevation, precipitation, spring vegetation greenness, summer soil brightness, geology, summer NDVI, shortwave radiation, spring NDVI, slope, and aspect. Decision tree modeling was rerun a third time using these 10 predictor variables, and the rules were implemented in a GIS.

The second iteration land cover map for AZ-2 has 27 land cover classes and a different probability of correct classification (Figure 4, right). The differences in the two land cover and associated probability maps can be attributed to the increase in data and the decrease in variables used in the decision tree modeling process. Because of the additional data, more ecological systems were mapped. The additional systems include two invasive classes and several more Intermountain Basins classes.

A comparison of the probability of correct classification between the second and first iteration maps shows a mean improvement of 2 percent. This change doesn't seem much of an improvement in classification, but the fact that many more classes were mapped without a decrease in correct classification is an improvement of the map.

However, overall accuracy determined from an error matrix of the second iteration map compared to withheld data was 86 percent. This is a decrease in accuracy from the first iteration map, but the increase in detail of map classes may justify the decrease in accuracy, depending on the application of the map.

Table 3. Usage of each predictor variable in the second iteration decision tree.

Predictor Variable	Avg %	Max %
Spring cluster map	0	1
Fall cluster map	0	1
Summer cluster map	0	1
Landform	1	1
Fall soil wetness	1	2
Terrestrial moisture index	2	5
Fall soil brightness	2	4
Summer soil wetness	2	3
Slope	2	6
Aspect	2	3
Spring soil wetness	2	5
Fall NDVI	2	3
Summer vegetation greenness	2	4
Fall vegetation greenness	3	3
Spring soil brightness	3	4
Summer soil brightness	3	6
Shortwave radiation	3	6
Spring NDVI	5	6
Summer NDVI	5	6
Spring vegetation greenness	6	9
Minimum temperature	8	10
Maximum temperature	14	18
Precipitation	14	16
Elevation	17	27

DISCUSSION

Land cover mapping using extensive field work, remote sensing, decision tree modeling, and GIS in an iterative process seems to improve the product. The improvement of the number of map classes from the first iteration to the second is tremendous. Although the accuracy may have decreased somewhat, the gain in detail by having more map classes makes the product better. In refining the map, the probability (of correct classification) map and error matrix results by map class were used to determine where to gather more field work. Since collecting field data is a costly endeavor, identification of the best areas for additional data collection is a valuable tool.

The methods presented in this paper are quantitative in nature. Expert knowledge guides the process, and the decision trees and GIS modeling provide a logical, repeatable, and traceable process.

Several land cover classes may not model well using the methods described. Certain classes are easy to identify or are already well mapped, such as water and recent forest burns. For urban and agriculture, we used a screen digitizing approach using the raw satellite imagery for each season plus ancillary data. Spectrally confusing classes, such as playas and sand dunes, may require other approaches. Similarly, important classes that have limited spatial extent such as riparian areas may benefit from additional work.

The final map for AZ-2 was subsequently improved upon with further iterations. Additional field data were obtained, bringing the total field samples to 4319. Twenty-three classes were modeled with 11 additional non-modeled classes. The final AZ-2 map was incorporated into the provisional land cover map of all five Southwest states (Arizona, Colorado, Nevada, New Mexico, and Utah) for the Southwest Regional Gap Analysis Project (map available at http://earth.gis.usu.edu/swgap/).

ACKNOWLEDGMENTS

This project relied heavily on the field data gathered by several teams of dedicated workers. It also incorporated the data analysis capabilities of Keith Pohs, Mar-Elise Hill, and Michael Tweiten.

LITERATURE CITED

Brown, D., C. Lowe, and C. Pase. 1979. A digitized classification system for the biotic communities of North America, with community (series) and association examples for the Southwest. Journal of the Arizona-Nevada Academy of Science 14 (Suppl. 1): 1–16.

Brown de Colstoun, E. C., M. H. Story, C. Thompson, K. Commisso, T. G. Smith, and J. R. Irons. 2003. National Park vegetation mapping using multitemporal Landsat 7 data and a decision tree classifier. Remote Sensing of Environment 85: 316–327.

Comer, P., D. Faber-Langendoen, R. Evans, S. Gawler, C. Josse, G. Kittel, S. Menard, M. Pyne, M. Reid, K. Schulz, K. Snow, and J. Teague. 2003. Ecological Systems of the United States: A Working Classification of U.S. Terrestrial Systems. NatureServe, Arlington VA.

Congalton, R. G., and K. Green. 1999. Assessing the Accuracy of Remotely Sensed Data: Principles and Practices. Lewis Publishers, Boca Raton FL.

Table 4. Correlations of all predictor variables in the second iteration decision tree.

	aspect	elev	fabrt	faclust 40	fagrn	fandvi	fawet	landf	precip	radiat	slope	spbrt	spclust 40	spgrn	spndvi	spwet	subrt	suclust 40	sugrn	sundvi	suwet	tmax	tmin	trmim
aspect	**1.00**	0.32	–0.18	–0.18	0.27	0.31	0.17	0.16	0.22	–0.06	0.67	–0.16	–0.18	0.28	0.28	0.19	–0.21	–0.20	0.32	0.32	0.19	–0.33	–0.20	–0.33
elev	0.32	**1.00**	–0.37	–0.31	0.58	0.65	0.20	0.12	0.74	0.16	0.36	–0.40	–0.43	0.59	0.44	0.43	–0.36	–0.29	0.59	0.63	0.14	–0.87	–0.83	–0.15
fabrt	–0.18	–0.37	**1.00**	0.95	–0.74	–0.68	–0.90	–0.09	–0.49	0.16	–0.27	0.87	0.84	–0.66	–0.36	–0.78	0.94	0.88	–0.70	–0.30	–0.80	0.40	0.38	0.08
faclust40	–0.18	–0.31	0.95	**1.00**	–0.75	–0.63	–0.93	–0.09	–0.45	0.11	–0.24	0.85	0.88	–0.67	–0.32	–0.83	0.90	0.92	–0.69	–0.28	–0.83	0.32	0.31	0.08
fagrn	0.27	0.58	–0.74	–0.75	**1.00**	0.90	0.72	0.15	0.60	0.06	0.32	–0.61	–0.66	0.88	0.62	0.68	–0.70	–0.68	0.89	0.64	0.61	–0.49	–0.41	–0.11
fandvi	0.31	0.65	–0.68	–0.63	0.90	**1.00**	0.58	0.14	0.63	–0.04	0.38	–0.54	–0.55	0.79	0.72	0.56	–0.64	–0.57	0.83	0.78	0.49	–0.61	–0.49	–0.14
fawet	0.17	0.20	–0.90	–0.93	0.72	0.58	**1.00**	0.12	0.34	–0.04	0.25	–0.78	–0.82	0.66	0.30	0.84	–0.89	–0.91	0.70	0.23	0.93	–0.17	–0.19	–0.06
landf	0.16	0.12	–0.09	–0.09	0.15	0.14	0.12	**1.00**	0.08	0.07	0.37	–0.08	–0.09	0.15	0.11	0.12	–0.10	–0.10	0.17	0.14	0.12	–0.07	–0.03	–0.43
precip	0.22	0.74	–0.49	–0.45	0.60	0.63	0.34	0.08	**1.00**	0.06	0.24	–0.43	–0.48	0.51	0.28	0.51	–0.41	–0.37	0.52	0.49	0.24	–0.65	–0.75	–0.10
radiat	–0.06	0.16	0.16	0.11	0.06	–0.04	–0.04	0.07	0.06	**1.00**	–0.07	0.00	–0.03	0.11	–0.10	0.10	0.19	0.12	0.03	–0.01	–0.07	0.29	–0.18	0.09
slope	0.67	0.36	–0.27	–0.24	0.32	0.38	0.25	0.37	0.24	–0.07	**1.00**	–0.21	–0.22	0.32	0.31	0.23	–0.27	–0.24	0.38	0.39	0.25	–0.37	–0.22	–0.32
spbrt	–0.16	–0.40	0.87	0.85	–0.61	–0.54	–0.78	–0.08	–0.43	0.00	–0.21	**1.00**	0.94	–0.62	–0.30	–0.79	0.85	0.83	–0.60	–0.21	–0.71	0.36	0.49	0.10
spclust40	–0.18	–0.43	0.84	0.88	–0.66	–0.55	–0.82	–0.09	–0.48	–0.03	–0.22	0.94	**1.00**	–0.68	–0.22	–0.92	0.84	0.87	–0.65	–0.26	–0.76	0.36	0.49	0.10
spgrn	0.28	0.59	–0.66	–0.67	0.88	0.79	0.66	0.15	0.51	0.11	0.32	–0.62	–0.68	**1.00**	0.71	0.72	–0.68	–0.67	0.88	0.61	0.60	–0.48	–0.41	–0.13
spndvi	0.28	0.44	–0.36	–0.32	0.62	0.72	0.30	0.11	0.28	–0.10	0.31	–0.30	–0.22	0.71	**1.00**	0.18	–0.37	–0.30	0.62	0.65	0.25	–0.45	–0.24	–0.14
spwet	0.19	0.43	–0.78	–0.83	0.68	0.56	0.84	0.12	0.51	0.10	0.23	–0.79	–0.92	0.72	0.18	**1.00**	–0.80	–0.85	0.69	0.30	0.80	–0.32	–0.43	–0.09
subrt	–0.21	–0.36	0.94	0.90	–0.70	–0.64	–0.89	–0.10	–0.41	0.19	–0.27	0.85	0.84	–0.68	–0.37	–0.80	**1.00**	0.94	–0.75	–0.29	–0.89	0.39	0.33	0.10
suclust40	–0.20	–0.29	0.88	0.92	–0.68	–0.57	–0.91	–0.10	–0.37	0.12	–0.24	0.83	0.87	–0.67	–0.30	–0.85	0.94	**1.00**	–0.72	–0.26	–0.91	0.30	0.26	0.10
sugrn	0.32	0.59	–0.70	–0.69	0.89	0.83	0.70	0.17	0.52	0.03	0.38	–0.60	–0.65	0.88	0.62	0.69	–0.75	–0.72	**1.00**	0.75	0.69	–0.52	–0.40	–0.12
sundvi	0.32	0.63	–0.30	–0.28	0.64	0.78	0.23	0.14	0.49	–0.01	0.39	–0.21	–0.26	0.61	0.65	0.30	–0.29	–0.26	0.75	**1.00**	0.19	–0.60	–0.42	–0.13
suwet	0.19	0.14	–0.80	–0.83	0.61	0.49	0.93	0.12	0.24	–0.07	0.25	–0.71	–0.76	0.60	0.25	0.80	–0.89	–0.91	0.69	0.19	**1.00**	–0.12	–0.11	–0.07
tmax	–0.33	–0.87	0.40	0.32	–0.49	–0.61	–0.17	–0.07	–0.65	0.29	–0.37	0.36	0.36	–0.48	–0.45	–0.32	0.39	0.30	–0.52	–0.60	–0.12	**1.00**	0.74	0.18
tmin	–0.20	–0.83	0.38	0.31	–0.41	–0.49	–0.19	–0.03	–0.75	–0.18	–0.22	0.49	0.49	–0.41	–0.24	–0.43	0.33	0.26	–0.40	–0.42	–0.11	0.74	**1.00**	0.10
trmim	–0.33	–0.15	0.08	0.08	–0.11	–0.14	–0.06	–0.43	–0.10	0.09	–0.32	0.10	0.10	–0.13	–0.14	–0.09	0.10	0.10	–0.12	–0.13	–0.07	0.18	0.10	**1.00**

Crist, E. P., and R. C. Cicone. 1984. Application of the tasseled cap concept to simulated thematic mapper data. Photogrammetric Engineering and Remote Sensing 50(3): 343–352.

Friedl, M. A, and C. E. Brodley. 1997. Decision tree classification of land cover from remotely sensed data. Remote Sensing of Environment 61: 399–409.

Grossman, D. H., D. Faber-Langendoen, A. S. Weakley, M. Anderson, P. Bourgeron, R. Crawford, K. Goodin, S. Landaal, K. Metzler, K. D. Patterson, M. Pyne, M. Reid, and L. Sneddon. 1998. International classification of ecological communities: Terrestrial vegetation of the United States. Volume I. The National Vegetation Classification System: Development, Status, and Applications. The Nature Conservancy, Arlington VA.

Halvorson, W., K. Thomas, and L. Graham. 2001. Arizona Gap Analysis Project Vegetation Map. U.S. Geological Survey, Sonoran Desert Field Station, University of Arizona, Tucson.

Homer, C., and P. Crist. 1999. Land cover characterization in Gap Analysis: Past, present, and future. In Gap Analysis Bulletin No. 8, edited by E. S. Brackney, P. J. Crist, and K. J. Gergely. USGS/BRD/Gap Analysis Program, Moscow ID.

Jacobs, S., K. Thomas, and C. Drost. 2001. Mapping land cover and animal species distributions for conservation planning: An overview of the Southwest Regional Gap Analysis Program in Arizona. In Proceedings of the Fifth Biennial Conference of Research on the Colorado Plateau, edited by C. van Riper, III, K. Thomas, and M. Stuart, pp. 159–172. U.S. Geological Survey/FRESC Report Series USGS FRESC/COPL/2001/24.

Jennings, M. 1997. Director's corner. In Gap Analysis Bulletin No. 6, edited by E. S. Brackney and M. D. Jennings. USGS/BRD/Gap Analysis Program, Moscow ID.

Kidwell, K. B. 1990. Global Vegetation Index User's Guide. U.S. Department of Commerce, National Oceanic and Atmospheric Administration, National Environmental Satellite Data and Information Service, National Climatic Data Center, Satellite Data Services Division.

Kirby, S. 2003. Unpublished map. Combined 1: 500 k Geology map for southwestern U.S., created for SWReGAP. Utah State University, Logan.

Manis, G. 2001. Unpublished map. Landform, created for SWReGAP. Utah State University, Logan.

Pal, M., and P. M. Mather. 2003. An assessment of the effectiveness of decision tree methods for land cover classification. Remote Sensing of Environment 86: 554–565.

Parker, A. J. 1982. The Topographic Relative Moisture Index: An approach to soil-moisture assessment in mountain terrain. Physical Geography 3(2): 160–168.

Quinlan, J. R. 1993. C4.5: Programs for Machine Learning. Morgan Kaufman, San Mateo CA.

Scott, J. M., F. Davis, B. Csuti, R. Noss, B. Butterfield, C. Groves, H. Anderson, S. Caicco, F. D'Erchia, T. Edwards, Jr., J. Ulliman, and R. Wright. 1993. Gap analysis: A geographic approach to protection of biological diversity. Wildlife Monographs 123: 1–41.

Thornton, P. E., S. W. Running, and M. A. White. 1997. Generating surfaces of daily meteorological variables over large regions of complex terrain. Journal of Hydrology 190: 214–251.

LAND-USE HISTORY OF THREE COLORADO PLATEAU LANDSCAPES: IMPLICATIONS FOR RESTORATION GOAL-SETTING

Gary P. Nabhan, Susan Smith, Marcelle Coder, and Zsuzsi Kovacs

In recent years, stand-replacing wildfires have burned catastrophically throughout the western United States, and society has invested an unprecedented amount of money in preventative forest treatment, creating heated policy debates about thinning, controlled burns, and other restoration activities (Snider et al. 2003). While researchers and resource managers generally agree that restoration of these forests and woodlands to more natural conditions is urgently needed (Covington et al. 1997; Allen et al. 2002; Baker and Shinneman 2004), they differ in the degree to which they attribute the apparently "unnatural" state of our forest ecosystems to fire exclusion and suppression, livestock grazing, climatic fluctuations, bark beetle infestations, or other factors (Allen 1996, 2001; Swetnam et al. 2001; Baker and Shinneman 2004). Activists, loggers, biologists, and recreationists on the Colorado Plateau differ in their opinions regarding the degree to which thinning or controlled burning can truly serve to restore wooded habitats, and what reference conditions or restoration goals are needed to guide such plans (Friederici 2003a). One proposed tool for guiding restoration has been the use of "pre-settlement conditions" to understand forest stand structure in the ponderosa pine forests before European settlement, and to use restored stand structure as a driver for restoration of ecosystem processes such as frequent, low-intensity fire (Covington and Moore 1994; Friederici 2003b). Policy makers have taken parts of the model and attempted to apply a one-size-fits-all forest thinning policy across the heterogeneous landscapes of the West (Healthy Forests 2002). Such restoration programs are susceptible to rigid prescriptions driven by an inferred ideal of what forests looked like just prior to Anglo settlement (Friederici 2003a).

There can be no doubt that understanding the shapers of our modern forests and woodlands is an important tool for restoration, and understanding stand structure and past fire frequencies is an important part of that picture. However, there are caveats associated with using pre-settlement conditions as the sole reference point for restoration. As will become clear through our use of case studies, cultural modifications of the landscape began well before European settlement of the West, and these important drivers of change must be taken into account. Second, it may be difficult or impossible to know with certainty what the "pre-settlement" conditions actually were. Using structure alone—or even in combination with fire frequency—to predict ecosystem function will likely miss such important pieces of the ecosystem as the role and importance of biodiversity, effects of the understory on hydrology, nutrient cycling dynamics, and the role of predators, birds, and other wildlife. Finally, using past conditions as a goal to restore modern forests also underestimates or ignores such new influences as pollution, habitat fragmentation and loss, exotic invasive species, and climate change. Restoration plans based on such goals may be able to replicate earlier stand structures, but may nevertheless fall short of addressing resilience, ecosystem processes such as hydrology and decadal climate fluctuation, or modern influences on forest and woodland ecology.

We suggest a new way of understanding how diverse land-use histories and natural heterogeneity impact the health of pine-dominated ecosystems in the West, one that acknowledges the importance of cultural and biological diversity. Using three case studies, we demonstrate the potential benefit of understanding natural and cultural complexity when shaping restoration plans, and we demonstrate how diverse land-use histories may have impacted the biodiversity and health of pine-dominated ecosystems in the western United States. This understanding can be used to investigate appropriate means for restoring and managing these dynamic and heterogeneous ecosystems in the future. Our research relies on published and unpublished research, documents, historic photographs, oral histories, and interviews with land managers in our three study areas. We evaluate the likely effects of prehistoric human occupation of these areas, the degree to which ungulate grazing, farming, and other land uses may have changed fire frequency and intensity, the degree of species richness, and the availability of non-timber forest products to local communities.

We suggest that managing for resilience—especially when faced with drought, rapid human population growth, invasive species, and other modern stressors—should include local stewardship of the forests, and that we should build upon local knowledge in our heterogeneous Western landscapes. Responsible harvesting of non-timber forest products, as opposed to large-scale commodity production for timber and cattle, may encourage such stewardship.

The Colorado Plateau is in many ways ideal for our study, as it demonstrates a remarkable degree of biological and cultural diversity. This area has been ranked in the top 4 of 109 ecoregions in North America for species richness in several taxonomic groups, and first for unique or endemic biota (Ricketts et al. 1999; Nabhan et al. 2002). The Colorado Plateau also contains the most extant Native American languages of all North American ecoregions—some 23 distinct languages and dialects—and 29 percent of the lands on the plateau are under tribal control (Nabhan et al. 2002). Each of these native cultures has influenced the vegetation history of its homeland through their land uses. In addition, this region has an exceptionally complete prehistoric and historic ethnobotanical record.

METHODS

We chose three case studies to highlight the range of variability, both past and present, on the Colorado Plateau: the Jemez Mountains and Bandelier National Monument in New Mexico; the Mesa Verde cuesta and Mesa Verde National Park in Colorado; and the San Francisco Volcanic Field and Wupatki and Sunset Crater National Monuments in Arizona (Figure 1). We surveyed published research in archaeology (including contract reports), paleoecology (including packrat midden, pollen, dendrochronological, and fire scar studies), ethnobotany, oral history, historic photographs, unpublished documents archived in the parks, and modern studies of floras, fire behavior, and ecology. We also conducted interviews with land managers in each region. Although our literature survey was not exhaustive, it did reveal patterns and insights that helped construct a more complete picture of the unique shapers of Colorado Plateau landscapes. By demonstrating the uniqueness of each of these landscapes, we provide a primer for how to build a site-specific reference envelope for guiding restoration and management practices.

San Francisco Volcanic Field: Wupatki and Sunset Crater National Monuments

Site Description

The San Francisco Volcanic Field is a chain of about 600 young volcanoes in northern Arizona, covering about 1900 square miles and including the Coconino National Forest and nearby national monuments, as well as the only stratovolcano in the region, the San

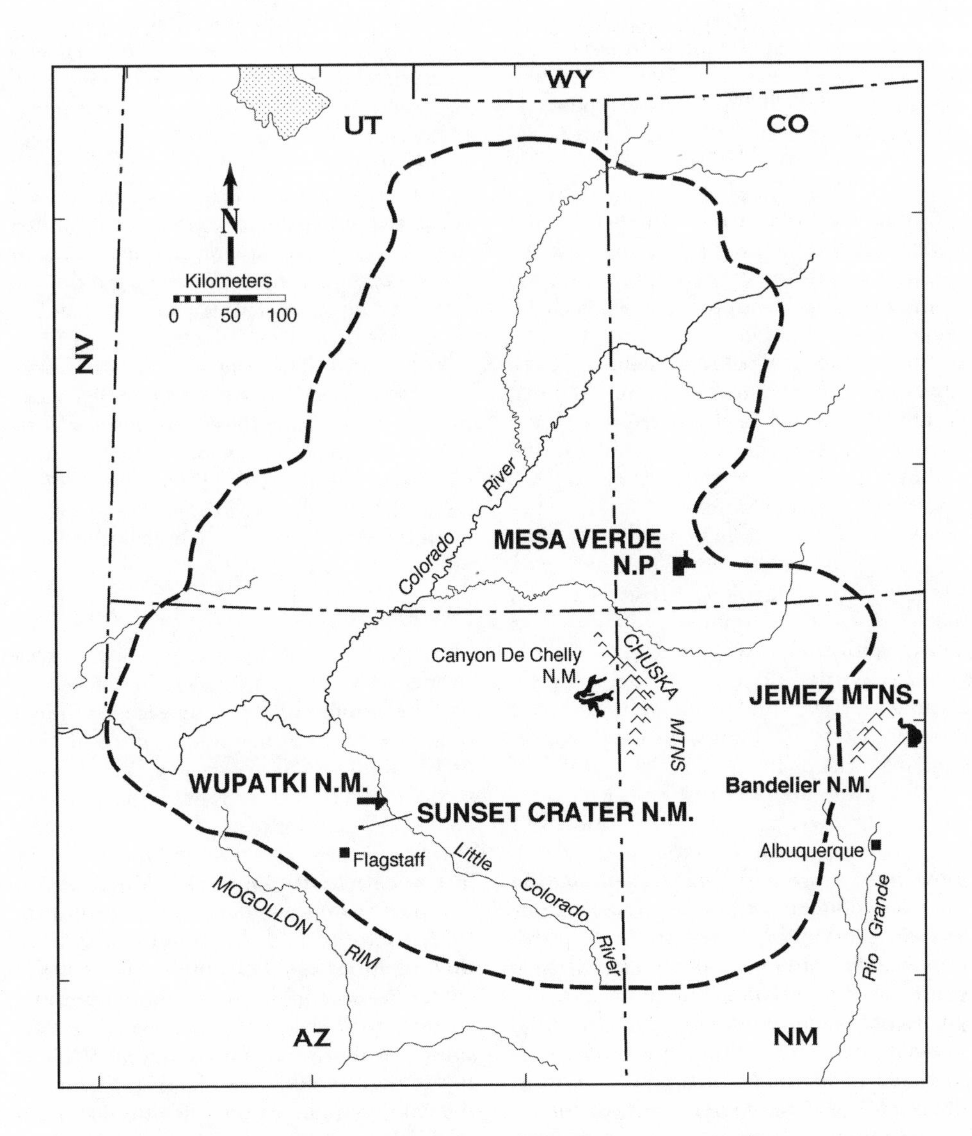

Figure 1. Location of study areas: Mesa Verde National Park, Wupatki and Sunset Crater National Monuments, and the Jemez Mountains.

Francisco Peaks, built by eruptions between 1 and 0.4 million years ago. The most recent volcanic feature in the chain is Sunset Crater, which erupted less than 1000 years ago (Shoemaker and Champion 1977). These late Miocene to Holocene volcanic deposits are predominantly basalt lava flows (Moore 1974; Holm 1986; Priest et al. 2001).

Wupatki and Sunset Crater National Monuments, where we focused our research, together encompass more than 153 sq km (59 sq mi) within the volcanic field. They are surrounded by a mix of private, state, and federal lands, including the Navajo Reservation and Coconino National Forest. Although these two monuments are almost geographically adjacent to one another, there are important geological, historical, and ecological differences between them. Elevation in this study area ranges between 6000 and 9000 ft at Sunset Crater to about 4280–5720 ft at Wupatki. Sunset Crater is dominated by black scoria tephra and cinders (Hooten et al. 2001), whereas Wupatki is more topographically complex, as it is underlain by the Kaibab and Moenkopi geological formations and contains a major fault—the Doney Fault—dividing the monument in half, each side having its own geology, elevation, and dominant vegetation (Thomas et al. 2003). Deposition of ash and cinders modified the hydrologic regime and ecotonal boundaries of the area (Colton 1932; Anderson 1990; Travis 1990; Blyth 1995). Wupatki and Sunset Crater are both in the rain-shadow of the San Francisco Peaks (Colton 1960; Anderson 1990). Precipitation at Wupatki averages 8.2 inches per year, and at Sunset Crater 16.7 inches (www.wrcc.dri.edu/).

Vegetation at Wupatki is dominated by juniper woodland and savanna, desert scrub, and open grassland biotic communities, which are currently being encroached upon by one-seed juniper (Cinnamon 1988; Paul Whitefield, National Park Service botanist, personal communication 2003). The grasslands at Wupatki are still dominated by native perennial bunchgrasses (National Park Service 2002a). Sunset Crater's vegetation, which may still be recovering from the most recent eruption, contains natural patches of pinyon pine, ponderosa, and aspen (National Park Service 2002b). Approximately 350 species of plants have been recorded within these two monuments (Anderson 1990).

Vegetative differences between the parks can be attributed to the differences in soil type, a major factor in vegetation community composition. Several studies compare the success of pinyon pines in the Sunset Crater cinder cone soils with those found in sandy-loam soils of limestone origin (e.g., Cobb et al. 1997; Gehring et al. 1998). Cinder soils tend to drain quickly and are therefore drier. Additionally, these soils are lower in the nutrients nitrogen and phosphorous (Selmants et al. 2003). Ectomycorrhizal associations with pinyon are more abundant in these soils to assist in nutrient accessibility (Gehring et al. 1998).

Land-Use History

More than 2600 sites of archaeological significance have been recorded in these two national monuments. Archaeologists have found evidence of nomadic cultures dating to 11,000 and 8,000 years ago. Year-round settlement commenced with the Sinagua culture in the ninth century (Colton 1932). Sinagua farmers, foragers, and hunters lived in the area until temporary abandonment caused by Sunset Crater's eruption (Colton 1932; Cinnamon 1988). Reoccupation after the eruption took advantage of the new cinder soils, and included more complex farming systems, and pithouse locations shifted to preferred soil conditions (Colton 1932; Short 1988; Anderson 1993; Pilles 1993). Large prehistoric agricultural fields—on the scale of 100 acres or more—have been identified at Wupatki, indicating a much more intensive use of this landscape prehistorically than at any time in the historic period (Cinnamon 1988; Anderson 1990). The region was abandoned around AD 1250–1300.

Historically, this landscape has served as a thoroughfare and hunting and collecting area for the Hopi, Apache, and Navajo (National Park Service 2002a, b; Christopher

Coder, personal communication 2004). While Hopi and Apache people continued to utilize the area for seasonal hunting and gathering, Navajos began seasonal use of the area for sheep in the 1820s (Short 1988), settling permanently in the region in 1860 (Roberts 1990). From 1870 to 1898, Mormons established cattle ranching settlements along the Little Colorado River, soon coming into conflict with Hopis and Navajos over water and land (Trimble 1982; Abruzzi 1985; Short 1988; Roberts 1990); the arrival of the Atlantic and Pacific Railroad in 1882 increased the numbers of settlers and cattle in the region (Wallace 1949; Roberts 1990). Livestock populations on the Navajo Reservation became quite large, with a peak of about 1,300,000 sheep around 1930; the resulting environmental degradation prompted the federal government to establish the Navajo Livestock Reduction Program, slaughtering nearly 70 percent of their sheep (Roberts 1990; National Park Service 2002a) and disrupting Navajo grazing patterns on a regional scale. Livestock grazing continued in the region and in some parts of Wupatki National Monument until it was finally phased out at the monument in 1989 (Cinnamon 1988; Abruzzi 1995). One Navajo family still retains grazing rights for 40 sheep (National Park Service 2002a; Whitefield, personal communication 2003).

The ecological impacts of livestock on the Little Colorado River basin include erosion and soil compaction and reduction of forage quantity and quality. Grazing also contributed to prominent shifts in vegetation communities from grassland to pinyon-juniper woodland (Abruzzi 1995), establishment of nonnative species, and increases in some desert species and noxious weeds (Cinnamon 1988; National Park Service 2002a; Whitefield, personal communication 2003).

Administrative History and Policies

Sunset Crater was designated a national monument in 1930 (National Park Service 2002a). Wupatki National Monument was established in 1924 and expanded in 1937 (Trimble 1982). At the time of designation, Navajo residency was declared a compatible use in the monument but only with a severe reduction of livestock grazing, wood cutting, and gathering of other natural resources (National Park Service 2002a, b). In 1960 and 1971, many Navajos were forced out of the monument, retaining only minimal grazing rights (Roberts 1990). Although gathering rights in both monuments are legally protected, there is disagreement as to whether or not such harvesting is compatible with the National Park Service mission to maintain the biological integrity of the parks (National Park Service 2002a; Whitefield, personal communication 2003).

Fire History

Fire suppression was a result of historic grazing, which also encouraged juniper encroachment; fire suppression policy reinforced these trends. Reliable data on fire frequencies are lacking for these grasslands, woodlands, and forests between 1911 and 1985, though the pinyon-juniper woodlands in this area are believed to have previously experienced natural fires every 15 to 30 years, restricting trees to rocky sites (Wright et al. 1979; Cinnamon 1988; Miller and Tausch 2001). Since the removal of most livestock from Wupatki's grasslands, fires of small spatial scale have dotted the landscape. The most recent event was the Antelope Fire of 2002, which scorched the southwestern end of Wupatki (Whitefield, personal communication 2003). National Park Service administrators believe that fire is necessary to restore and maintain biodiversity in the monument; a fire effects monitoring program was established in 2002 to implement that approach (National Park Service 2002a, b).

Environmental Consequences of Cumulative History

The plant communities existing today at Wupatki and Sunset Crater have developed in response to local topography, precipitation, and geology, prehistoric land use, and grazing (Cinnamon 1981, 1988; Whitefield, personal communication 2003). Between 1911 and 1918, climatic conditions suitable for juniper, pinyon, and ponderosa seed germination developed, and in the absence of natural fire, large cohorts were recruited (Covington and Moore 1994; Covington 2003). Juniper invaded the Wupatki grasslands, resulting in a loss of biodiversity in the grassland and loss of wildlife forage habitat.

Mesa Verde

Site Description

Mesa Verde National Park is in the southwestern corner of Colorado, in Montezuma County. The 21,074 ha (211 sq km) park is part of the Mesa Verde cuesta, an uplifted tableland that rises above the surrounding valley floor by about 610 m. The mesa, a labyrinth of narrow lateral canyons running north-south, is drained by the Mancos River and its tributaries. The park is bordered by the Mancos River on the east, by the Ute Mountain Ute Indian Reservation to the south and west, and by Montezuma Valley to the north (Torres-Reyes 1970).

The bedrock geology of the mesa is dominated by sedimentary formations: the Mancos and Menafee Shales and the Cliff House and Point Lookout Sandstones (Griffitts 2003). The soils in the park range from productive, red eolian soils in the higher elevations to thick deposits of colluvial soils in the canyon bottoms and fans (Ramsey 2003). Variability in soils supports isolated pockets of rare plant communities across the mesa (Floyd and Colyer 2003).

Mesa Verde, a cold, semi-arid steppe (Trewartha 1954), receives an average of 18.2 inches of precipitation annually, which is significantly higher than nearby areas (Erdman 1970; Omi and Emerick 1980; Ramsey 2003). It supports the largest natural reserve of the Upper Sonoran/Sierra Madrean Complex left in the world (Thomas et al. 2003). Vegetation is co-dominated by a pinyon-juniper woodland and a woodland scrub community or petran chaparral (Floyd et al. 2003a), with montane meadows and an estimated 200 hectares (2 sq km) of wetland and riparian habitat in the park (Floyd-Hannah and Romme 1995; Thomas et al. 2003). Mesa Verde's mature pinyon-juniper habitat is one of the oldest and largest on the Colorado Plateau, with most stands having trees that date to 400 years old or more (Erdman 1970; Floyd 2003). This mature woodland is home to some plants that may be endemic to this habitat type, as they do not occur in younger stands, even on Mesa Verde (Floyd and Colyer 2003).

Land-Use History

The pinyon-juniper woodland of the Colorado Plateau was probably established at about 8000 years ago, about the same time that human groups in the area became somewhat less mobile and more reliant on local wild animal and plant resources (Cordell 1994). The cultural history of Mesa Verde from AD 700 to 1350 is dominated by Puebloan farmers, who lived in cliff-house villages and pueblos near their agricultural fields in colluvial fans and valley bottoms. The final Puebloan occupation of Mesa Verde (AD 1150–1300) was characterized by an intensification of agriculture, decreased reliance on wild resources, and population aggregation (Wilshusen and Towner 1999). A generally cooler and wetter climate followed by a prolonged drought (van West and Dean 2000) apparently led to the abandonment of this lifeway on Mesa Verde around AD 1300–1350. In the years prior to abandonment, there is evidence of an overall decrease in health, increasing warfare (Wilshusen and Towner 1999), and possible deforestation (either human caused or natural in response to changing climate conditions; Petersen and Mehringer 1976; Kohler and Matthews 1988; Petersen 1994). The Mesa Verde area remained apparently unoccupied between AD 1350 and 1500 (Wilshusen and Towner 1999).

The area was reoccupied by Utes and proto-Navajo/Apache bands, who relied on wild food resources, about AD 1500 (Brown 1996; Wilshusen and Towner 1999). By the seventeenth century, Navajos living on the mesa were subsisting on a mix of hunting, upland gathering, and farming (Wilshusen and Towner 1999), with Utes continuing to rely seasonally on wild resources. After the

Pueblo Revolt in 1680, the Utes acquired horses from the Spanish and began applying pressure on the Navajos, driving them across the San Juan River to the south (Wilshusen and Towner 1999). For the next 200 years, Mesa Verde was used seasonally by the Utes for wild resources, with no farming or livestock grazing. Even until the late 1800s, the rugged landscape kept Europeans from coming to the area, which remained primarily Ute territory.

Beginning in the late 1870s and lasting until around 1930, several homesteads were established, and large herds of cattle, sheep, and horses grazed within what are now the park boundaries (O'Rourke 1980; Smith 2003). Homesteaders dug a total of 16 wells or pipe springs—primarily to water livestock—which impacted the hydrology of the canyon bottoms (M. Colyer, Mesa Verde National Park, personal communication October 6, 2003). Torres-Reyes (1970) indicated that "by the time the park was established Mesa Verde was already over grazed." Grazing leases were first issued in the park in 1911, but their restrictions were typically not observed (Torres-Reyes 1970). Grazing in the park officially ended in 1927 (Colyer, personal communication 2003), though cattle still grazed in private inholdings in canyon bottoms until 1935. After the exclusion of cattle from these inholdings in 1935 there was good recovery of springs, streams, and vegetation; the range and canyon bottomlands became "a veritable showplace for scientific and educational institutions" (Superintendent's annual report 1946, on file at Mesa Verde National Park). The park was the site of a range training school sponsored by the U.S. Forest Service in 1946, which focused on two of the last private inholdings to come under park ownership (Woodhead 1946). Repeat photography shows an impressive vegetation response, as do measurements documenting the recovery of water resources. The water table in Morefield Canyon had risen 17 feet in the 11 years after exclusion of grazing, in spite of several drought years in the intervening time (Woodhead 1946).

Administrative History

There was very little governmental involvement at Mesa Verde until establishment of the park in 1906. Until that time it had been Ute territory; the Ute reservation in southern Colorado was established in 1880 (www.utemountainute.com / chronology.htm). Reservation boundaries were altered drastically upon establishment of the park, and in 1913 there was a significant adjustment to the boundary of the park (Torres-Reyes 1970). Boundary disputes between the park and the Ute tribe continue to this day. Mesa Verde National Park was designated a wilderness on October 20, 1976, and a World Heritage site on September 6, 1978. The wilderness designation likely had an influence on the backcountry areas of the park; the lack of vehicle and foot traffic may have helped preserve biological soil crusts, decreased the establishment of invasive plants, and kept disturbance of wildlife to a minimum. Although wilderness status has provided protection for natural resources, the fact remains that humans had previously played a role in "managing" this landscape. Excluding such activities as hunting, plant resource gathering, and seasonal burning of some habitats has resulted in an ecosystem different from the prehistoric one, in which humans had a role.

Fire History

Wildfire in the pinyon-juniper habitat tends to be stand replacing (Rogers 2002). Mature trees, seedlings, and seed banks are destroyed, and reestablishment is reliant solely on reintroduction of seed (Floyd-Hanna and Romme 1995; Floyd et al. 2003a). It may take as long as 300 years for a severely burned mature pinyon-juniper forest to recover fully (Omi and Emerick 1980).

Most historic fires at Mesa Verde have been lightning caused (Rogers 2002), though there is evidence from oral histories, park administrative documents, and historic photos that the Utes deliberately used fire in the canyon bottoms (Torres-Reyes 1970:259–260; Colyer, personal communication October 6, 2003). This may have been done to control encroachment of pinyon-juniper into the oak scrub / petran chaparral and riparian communities to enhance wildlife habitat and range value, or as an aid for herding horses (Torres-Reyes 1970:259–260) and to increase abundance of some important wild plants (Adams 2002). In a 1969 study of fire in the pinyon-juniper habitat, deer and elk numbers were higher in burned areas than in adjacent unburned areas (Omi and Emerick 1980). Erdman et al. (1969) stated that "the prehistoric Indians deliberately burned [the] upper parts of the mesa ... in order to maintain the shrub vegetation, which supports a heavier game population than a pinyon-juniper forest approaching climax." Fire also increases the abundance of many plants used by the Utes and by the Puebloan farmers of the mesa; indeed, some rare plants such as wild tobacco (*Nicotiana attenuata*) only occur after fire (Adams 2002). Intentional use of fire by Utes may have shaped the distribution of the pinyon-juniper woodland in the park. Erdman (1970) estimated that 40 percent of the park was dominated by pinyon-juniper in 1970; today the estimate is around 50 percent. This increase may be due to a lack of fire (intentional or natural) in the petran chaparral habitats, which would have kept pinyons and junipers from invading.

Since its establishment in 1906, Mesa Verde National Park has had a total fire suppression policy because of concerns about damage to archaeological sites. However, in spite of nearly 100 years of active fire suppression, fire frequency has risen dramatically, beginning in the mid-1950s, and there has been a trend toward larger, hotter, stand-replacing wildfires (Rogers 2002; Floyd et al. 2003a). From 1906 until early 1934, the park experienced no large fires (Torres-Reyes 1970), though there is evidence of two possible large fires in the late 1880s (Erdman 1970). Floyd and colleagues (2003a) have proposed that this hiatus may have been the result of grazing during the preceding years. Since 1934, more than 80 percent of the park has burned, including five wildfires that burned more than half of the park in the past 7 years alone (Rogers 2002; Floyd et al. 2003a).

Environmental Consequences of Cumulative History

Mesa Verde has high levels of biodiversity and plant endemism when compared to other pinyon-juniper forests on the Colorado Plateau (Floyd and Colyer 2003). The park is a refuge from development and grazing, is largely wilderness, and is surrounded by undeveloped areas, creating natural wildlife corridors and a buffer zone. Endemism may also be due to the varied substrates on the mesa; protection of these soils from erosion should be a priority for managers (Floyd and Colyer 2003). Because much of this biodiversity and endemism is linked to old-growth pinyon-juniper forest, protecting it is also a high priority for resource managers. The greatest threats to these woodlands are climate change, fire, air pollution, and invasive exotic plant species (Romme et al. 2003a).

Fire is damaging to mature pinyon-juniper woodlands, even when used as a management tool (Romme et al. 2003a, b). Indeed, even thinning of dead trees in these habitats can be damaging because of increased risks of invasion by weedy species and removal of shade and protection for new pinyon and juniper seedlings (Romme et al. 2003b). In some cases, however, small-scale fire can enhance species richness, as well as the abundance and vigor of plant species with potential value to humans and wildlife. Adams (2002) observed rates of vegetative recovery in burned areas for 5 years after the 1989 Long Mesa fire. In the petran chaparral communities, several shrubs began to resprout almost immediately; many understory species also increased in both frequency and cover, most notably several grasses, wild tobacco, sego lily (*Calochortus* sp.), buckwheat (*Eriogonum* sp.), globemallow (*Sphaeralcea* sp.), goosefoot (*Chenopodium* sp.), and knotweed (*Polygonum* sp.). Fires can create edge habitats, which provide more diversity in potential resources for humans and wildlife. Recovery from fire in the petran chaparral community—where fire return intervals are about 100 years (Romme et al. 2003a)—was quick and resulted in greater biodiversity and vigor, whereas recovery from fire in pinyon-juniper woodlands—where fire return intervals are about 400 years (Floyd et al. 2003a)—was slower and accompanied by greater invasion of exotic plants (Adams 2002).

Jemez Mountains

Site Description

The Jemez Mountains in northern New Mexico are the southern extension of the Rocky Mountain geographical province. The heart of these mountains is the Sierra de los Valles, with three peaks higher than 10,000 feet (3048 m) forming a dramatic landscape west of Los Alamos National Laboratory. The Valles Caldera is located just west of the Sierra Crest, with the Rio Grande Valley to the east. Los Alamos and Bandelier National Monument are located on the Pajarito Plateau, a broad pediment sloping off east of the Jemez Mountains, carved by canyons into a series of long east-southeast trending mesas. The terrain is volcanic; a stack of dacites, andesites, rhyodacites, and quartz latites formed the mountains, and the Pajarito Plateau is cast from a series of massive ash-flow tuffs (ignimbrites; Griggs 1964; Smith et al. 1970; Goff et al. 1996; Reneau and McDonald 1996).

Modern vegetation is complex and forms mosaics of communities at different scales (Foxx 2002; Allen 2004). A broad classification of five main cover types is defined from the edge of the Pajarito Plateau to the mountain crest as follows: juniper savanna, pinyon-juniper woodland, ponderosa pine forest, mixed conifer forest, and spruce-fir forest (Balice et al. 1997). Boundaries between vegetation communities are diffuse transition zones moderated by slope, aspect, and the many canyons that carry narrow stringers of higher elevation species through lower elevation communities. The variety of habitats has created a remarkably diverse flora that has been studied by several researchers (Foxx and Tierney 1985; Allen 1989; Jacobs 1989; Hartman 2003). More than 700 plant species have been documented in Bandelier National Monument alone (Jacobs 1989).

Climate is semi-arid and continental with cool winters and moderately warm summers. Annual precipitation ranges from 25 cm at the lowest elevations to about 90 cm at higher elevations (Bowen 1990). The April through late June season is typically dry, and then, between July and September during "monsoons," almost half of the annual precipitation occurs as thunderstorms. The overall weather in the Jemez Mountains is extremely variable, not only year to year, but also by season. Over a 69-year period, annual precipitation extremes ranged from 17.8 to 77.1 cm (Bowen 1990). Three "events" are key to understanding the last 100 years: a severe drought between 1950 and 1956, a wet period that began in the 1970s (Grissino-Mayer and Swetnam 2000), and the current drought, which began in 1997. Some researchers predict that the next 20–30 years will be a protracted drought (Hereford et al. 2002), and possibly a century-scale dry period.

Land-Use History

Of the three landscapes that we examined, the Jemez Mountains has the longest continuous history of human occupation. Allen (2002a) believes that it may have the highest density of cultural sites in the entire National Park Service system, with more than 2400 archaeological sites recorded from the approximately 6500 hectares inventoried to date at Bandelier National Monument. The 10,000-year archaeological record is characterized by regional and local shifts in demography and land-use practices, a rhythm of boom and bust, occupation and abandonment. The first extensive farming on the Pajarito Plateau was about AD 1150 by Pueblo III peoples. The greatest populations (the most intense boom) occurred between AD 1200 and 1500 when settlements coalesced into large agriculture-based pueblos situated in the pinyon and juniper zone (Vierra et al. 2002). In the mid 1500s, the large pueblos were abandoned, possibly due to drought and soil depletion, and people moved into communities along the Rio Grande River. Small farms have shaped landscape patches in the Jemez Mountains and northern New Mexico since the AD 1100s. By the AD 1300s, Anasazi farmers were using a sophisticated dry-land farming technology that encompassed managing and conserving all manner of surface water (Toll 1995). By the 1700s, Spanish subsistence farmers and herders were setting up small settlements. In 1790, the Santa Fe census included more than "2500 souls," and almost everyone listed their profession as farmer (Dorman 1996). The Spanish imported their kitchen gardens to New Mexico and the local cuisine changed to include their imported favorite garden plants (Lopinot 1986); a variety of livestock was also introduced (Baydo 1970).

Americans began to travel through New Mexico in 1821, and the first American freight caravan reached Santa Fe in 1824. In the 1860s, Navajo and Apache Indians were confined to reservations, ending the raiding that had discouraged homesteaders and opening the land to Anglo, Hispanic, and Pueblo settlers. The railroads reached New Mexico in the 1880s and land use was transformed from subsistence to commercial activities across large areas (Rothman 1992). In 1937, 35 Anglo and Hispanic homesteads farmed about 15 sq km on the Pajarito Plateau (Foxx et al. 1997). Recent history is marked by the creation of the Los Alamos National Laboratory in 1942, which created almost overnight a new town of white, middle-class, highly educated professionals building nuclear bombs in the midst of a Hispanic and Indian Pueblo world (Allen 2002b).

During the Spanish occupation, sheep became the keystone of a subsistence economy (Carlson 1969). Cattle became more important than sheep in the 1800s, when Anglos started settling New Mexico. Economic policies and laws passed by the Texas legislature in 1879 and 1883 restricted cattle grazing on depleted Texas ranges, and cattlemen moved over to New Mexico (Foxx 2002). The most abusive overstocking, when the railroads opened the "frontier," started in the 1880s and extended into the 1900s. By 1895 virtually every acre of available grassland in New Mexico was overstocked with sheep or cattle—four times over what the range would carry (Foxx 2002). Grazing was not excluded in Bandelier until the National Park Service took over in 1932. Even after the amount of livestock was limited by the park, feral burros introduced by the Spanish continued to overgrazed prime habitats until they were removed in the 1970s (Allen 2004).

Administrative History

Mexico governed the territory from 1821 to 1846, the era of land grants. In 1848, the United States acquired the region during the Mexican-American war. Bandelier National Monument, which was established in 1917, encompasses 32,737 acres, 23,267 of which are designated wilderness (www.nps.gov/band/).

Fire History

Before the 1880s, there were frequent widespread lightning fires in the Jemez Mountains, but few "Indian" fires were recorded (Allen 2002a). The pre-1900 fire return interval was 6–15 years, with more frequent fires in ponderosa pine forests, and the seasonal timing of fires was during the dry spring and early summer months (Allen 2002a). In the higher elevation mixed conifer forests, fires are moderated by more mesic conditions and breaks in topography. Fires in the mixed conifer tend to be stand-replacing crown fires, but of limited areal extent (Allen 2002a; Falk and Swetnam 2003; Falk 2004). Tree-ring studies in the Jemez summarized by Touchan et al. (1996) show that in the ponderosa pine forests, from the 1600 to 1700s, fire regimes were characterized by high-frequency, low-intensity surface fires, with a sharp decrease about 1750. This presumed change in fire intervals may record overgrazing by sheep, which reduced grass and other fine fire-starting fuels (Touchan et al. 1996). An alternative theory, proposed by Swetnam and Baisan

(2003), is that a period of global El Niño climate and possibly major volcanic eruptions in the early 1800s caused a pronounced cooler period.

The net effect of livestock overgrazing and logging between the 1880s and 1900s, and the early 1900s policy of fire suppression, was fewer fires, which altered natural fire return intervals. In the Southwest, there were at least two significant wet intervals in this period, 1905–1941 and 1978–1998 (Hereford et al. 2002), and forests responded with surges in growth that built up dangerous fuel levels. In the past 20 years there have been catastrophic crown fires in the Jemez Mountains. More than 80,000 acres have burned on the east slope of the Jemez Mountains since the 1977 La Mesa fire. By far the largest and most dramatic was the 2000 Cerro Grande fire on the Pajarito Plateau, which burned 44,000 acres and 354 homes with a price tag of 5.1 million dollars (Griggs 2001).

Environmental Consequences of Cumulative History

The complex and at times intense land use in the Jemez Mountains has initiated a cascade of environmental responses. Most local vegetation types have undergone major changes in the past century from the cumulative and intertwined effects of livestock grazing, farming, logging, and fire suppression (Allen 1989), and the tempo of change is increasing with the pressures of modern development. Grasslands have been invaded by trees, and tree densities in mixed conifer and ponderosa pine forests have increased dramatically, escalating fire hazards to extreme conditions. Increased tree density in pinyon and juniper woodland has restricted herbaceous ground cover, creating bare ground, which in turn has accelerated erosion. In Bandelier National Monument, Craig Allen has documented 58–84 percent bare ground under pinyon juniper canopies. In the Frijolito watershed at Bandelier, sediment is eroding at approximately 2 mm/yr, which, at this rate, would strip all of the intercanopy soil from the watershed within 100–200 years (Allen 2004). Exotic species are becoming established over large areas and are taking over significant niches, especially the increasingly limited riparian and wetland communities (Allen 2004). At Bandelier National Monument, approximately 17 percent of the park flora are introduced plants (Brian Jacobs, Bandelier NM botanist, unpublished data), and at least 13 introduced species were released in the Jemez region before 1800 (Foxx 2002). Extreme drought conditions in 2002 and epidemic bark beetle infestations have devastated the unnaturally dense forests at Bandelier, and in some areas up to 100 percent of the pinyon trees have died.

DISCUSSION

The preceding case studies have described three unique landscapes within the greater Colorado Plateau ecoregion. Each area exhibits differences in the physical and cultural influences that shape their current ecosystems (Table 1). Differences in physical factors, such as substrate, bedrock geology, and climate were especially evident in our study areas. Mesa Verde, for example, has a predominantly sedimentary substrate, whereas the Jemez Mountains and Wupatki are primarily volcanic. These differences can have important implications for endemism, response to drought, and erosion.

Cultural influences on the landscape vary widely between our study sites. For example, the Jemez Mountains represent one of the richest cultural regions in the Southwest, with 14 long-established pueblos in the area, and with Hispanic people living in the region since 1590. Farming has been an important shaper of this landscape for at least 3000 years. Wupatki has been lightly occupied and heavily grazed in historic times, but before the 1300s it experienced more intensive land use (farming) than in any historic period. Mesa Verde has had an even less intensive history of human occupation since 1300, with about 250 years of intensive prehistoric farming. Farming affects western landscapes variably, depending on the scale and intensity of water control, ground disturbance, and clearing of native vegetation. Irrigation can have effects

Table 1. Natural and cultural heterogeneity at three study areas on the Colorado Plateau.

Jemez	Mesa Verde	SF Volcanic Field
Natural:		
Volcanic	Sedimentary	Recent volcanic, some sedimentary
Mesic to semi-arid	Cold semi-arid steppe	Cool, semi-arid
High biodiversity	High biodiversity, high endemics	Relatively low biodiversity
Dominated by ponderosa pine and pinyon-juniper	Dominated by old pinyon-juniper	Dominated by grassland-juniper savannah
Cultural:		
Long & culturally diverse occupation	Sporadic and less diverse occupation	Sporadic, culturally diverse presence
Very long grazing: more than 300 years	Short grazing history: 60 years	Long, intensive grazing: 175 years
Very long history of farming: 3000 years	Farming history short: 600 years	Farming history short, intensive: 500 years
Gathering, hunting	Gathering & hunting now prohibited, but formerly important to Ute, Hopi	Formerly important hunting and collecting area for Hopi, Apache, Navajo

on water table and soil salinity; annual plowing creates a soil disturbance that can prevent biological soil crusts, mycorrhizal associations, and microbes from establishing, increases erosion potential, and creates areas where weedy species thrive. Large-scale clearing of native vegetation for new fields can potentially decrease native biodiversity. A number of intentionally introduced species have escaped cultivation and become weedy invaders, for example Himalayan blackberry (*Rubus idaeus*), horehound (*Marrubium vulgare*), and tamarisk (*Tamarix* sp.); some of these naturalized species have become non-timber forest products (West et al. 2005), and *Corispermum*, native to the Colorado Plateau and common in the archaeological record but rare today, is thought to have been cultivated prehistorically (Hunter 1997).

Grazing has had a profound impact across the West, and although its influences on our study areas are recognized by land managers, these impacts have probably been underestimated (Fleischner 1994; Floyd et al. 2003b). At Jemez, early and continuous grazing of sheep, cattle, and horses influenced the landscape for more than 300 years, from the 1600s to the 1930s. Mesa Verde had a relatively short grazing history, primarily of cattle, limited to about 50–60 years. At Wupatki, Navajo and Mormon ranchers have grazed sheep and cattle, at times intensively, for 175 years, beginning around 1830, with limited grazing continuing to the present. Impacts of grazing include loss or severe reduction of native cool-season grasses, which are important for controlling erosion and as forage for wildlife, which was once very important as a food source for Native Americans (Bohrer 1974; Davenport et al. 1998). Interviews with Marilyn Colyer, natural resource specialist for more than 30 years at Mesa Verde National Park, support the idea that grazing exclusion coupled with reseeding has allowed some of these native cool-season grasses, especially Indian ricegrass (*Poa fendleriana*), to reestablish in the park and become common once again (Colyer, personal communication 2004). Indian ricegrass is also mentioned as one of the cool-season grasses most heavily impacted by grazing at Wupatki National Monument (P. Hogan, personal communication 2004). The disruptive impacts of cattle are particularly noteworthy in areas of the West where biological soil crusts are important for nutrient cycling, water retention, and native seedling establishment (Evans and Belnap 1999; Belnap 1995, 2003). Control of predators by ranchers has had an impact on populations of herbivores (Henke and Bryant 1999; Soulé et al. 2003; Wilmers et al. 2003), and the creation of tanks, wells, and pipe springs to benefit livestock has had an impact on hydrology (Woodhead 1946; Cole et al. 1997; Belsky et al. 1999). Mesa Verde National Park, with its livestock exclosure in place for 70 years, would make an excellent laboratory for studying some of these effects.

The disruption in natural fire frequencies resulting from grazing, logging, farming, and fire suppression has led to a decline in biodiversity in areas where frequent, low-intensity fires were once the norm. Some species require fire or smoke to germinate, or proliferate following fire, including plants which were important to the native inhabitants of the Colorado Plateau such as wild tobacco, goosefoot, sego lily, buckwheat, sumac (*Rhus* sp.), and several native grasses (Adams 2002; West et al. 2005). Conversely, frequent low-intensity fires inhibit the establishment of juniper and pinyon seedlings, which have become weedy invaders of grasslands in the Wupatki area, leading to a decline in native biodiversity (Jackson et al. 2002).

Long- and short-term studies of the effects of different forest restoration practices are important, and will help us better understand the relationships between land-use history, physical factors such as soil type and climate, and how a landscape responds to different treatment types. Such studies have begun in the ponderosa pine forests around Flagstaff, Arizona, though not all these studies take into account prehistoric land-use practices, non-timber forest products, the role and response of the understory, or the role of local stewardship of the forests (Friederici 2003a). The history of forest management in the pinyon-juniper woodlands across the Colorado Plateau is, however, much less self-critical. Furthermore, far from being a single habitat type, there are a wide variety of pinyon-juniper types, defined by soil type, predominance of one or another species of pinyon or juniper, or other factors. As demonstrated by the case studies here, some of these habitats may benefit from more frequent, low-intensity fire and/or release from grazing, as at Wupatki. Others, such as at Mesa Verde, thrive with an intact biological soil crust, and would be damaged by management using either thinning or burning, even in the face of a wildfire threat (Romme et al. 2003b). Additionally, where there is adequate survey data, it appears that prehistoric use of the pinyon-juniper habitat type was very intensive, and prehistoric land use was likely a strong influence on these ecosystems (Smith et al. 2005). Much of the pinyon-juniper woodland habitat type on the plateau is managed by the Bureau of Land Management (BLM); 39.3 percent of BLM lands on the Colorado Plateau are classified as pinyon-juniper (http://www.mpcer.nau.edu/pj/). The BLM has for decades used treatments such as chaining, hand thinning, burning, and chemical treatments to remove juniper and pinyon from grasslands, primarily to enhance range value for domestic livestock. Recently, the BLM has partnered with Northern Arizona University's Merriam Powell Center for Environmental Research (MPCER) to map the extent of different pinyon-juniper treatments in southern Utah. This is an exciting project, which has the potential to provide information which heretofore has been lacking in many current forest and woodland management plans, namely the associations between physical factors, land-use history, treatment types, and the results (http://www.mpcer.nau.edu/pj/; N. Cobb, personal communication April 2005) of treatment. Based on the results of our findings, we believe that pinyon-juniper restoration would greatly benefit by including ethnohistory, archaeology, and local knowledge into final restoration prescriptions.

The current threats facing our three study areas differ (Table 2). Surprisingly, not all of

Table 2. Current threats to our study areas, as identified by land managers.

Location / Threat
Bandelier National Monument
Urban growth and habitat fragmentation
Wildfire
Erosion
Invasive plant species
Mesa Verde National Park
Wildfire
Invasive plant species
Air pollution
Wupatki and Sunset Crater National Monuments
Juniper encroachment into grassland
Rural population growth and habitat fragmentation
Invasive plant species

the area managers identified wildfire as a major threat, though habitat loss and fragmentation and invasive species are identified as major threats in all three locations. At two of our study areas (Mesa Verde and Jemez Mountains), 10–17 percent of the floras are introduced species. In addition, climate change threatens the entire region, and is a factor that needs to be included in all management plans.

CONCLUSIONS

Each of our three study areas has had a very different land-use history, and each area is characterized by different factors affecting biodiversity and forest stands today. For some areas, it is clear that the last century of forest and woodland management has made the habitats within this ecoregion more homogeneous, more prone to high-severity, stand-replacing fires, and less diverse in understory non-timber forest products. These general trends have generated serious ecological, economic, and cultural consequences not only in the three landscapes we have studied, but for the ecoregion as a whole. We have summarized historic changes in forest stand structure and composition. We recognize that modern forests are imprinted with the different traditions of burning, foraging, farming, and hunting. The grazing effects from sheep, cattle, and horses have affected fire frequencies, sometimes decades before Anglos settled in the same areas and before official fire suppression policy was enacted. It is important to recognize that, to this day, fire and grazing patterns are varied, as these activities remain under distinctive cultural and administrative influence. Furthermore, these activities are and have been occurring in a place of geographic heterogeneity where the ecological ramifications remain as unique as the landscapes in which they occur.

Recognition of the potential cumulative effects of the various histories upon the landscape underscores the need for resource managers to cautiously base their restoration and management strategies on localized or site-specific scientific, historic, and cultural studies of reference conditions, rather than assuming that a one-size-fits-all "pre-settlement" formula will work for all pine-dominated ecosystems (Allen et al. 2002; Romme et al. 2003a; Baker and Shinneman 2004).

From comparing sites within and between the three landscapes analyzed here, we suggest that "pre-settlement conditions" in one region cannot be extrapolated completely to other regions or forest types over the Colorado Plateau. Instead, it is necessary to construct a "reference envelope" for each site, incorporating fire history, paleoecology, ethnohistory, archaeology, oral history, repeat photography, interviews with land stewards with a depth of local knowledge, and other information. It is important that managers not assume that thinning or shifting forest structure alone will be enough to regenerate understory species richness or non-timber forest products, because drought and arrested ecological processes produce time lags in recovery. In fact, thinning and prescribed burning can have negative effects on forest structure and biodiversity in some habitats (Romme et al. 2003a, b). The social importance of non-timber forest products and the effects of treatments on biodiversity need to be taken into consideration. An understanding of how specific cultural uses of non-timber forest resources affect and are affected by forest management policy changes, drought, and grazing can contribute valuable insights to management decisions and restoration goals.

ACKNOWLEDGMENTS

This work was funded by the National Commission on Science and Sustainable Forestry (NCSSF). We are grateful to many individuals and organizations who volunteered their time and expertise. At Northern Arizona University, we thank Bruce Hooper, Judy Springer, Patty West, Steve Buckley, Matt Clark, Ken Cole, and Thomas Whitham. At Mesa Verde National Park, we were assisted by Marilyn Colyer, George San Miguel, Liz Bauer, and Carolyn Landes. We thank Lisa Floyd, Prescott College, for information and guidance. Craig Allen, John Mack, and Brian Jacobs at Bandelier National Monument shared their valuable time and pointed us to great information. We thank Teralene Foxx (White Rock, New Mexico) and Randy Balice (Los Alamos National Laboratory, New Mexico) for providing us with their publications and reports, and Melissa Savage, Four Corners Institute, Santa Fe, New Mexico, for advice and information. And thanks to Paul Whitefield and Gwenn Gallenstein, Flagstaff Area National Monuments, Arizona. We also thank the editors of this volume and two anonymous reviewers for their help in refining our message.

REFERENCES

Abruzzi, W. S. 1985. Water and community development in the Little Colorado River basin. Human Ecology 12: 241–269.

Abruzzi, W. S. 1995. The social and ecological consequences of early cattle ranching in northeastern Arizona. Human Ecology 23: 75–98.

Adams, K. R. 2002. A five-year ecological and ethnobotanical study of vegetation recovery after the Long Mesa Fire of July 1989, Mesa Verde National Park, Colorado. Manuscript on file, Mesa Verde National Park.

Allen, C. D. 1989. Changes in the landscape of the Jemez Mountains, New Mexico. Unpublished Ph.D. dissertation, University of California, Berkeley.

Allen, C. D., editor. 1996. Fire Effects in Southwestern Forests: Proceedings of the Second La Mesa Fire Symposium. Los Alamos NM. USDA Forest Service General Technical Report RM-GTR-286.

Allen, C. D. 2001. Fire and vegetation history of the Jemez Mountains. In Water, Watersheds, and Land Use in New Mexico: The New Mexico Decision-makers Field Guide No. 1, edited by P. S. Johnson, pp. 29–33. New Mexico Bureau of Mines and Mineral Resources, Socorro NM.

Allen, C. D. 2002a. Lots of lightning and plenty of people. An ecological history of fire in the upland Southwest. In Fire, Native Peoples, and the Natural Landscape, edited by T. R. Vale, pp. 143–193. Island Press, Washington D.C.

Allen, C. D. 2002b. Rumblings in the Rio Arriba: Landscape changes in the southern Rocky Mountains of northern New Mexico. In Rocky Mountain Futures. An Ecological Perspective, edited by J. S. Baron, pp. 239–253. Island Press, Washington D.C.

Allen, C. D. 2004. Ecological patterns and environmental change in the Bandelier landscape. In Archaeology of Bandelier National Monument: Village Formation on the Pajarito Plateau, New Mexico, edited by T. A. Kohler. University of New Mexico Press, Albuquerque.

Allen, C. D., D. A. Falk, M. Hoffman, J. Klingel, P. Morgan, M. Savage, T. Schulke, P. Stacey, K. Suckling, and T. W. Swetman. 2002. Ecological restoration of southwestern ponderosa pine ecosystems: A broad framework. Ecological Applications 12: 1418–1433.

Anderson, B. A. 1990. Chapter 3. In The Wupatki Archaeological Inventory Survey Project: Final Report, edited by B. Anderson, pp. 1–38. Southwest Cultural Resources Professional Paper No. 35.

Anderson, B. A. 1993. Wupatki National Monument: Exploring into Prehistory. In Wupatki and Walnut Canyon: New Perspectives on History, Prehistory and Rock Art, edited by D. G. Noble, pp. 13–19. Ancient City Press, Santa Fe NM.

Bailey, R. G., P. E. Avers, T. King, and W. H. McNab, editors. 1994. Ecoregions and subregions of the United States (map). USDA Forest Service, Washington, D.C. 1: 7,5000,000. With supplementary table of map unit descriptions, compiled and edited by W. H. McNab and R. G. Bailey.

Baker, W. L., and D. J. Shinneman. 2004. Fire and restoration of pinyon-juniper woodlands in the western United States: A review. Forest Ecology and Management 189(1–3): 1–21.

Balice, R. G., S. G. Ferran, and T. S. Foxx. 1997. Preliminary vegetation and land cover classification for the Los Alamos region. Los Alamos National Laboratory, LA-UR-97-4627, Los Alamos NM.

Baydo, G. R. 1970. Cattle ranching in territorial New Mexico. Ph.D. dissertation, University of New Mexico, Albuquerque.

Belnap, J. 1995. Surface disturbances: Their role in accelerating desertification. Environmental Monitoring and Assessment 37: 39–57.

Belnap, J. 2003. Magnificent microbes: Biological soil crusts in pinyon-juniper communities. In Ancient Pinyon-Juniper Woodlands: A Natural History of Mesa Verde Country, edited by M. L. Floyd, pp. 75-88. University of Colorado Press, Boulder.

Belsky, A. J., A. Matzke, and S. Uselman. 1999. Survey of livestock influences on stream and riparian ecosystems in the western United States. Journal of Soil and Water Conservation 54(1): 419–431.

Blyth, C. P. 1995. The Cenozoic evolution of Wupatki National Monument. Master's thesis on file at Northern Arizona University, Flagstaff.

Bohrer, V. L. 1974. The prehistoric and historic role of the cool-season grasses in the Southwest. Economic Botany 29: 199–207.

Bowen, B. M. 1990. Los Alamos climatology. Los Alamos National Laboratory, LA-11735-MS UC-902. Los Alamos NM.

Brown, G. M. 1996. The protohistoric transition in the northern San Juan region. In The Archaeology of Navajo Origins, edited by R. H. Towner, pp. 47–69. University of Utah Press, Salt Lake City.

Carlson, A. W. 1969. New Mexico's sheep industry, 1850–1900: Its role in the history of the territory. New Mexico Historical Review XLIV 1: 25–49.

Cinnamon, S. K. 1981. Livestock grazing at Wupatki National Monument. Unpublished manuscript on file at Wupatki National Monument. 16 pp.

Cinnamon, S. K. 1988. The vegetation community of Cedar Canyon, Wupatki National Monument as influenced by prehistoric and historic environmental change. Master's thesis, Northern Arizona University, Flagstaff.

Cobb, N. S., S. Mopper, C. A. Gehring, M. Caouette, K. M. Christensen, and T. G. Whitham. 1997. Increased moth herbivory associated with environmental stress of pinyon pine at local and regional levels. Oecologia 109: 389–397.

Cole, K. L., N. Henderson, and D. S. Shafer. 1997. Holocene vegetation and historic grazing impacts at Capitol Reef National Park reconstructed from packrat middens. Great Basin Naturalist 57: 315–326.

Colton, H. S. 1932. Sunset Crater: The effect of a volcanic eruption on the ancient Pueblo people. Geographical Review 22: 582–590.

Colton, H. S. 1960. Black sand: Prehistory in northern Arizona. University of New Mexico Press, Albuquerque.

Cordell, L. S. 1994. Ancient pueblo peoples. Smithsonian Institution, Washington D.C.

Covington, W. W. 2003. The evolutionary and historical context. In Ecological Restoration of Southwestern Ponderosa Pine Forests, edited by P. Friederici, pp. 26–47. Island Press, Washington D.C.

Covington, W. W., P. Z. Fulé, M. M. Moore, S. C. Hart, T. E. Kolb, J. N. Mast, S. S. Sackett, and M. R. Wagner. 1997. Restoring ecosystem health in ponderosa pine forest of the Southwest. Journal of Forestry 95(4): 23–29.

Covington, W. W., and M. M. Moore. 1994. Southwestern ponderosa pine forest structure: Changes since Euro-American settlement. Journal of Forestry 92(1): 39–47.

Davenport, D. W., D. D. Breshears, B. P. Wilcox, and C. D. Allen. 1998. Viewpoint: Sustainability of pinyon-juniper ecosystems—a unifying perspective of soil erosion thresholds. Journal of Range Management 51: 231–240.

Dorman, R. L. 1996. The Chili Line and Santa Fe the City Different. R. D. Publications, Santa Fe NM.

Erdman, J. A. 1970. Pinyon-juniper succession after natural fires on residual soils of Mesa Verde, Colorado. Brigham Young University Science Bulletin, Biological Series 11(2).

Erdman, J. A., C. L. Douglas, and J. W. Marr. 1969. The environment of Mesa Verde, Colorado. Archaeological Research Series No. 7-B. National Park Service, Washington D.C.

Evans, R. D., and J. Belnap. 1999. Long-term consequences of disturbance on nitrogen dynamics in an arid ecosystem. Ecology 80(1): 150–160.

Falk, D. A. 2004. Scaling rules for fire regimes. Ph.D. dissertation, Department of Ecology and Evolutionary Biology, and Laboratory of Tree Ring Research, University of Arizona, Tucson.

Falk, D. A., and T. W. Swetnam. 2003. Scaling rules and probability models for surface fire regimes in ponderosa pine forests. In Fire, Fuel Treatments, and Ecological Restoration, edited by P. N. Omi and L. A. Joyce. RMRS-P-29: 301-317. U.S. Forest Service, Rocky Mountain Research Station, Ft. Collins CO.

Fleischner, T. L. 1994. Ecological costs of livestock grazing in western North America. Conservation Biology 8(3): 629–644.

Floyd, M. L., editor. 2003. Ancient Pinyon-Juniper Woodlands: A Natural History of Mesa Verde Country. University of Colorado Press, Boulder.

Floyd, M. L., and M. Colyer. 2003. Beneath the trees: Shrubs, herbs, and some surprising rarities. In Ancient Pinyon-Juniper Woodlands: A Natural History of Mesa Verde Country, edited by M. L. Floyd, pp. 31–60. University of Colorado Press, Boulder.

Floyd, M. L., W. H. Romme, and D. D. Hanna. 2003a. Fire history. In Ancient Pinyon-Juniper Woodlands: A Natural History of Mesa Verde Country, edited by M. L. Floyd, pp. 261–277. University of Colorado Press, Boulder.

Floyd, M. L., T. L. Fleischner, D. Hanna, and P. Whitefield. 2003b. Effects of historic livestock grazing on vegetation at Chaco Culture National Historic Park, New Mexico. Conservation Biology 17(6): 1703–1711.

Floyd-Hanna, L., and W. H. Romme. 1995. Fire history and fire effects of Mesa Verde National Park: Final Report, MEVE-R91-0160. Manuscript on file at Mesa Verde National Monument.

Foxx, T. 2002. Draft. Ecosystems of the Pajarito Plateau and East Jemez Mountains: Linking Land and People. Los Alamos National Laboratory Cultural Resources Management Plan.

Foxx, T. S., and G. D. Tierney. 1985. Status of the flora of the Los Alamos National Environmental Research Park. Checklist of vascular plants of the Pajarito Plateau and Jemez Mountains. Los Alamos National Laboratory, LA-8050-NERP, Vol. III. Los Alamos NM.

Foxx, T. S., G. Tierney, M. Mullen, and M. Salisbury. 1997. Old-field plant succession on the Pajarito Plateau. Los Alamos National Laboratory, LA-13350. Los Alamos NM.

Friederici, P., editor. 2003a. Ecological Restoration of Southwestern Ponderosa Pine Forests. Island Press, Washington D.C.

Friederici, P. 2003b. The "Flagstaff model." In Ecological Restoration of Southwestern Ponderosa Pine Forests, edited by P. Friederici, pp. 7–25. Island Press, Washington D.C.

Gehring, C. A., T. Theimer, T. G. Whitham, and P. Keim. 1998. Ectomycorrhizal fungal community structure of pinyon pines growing in two environmental extremes. Ecology 79(5): 1562–1572.

Goff, F., B. S. Kues, M. A. Rogers, L. D. McFadden, and J. N. Gardner, editors. 1996. The Jemez Mountains Region. New Mexico Geological Society Guidebook, 47th Field Conference. Albuquerque.

Griffitts, M. 2003. Bedrock geology. In Ancient Pinyon-Juniper Woodlands: A Natural History of Mesa Verde Country, edited by M. L. Floyd, pp. 183–196. University of Colorado Press, Boulder.

Griggs, A. B., managing editor. 2001. Cerro Grande canyons of fire, spirit of community. Prepared by Los Alamos National Laboratory, published by Los Alamos National Bank, NM.

Griggs, R. L. 1964. Geology and ground-water resources of the Los Alamos area, New Mexico. U.S. Geological Survey, Water Supply Paper 1753. Washington D.C.

Grissino-Mayer, H. D., and T. W. Swetnam. 2000. Century-scale climate forcing of fire regimes in the American Southwest. The Holocene 10.2: 213–220.

Hartman, R. L. 2003. Preliminary checklist of the vascular plants of Valles Caldera National Preserve, New Mexico. Unpublished report on file at Bandelier National Monument, NM.

Healthy Forests. 2002. An initiative for wildfire prevention and stronger communities. Office of the President of the United States, August 22, 2002.

Henke, S. E., and F. C. Bryant. 1999. Effects of coyote removal on the faunal community in western Texas. Journal of Wildlife Management 63(4): 1066–1081.

Hereford, R., R. H. Webb, and S. Graham. 2002. Precipitation history of the Colorado Plateau region, 1900–2000. U.S. Geological Survey Fact Sheet 119-02. U.S. Government Printing Office, Washington D.C.

Holm, R. F. 1986. Field guide to the geology of the Central San Francisco Volcanic Field, northern Arizona. In Geology of Central and Northern Arizona, edited by J. D. Nations, C. M. Conway, and G. A. Swann, pp. 27–41. Geological Society of America, Rocky Mountain Section Guidebook. Flagstaff, AZ.

Hooten, J. A., M. H. Ort, and M. D. Elson. 2001. The origin of cinders in Wupatki National Monument. Report prepared for Southwest Parks and Monuments Association Grant No. FY00-11. Technical Report 2001-12. Desert Archaeology, Tucson AZ.

Hunter, A. A. 1997. Seeds, cucurbits and corn from Lizard Man Village. Kiva 62(3): 221–244.

Jackson, R. B., J. L. Baner, E. G. Jobbagy, W. T. Pockmand, and D. H. Wall. 2002. Ecosystem carbon loss with woody plant invasion of grasslands. Nature (418): 623–626.

Jacobs, B. F. 1989. A flora of Bandelier National Monument. Unpublished report on file at Bandelier National Monument, NM.

Kohler, T. A., and M. H. Matthews. 1988. Long-term Anasazi land use and forest reduction: A case study from southwest Colorado. American Antiquity 53(3): 537–564.

Lopinot, N. H. 1986. The early Spanish introduction of new cultigens into the greater Southwest. Missouri Archaeologist 47: 61–84.

Miller, R. F., and R. J. Tausch. 2001. The role of fire in pinyon and juniper woodlands: A descriptive analysis. In Proceedings of the Invasive Species Workshop: The Role of Fire in the Control and Spread of Invasive Species, edited by K. E. M. Galley and T. P. Wilson, pp. 15–30. Fire Conference 2000: The First National Congress on Fire Ecology, Prevention, and Management. Miscellaneous Publication No. 11, Tall Timbers Research Station, Tallahassee FL.

Moore, R. B. 1974. Geology, petrology and geochemistry of the eastern San Francisco Volcanic Field, Arizona. Unpublished Ph.D. dissertation, Department of Geosciences, University of New Mexico, Albuquerque.

Nabhan, G. P., P. Pynes, and T. Joe. 2002. Safeguarding the uniqueness of the Colorado Plateau. Northern Arizona University Center for Sustainable Environments, Terralingua, Grand Canyon Wildlands Council, Flagstaff AZ.

National Park Service. 2002a. Sunset Crater Final Environmental Impact Statement, General Management Plan. U.S. Department of the Interior.

National Park Service. 2002b. Wupatki Final Environmental Impact Statement, General Management Plan. U.S. Department of the Interior.

Omi, P. N., and R. L. Emerick. 1980. Final report: Fire and resource management in Mesa Verde National Park. Contract report on file, Mesa Verde National Park, CO.

O'Rourke, P. M. 1980. Frontier in transition: A history of southwest Colorado. Colorado State Office, Department of the Interior, Bureau of Land Management, Denver CO.

Petersen, K. L. 1994. A warm and wet Little Climatic Optimum and a cold and dry Little Ice Age in the southern Rocky Mountains, USA. Climatic Change 26: 243–269.

Petersen, K. L., and P. J. Mehringer, Jr. 1976. Postglacial timberline fluctuations, La Plata Mountains, southwestern Colorado. Arctic and Alpine Research 8(3): 275–288.

Pilles, P. J., Jr. 1993. The Sinagua: Ancient people of the Flagstaff region. In Wupatki and Walnut Canyon: New Perspectives on History, Prehistory and Rock Art, edited by D. G. Noble, pp. 2–11. Ancient City Press, Santa Fe NM.

Priest, S. S., W. A. Duffield, K. Malis-Clark, J. W. Hendley II, and P. H. Stauffer. 2001. The San Francisco Volcanic Field, Arizona. U.S. Geological Survey Fact Sheet 017-01. U.S. Government Printing Office, Washington D.C.

Ramsey, D. 2003. Soils of Mesa Verde Country. In Ancient Pinyon-Juniper Woodlands: A Natural History of Mesa Verde Country, edited by M. L. Floyd, pp. 213–222. University of Colorado Press, Boulder.

Reneau, S. L., and E. V. McDonald. 1996. Landscape History and Processes on the Pajarito Plateau, Northern New Mexico: Rocky Mountain Cell. Friends of the Pleistocene, Field Trip Guidebook. Los Alamos NM.

Ricketts, T. H., E. Dinerstein, D. M. Olson, and C. Loucks. 1999. Who's where in North America? BioScience 49(5): 369–381.

Roberts, A. 1990. Navajo ethno-history and archaeology. In The Wupatki Archaeological Inventory Survey Project: Final Report, edited by B. A. Anderson. Southwest Cultural Resources Professional Paper No. 35.

Rogers, P. 2002. Mesa Verde fire history. Manuscript on file, Mesa Verde National Park, CO.

Romme, W. H., S. Olivia, and M. L. Floyd. 2003a. Threats to the pinyon-juniper woodlands. In Ancient Pinyon-Juniper Woodlands: A Natural History of Mesa Verde Country, edited by M. L. Floyd, pp. 339–260. University of Colorado Press, Boulder.

Romme, W. H., R. Balice, P. Brown, N. S. Cobb, T. DeGomez, L. Floyd-Hanna, P. Fulé, D. W. Huffman, B. A. Hungate, G. W. Koch, M. M. Moore, M. Savage, and E. W. Schupp. 2003b. Letter to U.S. Department of Agriculture, http://www.mpcer.nau.edu/direnet/files/Romme 2003.pdf.

Rothman, H. K. 1992. On Rims and Ridges: The Los Alamos Area Since 1880. University of Nebraska Press, Lincoln.

Ruel, J., and T. G. Whitham. 2002. Fast-growing juvenile pinyon suffer greater herbivory when mature. Ecology 83(10): 2691–2699.

Selmants, P. C., A. Elseroad, and S. C. Hart. 2003. Soils and nutrients. In Ecological Restoration of Southwestern Ponderosa Pine Forests, edited by P. Friederici, pp. 144–160. Island Press, Washington, D.C.

Shoemaker, E. M., and D. E. Champion. 1977. Eruption history of Sunset Crater, Arizona. Investigator's annual report. Manuscript on file, Area National Monuments Headquarters, Wupatki, Sunset Crater Volcano, and Walnut Canyon National Monuments, Flagstaff AZ.

Short, M. S. 1988. Walnut Canyon and Wupatki: A history. Master's thesis, Northern Arizona University, Flagstaff.

Smith, D. A. 2003. Only man is vile. In Ancient Pinyon-Juniper Woodlands: A Natural History of Mesa Verde Country, edited by M. L. Floyd, pp. 321–336. University of Colorado Press, Boulder.

Smith, R. L., R. A. Bailey, and C. S. Ross. 1970. Geologic map of the Jemez Mountains, New Mexico. Miscellaneous Geologic Investigations Map I-571. U.S. Geologic Survey, Washington D.C.

Smith, S. J., M. Coder, and G. P. Nabhan. 2005. Landscape history, climate change, and site setting: Ingredients for restoration of pinyon and juniper woodlands. Presented at the Colorado Plateau Chapter of the Society for Conservation Biology annual meeting, March 11–13, 2005, Prescott AZ.

Snider, G. B., D. B. Wood, and P. J. Dougherty. 2003. Analysis of costs and benefits of restoration-based hazardous fuel reduction. Treatment vs. no treatment. Progress Report No. 1, June 13, 2003. Unpublished report on file at Northern Arizona University, School of Forestry, Flagstaff.

Soulé, M. E., J. A. Estes, J. Berger, and C. M. Del Rio. 2003. Ecological effectiveness: Conservation goals for interactive species. Conservation Biology 17(5): 1238–1251.

Swetnam, T. W., and C. H. Baisan. 2003. Tree-ring reconstructions of fire and climate history in the Sierra Nevada and Southwestern United States. In Fire and Climate Change in Temperate Ecosystems of the Western Americas, edited by T. T. Veblen, W. L. Baker, G. Montenegro, and T. W. Swetnam. Ecological Studies 160. Springer Verlag.

Swetnam, T. W., C. H. Baisan, and J. M. Kaib. 2001. Forest fire histories of the Sky Islands of La Frontera. In Changing Plant Life of La Frontera, edited by G. L. Webster and C. J. Bahre, pp. 95–123. University of New Mexico Press, Albuquerque.

Thomas, L., J. Whittier, N. Tancreto, J. Atkins, M. Miller, and A. Cully, Preparers. 2003. Vital signs monitoring plan for the southern Colorado Plateau network: Phase I report, final draft. Southern Colorado Plateau Network, Flagstaff AZ.

Toll, H. W., editor. 1995. Soil, water, biology, and belief in prehistoric and traditional Southwestern agriculture. New Mexico Archaeological Council, Special Publication 2.

Torres-Reyes, R. 1970. Mesa Verde National Park: An Administrative History 1906–1970. U.S. Department of the interior, Office of History and Historic Architecture, Eastern Service Center, Washington D.C.

Touchan, R., C. D. Allen, and T. W. Swetnam. 1996. Fire history and climatic patterns in ponderosa pine and mixed-conifer forests of the Jemez Mountains, northern New Mexico. In Proceedings of the 1994 Symposium on the La Mesa Fire, edited by C. D. Alen, pp. 33–46. U.S. Department of Agriculture, Forest Service General Technical Report RM-286, Fort Collins CO.

Travis, S. E. 1990. Chapter 4. In The Wupatki Archaeological Inventory Survey Project: Final report, edited by B. Anderson, pp. 1–38. Southwest Cultural Resources Professional Paper 35. Santa Fe NM.

Trewartha, G. T. 1954. An Introduction to Climate. McGraw-Hill, New York.

Trimble, M. 1982. C O Bar. Bill Owen depicts the historic Babbitt Ranch. Northland Press, Flagstaff AZ.

U.S. Department of Agriculture, Forest Service. 2002. Forest insect and disease conditions in the Southwestern Region, 2002. http: / / www.fs.fed.us / r3 / publications / documents / fidc2002 .pdf.

van West, C. R., and J. S. Dean. 2000. Environmental characteristics of the A.D. 900–1300 period in the central Mesa Verde region. Kiva 66(1): 19–44.

Vierra, B., S. R. Hoagland, J. S. Isaacson, and A. L. Madsen. 2002. Department of Energy land conveyance data recovery plan and research design for the excavation of archaeological sites located within selected parcels to be conveyed to the incorporated County of Los Alamos, New Mexico. Los Alamos National Laboratory, LA-UR-02-1284.

Wallace, T. R. 1949. A brief history of Coconino County. Masters thesis, Arizona State College (Northern Arizona University), Flagstaff.

West, P., M. Coder, S. Smith, G. Nabhan, and Z. Kovacs. 2005. The Non-Timber Forest Products and Ethnobotanical Database for the Four Corners Region of the Southwestern United States. Center for Sustainable Environments, Northern Arizona University, Flagstaff.

Wilmers, C. C., R. L. Crabtree, D. W. Smith, K. M. Murphy, and W. M. Getz. 2003. Trophic facilitation by introduced top predators: Grey wolf subsidies to scavengers in Yellowstone National Park. Journal of Animal Ecology 72: 909–916.

Wilshusen, R. H., and R. H. Towner. 1999. Post-Puebloan occupation (A.D. 1300–1840). In Colorado prehistory: A Context for the Southern Colorado River Basin, edited by W. D. Lipe, M. D. Varien, and R. H. Wilshusen. Colorado Council of Professional Archaeologists, Denver.

Woodhead, P. V. 1946. Cooperative range training trip, Region 2 and Mesa Verde National Park. Manuscript on file, Mesa Verde National Park, CO.

Wright, H. A., L. F. Newenschwander, and C. M. Britton. 1979. The role and use of fire in sagebrush-grass and pinyon-juniper plant communities: A state-of-the-art review. U.S. Department of Agriculture, Forest Service General Technical Report INT-58. Intermountain Forest and Range Experiment Station, Ogden UT. 48 pp.

INDICATIONS OF LARGE CHANGES IN MIXED CONIFER FORESTS OF GRAND CANYON NATIONAL PARK

John L. Vankat, D. Coleman Crocker-Bedford, Don R. Bertolette, Paul Leatherbury, Taylor McKinnon, and Carmen L. Sipe

Mixed conifer forests of the Southwest are poorly understood because of sparse research and diverse stand structure and composition. Nevertheless, it is clear that mixed conifer forests are fundamentally different from the region's other coniferous forests in climate, landscape pattern, and fire regime.

Compared to lower elevation ponderosa pine (*Pinus ponderosa*) forests, mixed conifer forests have greater precipitation and cooler temperatures, and therefore greater moisture availability. In contrast with higher elevation spruce-fir (*Picea-Abies*) forests, mixed conifer forests have longer growing seasons. As a result, mixed conifer forests are the most productive coniferous forests of the Southwest (Moir 1993).

The landscape pattern in mixed conifer forests is heterogeneous (Moir 1993; White and Vankat 1993; Fulé et al. 2003a), and patterns were more pronounced before fire was excluded from these landscapes (White and Vankat 1993; Fulé et al. 2003a, 2004). In Grand Canyon National Park (GCNP), mixed conifer landscapes are typified by an array of topographically determined stands (patches) of varied forest structure and composition (White and Vankat 1993):

- Relatively dry sites such as ridge tops and south- and west-facing slopes have open to moderately dense stands dominated by ponderosa pine and Douglas-fir (*Pseudotsuga menziesii*). Before Euramerican influence, stands were more open, especially in the understory and midstory.
- More mesic sites such as north- and east-facing slopes have dense stands dominated by various combinations of ponderosa pine, Douglas-fir, white fir (*Abies concolor*), and quaking aspen (*Populus tremuloides*). Stands formerly were more open and had fewer small- and mid-diameter white firs.
- Relatively moist, forested valley bottoms have dense stands dominated by blue spruce (*Picea pungens*) and ponderosa pine, often with white fir and quaking aspen. Stands formerly were more open and had less white fir. Some sites have spruce-fir stands dominated by Engelmann spruce (*Picea engelmannii*) and subalpine fir (*Abies lasiocarpa*).

The climate and heterogeneity of mixed conifer landscapes produced a pre-Euramerican fire regime different from that of other coniferous forests in the Southwest. The fire regime of ponderosa pine forests was characterized by frequent, low-intensity surface fires (Swetnam and Baisan 1996; Dahms and Geils 1997). The pre-Euramerican fire regime of spruce-fir forests is not well documented for the Southwest (Moir 1993; Swetnam and Baisan 1996; Allen 2002). It may have included infrequent, large crown fires, as speculated by Merkle (1954) and White and Vankat (1993) for GCNP and as documented for Colorado (Veblen et al. 1994). However, there is evidence that the fire regime in GCNP included frequent, low-intensity surface fires (Fulé et al 2003a; see also Baker and Veblen 1990 for Colorado). In contrast, mixed conifer forests had a true mixed-severity fire regime (Jones 1974; Allen 1989; Allen et al. 1995; Touchan et al. 1996). Frequent, low-intensity surface fires likely characterized relatively dry sites (Fulé et al. 2003a; see also Wolf and Mast 1998). In dry years, these fires also burned mesic sites and occasionally crowned in denser stands with high fuel loads. Fire intervals also varied, apparently related to climate (Fulé et al. 2003a). Therefore, at the landscape scale, the fire regime of mixed conifer forests appears to have been complex, varying spatially according to stand structure and composition and varying temporally with climate. The pre-Euramerican mixed-severity fire regime appears to have both reflected and reinforced topography-based landscape heterogeneity.

Limited historical data from mixed conifer forests make it difficult to document long-term changes in forest structure and composition. Available historical data on uncut forests generally are summary statistics for different study sites across a broad region (e.g., Lang and Stewart 1910). We recently discovered field datasheets for sample plots from a quantitative vegetation study undertaken in GCNP in 1935. This study by the National Park Service's Branch of Forestry (BOF) included a portion of the mixed conifer forest zone of the park. This data set appears to be the earliest site-intensive, detailed documentation of coniferous forest structure and composition for anywhere in the Southwest. The BOF study included a small number of plots in the mixed conifer zone.

We could not relocate and resample the BOF plots in a timely manner, so instead we compared BOF data to data collected recently for forest inventory and monitoring. Our objective was to examine possible changes in forest structure and composition from 1935 to the present. Vegetation trends since 1935 can aid in estimating pre-Euramerican conditions, knowledge of which is critical for development of fire management goals and practices.

METHODS

Study Area

Grand Canyon National Park is located on the Colorado Plateau in northern Arizona. Forested areas of the park have been protected through a series of Presidential Proclamations and Congressional Acts, including Grand Canyon Forest Preserve (1893), Grand Canyon Game Preserve (1906), Grand Canyon National Monument (1908), GCNP (1919), and GCNP Expansion (1927).

Mixed conifer forests generally occur at 2500–2700 m elevation and occupy an area of about 110 sq km on the North Rim of GCNP, within N36°13' to N36°21' latitude and W111°59' to W112°14' longitude (Figure 1). The Bright Angel Ranger Station (2560 m) on the North Rim received an annual average of 64 cm precipitation, including 349 cm of snowfall, during 1948–2004 (http://www.wrcc.dri.edu/index.html; viewed on 2/19/2005). The average maximum and minimum temperatures for the warmest month (July) were 25.3 and 8.2° C; averages for the coldest month (January) were 3.4 and –8.3° C.

Euramerican impact in the region of the North Rim began with the introduction of livestock in the 1870s (Hughes 1967; Anderson 1998; Michael F. Anderson, GCNP archaeologist, personal communication). Livestock consumed herbs, which formerly fueled surface fires, and this decreased fire frequencies on the North Rim beginning in 1880 (Wolf and Mast 1998; Fulé et al. 2003a, b). Additional grazing occurred with elevated deer populations on the North Rim circa 1915–1935 (Rasmussen 1941; Mitchell and Freeman 1993). Active fire suppression began by 1924 (White and Vankat 1993), and livestock grazing ended with the fencing of the park's north boundary in the 1930s (cf. Hughes 1967). There is no record of timber harvesting in mixed conifer forests in the park, although small-scale cuttings occurred in areas of fires.

Field and Data Analysis Methods

We mapped the major forest zones of GCNP and located all forested plots from 1935 and from recent studies within the area of mixed conifer forest (Figure 1). Therefore, our analysis was at a landscape scale similar to that used in fire and vegetation management plans. At this scale, scattered stands of ponderosa pine and spruce-fir forests are included as part of the range of variability of mixed conifer forest; however, we excluded treeless meadows.

The 1935 BOF study employed field methods according to Coffman (1934), who directed that sample plots be placed in sites "representative of the average conditions" for a type of vegetation, to assist with preparation of a vegetation map. Nearly all plots were placed in areas away from lodges, campgrounds, and other facilities. The BOF used 0.2 acre (0.08 ha) plots with dimensions of 66 x 132 feet (20.1 x 40.2 m). In each plot, they tallied live trees by species according to diameter class: 4–11.9, 12–23.9, 24–35.9, and ≥ 36 inch (which we converted to 10–29.9, 30–60.9, 61–90.9, and ≥ 91 cm). For the purpose of calculating basal area, we assumed a range of 36–47.9 inches (91–122 cm) for the largest diameter class and used diameter-class midpoints (e.g., 20 cm for the 10–29.9 cm class). Only 12 BOF plots occurred in the mixed conifer forest zone, primarily in two broad-scale clusters (Figure 1).

We used recent data from White and Vankat (1993; W&V data collected in 1984), the Fire Effects Monitoring program of GCNP (FEM, data collected from 1990 to 2002), and the Ecological Restoration Institute of Northern Arizona University (ERI, data collected 1997 to 2001). All three sources used 0.1 ha, 20 x 50 m plots. We excluded plots having evidence of recent crown fire. The 20 W&V plots had been subjectively placed in clusters to sample sites representative of different slope aspects and elevations in Thompson Canyon watershed (Figure 1). The 7 FEM plots had been placed using a stratified random design within areas planned for prescribed burning. The objective was to compare pre- and post-burn conditions (we used only pre-burn data in our analyses). The 21 ERI plots had been placed according to a grid in the northwest part of the mixed conifer zone. Collectively, the 48 W&V, FEM, and ERI plots are widely scattered, but largely in areas not sampled during the 1935 BOF study.

We live-tallied trees from the W&V, FEM, and ERI plots according to the four diameter classes used in the 1935 BOF study and calculated basal area as we did for BOF data. We used Student's t-test (alpha = 0.10) to determine whether differences in mean densities (trees/ha) and basal areas (sq m/ha) were statistically significant between 1935 and recent studies. We present results for ponderosa pine, white fir, Douglas-fir, spruce (blue spruce and Engelmann spruce combined), and quaking aspen. We refer to individuals < 30 cm as small-diameter trees, 30–60.9 cm as mid-diameter trees, and ≥ 61 cm as large-diameter trees.

RESULTS

The present mean density of conifers 10–29.9 cm in diameter is 42 percent lower than measured in 1935 (Table 1), but no individual species exhibited a statistically significant difference. The present mean density of conifers 30–60.9 cm in diameter is 55 percent lower than measured in 1935, and both white fir (62% lower) and spruce (69% lower) exhibited significant differences. Conifers in the 61–90.9 cm and ≥ 91 cm diameter classes and quaking aspen ≥ 10 cm in diameter did not show significant differences in mean densities between 1935 and recent data. However, when aspens of 10–29.9 cm diameter were removed from analysis, the present mean density of the remaining quaking aspens (i.e., the 30–60.9 cm diameter class) is 73 percent lower than measured in 1935. Mean basal area of conifers is 48 percent lower today than measured in 1935.

DISCUSSION

Interpretation of our results is constrained by the small sample size of the 1935 BOF study. Unfortunately, that sample size is historically fixed. In addition, variances (standard deviations) are high; however, this is unavoidable given GCNP's heterogeneous mixed conifer landscapes. Although we used data from studies with different specific objectives, all studies shared the general objective of inventorying forest structure

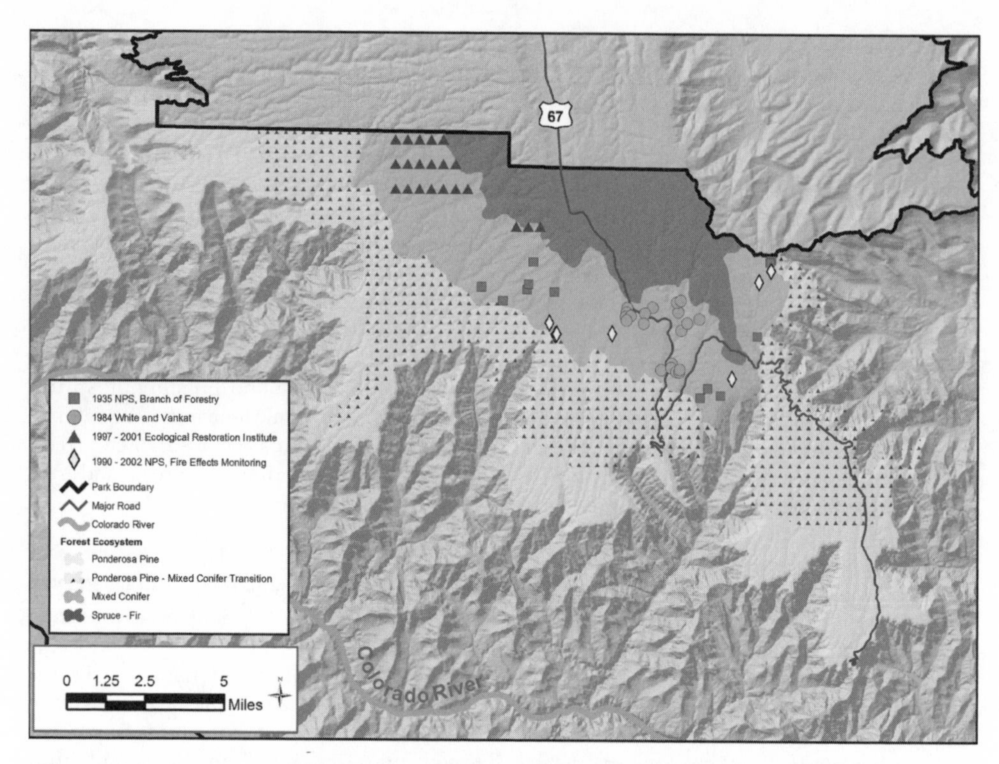

Figure 1. Distribution of major forest zones in GCNP, with plots from 1935 and recent studies superimposed on the mixed conifer forest zone (base map modified from Bertolette 2002).

and composition. Plots for the 1935 BOF and 1983 W&V studies were placed in representative areas; such non-random plots are commonly used in broad-scale vegetation studies. The recent plots are not a resample of the 1935 BOF plots. Despite these issues, we chose to use statistical analysis, which integrates the effects of sample size and variance, as a guide for assessing possible differences in mean values between 1935 and the present.

Several facts add credence to our findings. First, all statistically significant results involved large differences (42–73%). Second, all results have consistent direction of reported change. Third, results parallel field observations on permanent plots between 1984 and 2004 by the senior author and Mark A. White (The Nature Conservancy, Duluth MN, personal communication). Fourth, results are congruent with ecological theory of forest development (cf. Oliver and Larson 1996). Nevertheless, we suggest conservative interpretation of our results.

Mixed conifer forests appear to differ from lower elevation coniferous forests in GCNP in the direction and degree of apparent changes between 1935 and the present. For example, the large decrease in density of conifers 10–29.9 cm diameter suggested in mixed conifer forests differs from major increases in lower elevation forests (Crocker-Bedford et al., this volume). The apparent 48 percent decrease in total conifer basal area is larger than the approximately 27 percent decrease of ponderosa pine basal area in ponderosa pine forests (Crocker-Bedford et al., this volume).

Despite different responses in mixed conifer and lower elevation forests, the trigger

Table 1. Structure and composition of mixed conifer forests of GCNP based on 1935 and recent studies. Density units are trees/ha; basal area units are m^2/ha. Conifers included are ponderosa pine, white fir, Douglas-fir, spruce (Engelmann and blue spruce combined), and subalpine fir.

	1935 (n = 12)		Recent (n = 48)		
Attribute	Mean	SD	Mean	SD	*P*
Density of conifers 10–29.9 cm diameter	384.7	300.2	222.7	111.1	0.091
Density of white fir 10–29.9 cm diameter	134.8	205.4	71.1	83.5	0.314
Density of spruce 10–29.9 cm diameter	104.0	169.1	58.0	75.1	0.376
Density of Douglas-fir 10–29.9 cm diameter	69.9	93.1	22.2	30.6	0.105
Density of ponderosa pine 10–29.9 cm diameter	36.0	60.0	28.4	40.5	0.687
Density of conifers 30–60.9 cm diameter	222.2	94.1	99.0	55.6	< 0.001
Density of white fir 30–60.9 cm diameter	77.3	74.3	29.6	46.7	0.054
Density of spruce 30–60.9 cm diameter	57.5	71.4	17.8	27.9	0.081
Density of Douglas-fir 30–60.9 cm diameter	24.7	27.9	14.1	20.5	0.231
Density of ponderosa pine 30–60.9 cm diameter	49.4	73.1	24.2	26.2	0.262
Density of conifers 61–90.9 cm diameter	28.9	20.5	23.2	23.0	0.415
Density of white fir 61–90.9 cm diameter	10.4	14.8	4.9	12.6	0.270
Density of spruce 61–90.9 cm diameter	7.2	15.3	2.0	4.7	0.276
Density of Douglas-fir 61–90.9 cm diameter	4.2	8.1	3.2	6.4	0.758
Density of ponderosa pine 61–90.9 cm diameter	10.4	14.8	12.6	18.3	0.634
Density of conifers ≥ 91 cm diameter	3.2	7.4	0.7	2.7	0.337
Density of ponderosa pine ≥ 91 cm diameter	2.0	7.2	0.7	2.5	0.506
Basal area of conifers	64.9	24.0	34.0	14.7	< 0.001
Density of quaking aspen ≥ 10 cm diameter	201.7	172.3	128.1	117.3	0.184
Density of quaking aspen 30–60.9 cm diameter	77.3	89.9	20.7	27.2	0.052

factor for change appears to have been the same—a substantial reduction in fire frequencies when intensive livestock grazing began. Mean fire return interval (for fires recorded in ≥ 25% of fire-scarred trees) in mixed conifer forests in GCNP was likely about 7–30 years in the several decades before grazing (White, unpublished data; Wolf and Mast 1998; Fulé et al. 2003a, b). These fires essentially ended circa 1880 (Wolf and Mast 1998; Fulé et al. 2003a, b), soon after grazing began, although fires resulting in ≥ 10 percent scarring continued until active fire suppression began in the early twentieth century (Wolf and Mast 1998).

Densities of small- and mid-diameter conifers likely increased rapidly in mixed conifer forests following decreased frequency of fires, which formerly thinned stands. Shade-intolerant conifers such as ponderosa pine likely increased in open sites, and shade-tolerant conifers such as white fir increased in both open and denser sites (White and Vankat 1993). The increases in small- and mid-diameter conifers likely were much more rapid and pronounced than in lower elevation forests because mixed conifer forests had (a) higher densities of shade-tolerant, fire-intolerant trees, which provided seeds, (b) moister conditions, which favored germination and establishment, and (c) pulses of conifer regeneration immediately prior to circa 1870 (White, unpublished data; Fulé et al. 2002).

Increased densities in mixed conifer forests apparently became unsustainable around or after 1935. Self-thinning related to increased competition likely caused decreases in conifer density and basal area. Another probable factor was defoliation by western spruce budworm (*Choristoneura occidentalis*). This insect affects most tree species of mixed conifer forests (Linnane 1986; Lynch and Swetnam 1992), especially in dense, mature, low-vigor, multi-layered stands that have white fir and Douglas-fir as canopy dominants and shade-tolerant species in the understory (Linnane 1986). Many mixed conifer stands in GCNP likely fit this description throughout the twentieth century. Densities of quaking aspens apparently decreased because there was little post-1880 regeneration (Fulé et al. 2002) to replace individuals overtopped by conifers.

Post-1935 trends in densities and basal area in mixed conifer forest differ from trends at lower elevation (cf. Crocker-Bedford et al., this volume) likely because lower elevation forests did not reach a self-thinning stage of forest development during much of this period. In addition, lower elevation forests were less susceptible to western spruce budworm because they had open structure and were dominated by ponderosa pines.

The 1935 BOF data for mid- and large-diameter conifers in mixed conifer forests of GCNP are inconsistent with those of Lang and Stewart (1910) for the Kaibab Plateau (including what is now the North Rim of GCNP). For example, Lang and Stewart (1910) reported mean densities of 42.2 and 2.3 trees/ha for conifers 30–60.9 and ≥ 61 cm diameter, respectively, in their "mixed type" (which included mixed conifer forest). These densities are far less than the 222.2 and 32.1 trees/ha measured in these respective diameter classes in 1935 (Table 1). Possible increases from 1910 to 1935 may account for some of the higher densities of mid-diameter conifers in 1935, but it appears unlikely that the density of large-diameter conifers could have increased by nearly 30 trees/ha from 1910 to 1935. The differences may be partly an artifact of the 1935 BOF sample, but it also may represent the maturation of small- and mid-diameter conifers that began increasing in densities before circa 1870. In addition, mean densities reported by Lang and Stewart (1910) may be unrealistically low for characterizing mixed conifer forests in GCNP. Lang and Stewart's "mixed type" included spruce-fir forests, which have few large trees. Moreover, Lang and Stewart sampled long (805 m) transects arrayed systematically across the Kaibab Plateau, which has many treeless meadows in mixed conifer and spruce-fir zones (cf. their photo 86983). Therefore, despite the shortcomings of the 1935 BOF data, they may more accurately represent typical mixed conifer forests in the early twentieth century than Lang and Stewart's (1910) data, at least for mid- and large-diameter conifers in forests of GCNP.

Insight into changes in mixed conifer forests before 1935 is critical for management planning. Given our results and the likely dynamics of mixed conifer forests, we offer the following conjecture for likely changes during circa 1870–1935:

- Mean density of conifers 10–29.9 cm in diameter likely increased greatly circa 1870–1935, before decreasing to the present (Figure 2a). Shade-tolerant, fire-intolerant conifers such as white fir were important in these dynamics.
- Mean density of conifers 30–60.9 cm in diameter likely increased greatly circa 1870–1935, before decreasing to the present (Figure 2b). White fir was likely important in these dynamics.
- Mean density of quaking aspens 30–60.9 cm in diameter probably decreased slightly circa 1870–1935, before decreasing more rapidly to the present (Figure 2c).
- Mean basal area of conifers likely increased greatly circa 1870–1935, before decreasing to the present (Figure 2d).

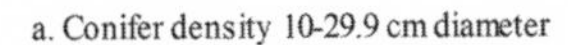

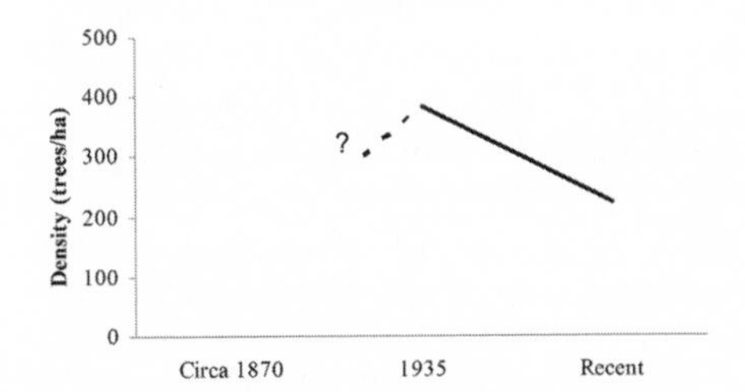

b. Conifer density 30-60.9 cm diameter

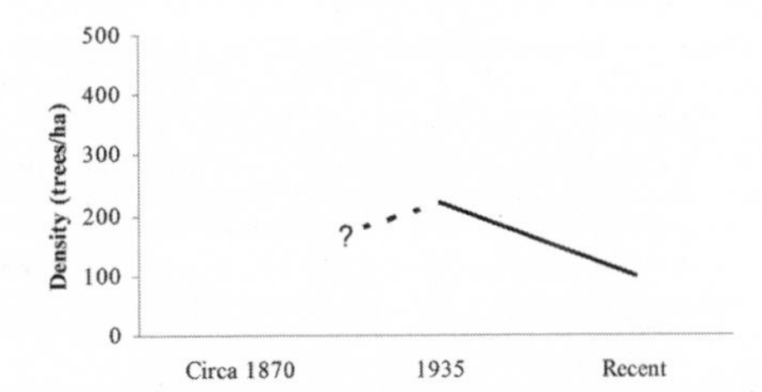

c. Quaking aspen density 30-60.9 cm diameter

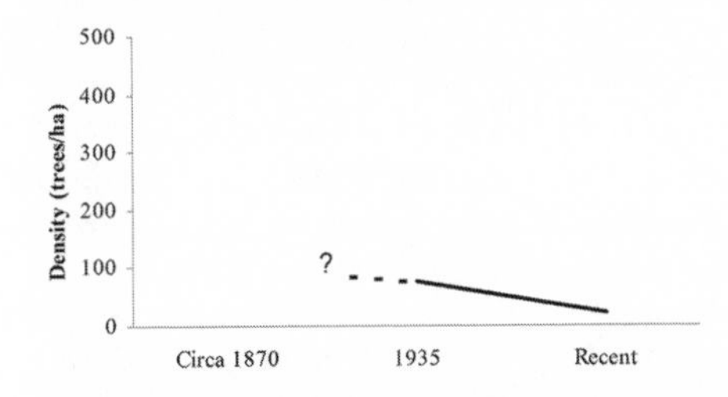

d. Conifer basal area

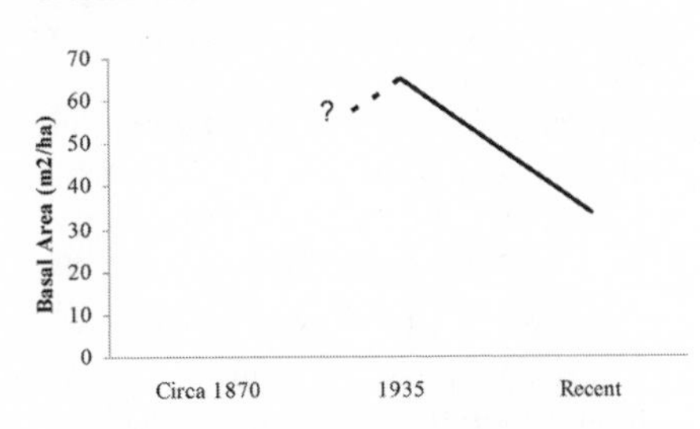

Figure 2. Trends in structure and composition of mixed conifer forests in GCNP from circa 1870 to recent studies: (a) density of conifers 10–29.9 cm diameter, (b) density of conifers 30–60.9 cm diameter, (c) density of quaking aspen 30–60.9 cm diameter, and (d) basal area of conifers. Solid lines are based on statistically significant differences; dashed lines are based on inferences.

The postulated widespread increases in white fir density circa 1870–1935 have important ramifications for today's landscape heterogeneity and fire behavior. As white firs matured in relatively open stands, these stands began to resemble the structure, composition, and fuel conditions of more closed stands on relatively moist sites, thereby decreasing landscape heterogeneity (White and Vankat 1993) and increasing the probability of crown fires (Fulé et al. 2004). With greater homogeneity of fuels and stand densities, recent fires in the mixed conifer zone of GCNP (Outlet Fire of 2000 and Poplar Fire of 2003) have included extensive (> 5 sq km) areas of contiguous crown fire. Such fires are undocumented for pre-Euramerican mixed conifer forests in GCNP (White and Vankat 1993; Fulé et al. 2003a) or elsewhere in the Southwest (Dieterich 1983; Lynch and Swetnam 1992; Brown et al. 2001).

CONCLUSIONS

The findings from our study, including our review of the literature on topography-based landscape heterogeneity, are important for the development of ecologically based fire management goals and practices for mixed conifer forests in GCNP. We recommend five specific goals, all within the overall goal of reestablishing ecosystem self-regulation by restoring forest structure, composition, and ecological processes without risking remnant features such as large-diameter conifers and vestiges of circa 1870 landscape heterogeneity:

1. Reestablish topography-based landscape heterogeneity. Although current management goals acknowledge the general concept of range of variability, our findings indicate the need to incorporate specifics of topography-based landscape heteroge-

neity, both as an individual goal and as context for other goals.

2. Reduce small-diameter white fir and ponderosa pine in formerly open stands on relatively dry sites (ridge tops and south- and west-facing slopes). Our findings suggest extending this long-standing management goal for ponderosa pine–dominated forests (National Park Service 1997) to mixed conifer forests, but at appropriate topographic positions.
3. Reduce small-diameter white fir in formerly moderately dense stands on mesic to moist sites (north- and east-facing slopes and valley bottoms). Our findings suggest applying this management goal to mixed conifer forests, but at appropriate topographic positions.
4. Promote quaking aspen primarily in mesic to moist sites (north- and east-facing slopes and valley bottoms). Our findings suggest topographic context for this long-standing management goal.
5. Restore a mixed-severity fire regime in accordance with small-scale topographic variation. The generic, long-standing management goal of restoring pre-Euramerican fire regimes, when applied to mixed conifer forests, needs to be expanded to incorporate interactions among fire, topography, vegetation, and fuels.

We also recommend additional research on mixed conifer forests to help refine management goals and practices, particularly research that focuses on differences among stands or ecological site types instead of overall means for mixed conifer forests. Resampling the 1935 BOF plots and 1984 White and Vankat plots in GCNP would clarify the past changes in vegetation suggested by this study. Additional inventories of existing vegetation would better document current conditions and provide for monitoring future changes. Additional analyses of pre-Euramerican mixed-severity fire regimes would clarify important details such as the size, distribution, and frequency of crown fires and site variability in fire return intervals. Geographic analysis of burn-severity patterns of recent large fires such as the Outlet and Poplar Fires in GCNP would provide insight into present fire regimes. Research on fire management would assist in developing methods that reestablish topography-based landscape heterogeneity without the loss of heterogeneity in extensive crown fires. Without the insight gained from research projects such as these, management in GCNP and other areas of the Southwest has the potential to leave unnatural footprints that may persist for centuries in mixed conifer landscapes.

ACKNOWLEDGMENTS

Julie Crawford, Lee Harrison, Jennifer McCollom, Lori Makarick, Niki Juarez-Cummings, Steven Mietz, Kara Leonard, Erin Uloth, and Sarah Falzarano of the National Park Service, Terence Arundel of the U.S. Geological Survey, and Michael Kearsley of Northern Arizona University's Department of Biological Sciences contributed to various aspects of this manuscript, particularly to data analyses. Peter Fulé of Northern Arizona University's Ecological Restoration Institute provided the ERI data used in this study. Thomas Heinlein and an anonymous reviewer provided helpful comments on an earlier version of the manuscript.

LITERATURE CITED

Allen, C. D. 1989. Changes in the landscape of the Jemez Mountains, New Mexico. Ph.D. dissertation, University of California, Berkeley.
Allen, C. D. 2002. Lots of lightning and plenty of people: An ecological history of fire in the upland Southwest. In Fire, Native Peoples, and the Natural Landscape, edited by T. R. Vale, pp. 143–193. Island Press, Covelo CA.
Allen, C. D., R. Touchan, and T. W. Swetnam. 1995. Landscape-scale fire history studies support fire management action at Bandelier. Park Science 15: 18.
Anderson, M. F. 1998. Living at the Edge. Grand Canyon Association, Grand Canyon AZ.
Baker, W. L., and T. T. Veblen. 1990. Spruce beetles and fire in the nineteenth-century subalpine forests of western Colorado, U.S.A. Arctic and Alpine Research 22: 65–80.
Bertolette, D. 2002. Wildfire hazard reduction research. Unpublished map, Grand Canyon National Park, Arizona.

Brown, P.M ., M. W. Kaye, L. S. Huckaby, and C. H. Baisan. 2001. Fire history along environmental gradients in the Sacramento Mountains, New Mexico: Influences of local patterns and regional processes. Ecoscience 8: 115–126.

Coffman, J. D. 1934. Suggestions for the mapping and study of vegetative cover types in areas administered by the National Park Service. Agency-wide directive, U.S. Department of the Interior, National Park Service, Branch of Forestry.

Dahms, C. W., and B. W. Geils. 1997. An assessment of forest ecosystem health in the Southwest. USDA Forest Service General Technical Report RM-GTR-295. Rocky Mountain Forest and Range Experiment Station, Fort Collins CO.

Dieterich, J. H. 1983. Fire history of southwestern mixed conifer: A case study. Forest Ecology and Management 6: 13–31.

Fulé, P. Z., W. W. Covington, M. M. Moore, T.A. Heinlein, and A.E.M. Waltz. 2002. Natural variability in forests of the Grand Canyon, USA. Journal of Biogeography 29: 31–47.

Fulé, P. Z., J. E. Crouse, T. A. Heinlein, M. M. Moore, W. W. Covington and G. Verkamp. 2003a. Mixed-severity fire regime in a high-elevation forest of Grand Canyon, Arizona, USA. Landscape Ecology 18: 465–486.

Fulé, P. Z., T. A. Heinlein, W. W. Covington and M. M. Moore. 2003b. Assessing fire regimes on Grand Canyon landscapes with fire-scar and fire-record data. International Journal of Wildland Fire 12: 129–145.

Fulé, P. Z., J. E. Crouse, A. E. Cocke, M. M. Moore, and W. W. Covington. 2004. Changes in canopy fuels and potential fire behavior 1880–2040: Grand Canyon, Arizona. Ecological Modelling 175: 231–248.

Hughes, J. D. 1967. The story of man at Grand Canyon. Grand Canyon Natural History Association Bulletin 14. K. C. Publications, Las Vegas NV.

Jones, J. R. 1974. Silviculture of southwestern mixed conifer and aspen: The status of our knowledge. USDA Forest Service Research Paper RM-122. Rocky Mountain Forest and Range Experiment Station, Fort Collins CO.

Lang, D. K., and S. S. Stewart. 1910. Reconnaissance of the Kaibab National Forest. Unpublished report, USDA Forest Service.

Linnane, J. P. 1986. Integrated forest management guide, western spruce budworm in the Southwest. USDA Forest Service, Southwestern Region Forest Pest Management Report R3-86-7. Albuquerque NM.

Lynch, A. M., and T. W. Swetnam. 1992. Old-growth mixed-conifer and western spruce budworm in the southern Rocky Mountains. In Old Growth Forests in the Southwest and Rocky Mountain Regions, Proceedings of a Workshop, Portal, AZ, March 9–13, 1992, technical coordinators M. R. Kauffman, W. H. Moir, and R. L. Bassett, pp. 66–80. USDA Forest Service General Technical Report RM-213. Rocky Mountain Forest and Range Experiment Station, Fort Collins CO.

Merkle, J. 1954. An analysis of the spruce-fir community on the Kaibab Plateau, Arizona. Ecology 35: 316–322.

Mitchell, J. E., and D. R. Freeman. 1993. Wildlife-livestock-fire interactions on the North Kaibab: A historical review. USDA Forest Service General Technical Report RM-222. Rocky Mountain Forest and Range Experiment Station, Fort Collins CO.

Moir, W. H. 1993. Alpine tundra and coniferous forest. In New Mexico Vegetation: Past, Present, and Future, edited by W. A. Dick-Peddie, pp. 47–84. University of New Mexico Press, Albuquerque.

National Park Service. 1997. Grand Canyon National Park Resource Management Plan. Grand Canyon National Park, Arizona.

Oliver, C. D. and B. C. Larson. 1996. Forest Stand Dynamics. John Wiley and Sons, New York.

Rasmussen, D. I. 1941. Biotic communities of Kaibab Plateau, Arizona. Ecological Monographs 11: 229–275.

Swetnam, T. W., and C. H. Baisan. 1996. Historical fire regime patterns in the southwestern United States since AD 1700. In Fire Effects in Southwestern Forests: Proceedings of the Second La Mesa Fire Symposium, Los Alamos NM, March 29–31, 1994, technical editor C. D. Allen, pp. 11–32. USDA Forest Service General Technical Report RM-GTR-286. Rocky Mountain Forest and Range Experiment Station, Fort Collins CO.

Touchan, R., C. D. Allen, and T. W. Swetnam. 1996. Fire history and climatic patterns in ponderosa pine and mixed-conifer forests of the Jemez Mountains, northern New Mexico. In Fire Effects in Southwestern Forests: Proceedings of the Second La Mesa Fire Symposium, Los Alamos NM, March 29–31, 1994, technical editor C. D. Allen, pp. 33–46. USDA Forest Service General Technical Report RM-GTR-286. Rocky Mountain Forest and Range Experiment Station, Fort Collins CO.

Veblen, T. T., K. S. Hadley, E. M. Nel, T. Kitzberger, M. Reid, and R. Villalba. 1994. Disturbance regime and disturbance interaction in a Rocky Mountain subalpine forest. Journal of Ecology 82: 125–135.

White, M. A., and J. L. Vankat. 1993. Middle and high elevation coniferous forest communities of the North Rim region of Grand Canyon National Park, Arizona, USA. Vegetatio 109: 161–174.

Wolf, J. J. and J. N. Mast. 1998. Fire history of mixed-conifer forests on the North Rim, Grand Canyon National Park, Arizona. Physical Geography 19: 1–14.

APPARENT INCREASES IN MIXED CONIFER CHARACTERISTICS SINCE 1935 IN PONDEROSA PINE–MIXED CONIFER TRANSITION FORESTS OF GRAND CANYON NATIONAL PARK

D. Coleman Crocker-Bedford, John L. Vankat, Don R. Bertolette, Paul Leatherbury, Taylor McKinnon, and Carmen L. Sipe

Coniferous forests of the Colorado Plateau traditionally have been classified as ponderosa pine (*Pinus ponderosa*), mixed conifer, and spruce-fir (*Picea-Abies*; Pase and Brown 1994a, 1994b). However, there is a broad band of forest transitional between ponderosa pine and mixed conifer forests on the Kaibab Plateau of northern Arizona (personal observation; Bertolette 2002). We define this ponderosa pine–mixed conifer transition forest as having (a) ≥ 70 percent overstory trees or basal area composed of ponderosa pine and (b) ≥ 20 percent understory trees composed of white fir (*Abies concolor*). Associated tree species include Douglas-fir (*Pseudotsuga menziesii*) and quaking aspen (*Populus tremuloides*). In comparison to this transition forest, ponderosa pine forests have greater dominance of the namesake species and mixed conifer forests have greater dominance of white fir, Douglas-fir, blue spruce (*Picea pungens*), or quaking aspen.

Similar forests have been described, although rarely explicitly labeled as transitional or ecotonal forests. Descriptions have treated these forests as high-elevation ponderosa pine forests (Hurst 1977; Pase and Brown 1994a), low-elevation mixed conifer forests (Wolf and Mast 1998; Mast and Wolf 2004), and fine-grained individual vegetation units (Brown et al. 1979; Warren et al. 1982). From a gradient perspective, the transition forest is an exceptionally wide ecotone in the gradient from ponderosa pine to mixed conifer forest. Descriptions of ponderosa pine–mixed conifer transition forests may be uncommon for several reasons. First, most vegetation classifications focus on zonal types and do not include transitions. Second, this transition forest may be uncommon elsewhere because most ponderosa pine forests on the Colorado Plateau are not adjacent to mixed conifer forests. Third, most other areas where these two forests meet are mountainous, where transitions may be narrow and less conspicuous. Fourth, the area of this transition forest may have been small in the past, expanding only when reductions in fire frequencies enabled white fir to invade high-elevation ponderosa pine forests.

Transition areas (ecotones) can be sensitive to environmental change (Neilson 1993; Risser 1995), especially in montane regions (Rusek 1993; Kupfer and Cairns 1996; Hessl and Baker 1997; Mast et al. 1998). However, few investigations of long-term environmental change have occurred in ponderosa pine–mixed conifer transition forests, not only because this vegetation type has lacked recognition, but also because historical data are uncommon for the Southwest in general and for the Colorado Plateau specifically. Moreover, available historical data on uncut forests generally are summary statistics for different study sites across a broad region (e.g., Lang and Stewart 1910).

We recently discovered field datasheets for sample plots from a quantitative vegetation study undertaken in Grand Canyon National Park (GCNP) in 1935. This study was conducted by the National Park Service's Branch of Forestry (BOF) and included nearly all areas of the park (1935 boundaries) with ponderosa pine–mixed conifer transition forests. This data set appears to be the earliest site-intensive, detailed documentation of coniferous forest structure and composition for anywhere in the Southwest.

We could not relocate and resample the BOF plots in a timely manner, so instead we compared BOF data to data recently collected for forest inventory and monitoring. Our objective was to examine possible changes in forest structure and composition from 1935 to the present. Vegetation trends since 1935 can aid in estimating pre-Euramerican conditions, knowledge of which is critical for development of fire management goals and practices.

METHODS

Study Area

Grand Canyon National Park is located on the Colorado Plateau in northern Arizona. Forested areas of the park were protected through a series of Presidential Proclamations and Congressional Acts, including the Grand Canyon Forest Preserve (1893), Grand Canyon Game Preserve (1906), Grand Canyon National Monument (1908), GCNP (1919), and GCNP Expansion (1927). Ponderosa pine–mixed conifer transition forests generally occur at 2300–2600 m elevation and occupy an area of about 130 sq km on the North Rim of the park, within N36°11' to N36°21' latitude and W111°55' to W112°17' longitude (Figure 1). The Bright Angel Ranger Station (2560 m) on the North Rim received an annual average of 64 cm of precipitation, including 349 cm of snowfall, during 1948–2004 (http://www.wrcc.dri.edu/index.html; viewed on 2/19/2005). The average maximum and minimum temperatures for the warmest month (July) were 25.3 and 8.2° C; averages for the coldest month (January) were 3.4 and –8.3° C.

Euramerican impact in the region of the North Rim began with the introduction of livestock in the 1870s (Hughes 1967; Anderson 1998; Michael F. Anderson, GCNP archaeologist, personal communication). Livestock consumed herbs, which formerly fueled surface fires, and this decreased fire frequencies on the North Rim beginning in 1880 (Wolf and Mast 1998; Fulé et al. 2003a, b). Additional grazing occurred with elevated deer populations on the North Rim ca. 1915–1935 (Rasmussen 1941; Mitchell and Freeman 1993). Active fire suppression began by 1924 (White and Vankat 1993), and livestock grazing ended with the fencing of the park's north boundary in the 1930s (cf. Hughes 1967). There is no record of timber harvesting in ponderosa pine–mixed conifer transition forests in the park, although small-scale cuttings occurred in areas of fires.

Field and Data Analysis Methods

We mapped the major forest zones of GCNP and located all forested plots from 1935 and recent studies within the area of ponderosa pine–mixed conifer transition forest (Figure 1). Therefore, our analysis was at a landscape scale similar to that used in fire and vegetation management plans. At this scale, scattered stands of ponderosa pine and mixed conifer forests are included as part of the range of variability of transition forest; however, we excluded treeless meadows.

The 1935 BOF study employed field methods according to Coffman (1934), who directed that sample plots be placed in sites "representative of the average conditions" for a type of vegetation, to assist with preparation of a vegetation map. Nearly all plots were placed in areas away from lodges, campgrounds, and other facilities. The BOF used 0.2 acre (0.08 ha) plots of 66 x 132 feet (20.1 x 40.2 m) dimensions. In each plot, they tallied live trees by species according to diameter class: 4–11.9, 12–23.9, 24–35.9, and ≥ 36 inch (which we converted to 10–29.9, 30–60.9, 61–90.9, and ≥ 91 cm). For the purpose of calculating basal area, we assumed a range of 36–47.9 inches (91–122 cm) for the largest diameter class and used diameter-class midpoints (e.g., 20 cm for the 10–29.9 cm class). The 95 BOF plots within the ponderosa pine–mixed conifer forest zone are widely distributed (Figure 1).

We used recent data from White and Vankat (1993; W&V, data collected in 1984) and the Fire Effects Monitoring program of GCNP (FEM, data collected from 1990 to 2002). Both sources used 0.1 ha plots of 20 x 50 m dimensions. We excluded plots having evidence of recent crown fire. The five W&V plots had been subjectively placed in one small cluster to sample sites representative of different slope aspects and elevations in the Thompson Canyon watershed (Figure 1). The 33 FEM plots had been placed using a stratified random design within areas planned for prescribed burning. The objective was to compare pre- and post-burn conditions (we used only pre-burn data in our analyses). In contrast to Crocker-Bedford et al. (this volume) and Vankat et al. (this volume), we did not use data from the Ecological Restoration Institute of Northern Arizona University because these data were collected from only one area which also included FEM plots. Collectively, the W&V and FEM plots are widely scattered.

We tallied live trees from the W&V and FEM plots according to the four diameter classes used in the 1935 BOF study and calculated basal area as we did for BOF data. We used the Student's t-test (alpha = 0.10) to determine whether differences in mean densities (trees/ha) and basal areas (sq m/ha) were statistically significant between 1935 and recent studies. We present results for tree taxa with the potential to reach overstory height: ponderosa pine, white fir, Douglas-fir, spruce (blue spruce and Engelmann spruce [*Picea engelmannii*] combined), subalpine fir (*Abies lasiocarpa*), and quaking aspen. We refer to individuals < 30 cm as small-diameter trees, 30–60.9 cm as mid-diameter trees, and ≥ 61 cm as large-diameter trees.

RESULTS

The present mean density of conifers 10–29.9 cm in diameter is 105 percent higher than measured in 1935 (Table 1). Within this diameter class, ponderosa pine, white fir, and spruce increased 61, 202, and 187 percent, respectively, and white fir surpassed ponderosa pine as the most abundant conifer. Mean densities of conifers in the 30–60.9 and 61–90.9 cm diameter classes and mean basal area of conifers are unchanged since 1935. In contrast, present mean densities of conifers ≥ 91 cm, ponderosa pine ≥ 91 cm, and quaking aspen ≥ 10 cm in diameter are lower by 82, 83, and 72 percent, respectively, than measured in 1935.

DISCUSSION

Interpretation of our results is constrained by high variance (standard deviations); however, this may be unavoidable given GCNP's diverse ponderosa pine–mixed conifer landscapes. In addition, we used data from studies with different specific objectives; however, all studies shared the general objective of inventorying forest structure and composition. Plots for the 1935 BOF and 1983 W&V studies were placed in representative areas; such non-random plots are commonly used in broad-scale vegetation studies. The recent plots are not a resample of the 1935 BOF plots. Despite these issues, we chose to use statistical analysis as a guide for assessing possible differences in mean values between 1935 and the present. The fact that all statistically significant results involved large differences (61–202%) and that results are congruent with trends reported in other studies adds credence to our findings. We nevertheless suggest conservative interpretation of our results.

Substantial changes have apparently occurred in both the structure and composition of ponderosa pine–mixed conifer transition forests of GCNP since 1935, a finding that matches the results of Mast and Wolf (2004). Much of this change is attributable to small-diameter conifers. Most of today's transition forest likely was ponderosa pine forest ca. 1870, as hypothesized by Mast and Wolf (2004). Densities of small-diameter conifers increased with livestock grazing (Mast and Wolf 2004), which reduced fire frequencies beginning ca. 1880. In the decades before 1880, mean fire return interval (for fires recorded in ≥ 25% of scarred trees) was 7–9 years (Wolf and Mast 1998; Fulé et

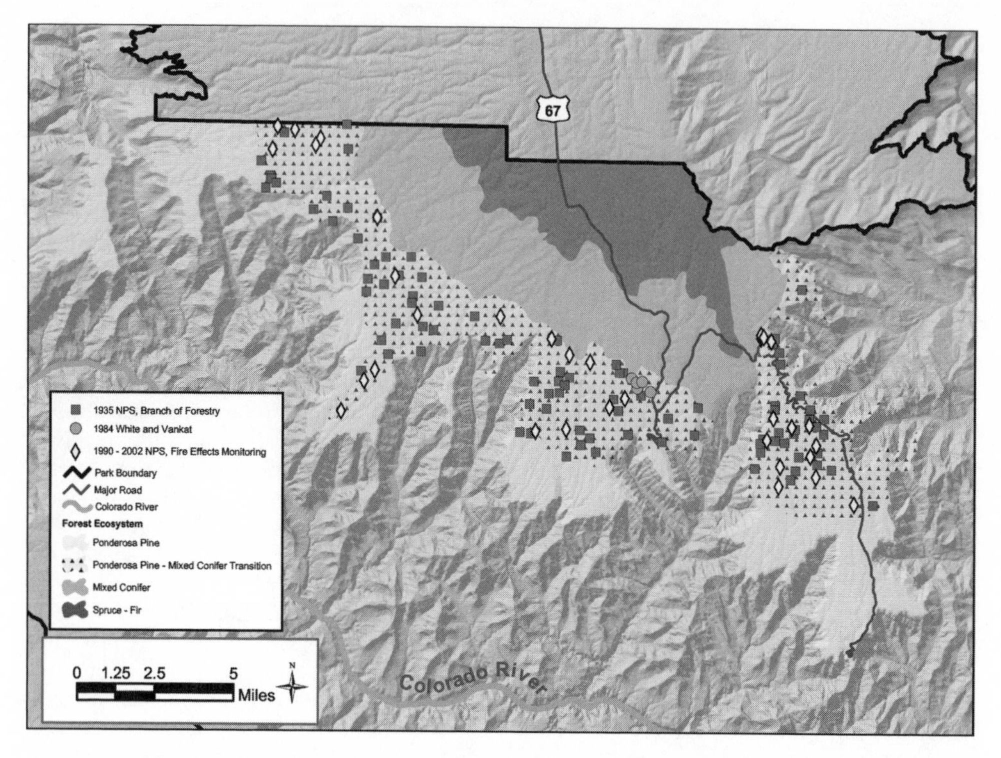

Figure 1. Distribution of major forest zones in GCNP, with plots from 1935 and recent studies superimposed on the ponderosa pine–mixed conifer transition forest zone (base map modified from Bertolette 2002).

al. 2003b). These fires were very uncommon after ca. 1880 (Wolf and Mast 1998; Fulé et al. 2003a, b), although fires resulting in ≥ 10 percent scarring continued in some areas until active fire suppression began in the early twentieth century (Wolf and Mast 1998).

Most post-1935 vegetation changes in transition forests differed from those in mixed conifer forests. Mean densities of small-diameter conifers increased in transition forests from 1935 to the present, but decreased in mixed conifer forests (Vankat et al., this volume). Mean densities of mid-diameter conifers and mean conifer basal areas were unchanged in transition forests from 1935 to the present, but decreased in mixed conifer forests. Mean densities of conifers ≥ 91 cm in diameter decreased in transition forests but were unchanged in mixed conifer forests. Most of these differences appear related to mixed conifer forests (a) having reached a self-thinning stage of forest development and (b) being more susceptible to outbreaks of western spruce budworm (*Choristoneura occidentalis*; Vankat et al., this volume).

Post-1935 changes in transition forests are more similar to changes in ponderosa pine forests, but there are important differences. For example, mean densities of small-diameter conifers increased from 1935 to the present in both forests, but the increases were larger in magnitude in transition forests (cf. Crocker-Bedford et al., this volume). White fir, ponderosa pine, and spruce increased in transition forests, but only ponderosa pine increased in ponderosa pine

Table 1. Structure and composition of ponderosa pine–mixed conifer transition forests of GCNP based on 1935 and recent studies. Density units are trees/ha; basal area units are sq m/ha. Conifers included are ponderosa pine, white fir, Douglas-fir, spruce (Engelmann and blue spruce combined), and subalpine fir.

	1935 (n = 95)		Recent (n = 38)		
Attribute	Mean	SD	Mean	SD	*P*
Density of conifers 10–29.9 cm diameter	148.1	194.8	303.5	166.7	< 0.001
Density of ponderosa pine 10–29.9 cm diameter	71.9	126.4	115.8	134.1	0.088
Density of white fir 10–29.9 cm diameter	49.6	101.5	149.9	171.1	0.001
Density of Douglas-fir 10–29.9 cm diameter	15.8	67.9	11.4	23.0	0.577
Density of spruce 10–29.9 cm diameter	7.9	24.4	22.7	49.4	0.085
Density of conifers 30–60.9 cm diameter	117.3	89.4	117.3	49.9	0.999
Density of ponderosa pine 30–60.9 cm diameter	70.4	78.5	69.6	53.1	0.960
Density of white fir 30–60.9 cm diameter	23.5	40.0	36.5	46.4	0.134
Density of conifers 61–90.9 cm diameter	32.6	34.8	36.5	25.2	0.469
Density of ponderosa pine 61–90.9 cm diameter	24.0	29.9	31.9	25.9	0.138
Density of white fir 61–90.9 cm diameter	4.7	12.3	3.0	6.4	0.280
Density of conifers ≥ 91 cm diameter	6.7	15.8	1.2	4.2	0.003
Density of ponderosa pine ≥ 91 cm diameter	5.9	15.6	1.0	4.0	0.005
Basal area of conifers	44.7	31.6	45.0	15.0	0.948
Density of quaking aspen ≥ 10 cm diameter	228.4	275.6	64.0	95.8	< 0.001

forests (authors' unpublished data). The greater increase in small-diameter conifers in transition forests likely resulted from (a) greater regeneration and faster growth with more soil moisture and (b) greater survivorship with fewer sizable post-1880 fires.

Although densities of small-diameter conifers increased from 1935 to the present in both transition and ponderosa pine forests, densities of conifers in the 61–90.9 cm diameter class significantly decreased only in ponderosa pine forests (Crocker-Bedford et al., this volume). This difference is likely attributable to greater availability of moisture in transition forests, which would mitigate the effects of increased competition. Densities of ponderosa pines ≥ 91 cm in diameter decreased in both forest types. Although absolute densities were different between the two forest types, the percentage decreases from 1935 to the present were similar (83 and 75% in transition and ponderosa pine forests, respectively). This loss of very large ponderosa pines in transition forests is likely the result of increased mortality and reduced recruitment caused by increased competition from small-diameter conifers (Mast and Wolf 2004; see also Biondi 1996; Feeney et al. 1998; Kaufmann and Covington 2001; Crocker-Bedford et al., this volume). Moreover, competition likely has acted synergistically with factors such as increased litter and duff, pathogens, drought, and bark beetle infestations (Covington et al. 1997; Feeney et al. 1998; Kaufmann and Covington 2001; Crocker-Bedford et al., this volume).

Decreased densities of quaking aspen in transition forests from 1935 to the present likely were related to several factors. Reduced fire frequencies probably resulted in increased densities of small-diameter white fir, which can be associated with decreased aspen regeneration (Fulé et al. 2002). In addition, quaking aspens probably were negatively impacted by herbivory from unnaturally elevated deer populations ca. 1915–1935 (Adams 1925; Rasmussen 1941; Merkle 1962; Mitchell and Freeman 1993; Fulé et al. 2002).

Perhaps the most fundamental difference between post-1935 changes in ponderosa pine and transition forests is the much greater increase in small-diameter white firs and spruces in transition forests, i.e., shifts toward mixed conifer characteristics. Such shifts also were reported by Mast and Wolf (2004). The increases of these taxa appear related to the presence of (a) occasional white fir in the overstory of transition stands and (b) white fir and spruce in the overstory of nearby mixed conifer stands. Abundant seed sources, reduced fire frequencies, and related subsequent increases in canopy cover enabled fire-intolerant, shade-tolerant white fir and spruce seedlings to increase in absolute and relative abundance.

Shifts in the species composition of small-diameter conifers may presage additional changes in transition forests. For example, the shift toward mixed conifer species composition in the understory has the potential to eventually influence the overstory, possibly converting transition forests to mixed conifer forests. Moreover, as understory white fir and spruce become sexually mature, they provide seed sources likely to affect changes in nearby ponderosa pine forests.

The 1935 BOF data for mid- and large-diameter conifers in transition forests of GCNP are inconsistent with those of Lang and Stewart (1910) for the Kaibab Plateau (including what is now the North Rim of GCNP). Lang and Stewart (1910) reported mean densities of 51.2 and 12.6 trees/ha for conifers 30–60.9 and ≥ 61 cm diameter, respectively, in their "yellow pine type" (which apparently included what are now transition forests). These densities are far less than the 117.3 and 39.3 trees/ha measured in these respective diameter classes in 1935 (Table 1). Likely increases in white fir from 1910 to 1935 may account for some of the higher densities of mid-diameter conifers in 1935, but it appears impossible that the densities of large-diameter conifers could have tripled. Fire exclusion would not have caused such an increase because surface fires before ca. 1870 rarely killed large conifers (Fulé et al. 2003b). We suggest that it is more plausible that densities of large-diameter conifers declined from 1910 to 1935. Mean densities reported by Lang and Stewart (1910) may be unrealistically low for characterizing what are now transition forests in GCNP. Lang and Stewart sampled long (805 m) transects arrayed systematically across the Kaibab Plateau, which includes many treeless meadows and openings, low-elevation ponderosa pine forests with open structure, and large areas transitional between ponderosa pine forests and pinyon-juniper woodlands (cf. their photo 91835). We conclude that the 1935 BOF data likely more accurately represent early twentieth century conditions in today's transition forests than Lang and Stewart's (1910) data, at least for mid- and large-diameter conifers in forests of GCNP.

Our results for apparent changes from 1935 to the present provide a basis for projecting backwards to ca. 1870 to gain insight critical for management planning. We estimated changes from ca. 1870 to 1935 based on three assumptions. First, we assumed that changes from ca. 1870 to the present were essentially unidirectional for each parameter. This assumption is congruent with forest reconstructions ca. 1880–present for GCNP (Fulé et al. 2002; see also Fulé et al. 1997). Second, for increases in densities of conifers 10–29.9 cm diameter, we assumed that percent changes were equal between ca. 1870–1935 and 1935–present because (1) conifers in GCNP "regenerated at a relatively consistent rate through most of the post-[Euramerican] settlement period" (Fulé et al. 2002) and (2) 1935 is approximately midway between 1870 and the present. Third, for decreases in densities of conifers ≥ 91 cm and quaking aspen ≥ 10 cm diameter, we assumed that percent changes ca. 1870–1935 were one-fourth that for 1935–present. We reasoned that (1) a lag period of approximately 50 years (to 1920) likely occurred before the negative effects of increased density of small-diameter conifers were substantial and (2) the 15 year period between 1920 and 1935 is approximately one-fourth of 1935–present. The 50-year lag period is a somewhat arbitrary position along a gradient of increased impacts on large conifers, but Biondi (1996) reported competition-based mortality of large ponderosa pines peaking at 50–60 years after a major regeneration year near Flagstaff, Arizona. Similar timeframes have been reported for reduced growth rates (Mast and Wolf 2004) and susceptibility to bark beetle outbreaks (Feeney et al. 1998).

Given these assumptions and our results for 1935–present, we offer the following inferences of likely conditions in GCNP ponderosa pine–mixed conifer transition forests ca. 1870:

- Mean density of conifers 10–29.9 cm diameter was approximately 70 trees/ha ca. 1870, resulting in an inferred average increase of about 300 percent from ca. 1870 to the present (Figure 2a).
- Mean density of ponderosa pines 10–29.9 cm diameter was approximately 45 trees/ha ca. 1870, resulting in an inferred average increase of about 150 percent from ca. 1870 to the present (Figure 2b).
- Mean density of white firs 10–29.9 cm diameter was approximately 15 trees/ha ca. 1870, resulting in an inferred average increase of about 800 percent from ca. 1870 to the present (Figure 2c).
- Mean density of conifers ≥ 91 cm diameter was approximately 8 trees/ha ca. 1870, giving an inferred average decrease of about 85 percent from ca. 1870 to the present (Figure 2d).
- Mean density of quaking aspens ≥ 10 cm diameter was approximately 275 trees/ha ca. 1870, giving an inferred average decrease of about 75 percent from ca. 1870 to the present (Figure 2e).

CONCLUSIONS

Our study is important for the development of ecologically based fire management goals and practices for ponderosa pine–mixed conifer transition forests in GCNP. We recommend four specific goals, all within the overall goal of reestablishing ecosystem self-regulation by restoring forest structure, composition, and ecological processes without risking remnant features such as large-diameter conifers:

1. Reduce small-diameter conifers, including white firs, ponderosa pines, and spruces. Our findings support the long-standing management goal of reducing small-diameter conifers (National Park Service 1997) and show which species need to be reduced in transition forests.
2. Maintain mid- and large-diameter conifers. Our findings not only support the long-standing management goal of maintaining large conifers, but also indicate the importance of focusing on trees ≥ 91 cm diameter while expanding the goal to include conifers 30–90.9 cm diameter to enable ongoing recruitment of larger trees.
3. Promote quaking aspen. This has not been perceived as a major management goal for GCNP forests of this elevation. However, our findings indicate the importance of applying this management goal to transition forests.
4. Restore a fire regime characterized by frequent, low-intensity surface fires. Our findings support this long-standing management goal for GCNP ponderosa pine–dominated forests to the degree that active management reduces small-diameter conifers while minimizing losses of mid- and large-diameter conifers.

We also recommend additional research on ponderosa pine–mixed conifer transition forests to help refine management goals and practices. Resampling the 1935 BOF plots would clarify past changes in vegetation and indicate whether the distribution of transition forests has expanded with post-1935 fire

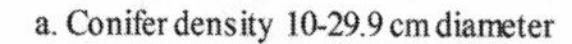

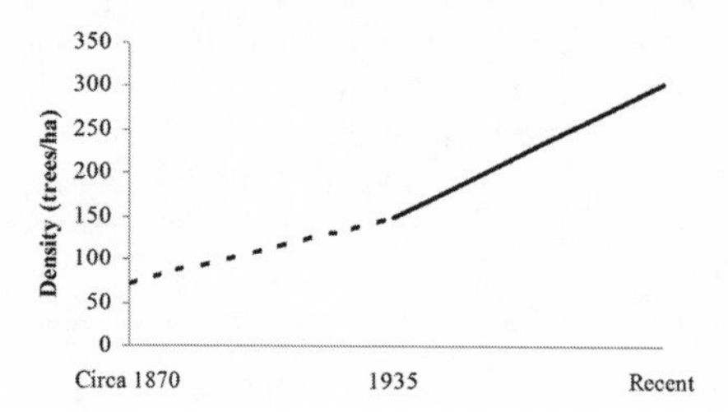

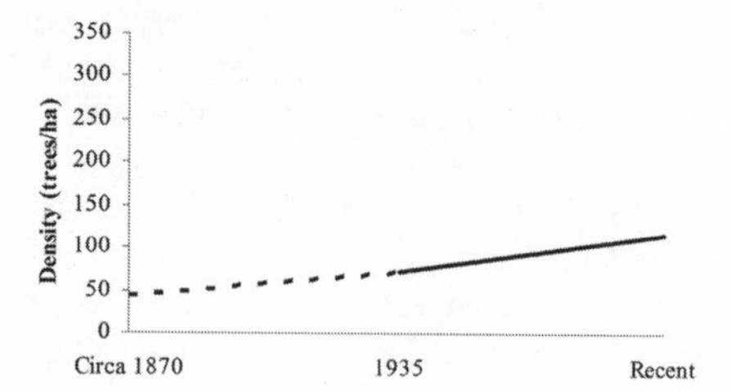

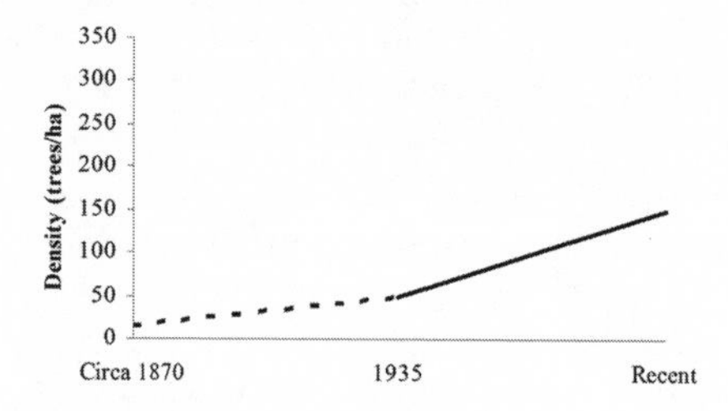

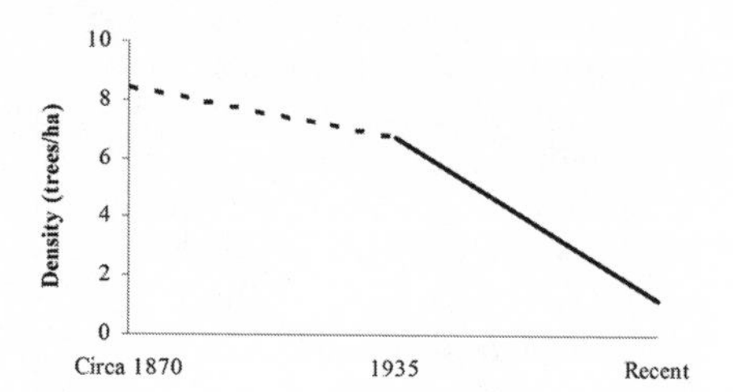

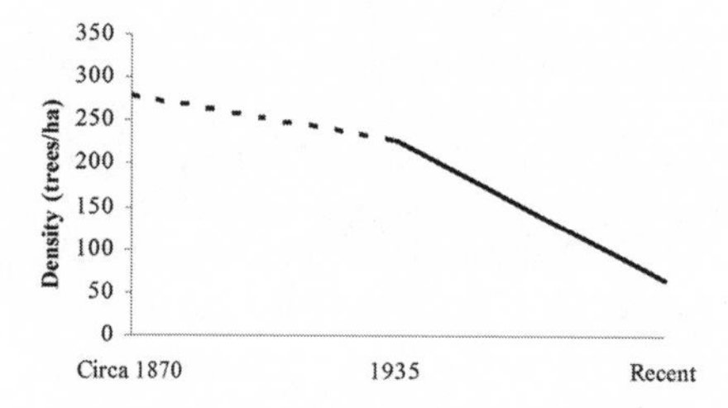

Figure 2. Trends in structure and composition of ponderosa pine–mixed conifer transition forests of GCNP from ca. 1870 to recent studies: (a) density of conifers 10–29.9 cm diameter, (b) density of ponderosa pine 10–29.9 cm diameter, (c) density of white fir 10–29.9 cm diameter, (d) density of conifers ≥ 91 cm diameter, and (e) density of quaking aspen ≥ 10 cm diameter. Solid lines are based on significant differences; dashed lines are based on inferences.

suppression. Additional inventories of existing vegetation in and near transition stands would better document current conditions and provide a stronger base for monitoring future changes, including possible expansion of transition forests. Investigations of causes of tree mortality would be insightful for maintaining mid- and large-diameter trees. Additional pattern analyses of tree ages across landscapes would provide insight into the size, distribution, and frequency of patches of pre-Euramerican crown fires, if present. Additional analyses of pre-Euramerican fire regimes would provide geographically broader understanding of fire return intervals. Geographic analysis of burn-severity patterns of recent fires would provide insight into present fire regimes. These and other research projects are needed so that ecologically appropriate management goals and practices can be developed for GCNP and other areas of the Southwest to reestablish ecosystem self-regulation, allowing ponderosa pine–mixed conifer transition forests to track future environmental changes with minimal active management.

ACKNOWLEDGMENTS

Julie Crawford, Lee Harrison, Jennifer McCollom, Lori Makarick, Niki Juarez-Cummings, Steven Mietz, Kara Leonard, Erin Uloth, and Sarah Falzarano of the National Park Service, Terence Arundel of the U.S. Geological Survey, and Michael Kearsley of Northern Arizona University's Department of Biological Sciences contributed to various aspects of this manuscript, particularly to data analyses. Two anonymous reviewers provided helpful comments on an earlier version of the manuscript.

LITERATURE CITED

Adams, C. C. 1925. Ecological conditions in National Forests and in National Parks. Scientific Monthly 20: 561–593.

Anderson, M. F. 1998. Living at the Edge. Grand Canyon Association, Grand Canyon, Arizona.

Bertolette, D. 2002. Wildfire hazard reduction research. Unpublished map, Grand Canyon National Park, Arizona.

Biondi, F. 1996. Decadal-scale dynamics at the Gus Pearson Natural Area: Evidence for inverse (a)symmetric competition? Canadian Journal of Forest Research 26: 1397–1406.

Brown, D. E., C. H. Lowe, and C. P. Pase. 1979. A digitized classification system for the biotic communities of North America, with community (series) and association examples for the Southwest. Journal of the Arizona-Nevada Academy of Science. 14 (Supplement 1): 1–16.

Coffman, J. D. 1934. Suggestions for the mapping and study of vegetative cover types in areas administered by the National Park Service. Agency-wide directive, U.S. Department of the Interior, National Park Service, Branch of Forestry.

Covington, W. W., P. Z. Fulé, M. M. Moore, S. C. Hart, T. E. Kolb, J. N. Mast, S. S. Sackett, and M. R. Wagner. 1997. Restoring ecosystem health in ponderosa pine forests of the Southwest. Journal of Forestry 95: 23–29.

Feeney, S. R., T. E. Kolb, W. W. Covington, and M. R. Wagner. 1998. Influence of thinning and burning restoration treatments on presettlement ponderosa pines at the Gus Pearson Natural Area. Canadian Journal of Forest Research 28: 1295–1306.

Fulé, P. Z., M. M. Moore, and W. W. Covington. 1997. Determining reference conditions for ecosystem management in southwestern ponderosa pine forests. Ecological Applications 7: 895–908.

Fulé, P. Z., W. W. Covington, M. M. Moore, T. A. Heinlein, and A. E. M. Waltz. 2002. Natural variability in forests of the Grand Canyon, USA. Journal of Biogeography 29: 31–47.

Fulé, P. Z., J. E. Crouse, T. A. Heinlein, M. M. Moore, W. W. Covington, and G. Verkamp. 2003a. Mixed-severity fire regime in a high-elevation forest of Grand Canyon, Arizona, USA. Landscape Ecology 18: 465–486.

Fulé, P. Z., T. A. Heinlein, W. W. Covington, and M. M. Moore. 2003b. Assessing fire regimes on Grand Canyon landscapes with fire-scar and fire-record data. International Journal of Wildland Fire 12: 129–145.

Hessl, A. E., and W. L. Baker. 1997. Spruce and fir regeneration and climate in the forest-tundra ecotone of Rocky Mountain National Park, Colorado, U.S.A. Arctic and Alpine Research 29: 173–183.

Hughes, J. D. 1967. The story of man at Grand Canyon. Grand Canyon Natural History Association Bulletin 14. K.C. Publications, Las Vegas NV.

Hurst, M. 1977. An ecological description of forest communities on the North Rim of Grand Canyon National Park. Unpublished report, Grand Canyon National Park, Arizona.

Kaufman, G. A., and W. W. Covington. 2001. Effect of prescribed burning on mortality of presettlement ponderosa pines in Grand Canyon National Park. In Ponderosa Pine Ecosystems Restoration and Conservation: Steps Toward Stewardship, Flagstaff AZ, April 25–27, 2000, compiled by R. K. Vance, C. B. Edminster, W. W. Covington, and J. A. Blake, pp. 36–42. USDA Forest Service Proceedings RMRS-P-22. Rocky Mountain Research Station, Ogden UT.

Kupfer, J. A., and D. M. Cairns. 1996. The suitability of montane ecotones as indicators of global climatic change. Progress in Physical Geography 20: 253–272.

Lang, D. K., and S. S. Stewart. 1910. Reconnaissance of the Kaibab National Forest. Unpublished report, U.S. Department of Agriculture, Forest Service.

Mast, J. N., and J. J. Wolf. 2004. Ecotonal changes and altered tree spatial patterns in lower mixed-conifer forests, Grand Canyon National Park, Arizona, USA. Landscape Ecology 19: 167–180.

Mast, J. N., T. T. Veblen, and Y. B. Linhart. 1998. Disturbance and climatic influences on age structure of ponderosa pine at the pine/grassland ecotone, Colorado Front Range. Journal of Biogeography 25: 743–755.

Merkle, J. 1962. Plant communities of the Grand Canyon area, Arizona. Ecology 43: 698–711.

Mitchell, J. E. and D. R. Freeman. 1993. Wildlife-livestock-fire interactions on the North Kaibab: A historical review. USDA Forest Service General Technical Report RM-222. Rocky Mountain Forest and Range Experiment Station, Fort Collins CO.

National Park Service. 1997. Grand Canyon National Park Resource Management Plan. Grand Canyon National Park, Arizona.

Neilson, R.P . 1993. Ecotone response to climate change. Ecological Applications 3: 385–395.

Pase, C. P., and D. E. Brown. 1994a. Rocky Mountain (Petran) and Madrean montane conifer forests. In Biotic Communities: Southwestern United States and Northwestern Mexico, edited by D. E. Brown, pp. 43–48. University of Utah Press, Salt Lake City.

Pase, C. P., and D. E. Brown. 1994b. Rocky Mountain (Petran) subalpine conifer forest. In Biotic Communities: Southwestern United States and Northwestern Mexico, edited by D. E. Brown, pp. 37–39. University of Utah Press, Salt Lake City.

Rasmussen, D. I. 1941. Biotic communities of Kaibab Plateau, Arizona. Ecological Monographs 11: 229–275.

Risser, P. G. 1995. The status of the science examining ecotones. BioScience 45: 318–325.

Rusek, J. 1993. Air-pollution mediated changes in alpine ecosystems and ecotones. Ecological Applications 3: 409–416.

Warren, P. L., K. L. Reichhardt, D. A. Mouat, B. T. Brown, and R. R. Johnson. 1982. Vegetation of Grand Canyon National Park. Technical Report No. 9, Cooperative National Park Resources Study Unit. University of Arizona, Tucson.

White, M. A., and J. L. Vankat. 1993. Middle and high elevation coniferous forest communities of the North Rim region of Grand Canyon National Park, Arizona, USA. Vegetatio 109: 161–174.

Wolf, J. J. ,and J. N. Mast. 1998. Fire history of mixed-conifer forests on the North Rim, Grand Canyon National Park, Arizona. Physical Geography 19: 1–14.

APPARENT REDUCTIONS IN LARGE-DIAMETER TREES SINCE 1935 IN PONDEROSA PINE FORESTS OF GRAND CANYON NATIONAL PARK

D. Coleman Crocker-Bedford, John L. Vankat, Don R. Bertolette, Paul Leatherbury, Taylor McKinnon, and Carmen L. Sipe

Knowledge of forest structure and composition prior to Euramerican influence is critical for the development of management plans by the National Park Service and other land management agencies responsible for forests. By comparing past with present-day forest conditions, land managers can identify changes that occurred during the period of Euramerican influence and possibly determine causative factors. Knowledge of changes and their causes provides the background necessary for the development of ecologically appropriate approaches to restoring forest ecosystems to their natural range of variability.

Previous studies have indicated that one major change in the ponderosa pine (*Pinus ponderosa*) forests of the Southwest was an increase in densities of small-diameter conifers, dating to the beginning of Euramerican influence (e.g., Weaver 1951; Harrington and Sackett 1990; Covington and Moore 1994a, 1994b; Dahms and Geils 1997; Swetnam et al. 1999; Allen 2002). Intensive livestock grazing reduced the herbs that had formerly fueled frequent surface fires. As fire frequencies decreased, fire-induced mortality of small-diameter conifers decreased and conifer densities increased.

In contrast, less information is available on whether densities of large-diameter trees (primarily ponderosa pines) changed in areas protected from logging. Early observers did not describe changes in densities of large trees, but any such changes would have involved few individuals and therefore would have been less visually obvious than increases in small trees. Recent observations in Grand Canyon National Park (GCNP) suggest losses of large ponderosa pines.

Changes in densities of large-diameter ponderosa pines are difficult to document quantitatively. Available historical data on uncut forests generally are summary statistics for different study sites across broad regions (e.g., Lang and Stewart 1910; Woolsey 1911). We recently discovered field datasheets for sample plots from a quantitative vegetation study undertaken in GCNP in 1935. This study was conducted by the National Park Service's Branch of Forestry (BOF) and included all areas of the park (1935 boundaries) with ponderosa pine forests. This data set appears to be the earliest site-intensive, detailed documentation of coniferous forest structure and composition for anywhere in the Southwest.

We could not relocate and resample the BOF plots in a timely manner, so instead we compared BOF data to data collected recently for forest inventory and monitoring. Our objective was to examine possible changes in forest structure—especially densities of large trees—from 1935 to the present. Vegetation trends since 1935 can aid in estimating pre-Euramerican conditions, knowledge of which is critical for development of fire management goals and practices.

METHODS

Study Area

Grand Canyon National Park is located on the Colorado Plateau in northern Arizona. Forested areas of the park have been protected through a series of Presidential Proclamations and Congressional Acts, including Grand Canyon Forest Preserve (1893), Grand Canyon Game Preserve (1906), Grand Canyon National Monument (1908), GCNP (1919), and GCNP Expansion (1927). Ponderosa pine forests generally occur at 2000–2400 m elevation and occupy a total area of about 140 sq km on the South and North Rims of the park, within N35°57' to N36°22' latitude and W111°54' to W112°28' longitude (Figure 1). The Grand Canyon National Park weather station 2, located in ponderosa pine forest at 2070 m on the South Rim, received an annual average of 43 cm of precipitation, including 118 cm of snowfall, during 1976–2004 (http://www.wrcc.dri.edu/index.html; viewed on 2/19/2005). Average maximum and minimum temperatures for the warmest month (July) were 29.2 and 9.9° C; averages for the coldest month (January) were 6.6 and –7.9° C.

Widespread Euramerican impact in the region of GCNP began with the introduction of livestock on the South Rim in the 1860s and the North Rim by the 1870s (Hughes 1967; Anderson 1998; Michael F. Anderson, GCNP archaeologist, personal communication). Livestock grazing decreased fire frequencies in GCNP beginning in 1880 (Fulé et al. 2003). Additional grazing occurred with elevated deer populations on the North Rim ca. 1915–1935 (Rasmussen 1941; Mitchell and Freeman 1993). Active fire suppression began by 1924 (White and Vankat 1993), and livestock grazing ended with the fencing of the park's north and south boundaries in the 1930s (cf. Hughes 1967). There are no records of timber harvesting in ponderosa pine forests of the park, but small-scale cuttings occurred near developed areas and in areas of fires. Insect-infected ponderosa pines were cut on the South Rim in the 1930s (cf. annual reports of the superintendent of GCNP).

Field and Data Analysis Methods

We mapped the major forest zones of GCNP and located all forested plots from 1935 and recent studies within the area of ponderosa pine forest (Figure 1). Therefore, our analysis was at a landscape scale similar to that used in fire and vegetation management plans. At this scale, scattered stands of mixed conifer forest are included as part of the range of variability of ponderosa pine forest; however, we excluded treeless meadows and areas of pinyon-juniper woodland lacking ponderosa pine.

The 1935 BOF study employed field methods according to Coffman (1934), who directed that sample plots be placed in sites "representative of the average conditions" for a type of vegetation, to assist with preparation of a vegetation map. Nearly all plots were placed in areas away from lodges, campgrounds, and other facilities. The BOF used 0.2 acre (0.08 ha) plots measuring 66 x 132 feet (20.1 x 40.2 m). In each plot, they tallied live trees by species according to diameter class: 4–11.9, 12–23.9, 24–35.9, and ≥ 36 inch (which we converted to 10–29.9, 30–60.9, 61–90.9, and ≥ 91 cm). To calculate basal area, we assumed a range of 36–47.9 inches (91–122 cm) for the largest diameter class and used diameter-class midpoints (e.g., 20 cm for the 10–29.9 cm class). The 72 BOF plots within the ponderosa pine forest zone are widely distributed (Figure 1).

We used recent data from the Fire Effects Monitoring program of GCNP (FEM, data collected from 1990 to 2002) and the Ecological Restoration Institute of Northern Arizona University (ERI, data collected 1997 to 2001). Both sources used 0.1 ha plots of 20 x 50 m dimensions. We excluded plots having evidence of recent crown fire. The 52 FEM plots had been placed using a stratified random design within areas planned for prescribed burning. The objective was to compare pre- and post-burn conditions (we used only pre-burn data in our analyses). The 148 ERI plots were clustered in five low-elevation areas, most of which were selected because they resembled likely pre-Euramerican conditions (Fulé et al. 2000, 2002). Plots

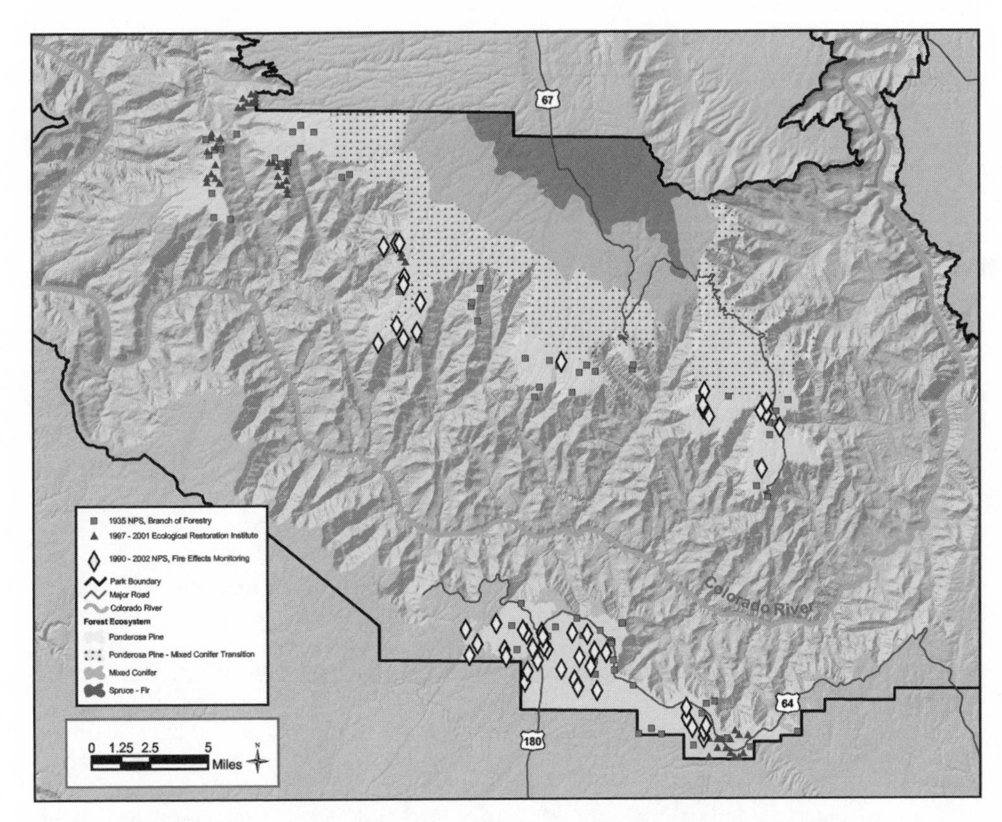

Figure 1. Distribution of major forest zones in GCNP, with plots from 1935 and recent studies superimposed on the ponderosa pine forest zone (base map modified from Bertolette 2002).

had been placed according to a grid in each area. We selected 52 of the 148 plots so that FEM and ERI plots would have equal importance in our analysis, offsetting ERI's intense sampling of restricted study areas. We included all six plots from a small ERI cluster and randomly selected a similar number of additional plots from each of the four larger clusters. Collectively, the 104 FEM and ERI plots used in this study are widely distributed (Figure 1).

We tallied live trees from the FEM and ERI plots according to the four diameter classes used in the 1935 BOF study and calculated basal area as we did for BOF data. We used Student's t-test (alpha = 0.10) to determine whether differences in mean densities (trees/ha) and basal areas (sq m/ha) were statistically significant between 1935 and recent studies. We present results only for ponderosa pine; other species with canopy potential, i.e., white fir (*Abies concolor*), Douglas-fir (*Pseudotsuga menziesii*), blue spruce (*Picea pungens*), and quaking aspen (*Populus tremuloides*), were minor elements of the three smallest diameter classes and were absent from the largest class. We refer to individuals < 30 cm as small-diameter trees, 30–60.9 cm as mid-diameter trees, and ≥ 61 cm as large-diameter trees.

RESULTS

The present mean density of ponderosa pines 10–29.9 cm in diameter is 45 percent higher than measured in 1935 (Table 1). In

Table 1. Structure of ponderosa pine forests of GCNP based on 1935 and recent studies. Density units are trees/ha; basal area units are sq m/ha. Other species with canopy potential, i.e., white fir (*Abies concolor*), Douglas-fir (*Pseudotsuga menziesii*), blue spruce (*Picea pungens*), and quaking aspen (*Populus tremuloides*), were minor elements of the three smallest diameter classes and absent from the largest class.

	1935 (n = 72)		Recent (n = 104)		
Attribute	Mean	SD	Mean	SD	*P*
Density of ponderosa pine 10–29.9 cm diameter	106.7	146.2	154.8	218.3	0.082
Density of ponderosa pine 30–60.9 cm diameter	92.6	89.8	77.8	51.6	0.209
Density of ponderosa pine 61–90.9 cm diameter	31.7	33.5	19.2	17.7	0.005
Density of ponderosa pine ≥ 91 cm diameter	2.0	6.2	0.5	2.2	0.042
Basal area of ponderosa pine	35.0	25.0	25.7	12.4	0.005

contrast, present mean densities of ponderosa pines are not significantly different in the 30–60.9 cm diameter class and are 39 and 75 percent lower in the 61–90.9 and ≥ 91 cm diameter classes, respectively. Mean basal area of ponderosa pines is 27 percent lower than measured in 1935.

DISCUSSION

Interpretation of our results is constrained by high variance (standard deviations); however, this may be unavoidable given GCNP's diverse ponderosa pine landscapes. In addition, we used data from studies with different specific objectives; however, all studies shared the general objective of inventorying forest structure and composition. Plots for the 1935 BOF study were placed subjectively in representative areas; such non-random plots are commonly used in broad-scale vegetation studies. The recent plots are not a resample of the 1935 BOF plots. Despite these issues, we chose to use statistical analysis as a guide for assessing possible differences in mean values between 1935 and the present. The fact that all statistically significant results involved large differences (27–75%) and that results are congruent with trends reported in other studies adds credence to our findings. Nevertheless, we suggest conservative interpretation of our results.

Density increases of small-diameter ponderosa pines likely began soon after fire frequencies decreased ca. 1880. Even in areas selected because they are relatively unchanged, mean fire return intervals (for fires recorded in ≥ 25% of fire-scarred trees) increased from 6–10 years before 1880 to 40–60 years after 1880 (Fulé et al. 2003). Outside of these areas, post-1880 mean fire return intervals were likely even longer. With reduced fire frequencies, fire-induced mortality of small-diameter ponderosa pines declined and densities increased. Moreover, establishment and growth of small pines were probably enhanced as grazing reduced competition from herbs (Belsky and Blumenthal 1997) and possibly as mycorrhizae became more abundant (Korb et al. 2003). Small-diameter ponderosa pines also increased (along with fire-intolerant, shade-tolerant taxa) after 1935 in ponderosa pine–dominated transition forests at higher elevations in GCNP (Crocker-Bedford et al., this volume). Similar increases are hypothesized to have occurred before 1935 in mixed conifer forests (Vankat et al., this volume).

Where densities of small-diameter ponderosa pines increased in ponderosa pine forests, the vigor of trees in all diameter classes likely decreased (Clary and Ffolliott 1969 in Harrington and Sackett 1990; Biondi 1996). Older, larger ponderosa pines may have been especially susceptible to increased competition, as predicted by Pearson (1950) for the Fort Valley Experiment Station near Flagstaff, Arizona:

> The reason for the good growth of large trees in the virgin stand is that most of them were partially isolated. But now that reproduction has come in, young trees will claim an ever increasing share of the limited moisture supply which the veterans were able to monopolize as long as fire and grazing prevented regeneration. It is expected that another 20 years will witness a marked decline in the growth of large trees.

This prediction has been realized (Biondi 1996; Feeney et al. 1998; Kaufmann and Covington 2001). Reduced vigor causes increased mortality rates among large ponderosa pines (Biondi 1996), decreasing density and overall basal area. Percent mortality is generally proportional to tree diameter (Avery et al. 1976; Biondi 1996). Decreased density of ponderosa pines ≥ 91 cm diameter from 1935 to the present has also been reported for higher elevation, ponderosa pine–dominated transition forests in GCNP (Crocker-Bedford et al., this volume). Losses of large conifers elsewhere in the Southwest have equaled 33–46 percent (Dahms and Geils 1997; Garrett and Soulen 1997).

Large ponderosa pines may also have been impacted by factors other than competition. Litter and duff likely increased in thickness with fewer surface fires. Although increased thickness reduces evaporation of soil moisture, it may also reduce infiltration of precipitation, particularly during the summer monsoon season (cf. Covington et al. 1997; Feeney et al. 1998). Pathogens, drought, bark beetles, and prescribed fires may also have contributed to reduced densities of large conifers (Feeney et al. 1998; Kaufmann and Covington 2001). Simultaneously with increased mortality, there has been decreased recruitment of large ponderosa pines (cf. Biondi 1996) because of fewer mid-sized, less vigorous trees. Factors involved in decreases in large-diameter ponderosa pines have likely acted synergistically (Kaufmann and Covington 2001).

The 1935 BOF data for mid- and large-diameter trees in ponderosa pine forests of GCNP are inconsistent with those of Lang and Stewart (1910) for the Kaibab Plateau (including what is now the North Rim of GCNP). Lang and Stewart (1910) reported mean densities of 46.3 and 11.7 trees/ha for ponderosa pines 30–60.9 and ≥ 61 cm diameter, respectively, in their "yellow pine type." These densities are far less than the 92.6 and 33.7 trees/ha measured in these respective diameter classes in 1935 (Table 1). Possible increases from 1910 to 1935 may account for some of the higher densities of mid-diameter ponderosa pines in 1935, but it appears impossible that the densities of large-diameter ponderosa pines could have nearly doubled or tripled. Fire exclusion would not have caused such an increase because surface fires before ca. 1870 rarely killed large trees, especially ponderosa pines (Fulé et al. 2003). We suggest that it is more plausible that trees of this size declined in density from 1910 to 1935. The mean density reported by Lang and Stewart (1910) may be unrealistically low for characterizing ponderosa pine forests in GCNP. Lang and Stewart sampled long (805 m) transects arrayed systematically across the Kaibab Plateau, which includes many treeless openings and large areas ecotonal between ponderosa pine forests and pinyon-juniper woodlands (cf. their photo 91835). We conclude that the 1935 BOF data likely more accurately represent typical ponderosa pine forests in the early twentieth century than do Lang and Stewart's 1910 data, at least for mid- and large-diameter ponderosa pines in forests of GCNP.

The 1935 BOF data are also inconsistent with average densities reported by Woolsey (1911) for areas south of GCNP in Coconino and Tusayan National Forests (most of Tusayan was later incorporated into Kaibab National Forest). Woolsey (1911) reported mean densities of 14.2–22.9 and 5.2–9.6 trees/ha for ponderosa pines 30–63 and > 63 cm diameter, respectively. These values are even less than Lang and Stewart's (1910) mean densities, suggesting that Woolsey's (1911) stands were on drier, less productive sites than present on the Kaibab Plateau. However, Woolsey also reported maximum densities of 62.0 and 17.3–65.4 trees/ha for the 30–63 and > 63 cm diameter classes, respectively, and indicated that areas of even higher densities had been logged. Therefore, there is greater congruence between the 1935

BOF data and the more productive sites described by Woolsey (1911).

Our results for apparent changes from 1935 to the present provide a basis for projecting backwards to ca. 1870 to gain insight critical for management planning. We based estimates for ca. 1870–1935 on four assumptions. First, we assumed that changes from ca. 1870 to the present were essentially unidirectional for each parameter. This assumption is congruent with ponderosa pine forest reconstructions ca. 1880–present for GCNP (Fulé et al. 2002; see also Fulé et al. 1997). Second, for increases in densities of ponderosa pines 10–29.9 cm diameter, we assumed that percent changes were equal between ca. 1870–1935 and 1935–present because (1) ponderosa pine "regenerated at a relatively consistent rate through most of the post- [Euramerican] settlement period" (Fulé et al. 2002) and (2) 1935 is approximately midway between 1870 and the present. Third, for decreases in densities of ponderosa pines ≥ 30 cm diameter, we assumed that percent changes ca. 1870–1935 were one-fourth that for 1935–present. We reasoned that (1) a lag period of approximately 50 years (to 1920) likely occurred before the negative effects of increased density of small-diameter ponderosa pines were substantial on larger trees and (2) the 15 year period between 1920 and 1935 is approximately one-fourth of 1935–present. The 50-year lag period is a somewhat arbitrary position along a gradient of increased impacts on large ponderosa pines, but Biondi (1996) reported competition-based mortality of large ponderosa pines peaking at 50–60 years after a major regeneration year near Flagstaff, Arizona. Similar time-frames have been reported for reduced growth rates (Mast and Wolf 2004) and susceptibility to bark beetle outbreaks (Feeney et al. 1998). Fourth, for changes in basal area, we applied our second and third assumptions.

Given these assumptions and our results for 1935–present, we offer the following inferences of likely conditions in GCNP ponderosa pine forests ca. 1870:

- Mean density of ponderosa pines 10–29.9 cm diameter was approximately 75 trees/ha ca. 1870, resulting in an inferred average increase of about 100 percent to the present (Figure 2a).
- Mean density of ponderosa pines 61–90.0 cm diameter was approximately 35 trees/ha ca. 1870, giving an inferred average decrease of about 50 percent from ca. 1870 to the present (Figure 2b).
- Mean density of ponderosa pines ≥ 91 cm diameter was approximately 2.5 trees/ha ca. 1870, giving an inferred average decrease of about 80 percent from ca. 1870 to the present (Figure 2c).
- Mean basal area of ponderosa pines was approximately 35 sq m/ha ca. 1870, giving an inferred average decrease of about 25 percent from ca. 1870 to the present (Figure 2d). Although basal areas ca. 1870 and 1935 were probably similar, inferences suggest a higher proportion of small-diameter trees and a slightly lower proportion of large-diameter trees in 1935.

CONCLUSIONS

Our study is important for the development of ecologically based fire management goals and practices for ponderosa pine forests in GCNP. We recommend three specific goals, all within the overall goal of reestablishing ecosystem self-regulation by restoring forest structure, composition, and ecological processes without risking remnant features such as large-diameter trees:

1. Reduce small-diameter ponderosa pines. Our findings support the long-standing management goal of reducing small-diameter trees (National Park Service 1997).
2. Maintain mid- and large-diameter ponderosa pines. Our findings not only support the long-standing management goal of maintaining large trees, but also indicate the importance of expanding the goal to include mid-diameter trees, which are needed for ongoing recruitment of large-diameter trees.

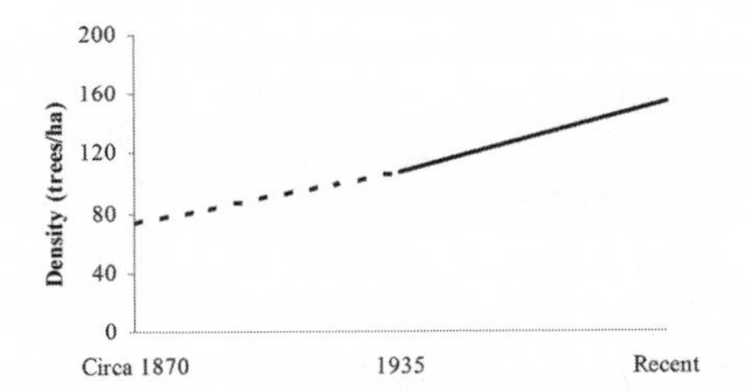

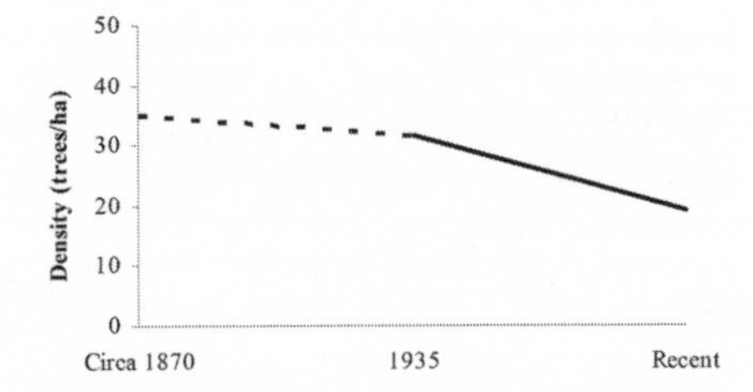

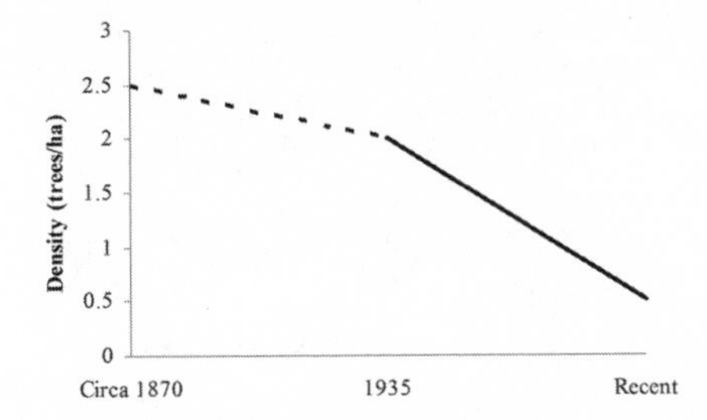

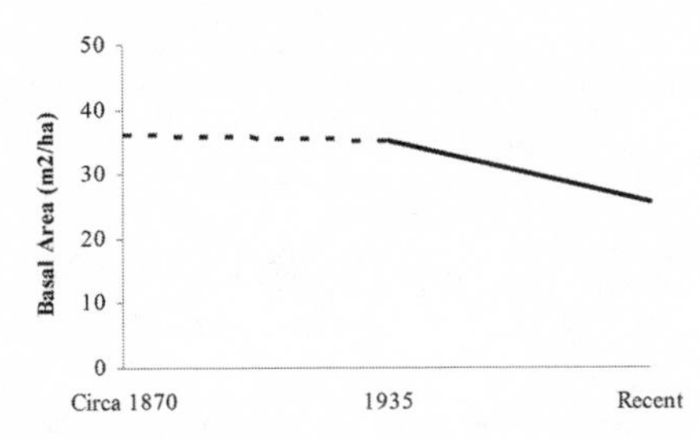

Figure 2. Trends in structure of ponderosa pine forests of GCNP from ca. 1870 to recent studies: (a) density of ponderosa pine 10–29.9 cm diameter, (b) density of ponderosa pine 61–90.9 cm diameter, (c) density of ponderosa pine ≥ 91 cm diameter, and (d) basal area of ponderosa pine. Solid lines are based on significant differences; dashed lines are based on inferences.

3. Restore a fire regime characterized by frequent, low-intensity surface fires. Our findings support this long-standing management goal for GCNP to the degree that active management reduces small-diameter ponderosa pines while minimizing losses of mid- and large-diameter trees.

We also recommend additional research on ponderosa pine forests to help refine management goals and practices. Resampling the 1935 BOF plots would clarify past changes in vegetation. Additional inventories of existing vegetation would better document current conditions and provide a stronger base for monitoring future changes. Investigations of causes of tree mortality would be insightful for maintaining mid- and large-diameter trees. Additional pattern analyses of tree ages across landscapes would provide insight into the size, distribution, and frequency of patches of pre-Euramerican crown fires, if present. Geographic analysis of burn-severity patterns of recent fires would provide insight into present fire regimes. These and other research projects are needed so that ecologically appropriate management goals and practices can be developed for GCNP and other areas of the Southwest to reestablish ecosystem self-regulation, allowing ponderosa pine forests to track future environmental changes with minimal active management.

ACKNOWLEDGMENTS

Julie Crawford, Lee Harrison, Jennifer McCollom, Lori Makarick, Niki Juarez-Cummings, Steven Mietz, Kara Leonard,

Erin Uloth, and Sarah Falzarano of the National Park Service, Terence Arundel of the U.S. Geological Survey, and Michael Kearsley of the Northern Arizona University's Department of Biological Sciences contributed to various aspects of this manuscript, particularly to data analyses. Peter Fulé of Northern Arizona University's Ecological Restoration Institute provided the ERI data used in this study. Thomas Heinlein and an anonymous reviewer provided helpful comments on an earlier version of the manuscript.

LITERATURE CITED

Allen, C. D. 2002. Lots of lightning and plenty of people: An ecological history of fire in the upland Southwest. In Fire, Native Peoples, and the Natural Landscape, edited by T. R. Vale, pp. 143–193. Island Press, Covelo CA.

Anderson, M. F. 1998. Living at the Edge. Grand Canyon Association, Grand Canyon, Arizona.

Avery, C. C., F. R. Larson, and G. H. Schubert. 1976. Fifty-year records of virgin stand development in southwestern ponderosa pine. USDA Forest Service General Technical Report RM-22. Rocky Mountain Forest and Range Experiment Station, Fort Collins CO.

Belsky, A. J., and D. M. Blumenthal. 1997. Effects of livestock grazing on stand dynamics and soils in upland forests of the Interior West. Conservation Biology 11: 315–327.

Bertolette, D. 2002. Wildfire hazard reduction research. Unpublished map, Grand Canyon National Park, Arizona.

Biondi, F. 1996. Decadal-scale dynamics at the Gus Pearson Natural Area: Evidence for inverse (a)symmetric competition? Canadian Journal of Forest Research 26: 1397–1406.

Clary, W. P., and P. F. Ffolliott. 1969. Water holding capacity of ponderosa pine forest floor layers. Journal of Soil and Water Conservation 24: 22–23.

Coffman, J. D. 1934. Suggestions for the mapping and study of vegetative cover types in areas administered by the National Park Service. Agency-wide directive, U.S. Department of the Interior, National Park Service, Branch of Forestry.

Covington, W. W., and M. M. Moore. 1994a. Postsettlement changes in natural fire regimes and forest structure: Ecological restoration of old-growth ponderosa pine forest. Journal of Sustainable Forestry 2: 153–181.

Covington, W. W., and M. M. Moore. 1994b. Southwestern ponderosa forest structure: Changes since Euro-American settlement. Journal of Forestry 92: 39–47.

Covington, W. W., P. Z. Fulé, M. M. Moore, S. C. Hart, T. E. Kolb, J. N. Mast, S. S. Sackett, and M. R. Wagner. 1997. Restoring ecosystem health in ponderosa pine forests of the Southwest. Journal of Forestry 95: 23–29.

Dahms, C. W., and B. W. Geils. 1997. An assessment of forest ecosystem health in the Southwest. USDA Forest Service General Technical Report RM-GTR-295. Rocky Mountain Forest and Range Experiment Station, Fort Collins CO.

Feeney, S. R., T. E. Kolb, W. W. Covington, and M. R. Wagner. 1998. Influence of thinning and burning restoration treatments on presettlement ponderosa pines at the Gus Pearson Natural Area. Canadian Journal of Forest Research 28: 1295–1306.

Fulé, P. Z., M. M. Moore, and W. W. Covington. 1997. Determining reference conditions for ecosystem management in southwestern ponderosa pine forests. Ecological Applications 7: 895–908.

Fulé, P. Z., T. A. Heinlein, W. W. Covington, and M. M. Moore. 2000. Continuing fire regimes in remote forests of Grand Canyon National Park. In Proceedings: Wilderness Science in a Time of Change Conference, Volume 5: Wilderness Ecosystems, Threats, and Management, compiled by D. N. Cole, S. F. McCool, W. T. Borrie, and F. O'Laughlin, pp. 242–248. Missoula MT, May 23–27, 1999. USDA Forest Service Proceedings RMRS-P-15-VOL-5, 307. Rocky Mountain Research Station, Ogden UT.

Fulé, P. Z., W. W. Covington, M. M. Moore, T. A. Heinlein, and A. E. M. Waltz. 2002. Natural variability in forests of the Grand Canyon, USA. Journal of Biogeography 29: 31–47.

Fulé, P. Z., T. A. Heinlein, W. W. Covington, and M. M. Moore. 2003. Assessing fire regimes on Grand Canyon landscapes with fire-scar and fire-record data. International Journal of Wildland Fire 12: 129–145.

Garrett, L. D., and M. H. Soulen. 1997. Changes in character and structure of Apache/Sitgreaves forest ecology: 1850–1990. In Proceedings of the Third Biennial Conference of Research on the Colorado Plateau, edited by C. van Riper III and E. T. Deshler, pp. 25–49. U.S. Department of the Interior, National Park Service Transactions and Proceedings Series NPS/NRNAU/NRTP-97/12.

Harrington, M. G., and S. S. Sackett. 1990. Using fire as a management tool in southwestern ponderosa pine forests. In Effects of Fire Management of Southwestern Natural Resources: Proceedings of the Symposium, technical coordinator J. S. Krammes, pp. 122–133. Tucson AZ, November 15–17, 1988. USDA Forest Service General Technical Report RM-191. Rocky Mountain Forest and Range Experiment Station, Fort Collins CO.

Hughes, J. D. 1967. The Story of Man at Grand Canyon. Grand Canyon Natural History Association Bulletin 14. K.C. Publications, Las Vegas NV.

Kaufman, G. A., and W. W. Covington. 2001. Effect of prescribed burning on mortality of presettlement ponderosa pines in Grand Canyon National Park. In Ponderosa Pine Ecosystems Restoration and Conservation: Steps Toward Stewardship, compiled by R. K. Vance, C. B. Edminster, W. W. Covington, and J. A. Blake, pp. 36–42. Flagstaff AZ, April 25–27,

2000. USDA Forest Service Proceedings RMRS-P-22. Rocky Mountain Research Station, Ogden UT.

Korb, J. E., N. C. Johnson, and W. W. Covington. 2003. Arbuscular mycorrhizal propagule densities respond rapidly to ponderosa pine restoration treatments. Journal of Applied Ecology 40: 101–110.

Lang, D. K., and S. S. Stewart. 1910. Reconnaissance of the Kaibab National Forest. Unpublished report, USDA Forest Service.

Mast, J. N., and J. J. Wolf. 2004. Ecotonal changes and altered tree spatial patterns in lower mixed-conifer forests, Grand Canyon National Park, Arizona, USA. Landscape Ecology 19: 167–180.

Mitchell, J. E., and D. R. Freeman. 1993. Wildlife-livestock-fire interactions on the North Kaibab: A historical review. USDA Forest Service General Technical Report RM-222. Rocky Mountain Forest and Range Experiment Station, Fort Collins CO.

National Park Service. 1997. Grand Canyon National Park Resource Management Plan. Grand Canyon National Park, Arizona.

Pearson, G. A. 1950. Management of ponderosa pine in the Southwest, as developed by research and experimental practice. U.S. Department of Agriculture, Agriculture Monographs 6. Washington D.C.

Rasmussen, D. I. 1941. Biotic communities of Kaibab Plateau, Arizona. Ecological Monographs 11: 229–275.

Swetnam, T. W., C. D. Allen, and J. L. Betancourt. 1999. Applied historical ecology: Using the past to manage for the future. Ecological Applications 9: 1189–1206.

Weaver, H. 1951. Fire as an ecological factor in the southwestern ponderosa pine forests. Journal of Forestry 49: 93–98.

White, M. A., and J. L. Vankat. 1993. Middle and high elevation coniferous forest communities of the North Rim region of Grand Canyon National Park, Arizona, USA. Vegetatio 109: 161–174.

Woolsey, T. S., Jr. 1911. Western yellow pine in Arizona and New Mexico. USDA Forest Service Bulletin 101. Washington D.C.

SPECIES-BASED VEGETATION MAPPING: AN EXAMPLE FROM THE GRAND CANYON

Kenneth L. Cole and John A. Cannella

National parks and other land management units have pursued two important goals: (1) developing vegetation association maps for use in Geographic Information Systems (GIS) (http://biology.usgs.gov/npsveg/index.html) and (2) establishing permanent vegetation plots as a baseline for monitoring change. These two goals are essential for many uses, but the information produced during the identification process on the occurrence of individual species is not always viewed as important enough to retain and distribute. These data, however, especially concerning the relevés sampled during the association mapping and ground-truthing processes, are indispensable for understanding plant distributions at the species level.

A relevé is a ground plot that provides a plant species list from a homogenous area of predetermined size and estimates of the coverage of each species (Mueller-Dombois and Ellenberg 1974). Although originally intended to be a fairly complete listing of species within a plant "community," this method has been abbreviated to record only a handful of dominant trees and shrubs for use in ground-checking maps of plant associations developed from remote sensing. The data from each relevé are far less quantitatively rigorous than those from a more intensively measured vegetation plot, being quickly estimated rather than actually measured. But because less time is required to record each plant, tens of relevés can be completed in the time required for each quantitative vegetation plot or transect. If a study requires measurements across a large and varied landscape, the relevé method can better represent this entire range of spatial variability than will a much smaller number of more detailed plots. For example, if a landscape contains 50 unique combinations of species, 100 relevés will encompass more of this variability than will 20 highly quantitative random plots.

FLEXIBLE CLASSIFICATION SYSTEMS

Vegetation maps are almost always based upon plant communities or associations defined by one to several dominant individual species. Alternatively they can be characterized by a vegetation structure or growth form (grassland vs. forest), or even a soil or hydrological condition (rock outcrop vs. wetland). These classification schemes are essential for lumping together distinct, yet similar microhabitats into units that can be mapped. Without these simplified units, polygons could not be developed to make up the final GIS map. Although vegetation maps are essential for many purposes, the near-exclusive focus on the plant community or association concept leads to misconceptions about the nature of plant species distributions. This is especially true of efforts to create regional or nation-wide vegetation maps using standardized classification schemes. Although a standardized classification system is essential for regional and inter-regional comparisons, it may not be the ideal system for particular needs within an individual park unit.

No single classification scheme serves all purposes for a vegetation map. For example, in 1986 one of us (Cole) participated in vegetation mapping and classification at Indiana Dunes National Lakeshore. The fire ecologist developed a scheme emphasizing fire frequencies, and the wetland ecologist developed a scheme based on flood frequency and depth to water table. Both schemes were valid, and each served certain National Park Service management needs. A compromise classification was eventually developed that was suboptimal for either fire or flood management planning. What was not evident at that time was that vegetation maps need not exist solely as fixed paper entities. Once in digital format, any hierarchical classification system can be reformulated to best address the issue at hand, providing that base-level raw data are still available.

Most of the polygons mapped in the Indiana Dunes project were consistent from one classification scheme to the next. That is, the placement of polygonal boundaries was independent from the specific association classified within that polygon. Thus, the hierarchical classification of associations was a secondary characteristic of the map that could be flexible. The primary characteristic of the map was the spatial orientation and boundaries of polygons representing fairly homogenous habitat areas. A map that efficiently represents homogenous vegetation units can be reclassified in many different ways, as long as the base-level raw data are available. This is especially true in the arid West, where plant distributions are usually controlled by such features as geologic substrate, elevation, slope aspect, and slope angle. Thus the raw data on occurrences of individual species obtained during the mapping process are just as important as the mapped polygons. Without these raw data, the map polygons cannot be reclassified for unanticipated future applications.

INDIVIDUALISTIC VS. COMMUNITY ECOLOGICAL MODELS

Vegetation can be viewed as an assortment of individual species, each distributed according to specific environmental requirements (Gleason 1926), or alternately as groups of species clustered together in larger ecological community, association, or biome units (Clements 1916). The latter view has worked well for characterizing vegetation across large regions, where the complex individual distributions of innumerable species would defy understandable classification and summarization. This community-based approach also works well where one or several species so modify the microhabitat that they are usually found with associates requiring this modification. For example, many shade-loving species occur only under a closed-canopy forest.

But these community concepts have limited utility in understanding the geographic range of a species, especially in open-canopy, arid to semi-arid systems. In a series of studies on plant species distributions along environmental gradients, Robert Whittaker (and many others) has demonstrated that species are usually distributed according to the environmental tolerances of each individual species, rather than in associated groups (Whittaker 1951; Whittaker and Niering 1965, 1968a, 1968b; Shipley and Keddy 1987; Bastow and Allen 1990). That is, the distribution for most species is independent of the distributions of other species; they rarely form reliable associations. Species occur along environmental gradients and are typically the most abundant near the central point of their range of tolerances. This concept has become the accepted model in plant ecology. For example, Begon et al. (1990:620) stated that "there may be communities that are separated by clear, sharp boundaries, where groups of species lie adjacent to, but do not intergrade into, each other. If they exist, they are exceedingly rare and exceptional." Yet these "exceptional" boundaries are precisely what is displayed along every polygon edge on a plant association map.

The notion of each species independently distributed according to its unique tolerances is often referred to as the "continuum concept" due to its view of the habitat as a gradually changing range of continuous variables rather than homogenous patches with discrete boundaries. The continuum concept is so well accepted by plant ecologists that it has passed beyond the testing stage and is instead the focus of refinements such as how to best mathematically represent the response functions of species along environmental gradients (e.g. Oksanen and Minchin 2002). Thus, the simplified community or association classifications used in most vegetation maps cannot be used to portray the ranges of individual species, except perhaps the dominant species for which the communities are named.

The difficulties with using a community-based map to describe a species range can be demonstrated using pinyon-juniper (P-J) woodland on the Colorado Plateau. This association is one of the most recognized of all western plant associations but it has little meaning in terms of species composition, even of its dominants. Throughout its range, P-J can contain one, but frequently neither, of two pinyon species (*Pinus edulis, P. monophylla*), which share the range with one (or none) of several different juniper species (*Juniperus californica, J. osteosperma, J. monosperma, J. deppeana, J. occidentalis*). Although other species may co-occur where these pinyons or junipers grow, there is probably no other plant species that conforms to the range of pinyons and junipers, aside from the species of mistletoe that infect them.

Rather than being a homogenous ecological unit, the P-J is perhaps better thought of as an entity visually recognizable to humans as small to medium sized dark green conifers, typically occurring at mid-elevations. As a term, P-J is useful for communication of human perceptions, but it does not necessarily convey much ecological meaning outside of a local area. P-J is more effectively understood as a general geographical term, much like "desert shrubland" or "montane forest." Unfortunately, the highly visible nature of these dark-colored conifers on a bare landscape, especially in aerial photographs, has led to the widespread impression that the P-J is an important ecological entity, homogenous from top to bottom and from range to range. Studying aspects of the P-J using a vegetation map, such as plotting seed availability for the pinyon jay (*Gymnorhinus cyanocephalus*), becomes an impossible exercise in guessing which parts of the P-J actually contain "P" rather than just "J."

The individualistic or continuum model of species distributions is further supported by the study of past vegetation change. Over the last 20,000 years, plant species migrated as individuals, rather than identifiable associations (Foster et al. 1990; Jackson and Overpeck 2000). These results are particularly evident in records of plant macrofossils from fossil packrat middens (Cole 1982, 1990), which allow the identification of individual plant species from specific locations in the past. These records demonstrate that the modern associations we use as ecological units are highly ephemeral entities. The record of change through time at any one locality is analogous to moving along a contemporaneous environmental gradient through space. Individual plant species appear or drop out as environmental tolerances for those species are met or exceeded.

Individualistic shifts of plant species are not restricted to ancient periods or even the historic past. This is how ongoing changes in plant distributions are described—as a change in the distribution of a particular species. Plant mortality during a drought is described as the death of an individual species, and almost never as a shift in plant associations. A plant invasion into a new area is described as an increase in that particular species, not as a change in plant associations. Using the P-J example, studies have shown that when these species increase or decrease, the shift usually involves just one or the other, and many times the pinyon shifts in the opposite direction of the juniper. For example, although overgrazing is widely believed to lead to range invasions by P-J, recent packrat middens from Capitol Reef National Park in Utah record an increase in only juniper, and a decrease in pinyon with grazing (Cole et al. 1997). Fossil pollen research from the Verde Valley of Arizona shows that juniper increased during the last 100 years, while pinyon remained stable (Davis and Turner 1986).

These results imply that maps of plant associations are inappropriate tools for measuring vegetation change (Davis 1989). A radical change in even one of the two dominant trees of the P-J (pinyon or juniper) would be difficult to detect using an association map. Detecting a change in a shrub or herb species would be impossible. We should expect the constituent members of associations to continue to respond in different ways, just as they have during the past and present.

SPECIES-BASED VEGETATION MAPPING

Plant species are the most stable entities for mapping vegetation. Despite continuing disagreements among plant taxonomists, and the gray areas caused by hybridization and ecotypes, species are scientifically well documented and defined. Even though the taxonomic rank of an individual species may undergo periodic shifting through splitting and lumping, it usually remains an identifiable entity, perhaps continuing as a variety or species-complex, if not a species. In contrast, definitions of associations, communities, and formations will change between times, classifiers, and regions. These definitions, usually defined by one or a few dominant species, are not suited for monitoring change in the secondary plant species. Although these secondary plant species may not individually dominate the community in terms of coverage or abundance, they nevertheless can be critical components of the habitat.

Despite the utility of the species concept, the spatial distributions of individual species are typically poorly documented in western North America. Almost all state floras report distributions only by county of occurrence. Although many herbariums are currently digitizing their collections and making them available on-line, they are using the county as their spatial unit. These county records are adequate for states with many small counties spread over homogenous landscapes, but they are particularly poor for use in western states where counties can be larger than entire northeastern states, and the range of climatic conditions within a mountainous county can exceed the range found within an entire region of several more homogenous states. County-based records are particularly problematic for parts of California, Oregon, Arizona, New Mexico, Nevada, Utah, and Wyoming.

There are some exceptions to this paucity of data on individual species distributions. For example, Turner et al. (1995) published an atlas of 414 species ranges based upon 100,000 records of occurrence in the Sonoran Desert (http://wwwpaztcn.wr.usgs.gov/atlas/). This atlas, compiling 30 years of field observations, has no peer elsewhere in the western United States. Charlet (1996) has mapped the locations of herbarium collections for 24 species of conifer in Nevada. The Southwest Exotic Mapping Program (SWEMP; http://www.usgs.nau.edu/SWEPIC/swemp/maps.html) also uses an individualistic approach in plotting the occurrence of weed species, rather than attempting to assign them to association types.

The most unfortunate aspect of this sparse information is not that the data have not been collected, but rather that the data have not been synthesized and made available for widespread use. Individual scientists who have conducted field investigations have little incentive to expend the considerable effort required to edit their raw data to make it available to other scientists who are their potential competitors. Large agencies that have collected data for other uses often have barriers to open data sharing. For example, the U.S. Forest Service Inventory and Analysis Program's (http://fia.fs.fed.us/) target is to sample one site for every 6000 acres of forest every 5 years. Although these data have recently become available at the county and plant association level, information on occurrences of species and vegetation change is still not available even to most scientists within the U.S. Forest Service itself. Research projects coordinated through the USGS GAP Analysis Program (http://www.gap.uidaho.edu/) have used data from tens of thousands of relevés to classify and assess vegetation maps, but these raw data were never considered to be products and are not generally available except to the original project investigators.

In contrast, the NPS-USGS vegetation mapping program (http : / / biology . usgs . gov / npsveg / index.html) has fortunately made detailed vegetation plot data available for the parks that have been completed. VegBank, developed by the Ecological Society of America (http: / / vegbank.org / vegbank / index.jsp), represents an important continuing effort to compile plot data, but so far it contains data from relatively few plots.

Species-based mapping efforts will not replace either permanent monitoring plots or plant association (land cover) maps, both of which fill critical needs. But plot-specific data produced by these other efforts can be included in spatial databases for individual species. Table 1 summarizes some of the strengths, weaknesses, and features of these three approaches to plant monitoring.

APPLYING SPECIES-BASED DATA: AN EXAMPLE FROM THE GRAND CANYON

Between 1978 and 1981, 1420 plant relevés were sampled throughout Grand Canyon National Park (GRCA) to produce a vegetation map (Warren et al. 1982). An additional 130 relevés surveyed by Cole (1981) in a gradient analysis of individual species distributions were added to that total and are included in the Warren report. These relevé data were especially valuable as many had been collected from remote locations where water is unavailable. The collectors did extensive hiking, sometimes starting from remote helicopter landing sites. Since that time, helicopter access within the GRCA wilderness has become much more sensitive and expensive. These relevé locations can still be reached without mechanical assistance, but only at tremendous expense in time, logistics, and safety. Thus, these Grand Canyon relevé data are a highly valuable asset.

Warren et al. (1982) used relevés and aerial reconnaissance data to produce an excellent vegetation map, applying the vegetation classification system of Brown et al. (1979). Although this review may seem unfavorable for the map compiled by Warren et al. (1982), our criticism applies only to the plant association paradigm mandated for the mapping effort, not to the data themselves, or to the field effort.

In 1998, we became interested in analyzing the GRCA relevés to estimate the climatic tolerances for individual plant species for a USGS paleoclimatology project. Unfortunately, the final vegetation map only displayed plant associations classified according to the Brown et al. (1979) system. The problem was not in the system itself, but with the underlying assumption that individual plant species were constrained within any association. We needed the raw, unclassified field data from the relevés in order to extract the distributions of individual species. The FORTRAN punch cards containing these data had been discarded long ago. After several years of searching, Peter Warren was able to locate photocopies of the original field data sheets and he generously made them available for our effort. We entered the field data into a relational database for use in GIS.

UTAH AGAVE WITHIN THE GRAND CANYON

Utah agave (*Agave utahensis*) is one of the most common plant species in the Grand Canyon. We will use it to demonstrate an application of species-based data. Although Utah agave is a very distinctive plant, it is too small (30–60 cm broad) to have been incorporated in remote sensing mapping efforts. Utah agave is an important food source for many species, especially packrats. Pack rats require succulent plant food (at least 50% water by weight), often *Juniperus, Opuntia*, or *Agave*, to supply them with adequate moisture (Vaughn 1990). Pack rat populations, in turn, can affect the populations of other species, such as rattlesnakes (Repp 1998). Utah agave was also an important food source for Native Americans, as evidenced by the abundant agave roasting pits distributed over the Tonto Platform of the Grand Canyon. Thus, the distribution of this fairly insignificant plant species has

Table 1. Three different types of spatial vegetation data and their uses.

	Plant Association Map	Species Occurrence Data	Permanent Vegetation Plot
Coverage type	Polygon coverage.	Point data that can be used to model polygonal distributional rangemaps.	Usually point data.
Primary Use	Analyzing/comparing areas of different habitat type.	Documenting distributions of and co-occurrences of individual species.	Documenting current resources and vegetation change.
Strengths	Classifies entire landscape. Simplifies vegetation into convenient units.	Essential for use in studies of individual species. Can be used to reclassify association map for unanticipated needs. Ideal for use with well-distributed threatened and exotic species.	Can be a convincing documentation of change, especially when combined with replicate photographs. Ideal for monitoring change with specific populations of threatened, endangered, and exotic species.
Weaknesses	Assumes that species co-occur in associations through time and space. Allows only the one classification scheme anticipated by the authors. Statistical analysis of change is impossible.	Data are rarely collected in a systematic manner, making local statistical analysis problematic.	Time-consuming sampling procedures produce too few plots to statistically characterize entire spatial variability of landscape. Statistical analysis complicated by the pseudoreplication.
Location Accuracy	Mapped polygon boundaries may be misleadingly accurate due to the random diffusion of species across association boundaries.	Variable, depending upon the mapping and technological resources of the scientists.	Highly accurate. Hopefully documented on the ground at each plot.
Taxonomic Accuracy	Poor (association level). Assumes constant associations in time and space.	Detailed (species level observations).	Precise (species, subspecies, or variety level with voucher specimens).
Optimal Spatial Extent	Medium range—the size of the management unit being mapped.	Large—possibly the entire range of the species.	Small—hopefully characterizes entire landscape, but more likely only characterizes area within plot.

Table 2. Percent occurrence of Utah agave in all plant associations that contain it in at least half of their relevés from the eastern Grand Canyon. Importance rankings for Utah agave are from Warren et al. (1982). Percent frequency in relevés was calculated by overlaying the digital vegetation map with the relevé locations.

Plant Association	Importance of Utah agave	Percent of Relevés
Scrub oak–snakeweed–beargrass–blackbush	Not listed	83
Blackbrush–pinyon–juniper	Characteristic	78
Saltbush–banana yucca–snakeweed	Occasional	67
Juniper–pinyon–Mormon tea–scrub oak	Characteristic	67
Pinyon–juniper–scrub oak–little leaf mtn mahogany	Characteristic	65
Cottonwood–brickellia–acacia–Apache plume	Associated	62
Snakeweed–Mormon tea–Utah agave	Characteristic	61
Juniper–pinyon–Mormon tea–greasebush	Characteristic	61
Pinyon–scrub oak–manzanita	Associated	61
Blackbrush–Mormon tea–banana yucca	Characteristic	51
Mormon tea–big galleta–catclaw acacia	Not listed	50

implications for mammalogists, herpetologists, and archaeologists, as well as for botanists and paleoclimatologists.

Because Utah agave could also be an important indicator of climate, it became of interest to the USGS Global Change program. During the ice ages, Utah agave distributions fluctuated in elevation from the bottom to the top of the Grand Canyon. So, understanding the modern distribution of Utah agave was essential to understanding past climate change. We produced a range map for Utah agave (Figure 1) through a compilation of all available records (Cole et al. 2004). These GRCA relevé data were invaluable in defining its eastern distribution. We discovered that its upper range limits are controlled by minimum winter temperatures. This information then allowed us to use past Utah agave distribution as an indicator of paleoclimates (Cole and Arundel, unpublished manuscript).

The vegetation classification produced by Warren et al. (1982) includes Utah agave in the name of only one of 63 associations—the snakeweed–Mormon tea–Utah agave association (153.11011; Figure 2). But Utah agave actually occurs in only 61 percent of the relevés obtained from areas mapped as this unit (Table 2). From this, one might conclude that Utah agave is only an occasional species within the canyon. But the relevé database shows that it actually occurred in 360 (35%) of all eastern Grand Canyon relevés, and was listed as occurring within 27 (43%) plant associations (Warren et al. 1982). Thus, even though it was not found in many of the relevés named for it, it is one of the most widely distributed species within GRCA.

The snakeweed–Mormon tea–Utah agave association is one of exclusion rather than inclusion. All three plant species actually grow throughout the Grand Canyon and are not particularly more abundant within any one particular association than elsewhere, but they had been selected to typify the association of extremely arid, barren rockscapes because little else can grow there. That is, the environment on these low-elevation rockscapes is so stressful that only these three extremely well adapted species remain; most other species are excluded.

Utah agave occupies an extreme range of elevations. Phillips et al. (1987) gave a range of 366–2195 m (1200–7200 feet), whereas the Warren et al. (1982) database records it in relevés from 427 m to 2438 m (1400–8000 feet). Its frequency throughout this range

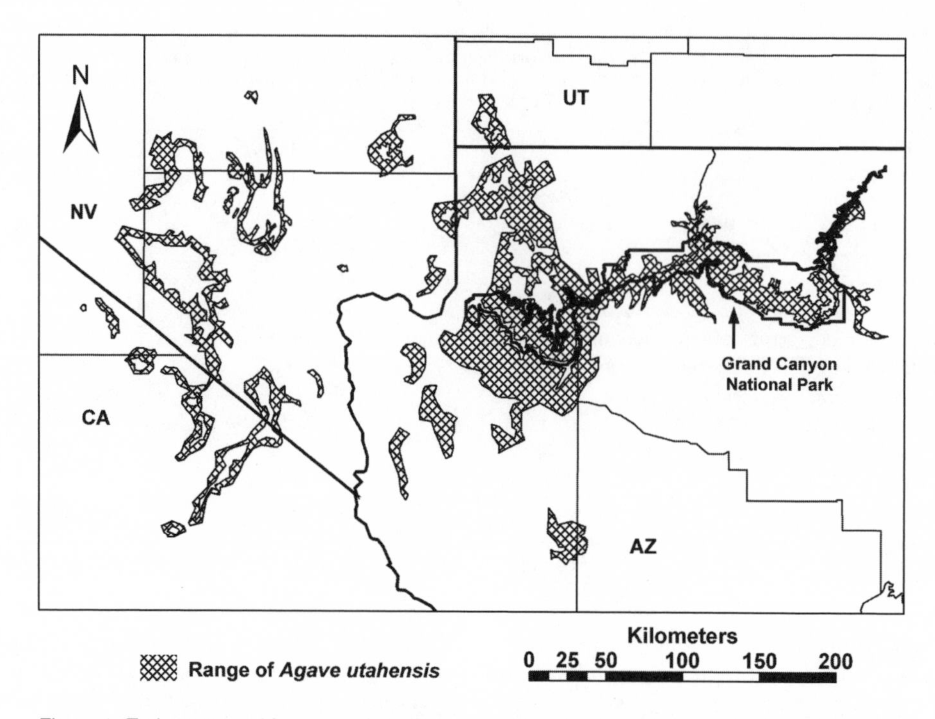

Figure 1. Entire geographic range of Utah agave modeled using data from the Grand Canyon relevé database, a Digital Elevation Model, herbarium records, published regional estimates of elevational range, and field experience of several regional specialists.

can be estimated by calculating the percentage of relevés containing Utah agave in elevational classes (Figure 3). This analysis shows that it is most abundant in the eastern Grand Canyon at middle elevations, between about 1000 and 2000 m elevation.

Calculating the frequency of Utah agave on different rock types reveals that it is concentrated on rocky substrates, especially limestone, sandstone, and schist (Figure 4). These data somewhat contradict the usual description of Utah agave being abundant on limestone substrates (Gentry 1982). Although it frequent grows on limestone, it is also abundant on other steep rocky substrates with minimal soil. Because rocky substrates with the highest frequency of Utah agave are also those at middle elevations in the Grand Canyon, it is difficult to discriminate between elevational and substrate control of its distribution.

The distribution of Utah agave in the Grand Canyon could be determined through the circular process of classifying and mapping associations, then creating a map of those associations known to contain Utah agave. Although this approach is possible, it is entirely dependent upon the assumption that species limits co-occur along association boundaries; this assumption has been refuted in many studies. Applying this circular process to the Grand Canyon data of Warren et al. (1982) would map Utah agave as growing at up to 2680 m (8800 ft), where it occasionally occurs within the Douglas fir–white fir–New Mexico locust association. A check of the raw data reveals that the co-occurrence of Utah agave in this association type happened only around 1830 m (6000 ft) elevation, where fir and locust trees grow on north-facing cliffs. The two highest documented occurrences of Utah agave, at 2438

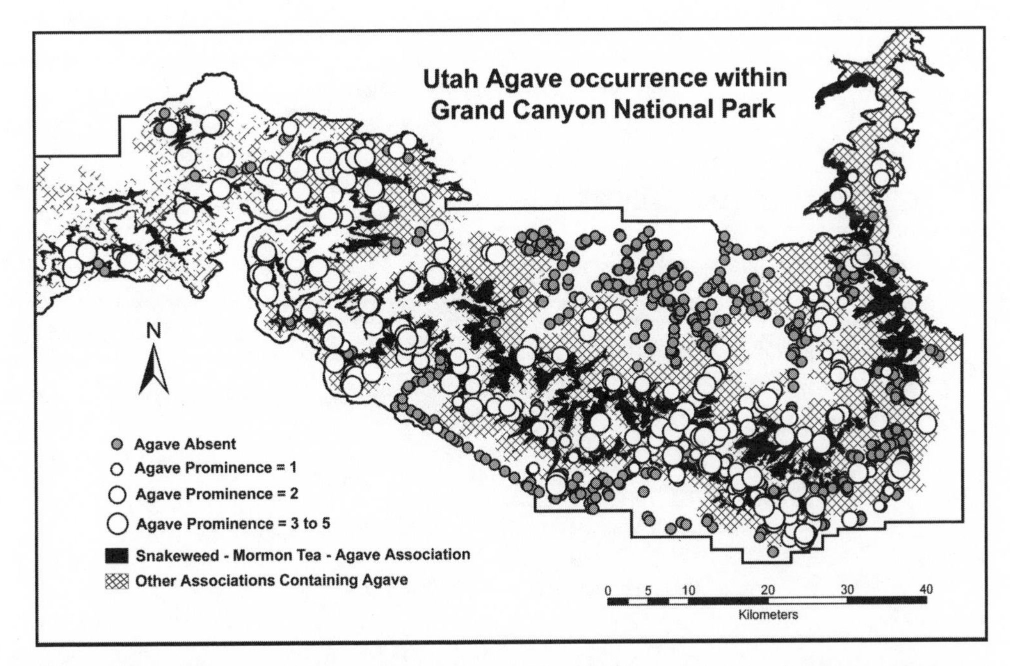

Figure 2. Map of Utah agave records within the eastern Grand Canyon showing the actual presence and absence data for *Agave utahensis* in the eastern Grand Canyon using data from 1200 relevés from the GRCA database; the distribution of the snakeweed–Mormon tea–Utah agave association; and the mapped distribution of associations in which *Agave utahensis* is a "characteristic species" or "associated species." Associations where Utah agave is an "occasional species" are not mapped.

m (8000 ft) within the pinyon–serviceberry–Gambel oak association, actually occur on exposed limestone cliffs within the canyon, rather than above the rim. Our site-specific analysis of Utah agave reveals that it has substrate and thermal requirements much more specific than would be apparent from the association map. It is located inside the confines of the canyon rim where cloud formation is less frequent and the steep rocky slopes provide open ground for stress-tolerant species (Grime 1979).

BUILDING A VIRTUAL SPECIES DISTRIBUTION MAP

The occurrence of Utah agave in a relevé database was used as an example of how species-specific data can assist an unanticipated research project many years after the data were collected. In this example, the raw field data were required for our effort. Any attempt to use the classified plant association map would have produced extremely misleading results about the distribution of Utah agave. These same types of plot or relevé data could be used in projects involving any major plant taxa.

The techniques used here for Utah agave could be applied to all the major plant taxa within an area, creating a true species-based vegetation map. This map would exist only as a virtual (digital) entity as the polygonal boundaries circumscribing each species range would not necessarily coincide as they are forced to on an association map. From this virtual map, individual plant species could then be selected for inclusion on numerous printed versions. Such an approach, and its ability to generate customized maps, could be especially useful for informing land management decisions related to individual species distributions.

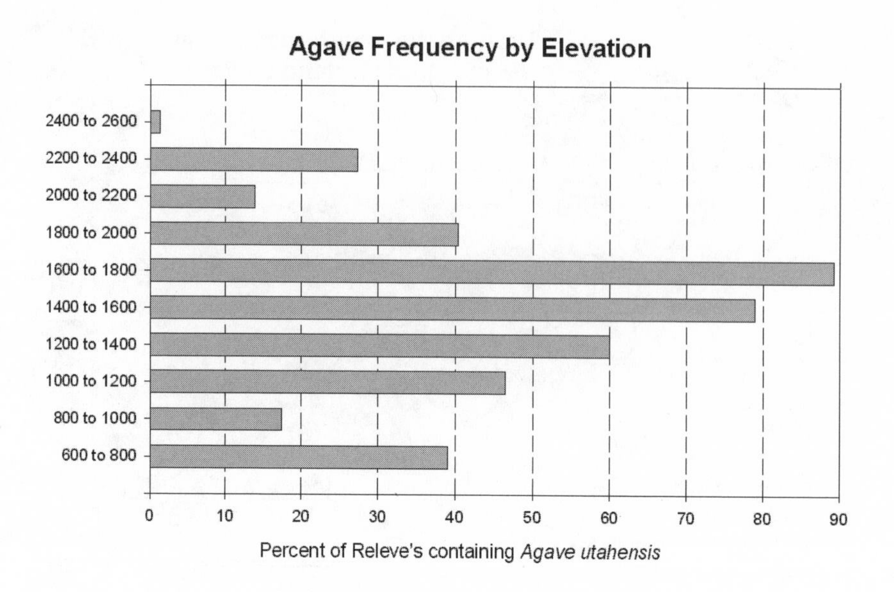

Figure 3. Distribution of Utah agave within eastern Grand Canyon by elevation (m).

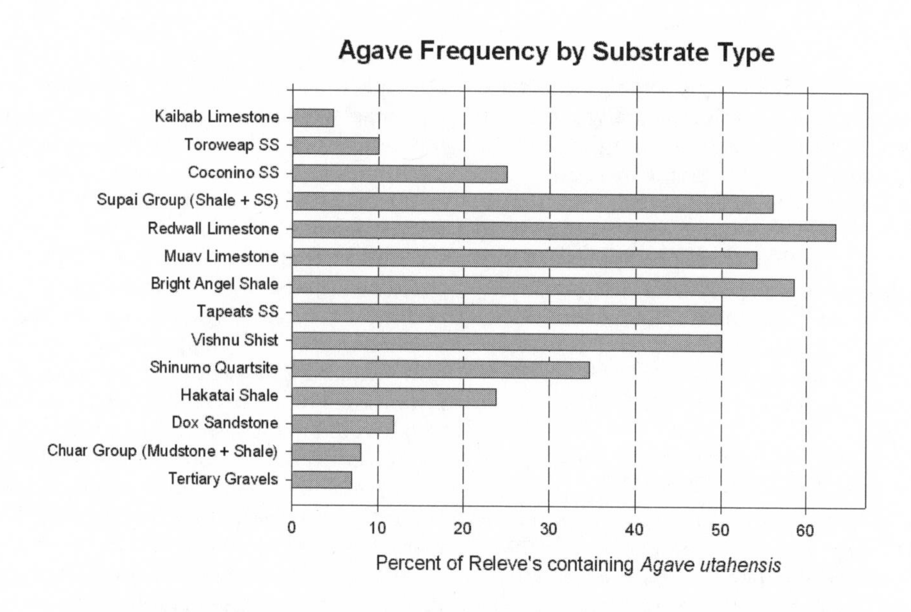

Figure 4. Distribution of Utah agave in the eastern Grand Canyon by substrate type. Substrate types are shown roughly in elevational order as they occur within the canyon.

Creating species-based maps requires a reordering of priorities in vegetation mapping efforts. Ground-based observations such as relevés would be given a much greater emphasis as the focus shifted from spatial accuracy to taxonomic accuracy. Because only a few plant species can be identified to the species level from remote sensing data, such data would be less emphasized.

Species-based data collected as part of other efforts can be invaluable. The importance of these data should be recognized in their preservation and distribution. Coordinated efforts should be undertaken to compile, archive, and distribute available species-based data before such data are lost.

ACKNOWLEDGEMENTS

We appreciate comments received from many individuals that greatly improved this paper—Dave Mattson, Denny Fenn, Mark Sogge, and reviewers. Special thanks go to Terry Arundel for discussions on species mapping, and to Peter Warren, who saved boxes of raw data sheets for more than 20 years in his garage and generously made them available to us. Development of the Grand Canyon digital relevé database was funded through the USGS Global Change Research Program.

REFERENCES CITED

Arundel, S. T. In press. Assessing spatial models to establish climatic limiters of plant species' distributions. Ecological Modeling.

Bastow, J., and R. B. Allen. 1990. Deterministic versus Individualistic community structure: A test from invasion by *Nothofagus menziesii* in southern N. Zealand. Journal of Vegetation Science 1: 467.

Begon, M., J. L. Harper, and C. R. Townsend. 1990. Ecology: Individuals, populations, and communities. Blackwell Scientific, Boston.

Brown, D. E., C. H. Lowe, and C. P. Pase. 1979. A digitized classification system for the biotic communities of North America, with community (series) and association examples for the southwest. Journal of the Arizona-Nevada Academy of Science, Vol. 14, supplement 1.

Charlet, D. A. 1996. Atlas of Nevada Conifers: A Phytogeographic Reference. University of Nevada Press, Las Vegas.

Cole, K. L. 1981. Late Quaternary vegetation of the Grand Canyon: Vegetational gradients over the last 25,000 years. PhD dissertation, University of Arizona, Tucson.

Cole, K. L. 1982. Late Quaternary zonation of vegetation in the eastern Grand Canyon. Science 217: 1142–1145.

Cole, K. L. 1985. Past rates of change, species richness, and a model of vegetational inertia in the Grand Canyon, Arizona. American Naturalist 125: 289–303.

Cole, K. L. 1990. Late Quaternary vegetation gradients through the Grand Canyon. In Fossil Packrat Middens: The last 40,000 Years of Biotic Change in the Arid West, edited by J. L. Betancourt, P. S. Martin, and T. R. Van Devender. University of Arizona Press, Tucson.

Cole, K. L., N. Henderson, and S. Shafer. 1997. Holocene vegetation and historic grazing impacts at Capitol Reef National Park Reconstructed using packrat middens. Great Basin Naturalist 57: 315–326.

Cole, K. L., W. G. Spaulding, W. Hodgson, and K. Thomas. 2004. A digital rangemap for Utah agave (Agave utahensis). Available online at http://www.usgs.nau.edu/global_change/RangeMaps.html (map, metadata, and digital coverage).

Davis, M. B. 1989. Insights from paleoecology on global change. Bulletin of the Ecological Society of America 70: 222–228.

Davis, O. K., and R. M. Turner. 1986. Palynological evidence for the historic expansion of juniper and desert shrubs in Arizona, USA. Review of Palaeobotany and Palynology 49: 177–193.

Foster, D. R., P. K. Schoonmaker, and S. T. A. Pickett 1990. Insights from paleoecology to community ecology. Tree 5: 119–122.

Gentry, H. S. 1982. Agaves of continental North America. University of Arizona Press, Tucson.

Gleason, H. A. 1926. The individualistic concept of the plant association. Bull. Torrey Botan. Club 53: 7–26.

Grime, J. P. 1979. Plant Strategies and Vegetation Processes. Wiley and Sons, New York.

Jackson, S., and J. Overpeck. 2000. Responses of plant species, populations and communities to long-term environmental change. In Deep Time: Paleobiology's Perspective. Paleobiology 26 (Supplement to No. 4): 194–220..

Mueller-Dombois, D., and H. Ellenberg. 1974. Aims and Methods of Vegetation Ecology. Wiley and Sons, New York.

Oksanen, J., and P. R. Minchin. 2002. Continuum theory revisited: What shape are species responses along ecological gradients? Ecological Modelling 157: 119–129.

Phillips, B. G., A. M. Phillips, and M. A. Schmidt Bernzott. 1987. Annotated checklist of vascular plants of Grand Canyon National Park. Grand Canyon Natural History Association, Monograph Number 7.

Repp, R. 1998. Wintertime observations on five species of reptiles in the Tucson Area: Sheltersite selections/fidelity to sheltersites/notes on Behavior. Bull. Chicago Herp. Soc. 33: 49–56.

Shipley, B., and P. A. Keddy. 1987. The individualistic and community-unit concepts as falsifiable hypotheses. Vegetatio 69: 47–55.
Turner, R. M.., J. E. Bowers, and T. L. Burgess. 1995. Sonoran desert plants: An ecological atlas. University of Arizona Press, Tucson.
Vaughn, T. A. 1990. Ecology of living packrats. In Packrat Middens: The Last 40,000 Years of Biotic Change, edited by J. Betancourt, T. R. Van Devender, and P. S. Martin. University of Arizona Press, Tucson.
Warren, P. L., K. L. Reichhardt, D. Mouat, B. T Brown, and R. R. Johnson. 1982. Technical Report No. 9, Vegetation of Grand Canyon National Park. Cooperative National Park Resources Studies Unit, University of Arizona, Tucson.
Whittaker, R. H. 1951. A criticism of the plant association and climatic climax concepts. Northwest Science 25: 17–31.
Whittaker, R. H., and W. A. Niering. 1965. Vegetation of the Santa Catalina Mountains, Arizona: A gradient analysis of the south slope. Ecology 46: 429–452.
Whittaker, R. H., and W. A. Niering. 1968a. Vegetation of the Santa Catalina Mountains, Arizona, IV. Limestone and acid soils. Journal of Ecology 56: 523–544.
Whittaker, R. H., and W. A. Niering. 1968b. Vegetation of the Santa Catalina Mountains, Arizona, III. Species distribution and floristic relations of the north slope. Journal of the Arizona Academy of Sciences 5: 3–21.

A QUANTITATIVE MODEL OF AVIAN COMMUNITY AND HABITAT RELATIONSHIPS ALONG THE COLORADO RIVER IN THE GRAND CANYON

Mark K. Sogge, David Felley, and Mark Wotawa

Riparian habitats in the western United States, especially those dominated by native vegetation, typically support a disproportionately large number of birds as compared to adjacent non-riparian habitats, in terms of both bird abundance and species richness. The Colorado River corridor in the Grand Canyon is no exception, as it provides important habitat to a large number of wintering, migrant, and breeding birds (Brown et al. 1987; Carothers and Brown 1991; Felley and Sogge, this volume), even though much of the habitat is dominated by introduced tamarisk (*Tamarix ramosissima*).

Knopf et al. (1988) calculated that riparian habitat covers less than one percent of the land area of the western United States, and much of this is subject to the influence of water management activities (Collier et al. 1996). The rarity and importance of riparian habitats highlights the uniqueness and value of the Colorado River corridor through the Grand Canyon, one of the largest protected riparian areas in the West. The operation of Glen Canyon Dam has affected riparian habitats in the Grand Canyon (Turner and Karpiscak 1980; Anderson and Ruffner 1988; Stevens and Ayers 1994) and the associated bird communities in the past (Brown et al. 1987; Brown 1988b; Carothers and Brown 1991), and will continue to do so in the future.

Understanding the relationship between riparian vegetation and the habitat of a bird species or community is central to informed river management and conservation of avian resources. This is especially true of the breeding bird community in the Grand Canyon, which includes many riparian-obligate breeding species (Brown et al. 1987; Felley and Sogge, this volume). The avifauna of the riparian corridor in the Grand Canyon has received much study, and a number of researchers have looked at bird-habitat relationships in the past (Brown and Johnson 1985, 1987, 1988; Brown et al. 1987; Brown and Trosset 1989). These studies have provided much useful information for some research and management questions, but are generally hard to apply to broader questions regarding bird communities. For example, studies of single species (e.g., Sogge et al. 1997; Brown 1988a) or indicator species (Brown and Johnson 1987; Brown 1988b) have limited application to community-level questions. The use of avian indicator species has been the subject of much debate and criticism in general (Verner 1985; Morrison 1986), and Brown (1988b) found that indicator species were not useful for predicting long-term trends in the breeding bird community along riparian areas in the Grand Canyon. Therefore, bird community studies should include all species within the community of interest (e.g., migrants, breeders, insectivores).

Furthermore, bird-habitat relationship models can be greatly affected by the type and scale of habitat measurements (Wiens 1989a; Morrison et al. 1992). Studies involving a small number of study sites (e.g., Brown and Johnson 1987; Brown 1988b) can provide accurate information for these particular sites but may not include a variety of sites representative of the diversity within a region (Wiens 1989a). Habitat measurements that are fine-scale and/or focus on nest-centered measurements (e.g., Brown and Trosset 1989) may provide different results or models than studies conducted at different habitat scales (Morrison et al. 1992; Block and Brennan 1993). In addition, the nature of the bird survey techniques used can affect estimates of abundance and species richness (Wilson et al. 2000; Dieni and Jones 2002), and thereby influence bird-habitat models (Verner 1985; Wiens 1989a; Block and Brennan 1993; Felley and Sogge 1997).

Our objective was to quantify the relationship between riparian vegetation/habitat characteristics and measures of the resident breeding bird community, which in turn could be used to evaluate possible indirect impacts from dam operations. We wanted to base our habitat measurements on protocols that could be applied quickly and parameters that could be quantified easily, using remote sensing (aerial photography) to the greatest degree possible. We also emphasized patch-scale parameters, rather than nest-based measurements, in order to more easily relate our bird-habitat models to larger-scale patterns and future dam-induced habitat changes.

METHODS

Study Sites

We selected study sites to form a representative sample of riparian habitats between Lees Ferry (River Mile 0) and Diamond Creek (RM 225; river mile designations per Stevens 1983). At the largest geographic scale, the river corridor follows a 540 m elevation gradient with changing vegetation communities from Colorado Plateau desert-scrub (elevation 950 m at Lees Ferry) to Sonoran desert-scrub (elevation 410 m at Diamond Creek; Turner and Karpiscak 1980), and characteristics of riparian vegetation relate naturally to the geomorphology of the 12 reaches of the river (as defined by Schmidt and Graf 1988). For these reasons, we located sites in a variety of reaches, within the constraints of river logistics. Riparian patches that supported breeding Southwestern Willow Flycatchers (*Empidonax traillii extimus*) were purposely omitted from consideration, due to potential disturbance impacts.

We conducted bird surveys and measured habitat parameters at 50 study sites in 1995. We had conducted bird surveys at 32 of these sites in 1994. Each site was a discrete patch of riparian vegetation, separated from other riparian habitat by open water, or more than 50 m of non-riparian vegetation or unvegetated ground (see site aerial photographs in Sogge et al. 1998). Most sites were located above RM 75 or below RM 150 (Figure 1). Because of the constraints of river-based logistics, random site selection was not possible. Therefore, sites were chosen from among habitat patches that were downstream of suitable camping locations, that could be reached and surveyed during morning hours. Sites were selected to include a wide range of the sizes, shapes, geomorphology, and floristic composition of patches of riparian vegetation found along the river. All surveys were conducted between half an hour before sunrise and 10:00 AM.

Bird Surveys

We conducted monthly bird surveys from April through July in 1994, and from March through June in 1995. We integrated information from walking count, floating count, and point count techniques in order to develop a composite data set on the numbers and types of birds detected in each patch.

Walking surveys consisted of one observer walking slowly through the riparian habitat, recording species and numbers of all birds observed at the study site. Walking surveys were made at all study sites surveyed during 1994–95.

Floating surveys consisted of observers floating past each study site in a raft ahead of the walking surveyor and recording all birds heard or observed. Floating surveys were conducted only in 1994.

We conducted point count surveys at 11 study sites (37 total point count stations) where walking counts were also conducted.

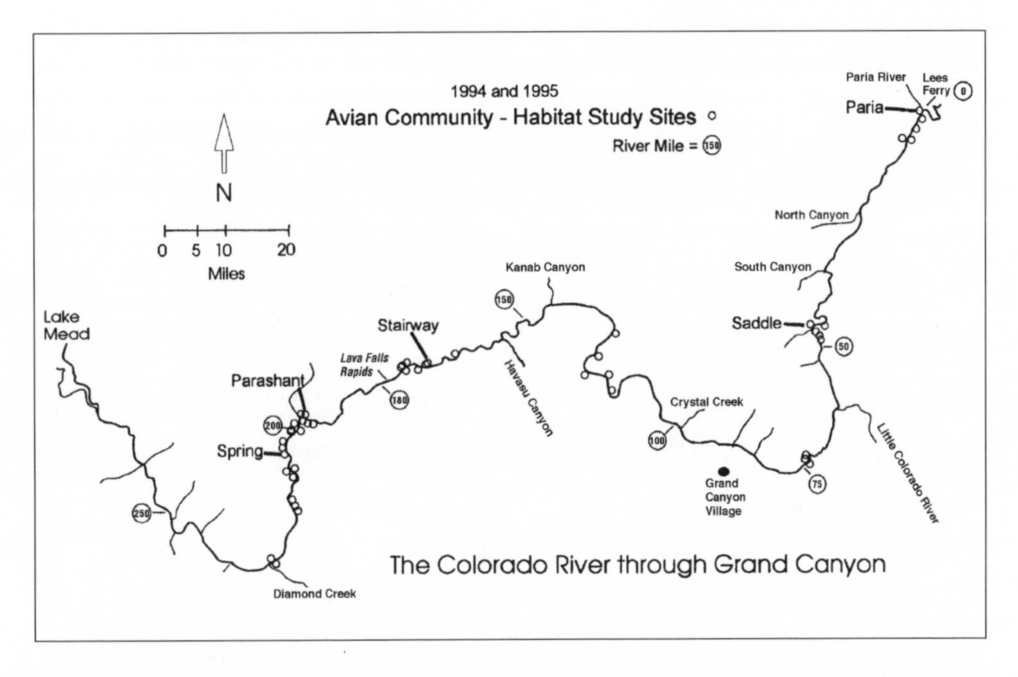

Figure 1. Location of Grand Canyon avian community-habitat study sites, 1994–1995. Open circles are sites where bird surveys and habitat measurements were conducted.

Point counts were made once per month at each site, March–June 1995. Point count survey methods generally followed the recommendations of Ralph et al. (1993). Points were placed systematically 125–150 m apart, within the riparian habitat, halfway between the upland habitat and the river's edge. The number of points varied from one to five per site depending on patch size.

We integrated the data from all survey techniques into a composite list of species detections and abundance estimates for each day of survey for each patch. This was done by pooling species lists (e.g., species were included as present in the patch if they were found during any of the surveys on that day) and using the highest abundance estimate for each species from any one technique. These results, in turn, were aggregated to provide a composite data set of species presence and abundance for each patch, by month. These data sets were then fine-tuned through the decision rule process outlined below.

Decision Rules for Determining Breeding Status and Abundance

Raw data from surveys provide information on the total number of species and birds detected; however, even during the breeding season, these totals include migrants as well as local breeders (Felley and Sogge, this volume). Therefore, survey totals and/or high counts are subject to potentially high variability due to yearly or seasonal changes in the number of migrants, even if the number of local breeders remains nearly constant. Inclusion of migrants can also inflate calculated values for species richness and abundance, and alter avian community/habitat relationship models (Wiens 1989a; Morrison et al. 1992). These considerations led us to establish a set of "decision rules" (Table 1), tailored to each species, that

Table 1. Decision rules used to determine whether a species was considered a breeder at a particular patch and to estimate the number of breeding pairs of that species.

Species	Criteria for Breeder Status in Patch	Calculation of Number of Breeding Pairs
Ash-throated Flycatcher	Observed on two visits, not in July	Highest count from April or May divided by two
Brown-crested Flycatcher	Observed	Highest count divided by two
Black-chinned Hummingbird	Observed	Highest count of any sex or highest count divided by two
Bell's Vireo	Observed on two visits	Highest count of singing males, obvious pairs, or nests found
Bewick's Wren	Observed on two visits	Highest count of singing males, obvious pairs, or nests found
Blue-gray Gnatcatcher	Observed on two visits	Highest count from April or May, divided by two
Brown-headed Cowbird	Observed	Highest count of males or females
Blue Grosbeak	Observed on two visits	Highest count of singing males or obvious pairs
Costa's Hummingbird	Observed	Highest number of males observed
Common Yellowthroat	Observed on two visits	Highest count of singing males or obvious pairs common to two months
House Finch	Observed on two visits	Highest count of singing males or obvious pairs common to two months
Hooded Oriole	Observed on two visits, not in July	Highest number of males or females from any visit
Lesser Goldfinch	Observed in April or May	Highest number of singing males or total count, divided by two
Lucy's Warbler	Observed on two visits	Highest number of singing males, obvious pairs, or count divided by two, except June or July
Mourning Dove	Observed on two consecutive visits	Highest number of singing males or obvious pairs
Northern Mockingbird	Observed on two consecutive visits	Highest number of singing males or obvious pairs from two visits
Northern Oriole	Observed on two visits, not in July	Highest number of males or females from any visit, not July
Phainopepla	Observed on two visits	Highest number of males or obvious pairs from any visit
Rufous-crowned Sparrow	Observed on two visits	Highest number of males or obvious pairs from any visit
Say's Phoebe	Observed on two consecutive visits	Highest count common to two visits, divided by two
Summer Tanager	Observed on two visits	Highest number of singing males or obvious pairs
Yellow-breasted Chat	Observed on two visits	Highest number of singing males common to two visits
Yellow Warbler	Observed on two visits	Highest number of singing males common to two visits, excluding May

allowed us to develop a more conservative and accurate estimate of the type of breeding species and the number of breeding pairs in each patch. In some cases, these decision rules involved using information gathered from nest searches and/or banding. We used the maximum count for each species (adjusted per the decision rules), which provides a better approximation of actual species abundance than does an average count value (Dieni and Jones 2002). General bird survey techniques do not allow for confirmation of pairing status for all individuals at every study site; therefore, we estimated the number of pairs in a patch by integrating data on numbers of males, females, and nests.

Habitat Mapping and Measurements

The standardized habitat measurement techniques often used for vegetation studies are not always the most appropriate protocols for characterizing habitat parameters important for animals (Morrison et al. 1992). This is particularly true in avian habitat studies, which have generally found that most birds are not responding to fine-scale habitat features (Wiens 1989a). Therefore, it is best to select or develop habitat variables and measurement protocols tailored to the objectives of a particular study. We developed our own methodology to measure macro-habitat scale features such as patch size, vegetative structure, and composition. In order to do so, we delineated vegetation types and structure on aerial photographs of each patch, estimated density and average height of shrub and tree vegetation, georeferenced the aerial photographs using GPS and GIS, and determined the aerial extent and/or volume of different vegetation and structural types.

We procured color xerox copies of 1:2400 aerial photographs (GCES photo series 5-29-95) for each survey site. While at each site, we delineated and labeled polygons of visually distinct vegetation types and non-vegetated areas. The classifications used to delineate vegetation/habitat categories are described in Table 2.

For each vegetation/habitat polygon, we recorded Braun-Blanquet cover estimates (Bonham 1989) for the dominant plant species in each of three structural layers: tree (woody vegetation > 2 m tall), shrub (woody vegetation < 2 m tall), and herb (nonwoody vegetation). We also estimated the mean height of the top and bottom of the tree and shrub layers.

While at each site, we used Trimble Pathfinder GPS units to record (as waypoints) the coordinates of three to four points at or near each patch. These coordinates were taken at sites that could be easily recognized on each aerial photograph. At each georeference point, we collected a minimum of 100 data points, to provide for effective differential correction once we returned from the field. Georeference locations were dispersed across the patch and/or photograph in order to minimize the proportional spatial error during georeferencing. The GPS points were later differentially corrected using data from a Trimble base station located in Flagstaff, Arizona. We used a digitizing pad to trace the delineated vegetation polygons and created ArcView coverages of the vegetation at each study site. The differentially corrected GPS points were transferred to these coverages, creating a georeferenced habitat theme.

We used ArcView to calculate the aerial extent (in square meters) of each of the delineated polygons for a patch. In order to calculate volume indices, we multiplied the areal extent of the vegetation or structural type (e.g., tamarisk, tree) by the vertical thickness (mean height of top minus mean height of bottom) of the vegetation layer. This was weighted by the Braun-Blanquet cover estimate (weighting factor ranging from 0 to 1). This produced a unitless volume index that increased with increasing area, density, and/or vertical thickness of that habitat type. In order to characterize different habitat and structural types, we combined the habitat types listed in Table 2 into a variety of categories, which were used as covariates in further analyses (Table 3).

Table 2. Description of habitat types used to classify and characterize vegetation components at avian community–habitat study sites along the riparian corridor in the Grand Canyon during 1995. (S = shrub, T = tree, H = herb; N = new high water zone, O = old high water zone.)

Code	Description	Life Form	Water Zone
a	Arrowweed: Monotypic arrowweed. Pioneer on beach/sandy sites. Always a shrub. Canyon-wide.	S	N
abt	Arrowweed-Baccharis-Tamarisk: Mixture of a, b, and t; shrub form, linear along river, no obvious dominant.	S	N
ag	Arrowweed-Grass: Mix of a & g, seral stage where grass moves into arrowweed, shrub form, relatively open with % cover dominated by grass, linear along river.	S	N
ax	Arrowweed–Desert Broom: Shrub form, usually a linear stretch between OHWZ and NHWZ, usually dominated by arrowweed but relative amounts vary.	S	N
b	Baccharis: Monotypic or dominant seepwillow, tall shrubby form, deep and relatively wet soil.	S	N
bmt	Baccharis-Mesquite-Tamarisk: Large area of shrubby b and t, with scattered larger mesquites; mesquite is minor component. Everything but mesquite is shrub form.	S	N
bt	Baccharis-Tamarisk: Mixture of b and t, no obvious or consistent dominant; shrub form, linear along river in wet sites.	S	N
btw	Baccharis-Tamarisk-Willow: Mixture of b, w, and t, no obvious or consistent dominant—usually an even mix (at least 20% of any one); shrub form, linear along river in wet sites.	S	N
bw	Baccharis-Willow: Mixture of b and w, no obvious or consistent dominant—usually an even mix; shrub form, linear along river in wet sites.	S	N
btx	Baccharis-Tamarisk–Desert Broom: Mixture of b, t, and x, usually a mix with no obvious dominant; shrub form, linear along river, only along lower river.	S	N
c	ACACIA: Basically OHWZ with northern exposure where mesquite does not grow. Acacia dominated, but never monotypic (usually with some upland species). Rocky substrate, dry sites. Varies from shrub to low trees. Open stands, low density.	T	O
d	Sand, Beach, Dune, Rock, Talus: No vegetation.	n/a	n/a
e	*Equisetum*-Sedge-Shore: No woody vegetation, very wet sites within inundation zone. Generally dominated by *Equisetum*, but sometimes with other wetland species.	H	n/a
f	Debris Fan: Rocky/boulder substrate with very sparse, dispersed shrubby vegetation (usually a mix of species such as desert broom, tamarisk, seepwillow, willow).	S	n/a
g	Grass: Monotypic grass (predominantly Bermuda grass and red brome).	H	O
h	Hackberry (*Celtis reticulata*): Monotypic hackberry, small stands in shady sites of OHWZ, tree form, possibly important for some birds such as Summer Tanager.	T	O
k	Cliffrose: Monotypic cliffrose, shady sites in OHWZ along upper canyon, shrub form.	S	O
m	Mesquite: Monotypic or dominant mesquite, often in savannah landscape; tree forms typically in deeper well-watered soil; shrub form usually higher in slopes in dryer/rockier sites with higher stem density (but high canopy cover) than tree form.	T	O
p	Marsh, Reeds, Cattail: Wetland vegetation dominated by or with monotypic emergent vegetation; dominated by phragmites (usually) or cattail; often in return channel configuration.	H	N
t	Tamarisk: Monotypic tamarisk, almost always a tree form when monotypic. Overwhelmingly NHWZ.	T	N
tw	Tamarisk-Willow: Mixture of t and w, no consistent dominant, only one site.	S	N
tx	Tamarisk–Desert Broom: Mixture of t and x, only in lower canyon. Shrub form. No consistent dominant. Usually linear between NHWZ and OHWZ. Structurally similar to ax.	S	N
w	Willow: Monotypic coyote willow, usually shrub form. In wetter sites, usually not linear.	S	N
x	Desert Broom: Monotypic but not closed canopy/total cover. Intermediate band between OHWZ and NHWZ much like ax and tx. Shrub form.	S	N
xy	Desert Broom–Shrub: Desert broom component but not overwhelmingly dominant; intermediate band, similar to x category.	S	N
y	Shrub-*Baccharis*?: Very short shrub form, similar to *Baccharis*, dispersed in sandy/rocky substrate.	S	N

Table 3. Response variables, factors, and covariates used in analyzing breeding bird community and habitat relationships along the riparian corridor in the Grand Canyon, 1994 and 1995. Refer to Table 2 for descriptions of habitat type codes noted under "Covariates" below.

Name	Description
Response Variables	
ABUND	Number of breeding pairs for each patch.
RICH	Number of breeding species for each patch.
SDI	Shannon Diversity Index for each patch, calculated per MacArthur and MacArthur (1961).
Factors	
YEAR	1994 or 1995.
SEG	River segment: A = RM 0–77; B = RM 117–132; C = RM 171–224 (river miles per Stevens 1983).
Covariates	
AREA	Total patch area (sq m).
VEG	Total vegetation area (sq m) for each patch; AREA minus area of habitat type d.
TREE	Tree area (sq m) for each patch; sum of the areas for vegetation types c + h + m + t.
SHRB	Shrub area (sq m) for each patch; sum of the areas for vegetation types shrub = a + ag + ax + abt + bw + btw + btx + bt + b + tw + tx + x + xy + y + w + bmt + f + k.
HERB	Herb area (sq m) for each patch; sum of the areas for vegetation types e + g + p.
NHWZ	Total tree area (sq m) for each patch; sum of the areas for vegetation types a + ag + ax + abt + bw + btw + btx + bt + b + tw + tx + x + xy + y + w + bmt + p + t.
OHWZ	Total tree area (sq m) for each patch; sum of the areas for vegetation types c + h + g + k + m.
TREEVOL	Tree volume per patch calculated as per text.
SHRBVOL	Shrub volume per patch calculated as per text.
TAMVOL	Tamarisk volume per patch calculated as per text.
MESQVOL	Mesquite volume per patch calculated as per text.
ARROW	Patch area (sq m) that contained arrowweed; sum of the areas for vegetation types a + abt + ag + ax.
BACH	Patch area (sq m) that contained *Baccharis* spp.; sum of the areas for vegetation types abt + b + bmt + bt + btw + bw + btx.
GRASS	Patch area (sq m) that contained grass; sum of the areas for vegetation types ag + g.
MESQ	Patch area (sq m) that contained mesquite; sum of the areas for vegetation types bmt + m.
TAM	Patch area (sq m) that contained tamarisk; sum of the areas for vegetation types abt + bmt + bt + btw + btx + t + tw + tx.
WILLOW	Patch area (sq m) that contained willow species.; sum of the areas for vegetation types btw + bw + tw + w.

Data Integration

Integrating two years (1994 and 1995) of bird survey data with the one year (1995) of vegetation data required several assumptions: (1) There were no between-year differences in the area of a given surveyed patch (e.g. due to changes in the bird survey route between years). This assumption was met because survey patches were relocated during each visit, such that the same patch was surveyed each visit and / or year.

(2) There were no between-year differences in the size or location of any given patch (e.g. due to different observers delineating boundaries differently between years). This assumption was met because survey boundaries were standardized and indicated on aerial photographs of each site, and we surveyed the entire patch area on each visit.

(3) There were no between-year differences in the proportion or location of the 1995 vegetation classes (e.g. due to major changes in the actual vegetation, such as a new moist-soil vegetation type emerging between years in response to water level fluctuations or bank erosion). We based this assumption on our familiarity with each survey patch, and on comparisons of aerial photographs between 1994 and 1995.

General Analysis Methods

All data manipulations and analyses were conducted using SPSS version 7.5, following statistical methods and tests described in Sokal and Rolf (1995). Variables of interest include three response variables, two factors, and 17 covariates (Table 3). The number of breeding pairs per patch (ABUND) was $\log_{10}$ transformed and the Shannon Diversity Index (SDI) was squared to meet the assumptions of general linear models. No transformations were necessary for the number of breeding species (RICH). Covariates were calculated from original vegetation type classifications as noted in Table 3.

The non-parametric Spearman's correlation coefficient was used for all bivariate correlations of untransformed variables. Techniques associated with general linear models (analysis-of-variance, ANOVA and analysis-of-covariance, ANCOVA) were used to evaluate factors and covariates. Similar analyses were conducted separately for each response variable. The factors were evaluated first using ANOVA in a step-down approach beginning with a full-factorial model that included YEAR, SEG, and their interaction. The covariates were then added to the full-factorial model and evaluated using ANCOVA in the presence of the factors.

Analyses were conducted separately for each covariate as follows: The data were fit to a series of seven models beginning with the most general model (Model 1: YEAR, SEG, the covariate, and their interactions), and proceeding in a step-down approach analogous to backwards multiple regression. The three-way interaction was tested first and dropped from the model if it was not significant ($P < 0.05$). Next, each two-way interaction was sequentially dropped from the model (Models 2–4), beginning with the least significant (highest P-value). Then, each main effect was dropped from the model (Models 5–6), beginning with the least significant (highest P-value), until all main effects were significant ($P < 0.05$). Finally, the interaction of the remaining main effects were added back to the model (Model 7). A "best" model was determined for each covariate. Consequently, the relative importance of each covariate (in the presence of the factors) was noted for each response variable. This approach was feasible because most of the covariates were significantly and highly correlated with each other.

RESULTS

Overall Bird Community Parameters

Because simply detecting a species in a patch during the breeding season does not prove that it is a local breeder, we used species-specific decision rules (refer to Methods section) to more accurately determine what species and how many pairs were breeding in each patch. This minimizes the likelihood of including non-breeding species or migrant individuals, and thereby inflating resident breeding population estimates.

Abundance (number of breeding pairs of all species combined) ranged from 0 to 57 (Table 4), and was generally lower at smaller patches. Only five of the patches we surveyed supported more than 20 breeding pairs, and most had 10 or fewer. Spring Canyon had the greatest number of breeding pairs (50 in 1994 and 57 in 1995). The number of breeding pairs was similar between years at some patches, but varied widely at others (Table 4).

Richness (number of breeding species in a patch) ranged from 0 (at the smallest patches) to 14 (Table 4), with most patches having between two and nine breeding species. Spring Canyon, one of our largest patches, had the highest species richness. The number of breeding species detected was relatively stable among years at many patches, but varied substantially at others (Table 4).

Shannon Diversity Indices varied from 0 to 2.39, though most were between 1.0 and 2.0 (Table 4). The dominance of values in the range of 1.0 to 2.0 suggests that, in most patches, there were similar numbers of breeding pairs among the different breeding species. At many of the patches with high SDI values (e.g., Parashant, Spring Canyon), a small number of species contributed a large proportion of the number of breeding pairs.

Overall Vegetation/Habitat Parameters

The characteristics of the habitat patches used in this study varied widely (Table 5). Patches ranged from very small (0.002 ha) to large (7.4 ha), with most between 0.5 and 2 ha. The absolute and relative amounts of different habitat types (e.g., Old High Water Zone [OHWZ], New High Water Zone [NHWZ], tamarisk, mesquite) and vegetative structure (e.g., area and volume of tree, shrub) also varied widely between patches. Larger patches in the upper portion of the river tended to be dominated by tamarisk vegetation, whereas many larger patches in the lower segment included a major mesquite component.

All but one patch included tree or shrub habitat, usually tamarisk and/or mesquite. Herbaceous understory was present to some degree in every patch. Overall, the variety of habitats, configurations, sizes, and geographic locations provides for a wide range of habitat characteristics to consider in modeling bird community–habitat relationships.

Bird Community and Habitat Relationships: Breeding Pair Abundance and Habitat

Differences between years. Overall mean breeding pair abundance at sites did not differ (one-way ANOVA, $F = 0.425$, $df = 1$, $P = 0.516$) between 1994 (10.8 ± 1.6) and 1995 (10.7 ± 1.3). Additionally, the lack of year differences was consistent among river segments (test of YEAR*SEG interaction in full factorial model, $F = 0.596$, $df = 2$, $P = 0.554$).

There was also no evidence of year differences in the abundance of breeding pairs per patch for any analysis that included a single covariate. For each series of the seven models that we ran for each covariate, there was no significant three-way interaction ($P \geq 0.217$) among YEAR, SEG, and the covariate (see Model 1 of Table 6). Likewise, we found no significant two-way interactions ($P \geq 0.061$) between YEAR and SEG, or YEAR and the covariate (see Models 2–4 of Table 6). Finally, there was no significant contribution by YEAR ($P \geq 0.319$) in the main effects only models (see Model 5 of Table 6). We therefore conclude that the abundance of breeding pairs per patch did not differ between 1994 and 1995 for any segment of the river.

Differences among river segments. Bird abundance differed by river segment (one-way ANOVA, $F = 9.6$, $df = 2$, $P < 0.001$). Abundance for river segment B (4.5 ± 1.1) was lower than for river segment A (9.0 ± 1.0) and river segment C (13.2 ± 1.7; Bonferroni's multiple comparisons, $P = 0.021$ and < 0.001 respectively), while river segment A did not differ from river segment C ($P = 0.084$).

Table 4. Total number of breeding species, number of breeding pairs (all species combined), and Shannon Diversity Index (SDI) for each avian community–habitat survey patch, 1994–1995. Determination of status as a breeding species and calculation of the number of breeding pairs based on a set of "decision rules" (refer to text). SDI calculated per MacArthur and MacArthur (1961). River miles follow Stevens (1983). River segments A–C responded differently in bird community–habitat models (refer to text).

	1994			1995		
Site	# Breeding Species	# Breeding Pairs	SDI	# Breeding Species	# Breeding Pairs	SDI
River Segment A						
1.0 R (Paria)	11	14	2.34	9	10	2.16
1.6 R	6	6	1.79	2	2	0.69
2.0 L	–	–	–	3	4	1.04
3.7 L	–	–	–	3	3	1.10
5.1 L	5	5	1.61	3	3	1.10
5.2 R	5	7	1.55	8	9	2.04
46.0 L	–	–	–	8	13	1.93
46.7 R (Saddle)	11	22	2.24	11	23	2.12
47.5 L	5	6	1.56	2	6	0.69
48.5 L	3	5	1.05	4	5	1.33
49.1 R	6	13	1.52	8	16	1.81
49.2 L	4	6	1.24	8	9	2.04
50.0 R	10	15	2.15	7	20	1.73
73.9 R	8	10	2.02	3	4	1.04
74.1 R	5	6	1.56	3	4	1.04
74.3 R	4	5	1.33	3	3	1.10
74.4 L	–	–	–	9	15	2.06
76.5 L	–	–	–	8	9	2.04
River Segment B						
117.5 R	–	–	–	3	3	1.1
119.5 R	–	–	–	1	1	0
122.8 L	6	9	1.67	8	8	2.08
125.5 R	1	2	0	2	4	0.69
131.3 R	5	7	1.47	1	5	0
River Segment C						
167.0 R	5	5	1.61	5	5	1.61
171.0 R (Stairway)	11	14	2.34	8	13	1.99
171.1 R	10	13	2.24	6	7	1.75
172.2 L	6	9	1.74	9	13	2.10
173.1 R	7	10	1.75	6	8	1.67
174.4 R	6	9	1.68	7	11	1.85
174.5 R	–	–	–	5	7	1.55
174.7 R	5	5	1.61	4	4	1.39
197.3 L	–	–	–	11	17	2.26
197.6 L	–	–	–	9	12	2.09
198.0 R (Parashant)	13	29	2.34	14	30	2.34
198.2 L	–	–	–	12	20	2.32
198.3 R	–	–	–	9	12	2.14
199.5 R	10	10	2.30	5	5	1.61
200.0 L	5	7	1.55	6	9	1.74
200.4 R	5	6	1.56	8	12	1.98
200.5 R	12	16	2.39	10	16	2.19
202.5 R	7	10	1.83	7	12	1.82
204.1 R	–	–	–	13	30	2.37
204.5 R (Spring)	18	50	2.51	15	57	2.36
206.5 L	3	5	1.05	6	8	1.67
208.7 R	7	11	1.85	9	14	2.07
213.6 L	–	–	–	7	10	1.83
214.0 L	–	–	–	10	13	2.25
214.2 L	–	–	–	2	2	0.69
224.0 L	–	–	–	7	8	1.91
224.1 R	–	–	–	3	4	1.04

Table 5. Select habitat variables (see text) for each avian community–habitat survey patch (in sq m) in 1994–1995. River mile designations follow Stevens (1983). River segments A–C responded differently in the bird community–habitat models (refer to text).

Site	Total Vegetated Area	Area NHWZ	Area OHWZ	Tree Area	Shrub Area	Tree Volume Index	Shrub Volume Index	Tamarisk Volume Index	Mesquite Volume Index
River Segment A									
1.0 R (Paria)	19,004	18,552	0	15,665	2,857	40,154	17,685	53,261	0
1.6 R	7,692	7,321	153	0	7,321	11,257	6,617	0	0
2.0 L	9,804	1,614	7,644	1,110	3,391	1,624	4,049	1,776	0
3.7 L	10,575	4,111	2,078	3,728	2,016	9,472	1,911	10,345	0
5.1 L	4,095	1,807	571	1,807	0	7,228	367	7,589	0
5.2 R	7,891	6,910	346	4,567	2,343	12,960	1,274	9,796	0
46.0 L	24,914	13,023	11,891	13,768	11,146	33,226	10,140	6,710	11,296
46.7 R (Saddle)	41,346	25,045	14,237	29,629	7,405	80,794	14,259	65,84	13,528
47.5 L	6,982	2,530	2,842	3,854	2,974	8,219	2,349	3,127	0
48.5 L	6,556	3,093	3,396	3,396	3,093	11,956	1,916	0	7,556
49.1 R	38,978	10,337	25,298	29,168	7,969	51,587	18,180	15,056	39,148
49.2 L	16,714	4,342	10,855	9,433	5,794	20,512	7,939	0	18,394
50.0 R	55,303	20,303	26,254	39,865	15,276	106,086	23,142	58,868	45,944
73.9 R	3,485	3,177	0	896	2,282	5,514	2,053	2,890	0
74.1 R	6,173	2,003	0	1,684	4,489	7,478	2,245	4,842	0
74.3 R	2,102	2,009	0	0	2,009	1,356	2,232	0	0
74.4 L	37,882	19,431	15,354	18,217	19,369	47,138	27,964	8,088	31,476
76.5 L	16,842	3,917	12,925	12,925	3,917	14,727	7,520	0	14,217
River Segment B									
117.5 R	217	203	0	0	161	326	74	0	0
119.5 R	3,471	734	2,556	2,556	734	2,761	984	0	2,556
122.8 L	8,424	4,703	0	935	7,195	8,456	3,796	2,688	0
125.5 R	1,171	305	0	305	865	1103	612	831	0
131.3 R	6,973	2,257	4,198	4,198	2,257	3,830	3,555	0	0
River Segment C									
167.0 R	5,628	2,848	2,634	2,634	2,848	937	2,982	0	1,870
171.0 R (Stairway)	29,985	7,822	7,220	8,426	20,614	25,303	17,175	3,437	19,675
171.1 R	12,478	5,305	2,092	2,092	10,046	9,825	2,861	0	3,400
172.2 L	14,970	10,154	4,622	6,122	8,653	13,702	9,215	4,050	7,626
173.1 R	10,487	2,845	2,758	3,078	7,296	17,043	6,717	672	9,308
174.4 R	22,775	5,713	7,288	11,037	11,738	26,826	12,768	10,558	13,665
174.5 R	14,594	5,457	6,819	6,267	4,280	12,849	7,693	9,087	4,089
174.7 R	6,119	1,735	4,203	4,203	1,735	6,815	3,984	0	8,196
197.3 L	9,727	3,692	5,579	7,251	2,020	16,341	4,424	10,654	6,805
197.6 L	7,112	2,873	4,005	5,049	1,829	9,153	3,413	2,798	6,174
198.0 R (Parashant)	23,576	6,895	16,267	16,214	3,400	41,689	15,918	18,853	34,800
198.2 L	4,828	2,381	2,593	3,978	596	9,970	2,759	4,819	6,872
198.3 R	8,829	3,079	4,872	6,708	1,243	13,056	6,781	6,426	11,206
199.5 R	10,204	3,260	4,850	4,850	5,354	16,666	7,343	9,393	8,730
200.0 L	21,247	11,658	9,588	11,985	9,262	14,360	19,486	6,829	15,341
200.4 R	10,195	3,288	5,719	6,593	2,864	10,400	4,466	744	8,864
200.5 R	17,453	9,182	6,124	10,217	6,268	23,155	13,091	15,758	10,717
202.5 R	17,614	5,024	12,297	12,297	5,024	25,024	15,437	0	34,432
204.1 R	41,429	16,256	23,699	29,120	10,834	48,560	22,526	17,834	48,857
204.5 R (Spring)	74,133	24,868	46,924	48,961	19,252	112,860	41,835	12,110	120,315
206.5 L	14,993	7,686	7,308	7,308	7,686	14,889	10,711	0	14,249
208.7 R	8,785	4,050	4,337	6,086	2,302	15,180	3,469	6,549	9,108
213.6 L	5,654	3,374	2,280	3,453	2,135	12,270	3,416	6,041	4,332
214.0 L	5,993	3,050	2,943	5,609	384	13,543	4,151	11,597	5,886
214.2 L	777	396	382	382	396	465	544	0	745
224.0 L	1,278	246	48	0	340	396	289	228	0
224.1 R	1,390	508	103	279	1,057	710	347	739	0

Table 6. Response variable $\log_{10}$(ABUND). Correlation coefficients and results of hypothesis tests of the effective contribution of terms used during the step-down modeling approach to evaluate the significance of individual covariates in the presence of factors YEAR and SEG on the transformed response variable $\log_{10}$(ABUND). The first line in each cell is the model correlation coefficient. The second line is the term in the model that had the highest P-value within the class of interaction (3-way, 2-way, or main effect), and this was the term that was dropped prior to running the model in the next column. The last line in each cell is the P-value of the specified term. Model 7 contains two main effects and their interaction, and differs from Model 4 which has all three main effects and the single interaction. Bolded entries contain P values < 0.1, while underlined entries indicate a good model for inference for each covariate. Comparisons within columns allow selection of the best covariates for a given model.

		Models with 2-Way Interactions			Models with Only Main Effects		
Covariate	Model 1: Full Factorial Model - 7 Parameters	Model 2: Three 2-way Interactions 6 Parameters	Model 3: Two 2-way Interactions 5 Parameters	Model 4: One 2-way Interaction 4 Parameters	Model 5: All 3 Main Effects Only	Model 6: Two Main Effects	Model 7: Model 6 with Interaction Added
AREA	$R^2 = 0.673$ 3-way $P = 0.493$	$R^2 = 0.666$ YEAR*SEG $P = 0.293$	$R^2 = 0.655$ YEAR*AREA $P = 0.205$	**$R^2 = 0.647$** **SEG*AREA** **$P = 0.017$**	$R^2 = 0.607$ YEAR $P = 0.576$	**$R^2 = 0.605$** **SEG & AREA** **$P < 0.001$**	**$R^2 = 0.646$** **SEG*AREA** **$P = 0.015$**
VEG	$R^2 = 0.688$ 3-way $P = 0.404$	$R^2 = 0.679$ YEAR*SEG $P = 0.269$	$R^2 = 0.667$ YEAR*VEG $P = 0.162$	**$R^2 = 0.658$** **SEG*VEG** **$P = 0.019$**	$R^2 = 0.620$ YEAR $P = 0.563$	**$R^2 = 0.619$** **SEG & VEG** **$P < 0.001$**	**$R^2 = 0.658$** **SEG*VEG** **$P = 0.017$**
TREE	$R^2 = 0.668$ 3-way $P = 0.368$	$R^2 = 0.658$ SEG*TREE $P = 0.934$	$R^2 = 0.657$ YEAR*SEG $P = 0.267$	**$R^2 = 0.645$** **YEAR*TREE** **$P = 0.061$**	$R^2 = 0.628$ YEAR $P = 0.468$	**$R^2 = 0.625$** **SEG & TREE** **$P < 0.001$**	$R^2 = 0.627$ SEG*TREE $P = 0.896$
SHRB	$R^2 = 0.454$ 3-way $P = 0.917$	$R^2 = 0.453$ YEAR*SHRB $P = 0.886$	$R^2 = 0.452$ YEAR*SEG $P = 0.348$	**$R^2 = 0.436$** **SEG*SHRB** **$P = 0.076$**	$R^2 = 0.396$ YEAR $P = 0.673$	**$R^2 = 0.395$** **SEG & SHRB** **$P \leq 0.001$**	**$R^2 = 0.435$** **SEG*SHRB** **$P = 0.076$**
HERB	$R^2 = 0.474$ 3-way $P = 0.831$	$R^2 = 0.472$ YEAR*SEG $P = 0.499$	$R^2 = 0.461$ YEAR*HERB $P = 0.252$	**$R^2 = 0.452$** **SEG*HERB** **$P = 0.025$**	$R^2 = 0.395$ YEAR $P = 0.319$	**$R^2 = 0.387$** **SEG & HERB** **$P < 0.001$**	**$R^2 = 0.446$** **SEG*HERB** **$P = 0.021$**
NHWZ	$R^2 = 0.655$ 3-way $P = 0.481$	$R^2 = 0.655$ YEAR*NHWZ $P = 0.343$	$R^2 = 0.651$ YEAR*SEG $P = 0.307$	**$R^2 = 0.639$** **SEG*NHWZ** **$P = 0.012$**	$R^2 = 0.594$ YEAR $P = 0.571$	**$R^2 = 0.593$** **SEG & NHWZ** **$P < 0.001$**	**$R^2 = 0.639$** **SEG*NHWZ** **$P = 0.010$**
OHWZ	$R^2 = 0.591$ 3-way $P = 0.324$	$R^2 = 0.578$ SEG*OHWZ $P = 0.882$	$R^2 = 0.577$ YEAR*SEG $P = 0.307$	**$R^2 = 0.563$** **YEAR*OHWZ** **$P = 0.094$**	$R^2 = 0.546$ YEAR $P = 0.414$	**$R^2 = 0.542$** **SEG & OHWZ** **$P < 0.001$**	$R^2 = 0.544$ SEG*OHWZ $P = 0.906$
TREEVOL	$R^2 = 0.707$ 3-way $P = 0.787$	$R^2 = 0.705$ YEAR*SEG $P = 0.314$	$R^2 = 0.695$ YEAR*TREEVOL $P = 0.122$	**$R^2 = 0.685$** **SEG*TREEVOL** **$P = 0.004$**	$R^2 = 0.635$ YEAR $P = 0.582$	**$R^2 = 0.634$** **SEG & TREEVOL** **$P < 0.001$**	**$R^2 = 0.685$** **SEG*TREEVOL** **$P = 0.003$**

Table 6 (continued)

Covariate	Model 1: Full Factorial Model - 7 Parameters	Models with 2-Way Interactions: Model 2: Three 2-way Interactions 6 Parameters	Models with 2-Way Interactions: Model 3: Two 2-way Interactions 5 Parameters	Models with 2-Way Interactions: Model 4: One 2-way Interaction 4 Parameters	Models with Only Main Effects: Model 5: All 3 Main Effects Only	Models with Only Main Effects: Model 6: Two Main Effects	Model 7: Model 6 with Interaction Added
SHRBVOL	$R^2 = 0.3.7$ 3-way $P = 0.783$	$R^2 = 0.634$ YEAR*SHRBVOL $P = 0.529$	$R^2 = 0.632$ YEAR*SEG $P = 0.251$	**$R^2 = 0.618$** **SEG*SHRBVOL** **$P = 0.008$**	$R^2 = 0.565$ YEAR $P = 0.628$	**$R^2 = 0.564$** **SEG & SHRBVOL** **$P \leq 0.003$**	**$R^2 = 0.617$** **SEG*SHRBVOL** **$P = 0.007$**
TAMVOL	$R^2 = 0.568$ 3-way $P = 0.790$	$R^2 = 0.565$ YEAR*SEG $P = 0.667$	$R^2 = 0.560$ YEAR*TAMVOL $P = 0.612$	**$R^2 = 0.559$** **SEG*TAMVOL** **$P < 0.001$**	$R^2 = 0.437$ YEAR $P = 0.482$	**$R^2 = 0.434$** **SEG & TAMVOL** **$P < 0.001$**	**$R^2 = 0.555$** **SEG*TAMVOL** **$P < 0.001$**
MESQVOL	$R^2 = 0.591$ 3-way $P = 0.217$	$R^2 = 0.582$ YEAR*SEG $P = 0.620$	$R^2 = 0.577$ YEAR*MESQVOL $P = 0.140$	**$R^2 = 0.564$** **SEG*MESQVOL** **$P = 0.012$**	$R^2 = 0.510$ YEAR $P = 0.529$	**$R^2 = 0.507$** **SEG & MESQVOL** **$P \leq 0.002$**	**$R^2 = 0.563$** **SEG*MESQVOL** **$P = 0.011$**
ARROW	$R^2 = 0.371$ 3-way $P = 0.546$	$R^2 = 0.360$ YEAR*ARROW $P = 0.976$	$R^2 = 0.360$ SEG*ARROW $P = 0.530$	$R^2 = 0.349$ YEAR*SEG $P = 0.327$	$R^2 = 0.329$ YEAR $P = 0.506$	**$R^2 = 0.325$** **SEG & ARROW** **$P \leq 0.004$**	$R^2 = 0.341$ SEG*ARROW $P = 0.402$
BACH	$R^2 = 0.365$ 3-way $P = 0.976$	$R^2 = 0.364$ YEAR*BACH $P = 0.776$	$R^2 = 0.364$ YEAR*SEG $P = 0.524$	$R^2 = 0.352$ SEG*BACH $P = 0.140$	$R^2 = 0.317$ YEAR $P = 0.459$	**$R^2 = 0.312$** **SEG & BACH** **$P \leq 0.001$**	$R^2 = 0.349$ SEG*BACH $P = 0.126$
GRASS	$R^2 = 0.297$ 3-way $P = 0.998$	$R^2 = 0.297$ SEG*GRASS $P = 0.763$	$R^2 = 0.296$ YEAR*SEG $P = 0.455$	$R^2 = 0.281$ YEAR*GRASS $P = 0.265$	$R^2 = 0.269$ YEAR $P = 0.358$	**$R^2 = 0.261$** **SEG & GRASS** **$P \leq 0.011$**	$R^2 = 0.263$ SEG*GRASS $P = 0.648$
MESQ	$R^2 = 0.604$ 3-way $P = 0.306$	$R^2 = 0.598$ YEAR*SEG $P = 0.541$	$R^2 = 0.591$ YEAR*MESQ $P = 0.100$	**$R^2 = 0.575$** **SEG*MESQ** **$P = 0.018$**	$R^2 = 0.527$ YEAR $P = 0.440$	**$R^2 = 0.524$** **SEG & MESQ** **$P \leq 0.001$**	**$R^2 = 0.574$** **SEG*MESQ** **$P = 0.015$**
TAM	$R^2 = 0.663$ 3-way $P = 0.892$	$R^2 = 0.662$ YEAR*TAM $P = 0.412$	$R^2 = 0.658$ YEAR*SEG $P = 0.350$	**$R^2 = 0.648$** **SEG*TAM** **$P <0.001$**	$R^2 = 0.558$ YEAR $P = 0.455$	**$R^2 = 0.555$** **SEG & TAM** **$P < 0.001$**	**$R^2 = 0.647$** **SEG*TAM** **$P < 0.001$**
WILLOW	$R^2 = 0.347$ 3-way $P = 0.852$	$R^2 = 0.344$ YEAR*WILLOW $P = 0.684$	$R^2 = 0.343$ YEAR*SEG $P = 0.646$	$R^2 = 0.335$ SEG*WILLOW $P = 0.104$	$R^2 = 0.293$ YEAR $P = 0.386$	**$R^2 = 0.286$** **SEG & WILLOW** **$P \leq 0.002$**	**$R^2 = 0.330$** **SEG*WILLOW** **$P = 0.094$**

Among the series of seven models we ran for each covariate, SEG was a significant factor in the best model for every covariate. The main effect SEG and/or the SEG*covariate interaction was always highly significant ($P \leq 0.021$; see underlined values in Table 6). We therefore conclude that the abundance of breeding pairs per patch differed by river segment.

Covariates. All covariates were positively correlated with ABUND; all except BACH, GRASS, and WILLOW were highly and significantly so (Spearman's rho ≥ 0.426, $P < 0.01$). Each individual covariate that we considered, in the presence of SEG, accounted for a significant proportion of the variation in abundance ($P \leq 0.021$; see underlined values in Table 6). For seven of the covariates (TREE, SHRB, OHWZ, ARROW, BACH, GRASS, WILLOW), the best model included the main effects for SEG and the respective covariates (see Model 6 in Table 6). For the remaining 10 covariates, the best model also included the SEG*covariate interaction (see Model 7 in Table 6). Covariates AREA, VEG, TREE, NHWZ, TREEVOL, SHRBVOL, and TAM (in the presence of SEG) accounted for the greatest proportion of the variation in abundance ($0.605 \leq R^2 \leq 0.685$; see Models 6 and 7 in Table 6). Among the significant interactions, abundance increased with AREA, VEG, HERB, NHWZ, TREEVOL, and SHRBVOL at a greater rate for river segment B than for river segments A and C (Figure 2A). Abundance increased with TAM and TAMVOL at a greater rate for river segments B and C than for segment A (Figure 2B). For covariates MESQ and MESQVOL, the interaction term was not meaningful because mesquite was found at only one patch in segment B.

We therefore conclude that covariates associated with total patch size (total area and vegetated area) and covariates associated with areas and volumes composed primarily of large vegetation structures (tree area, NHWZ area, tamarisk area, tree volume, and shrub volume) have the greatest potential, among those considered, to predict abundance. Additionally, abundance increased at the greatest rate with most of the vegetation variables for river segment B, but increased at a lesser rate for tamarisk area and volume for river segment A.

Bird Species Richness and Habitat

Differences between years. Overall mean bird species richness per patch did not differ (one-way ANOVA, $F = 0.300$, $df = 1$, $P = 0.585$) between 1994 (8.0 ± 0.6) and 1995 (6.6 ± 0.5). Additionally, the lack of year differences was consistent among river segment (test of YEAR*SEG interaction in full factorial model, $F = 0.073$, $df = 2$, $P = 0.929$).

There was also no evidence of year differences in the richness of breeding pairs per patch for any analysis that included a single covariate. For each series of the seven models we ran for each covariate, there was no significant three-way interaction ($P \geq 0.271$) among YEAR, SEG, and the covariate (see Model 1 of Table 7). Likewise, we found no significant two-way interactions ($P \geq 0.099$) between YEAR and SEG, or YEAR and the covariate (see Models 2–4 of Table 7). Finally, there was no significant contribution by YEAR ($P \geq 0.371$) in the main effects only models (see Model 5 of Table 7). We therefore conclude that the richness of breeding pairs per patch did not differ between 1994 and 1995 for any segment of the river.

Differences among river segments. Species richness differed by river segment (one-way ANOVA, $F = 8.7$, $df = 2$, $P < 0.001$). Species richness for river segment C (8.0 ± 0.5) was greater than for river segment A (6.0 ± 0.5) and river segment B (3.4 ± 0.9; Bonferroni's multiple comparisons, $P = 0.026$ and 0.001 respectively), whereas river segment A did not differ from segment B ($P = 0.128$).

Among the series of seven models run for each covariate, SEG was a significant factor in the best model for every covariate. The main effect SEG and/or the SEG*covariate interaction was always highly significant ($P \leq 0.019$; see underlined values in Table 7). We therefore conclude that species richness of breeding pairs per patch differed by river segment.

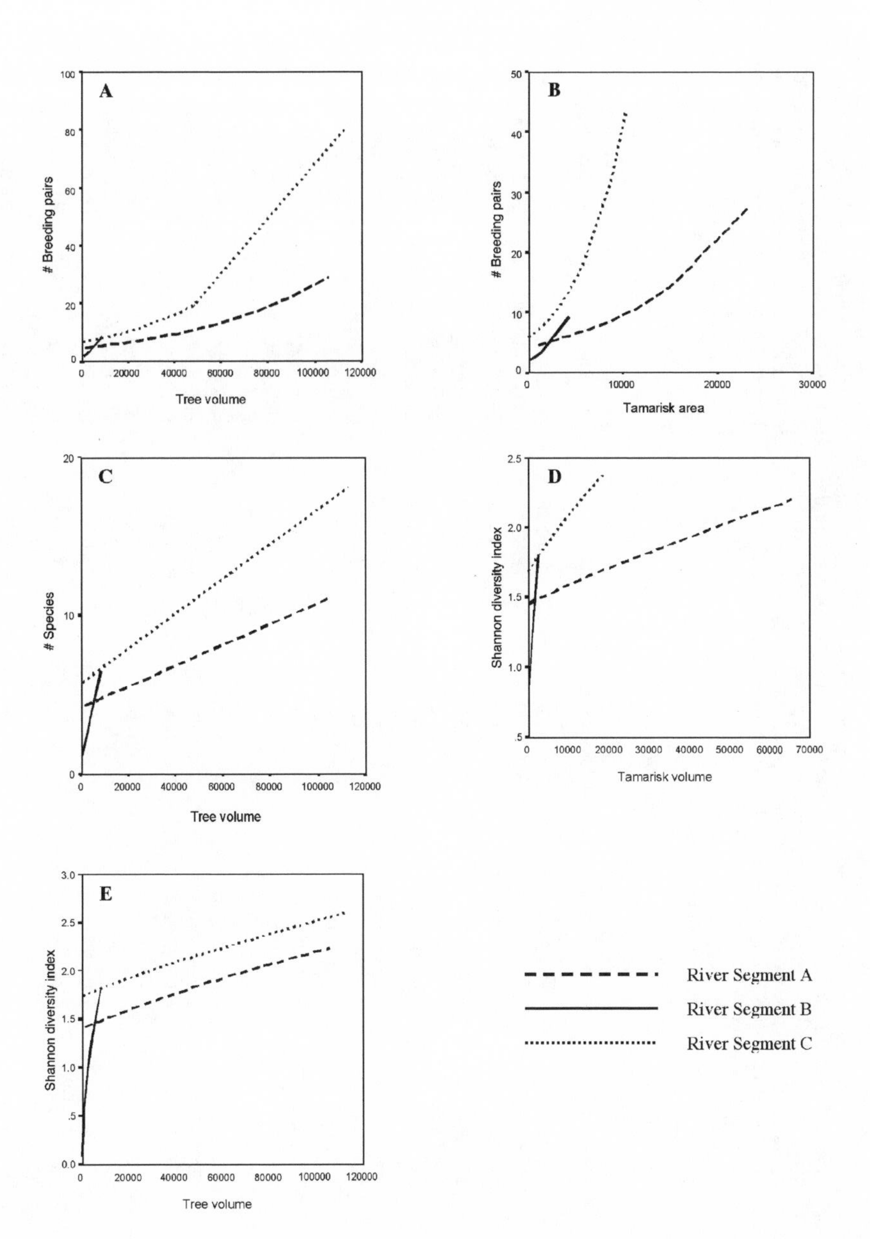

Figure 2. (A) Predicted values of model $\log_{10}$ (ABUND) = SEG + TREEVOL + (SEG*TREEVOL). Tree volume index calculations are described in the text. (B) Predicted values from model $\log_{10}$ (ABUND) = SEG + TAM + (SEG*TAM). Tamarisk area is in square meters. (C) Predicted values for model RICH = SEG + TREEVOL + (SEG*TREEVOL). Tree volume index calculations are described in the text. RICH showed similar patterns with TAM and TAMVOL. (D) Predicted values of model SDI2 = SEG + TAMVOL + (SEG*TAMVOL). Tamarisk volume index calculations are described in the text. (E) Predicted values for model SDI2 = SEG + TREEVOL + (SEG*TREEVOL). Tree volume index calculations are described in the text.

Table 7. Response variable RICH. Correlation coefficients and results of hypothesis tests of the effective contribution of terms used during the step-down modeling approach to evaluate the significance of individual covariates in the presence of factors YEAR and SEG on the response variable RICH. The first line in each cell is the model correlation coefficient. The second line is the term in the model that had the highest P-value within the class of interaction (3-way, 2-way, or main effect), and this was the term that was dropped prior to running the model in the next column. The last line in each cell is the P-value of the specified term. Model 7 contains two main effects and their interaction, and differs from Model 4 which has all three main effects and the single interaction. Bolded entries contain P values < 0.1, while underlined entries indicate a good model for inference for each covariate. Comparisons within columns allow for selecting the best covariates for a given model.

		Models with 2-Way Interactions			Models with Only Main Effects		
Covariate	Model 1: Full Factorial Model - 7 Parameters	Model 2: Three 2-way Interactions 6 Parameters	Model 3: Two 2-way Interactions 5 Parameters	Model 4: One 2-way Interaction 4 Parameters	Model 5: All 3 Main Effects Only	Model 6: Two Main Effects	Model 7: Model 6 with Interaction Added
AREA	$R^2 = 0.522$ 3-way $P = 0.547$	$R^2 = 0.545$ YEAR*AREA $P = 0.658$	$R^2 = 0.543$ YEAR*SEG $P = 0.630$	$R^2 = 0.538$ SEG*AREA $P = 0.182$	$R^2 = 0.516$ YEAR $P = 0.660$	**$R^2 = 0.515$** **SEG & AREA** **$P < 0.001$**	$R^2 = 0.537$ SEG*AREA $P = 0.167$
VEG	$R^2 = 0.557$ 3-way $P = 0.472$	$R^2 = 0.548$ YEAR*VEG $P = 0.720$	$R^2 = 0.547$ YEAR*SEG $P = 0.614$	$R^2 = 0.541$ SEG*VEG $P = 0.291$	$R^2 = 0.526$ YEAR $P = 0.650$	**$R^2 = 0.524$** **SEG & VEG** **$P < 0.001$**	$R^2 = 0.540$ SEG*YEAR $P = 0.271$
TREE	$R^2 = 0.568$ 3-way $P = 0.345$	$R^2 = 0.554$ YEAR*TREE $P = 0.884$	$R^2 = 0.554$ YEAR*SEG $P = 0.713$	$R^2 = 0.550$ SEG*TREE $P = 0.301$	$R^2 = 0.535$ YEAR $P = 0.568$	**$R^2 = 0.533$** **SEG & TREE** **$P < 0.001$**	$R^2 = 0.548$ SEG*TREE $P = 0.292$
SHRB	$R^2 = 0.360$ 3-way $P = 0.764$	$R^2 = 0.355$ YEAR*SEG $P = 0.596$	$R^2 = 0.346$ HEAR*SHRB $P = 0.450$	$R^2 = 0.341$ SEG*SHRB $P = 0.366$	$R^2 = 0.323$ YEAR $P = 0.708$	**$R^2 = 0.322$** **SEG & SHRB** **$P \leq 0.002$**	$R^2 = 0.340$ SEG*SHRB $P = 0.360$
HERB	$R^2 = 0.462$ 3-way $P = 0.745$	$R^2 = 0.458$ YEAR*SEG $P = 0.870$	**$R^2 = 0.456$** **YEAR*HERB** **$P = 0.099$**	**$R^2 = 0.435$** **SEG*HERB** **$P = 0.044$**	$R^2 = 0.386$ YEAR $P = 0.371$	**$R^2 = 0.380$** **SEG & HERB** **$P < 0.001$**	**$R^2 = 0.430$** **SEG*HERB** **$P = 0.040$**
NHWZ	$R^2 = 0.574$ 3-way $P = 0.900$	$R^2 = 0.573$ YEAR*NHWZ $P = 0.715$	$R^2 = 0.572$ YEAR*SEG $P = 0.678$	$R^2 = 0.568$ SEG*NHWZ $P = 0.146$	$R^2 = 0.545$ YEAR $P = 0.659$	**$R^2 = 0.544$** **SEG & NHWZ** **$P < 0.001$**	$R^2 = 0.567$ SEG*NHWZ $P = 0.134$
OHWZ	$R^2 = 0.497$ 3-way $P = 0.271$	$R^2 = 0.478$ YEAR*OHWZ $P = 0.860$	$R^2 = 0.478$ YEAR*SEG $P = 0.675$	$R^2 = 0.472$ SEG*OHWZ $P = 0.274$	$R^2 = 0.454$ YEAR $P = 0.508$	**$R^2 = 0.451$** **SEG & OHWZ** **$P \leq 0.001$**	$R^2 = 0.470$ SEG*OHWZ $P = 0.259$
TREEVOL	$R^2 = 0.618$ 3-way $P = 0.782$	$R^2 = 0.615$ YEAR*TREEVOL $P = 0.817$	$R^2 = 0.615$ YEAR*SEG $P = 0.740$	**$R^2 = 0.612$** **SEG*TREEVOL** **$P = 0.012$**	$R^2 = 0.563$ YEAR $P = 0.673$	**$R^2 = 0.562$** **SEG & TREEVOL** **$P < 0.001$**	**$R^2 = 0.611$** **SEG*TREEVOL** **$P = 0.011$**

Table 7 (continued)

Covariate	Model 1: Full Factorial Model - 7 Parameters	Model 2: Three 2-way Interactions 6 Parameters	Model 3: Two 2-way Interactions 5 Parameters	Model 4: One 2-way Interaction 4 Parameters	Model 5: All 3 Main Effects Only	Model 6: Two Main Effects	Model 7: Model 6 with Interaction Added
		Models with 2-Way Interactions			Models with Only Main Effects		
SHRBVOL	$R^2 = 0.525$ 3-way P = 0.794	$R^2 = 0.522$ YEAR*SEG P = 0.528	$R^2 = 0.513$ YEAR*SHRBVOL P = 0.613	$R^2 = 0.512$ SEG*SHRBVOL P = 0.382	$R^2 = 0.499$ YEAR P = 0.704	**$R^2 = 0.498$** **SEG & SHRBVOL** **$P \leq 0.005$**	$R^2 = 0.511$ SEG*SHRBVOL P = 0.367
TAMVOL	$R^2 = 0.579$ 3-way P = 0.621	$R^2 = 0.573$ YEAR*SEG P = 0.994	$R^2 = 0.573$ YEAR*TAMVOL P = 0.583	**$R^2 = 0.571$** **SEG*TAMVOL** **$P < 0.001$**	$R^2 = 0.420$ YEAR P = 0.554	**$R^2 = 0.417$** **SEG & TAMVOL** **$P < 0.001$**	**$R^2 = 0.569$** **SEG*TAMVOL** **$P < 0.001$**
MESQVOL	$R^2 = 0.476$ 3-way P = 0.274	$R^2 = 0.467$ YEAR*MESQVOL P = 0.946	$R^2 = 0.467$ YEAR*SEG P = 0.758	$R^2 = 0.463$ SEG*MESQVOL P = 0.598	$R^2 = 0.456$ YEAR P = 0.607	**$R^2 = 0.454$** **SEG & MESQVOL** **$P \leq 0.006$**	$R^2 = 0.462$ SEG*MESQVOL P = 0.568
ARROW	$R^2 = 0.346$ 3-way P = 0.277	$R^2 = 0.322$ SEG*ARROW P = 0.985	$R^2 = 0.322$ YEAR*SEG P = 0.538	$R^2 = 0.310$ YEAR*ARROW P = 0.700	$R^2 = 0.309$ YEAR P = 0.572	**$R^2 = 0.306$** **SEG & ARROW** **$P \leq 0.013$**	$R^2 = 0.307$ SEG*ARROW P = 0.960
BACH	$R^2 = 0.297$ 3-way P = 0.856	$R^2 = 0.294$ YEAR*BACH P = 0.974	$R^2 = 0.294$ YEAR*SEG P = 0.815	$R^2 = 0.290$ SEG*BACH P = 0.365	$R^2 = 0.271$ YEAR P = 0.523	**$R^2 = 0.267$** **SEG & BACH** **$P \leq 0.003$**	$R^2 = 0.288$ SEG*BACH P = 0.336
GRASS	$R^2 = 0.258$ 3-way P = 0.752	$R^2 = 0.257$ YEAR*SEG P = 0.898	$R^2 = 0.255$ SEG*GRASS P = 0.557	$R^2 = 0.251$ YEAR*GRASS P = 0.345	$R^2 = 0.243$ YEAR P = 0.419	**$R^2 = 0.236$** **SEG & GRASS** **$P \leq 0.019$**	$R^2 = 0.241$ SEG*GRASS P = 0.482
MESQ	$R^2 = 0.482$ 3-way P = 0.369	$R^2 = 0.473$ YEAR*MESQ P = 0.821	$R^2 = 0.472$ YEAR*SEG P = 0.741	$R^2 = 0.468$ SEG*MESQ P = 0.365	$R^2 = 0.453$ YEAR P = 0.526	**$R^2 = 0.450$** **SEG & MESQ** **$P \leq 0.002$**	$R^2 = 0.466$ SEG*MESQ P = 0.332
TAM	$R^2 = 0.622$ 3-way P = 0.526	$R^2 = 0.615$ YEAR*SEG P = 0.828	$R^2 = 0.613$ YEAR*TAM P = 0.602	**$R^2 = 0.612$** **SEG*TAM** **$P < 0.001$**	$R^2 = 0.522$ YEAR P = 0.539	**$R^2 = 0.520$** **SEG & TAM** **$P < 0.001$**	**$R^2 = 0.611$** **SEG*TAM** **$P < 0.001$**
WILLOW	$R^2 = 0.286$ 3-way P = 0.784	$R^2 = 0.281$ YEAR*WILLOW P = 0.757	$R^2 = 0.278$ YEAR*SEG P = 0.845	$R^2 = 0.275$ SEG*WILLOW P = 0.328	$R^2 = 0.253$ YEAR P = 0.454	**$R^2 = 0.247$** **SEG & WILLOW** **$P \leq 0.010$**	$R^2 = 0.270$ SEG*WILLOW P = 0.320

Covariates. All covariates were positively correlated with bird species richness and all except BACH, GRASS, and WILLOW were highly and significantly correlated with richness (Spearman's rho ≥ 0.401, $P < 0.01$). Each individual covariate considered, in the presence of SEG, accounted for a significant proportion of the variation in richness ($P \leq 0.019$; see underlined values in Table 7). For 14 of the covariates (all except TREEVOL, TAMVOL, and TAM) the best model included the main effects for SEG and the respective covariate (see Model 6 in Table 7). For TREEVOL, TAMVOL, and TAM, the best model also included the SEG*covariate interaction (see Model 7 in Table 7). Among the series of models considered for each covariate, the best model contained the main effects SEG and the covariate for 14 of the covariates (see Model 6 in Table 7), plus the additional SEG*covariate interaction term for the remaining three covariates (see Model 7 in Table 7).

Covariates TREEVOL, TAM, and TAMVOL (in the presence of SEG) accounted for the greatest proportion of the variation in richness for the interaction models ($0.569 \leq R^2 \leq 0.611$; see Model 7 in Table 7), and covariates TREEVOL, TREE, and NHWZ accounted for the greatest proportion of the variation in richness for the main effects models ($0.533 \leq R^2 \leq 0.562$; see Model 6 in Table 7). Richness increased with TREEVOL, TAM, and TAMVOL at the greatest rate for river segment B, then C, and at the lowest rate for river segment A (Figure 2C).

We therefore conclude that covariates associated with tamarisk and trees (tree area and volume, tamarisk area and volume, and NHWZ area) have the greatest potential, among those considered, to predict species richness. Additionally, species richness increased at the greatest rate with tamarisk area and volume, and tree volume for river segment B.

Shannon Diversity Index and Habitat

Differences between years. As with bird abundance and species richness, the overall mean Shannon Diversity Index per site did not differ between 1994 (1.7 ± 0.09) and 1995 (1.6 ± 0.08; one-way ANOVA, $F = 0.449$, df = 1, $P = 0.505$). Additionally, the lack of year differences was consistent among river segment (test of YEAR*SEG interaction in full factorial model, $F = 0.157$, df = 2, $P = 0.855$).

There was also no evidence of year differences in the SDI per patch for any analysis that included a single covariate. For each series of the seven models we ran for each covariate, there was no significant three-way interaction ($P \geq 0.271$) among YEAR, SEG, and the covariate (see Model 1 of Table 8). Likewise, we found no significant two-way interactions ($P \geq 0.099$) between YEAR and SEG, or YEAR and the covariate (see Models 2–4 of Table 8). Finally, we found no significant contribution by YEAR ($P \geq 0.371$) in the main effects only models (see Model 5 of Table 8). We therefore conclude that diversity per patch, as measured by the Shannon Diversity Index, did not differ between 1994 and 1995 for any segment of the river.

Differences among river segments. The Shannon Diversity Index differed by river segment (one-way ANOVA, $F = 10.193$, df = 2, $P < 0.001$). The SDI for river segment C (1.9 ± 0.06) was greater than for river segment A (1.6 ± 0.09) and river segment B (0.9 ± 0.3; Bonferroni's multiple comparisons, $P = 0.018$ and < 0.001 respectively), while river segment A did not differ from river segment B ($P = 0.066$).

Among the series of seven models run for each covariate, SEG was a significant factor in the best model for every covariate. The main effect SEG and/or the SEG*covariate interaction was always highly significant ($P \leq 0.036$; see underlined values in Table 8). We therefore conclude that avian diversity per patch, as measured by the Shannon Diversity Index, differed by river segment.

Covariates. Interpretation of covariates for the Shannon Diversity Index was almost identical to that for species richness. All covariates were positively correlated with SDI and all except BACH, GRASS, and WILLOW were highly and significantly correlated with SDI (Spearman's rho ≥ 0.406, $P < 0.01$). In all but one case (GRASS), each individual covariate we considered (in the presence of SEG) accounted for a significant proportion of the variation in SDI ($P \leq 0.013$; see underlines values in Table 8). For 13 of the covariates (all except TREEVOL, TAMVOL, TAM, and GRASS), the best model included the main effects for SEG and the respective covariate (see Model 6 in Table 8). For TREEVOL, TAMVOL, and TAM, the best model also included the SEG*covariate interaction (see Model 7 in Table 8). No good model was identified for GRASS. Covariates TAM, TAMVOL, and NHWZ (in the presence of SEG) accounted for the greatest proportion of the variation in SDI for the interaction models ($0.504 \leq R^2 \leq 0.545$; see Model 7 in Table 8), while covariates TAM, NHWZ, and TREEVOL accounted for the greatest proportion of the variation in SDI for the main effects models ($0.435 \leq R^2 \leq 0.486$; see Model 6 in Table 8). SDI increased with TREEVOL, TAM, and TAMVOL at the greatest rate for river segment B, then C, and at the lowest rate for river segment A (Figures 2D and 2E).

We therefore conclude that covariates associated with tamarisk and trees (tamarisk area and volume, tree volume, and NHWZ area) have the greatest potential, among those considered, to predict species diversity as measured by the Shannon Diversity Index. Additionally, the SDI increased at the greatest rate with tamarisk area and volume, and tree volume, for river segment B.

Correlation among covariates

Among the 136 bivariate correlations from the 17 covariates, only two were negative (BACH vs. TAMVOL and WILLOW vs. ARROW); neither were significant ($P > 0.05$). Most covariates were significantly ($P < 0.01$ for 106 pairs, $P < 0.05$ for 9 pairs) and highly (Spearman's rho > 0.5 for 86 pairs) correlated with one another. Non-significant correlations included covariates HERB (5 pairs), BACH (6), GRASS (5), and/or WILLOW (6) and the ARROW vs. TAMVOL pair. Consequently, the covariates BACH, GRASS, and WILLOW represented a group that tended to be correlated with each other, but were uncorrelated with most other covariates (including HERB). Small significant correlations (Spearman's rho < 0.5) were primarily associated with HERB (10 pairs), ARROW (10), BACH (6), GRASS (10), WILLOW (8), TAMVOL (7), TAM (6), and MESQ (3).

DISCUSSION

Habitat and Habitat Selection Theory

Birds and bird communities have played a major role in the development of the concept of habitat, yet specific definitions of the term "habitat" are often vague and/or differ from one another (Block and Brennan 1993). However, a common theme among the various definitions and terms is that "habitat" includes the physical and biological environmental attributes that influence the presence or absence of a bird species (Morrison et al. 1992). A bird community is an assemblage of individuals of several species that occur together, and community ecology is concerned with identifying patterns in species assemblages, understanding the causes of these patterns, and determining how general they are (Wiens 1989a).

Habitat involves many components other than the vegetation composition and structure at a site. Environmental features (climate, food, size/area), predation, competition, parasitism, disease, disturbance, past history, and even chance influence the distribution and abundance of species, and thus community attributes (Wiens 1989b). Research is usually focused on the habitat components that are most easily or reliably quantified and/or considered most likely to influence the bird community, and no single study can address all of the factors that may influence bird species or community use in a system.

Bird communities are dynamic and can vary temporally, even in the same region and/or site (Wiens 1989a; Morrison et al. 1992). Some of these changes may be due to regional and local changes in habitat distribution or characteristics, but others may be due to factors intrinsic to the bird species (such as increasing abundance or dispersal). Riparian corridors in particular facilitate faunal mixing on a broad, regional level (Knopf and Samson 1994), especially at the

Table 8. Response variable SDI2. Correlation coefficients and results of hypothesis tests of the effective contribution of terms used during the step-down modeling approach to evaluate the significance of individual covariates in the presence of factors YEAR and SEG on the transformed response variable SDI2. The first line in each cell is the model correlation coefficient. The second line is the term in the model that had the highest P-value within the class of interaction (3-way, 2-way, or main effect), and this was the term that was dropped prior to running the model in the next column. The last line in each cell is the P-value of the specified term. Model 7 contains two main effects and their interaction, and differs from Model 4 which has all three main effects and the single interaction. Bolded entries contain P values < 0.1, while underlined entries indicate a good model for inference for each covariate. Comparisons within columns allow for selecting the best covariates for a given model.

	Models with 2-Way Interactions				Models with Only Main Effects		
Covariate	Model 1: Full Factorial Model - 7 Parameters	Model 2: Three 2-way Interactions 6 Parameters	Model 3: Two 2-way Interactions 5 Parameters	Model 4: One 2-way Interaction 4 Parameters	Model 5: All 3 Main Effects Only	Model 6: Two Main Effects	Model 7: Model 6 with Interaction Added
AREA	$R^2 = 0.446$ 3-way P = 0.777	$R^2 = 0.446$ YEAR*AREA P = 0.777	$R^2 = 0.442$ YEAR*SEG P = 0.614	$R^2 = 0.435$ SEG*AREA P = 0.153	$R^2 = 0.406$ YEAR P = 0.519	**$R^2 = 0.403$** **SEG & AREA** **P < 0.001**	$R^2 = 0.433$ SEG*AREA P = 0.139
VEG	$R^2 = 0.449$ 3-way P = 0.682	$R^2 = 0.443$ YEAR*VEG P = 0.857	$R^2 = 0.443$ YEAR*SEG P = 0.612	$R^2 = 0.435$ SEG*VEG P = 0.193	$R^2 = 0.410$ YEAR P = 0.512	**$R^2 = 0.407$** **SEG & VEG** **P < 0.001**	$R^2 = 0.433$ SEG*VEG P = 0.180
TREE	$R^2 = 0.440$ 3-way P = 0.376	$R^2 = 0.424$ SEG*TREE P = 0.784	$R^2 = 0.420$ YEAR*SEG P = 0.725	$R^2 = 0.415$ YEAR*TREE P = 0.542	$R^2 = 0.412$ YEAR P = 0.462	**$R^2 = 0.408$** **SEG & TREE** **P < 0.001**	$R^2 = 0.412$ SEG*TREE P = 0.797
SHRB	$R^2 = 0.367$ 3-way P = 0.727	$R^2 = 0.362$ YEAR*SHRB P = 0.576	$R^2 = 0.359$ YEAR*SEG P = 0.584	**$R^2 = 0.349$** **SEG*SHRB** **P = 0.096**	$R^2 = 0.307$ YEAR P = 0.565	**$R^2 = 0.304$** **SEG & SHRB** **P = 0.001**	**$R^2 = 0.346$** **SEG*SHRB** **P = 0.093**
HERB	$R^2 = 0.345$ 3-way P = 0.767	$R^2 = 0.340$ YEAR*SEG P = 0.833	$R^2 = 0.337$ SEG*HERB P = 0.296	$R^2 = 0.315$ YEAR*HERB P = 0.212	$R^2 = 0.300$ YEAR P = 0.344	**$R^2 = 0.292$** **SEG & HERB** **P ≤ 0.003**	$R^2 = 0.318$ SEG*HERB P = 0.242
NHWZ	$R^2 = 0.513$ 3-way P = 0.968	$R^2 = 0.513$ YEAR*NHWZ P = 0.937	$R^2 = 0.513$ YEAR*SEG P = 0.616	**$R^2 = 0.506$** **SEG*NHWZ** **P = 0.062**	$R^2 = 0.468$ YEAR P = 0.522	**$R^2 = 0.465$** **SEG & NHWZ** **P < 0.001**	**$R^2 = 0.504$** **SEG*NHWZ** **P = 0.057**
OHWZ	$R^2 = 0.378$ 3-way P = 0.260	$R^2 = 0.354$ YEAR*SEG P = 0.735	$R^2 = 0.348$ SEG*OHWZ P = 0.550	$R^2 = 0.337$ YEAR*OHWZ P = 0.461	$R^2 = 0.333$ YEAR P = 0.426	**$R^2 = 0.327$** **SEG & OHWZ** **P < 0.001**	$R^2 = 0.338$ SEG*OHWZ P = 0.541
TREEVOL	$R^2 = 0.491$ 3-way P = 0.927	$R^2 = 0.489$ YEAR*TREEVOL P = 0.882	$R^2 = 0.489$ YEAR*SEG P = 0.714	**$R^2 = 0.485$** **SEG*TREEVOL** **P = 0.040**	$R^2 = 0.438$ YEAR P = 0.527	**$R^2 = 0.435$** **SEG & TREEVOL** **P < 0.001**	**$R^2 = 0.483$** **SEG*TREEVOL** **P = 0.036**

Table 8 (continued)

Covariate	Model 1: Full Factorial Model - 7 Parameters	Model 2: Three 2-way Interactions 6 Parameters	Model 3: Two 2-way Interactions 5 Parameters	Model 4: One 2-way Interaction 4 Parameters	Model 5: All 3 Main Effects Only	Model 6: Two Main Effects	Model 7: Model 6 with Interaction Added
		Models with 2-Way Interactions			Models with Only Main Effects		
SHRBVOL	$R^2 = 0.447$ 3-way $P = 0.850$	$R^2 = 0.444$ YEAR*SHRBVOL $P = 0.930$	$R^2 = 0.444$ YEAR*SEG $P = 0.562$	**$R^2 = 0.435$** **SEG*SHRBVOL** **$P = 0.099$**	$R^2 = 0.400$ YEAR $P = 0.548$	**$R^2 = 0.397$** **SEG & SHRBVOL** **$P \leq 0.002$**	**$R^2 = 0.433$** **SEG*SHRBVOL** **$P = 0.095$**
TAMVOL	$R^2 = 0.550$ 3-way $P = 0.449$	$R^2 = 0.540$ YEAR*SEG $P = 0.973$	$R^2 = 0.539$ YEAR*TAMVOL $P = 0.494$	**$R^2 = 0.536$** **SEG*TAMVOL** **$P < 0.001$**	$R^2 = 0.418$ YEAR $P = 0.446$	**$R^2 = 0.414$** **SEG & TAMVOL** **$P < 0.001$**	**$R^2 = 0.532$** **SEG*TAMVOL** **$P < 0.001$**
MESQVOL	$R^2 = 0.358$ 3-way $P = 0.777$	$R^2 = 0.347$ YEAR*SEG $P = 0.792$	$R^2 = 0.343$ SEG*MESQVOL $P = 0.540$	$R^2 = 0.332$ YEAR*MESQVOL $P = 0.503$	$R^2 = 0.328$ YEAR $P = 0.480$	**$R^2 = 0.324$** **SEG & MESQVOL** **$P \leq 0.001$**	$R^2 = 0.336$ SEG*MESQVOL $P = 0.490$
ARROW	$R^2 = 0.319$ 3-way $P = 0.170$	$R^2 = 0.284$ SEG*ARROW $P = 0.796$	$R^2 = 0.279$ YEAR*ARROW $P = 0.798$	$R^2 = 0.278$ YEAR*SEG $P = 0.678$	$R^2 = 0.271$ YEAR $P = 0.458$	**$R^2 = 0.266$** **SEG & ARROW** **$P \leq 0.013$**	$R^2 = 0.267$ SEG*ARROW $P = 0.913$
BACH	$R^2 = 0.318$ 3-way $P = 0.874$	$R^2 = 0.315$ YEAR*BACH $P = 0.833$	$R^2 = 0.315$ YEAR*SEG $P = 0.719$	$R^2 = 0.309$ SEG*BACH $P = 0.142$	$R^2 = 0.272$ YEAR $P = 0.428$	**$R^2 = 0.266$** **SEG & BACH** **$P \leq 0.013$**	$R^2 = 0.304$ SEG*BACH $P = 0.131$
GRASS	$R^2 = 0.245$ 3-way $P = 0.760$	$R^2 = 0.244$ YEAR*SEG $P = 0.847$	$R^2 = 0.241$ SEG*GRASS $P = 0.750$	$R^2 = 0.240$ YEAR*GRASS $P = 0.456$	$R^2 = 0.234$ YEAR $P = 0.368$	$R^2 = 0.226$ GRASS $P = 0.153$	$R^2 = 0.228$ SEG*GRASS $P = 0.672$
MESQ	$R^2 = 0.368$ 3-way $P = 0.327$	$R^2 = 0.359$ YEAR*SEG $P = 0.748$	$R^2 = 0.354$ SEG*MESQ $P = 0.547$	$R^2 = 0.343$ YEAR*MESQ $P = 0.406$	$R^2 = 0.337$ YEAR $P = 0.436$	**$R^2 = 0.332$** **SEG & MESQ** **$P = \leq 0.001$**	$R^2 = 0.344$ SEG*MESQ $P = 0.488$
TAM	$R^2 = 0.556$ 3-way $P = 0.765$	$R^2 = 0.553$ YEAR*TAM $P = 0.998$	$R^2 = 0.553$ YEAR*SEG $P = 0.652$	**$R^2 = 0.548$** **SEG*TAM** **$P = 0.011$**	$R^2 = 0.490$ YEAR $P = 0.432$	**$R^2 = 0.486$** **SEG & TAM** **$P < 0.001$**	**$R^2 = 0.545$** **SEG*TAM** **$P = 0.009$**
WILLOW	$R^2 = 0.341$ 3-way $P = 0.696$	$R^2 = 0.334$ YEAR*SEG $P = 0.556$	$R^2 = 0.323$ YEAR*WILLOW $P = 0.699$	**$R^2 = 0.322$** **SEG*WILLOW** **$P = 0.081$**	$R^2 = 0.275$ YEAR $P = 0.369$	**$R^2 = 0.267$** **SEG & WILLOW** **$P \leq 0.012$**	**$R^2 = 0.314$** **SEG*WILLOW** **$P = 0.079$**

interface of different biomes or ecoregions (as occurs between the upper and lower portions of the Grand Canyon).

Study Design Considerations

As evidenced by the preceding discussion, when embarking on this Grand Canyon avian community–habitat relationship study we were faced with a concept of considerable complexity. Study design challenges included deciding what spatial scales to use, what variables to consider, and what methods to use. For scale, we selected a macrohabitat approach that included patch characteristics (e.g., size, structure, composition) and landscape factors (e.g., river segment). We used a larger number of study sites than previous Grand Canyon bird studies in order to increase confidence in our model and to capture a wide variety of patch/habitat attributes. There is almost always an increasing relationship between the number of bird species and individuals and the amount of habitat present or surveyed (Wiens 1989a; Morrison et al. 1992), so we included a wide range of patch sizes rather than focus only on larger patches.

In terms of response variables, we selected bird species richness, abundance, and diversity for the entire riparian breeding community, rather than for a subset of species (e.g., indicator species). These variables are readily determined, reliably measured, and easily understood, and they can be used when developing or measuring management objectives and actions. Use of more basic data such as comparisons of simple species lists among sites is often flawed and inappropriate (Remsen 1994), while more complex and derived information such as density estimates are subject to major methodological problems (Verner 1985; Norvell et al. 2003) and may be misleading in terms of habitat quality (Vickery et al. 1992). Although virtually all bird census techniques are subject to biases and errors, we used and integrated a variety of methods in order to maximize the likelihood of detecting the species that were present, and we used decision rules to avoid overestimating the abundance of local breeders. Observers and methods were standardized in order to minimize sampling inconsistencies.

With regard to habitat, we selected from among an almost limitless number of parameters and sampling methods. Given our objective of quantifying the relationship between the riparian breeding bird community and vegetation/habitat characteristics, we focused on vegetative structure and composition, both of which have and will respond directly to dam operations. Factors such as predators, competitors, or disease, for example, are difficult to study and impossible to directly relate to river flow, and hence are not appropriate for this study. Given our scale of interest (macro-habitat), we focused on patch-scale vegetation area and structure, and we de-emphasized fine-scale floristics (other than major tree/shrub community dominants and components) and vegetation measurements (e.g., number of stems). Relevé, total vegetation volume (TVV), and/or point-based vegetation measurements are often very time consuming (and hence expensive in terms of river-based logistics) and can negatively impact the vegetation at a site, especially where they require entry into dense or fragile habitats. To avoid this, we used aerial photographs and rapid ocular-estimation techniques to estimate habitat parameters we thought (based on a review of the literature) would influence the bird community along the river corridor.

Bird Community–Habitat Patterns in the Grand Canyon

The discussion that follows relates only to the breeding bird community, and excludes migrants, wintering species, and vagrants. Note that our decision rules could not account for the possible presence of unpaired males or non-territorial "floater" individuals; no data exist on the degree to which either of these occur in the habitats along the Colorado River corridor. Also, the inclusion of a species in our model does not imply that it is a riparian obligate; some species in Table 1 nest in other habitats as well. In addition, some species like the Rufous-crowned Sparrow (*Aimophila ruficeps*) typically place their nests outside the riparian zone, but regularly use riparian patches for foraging and as post-fledgling habitat for family groups. Their frequency of detection and use of the riparian zone was extensive enough to warrant their inclusion in our modeling of the bird community.

Bird species abundance, species richness, and diversity (SDI) per patch did not differ between 1994 and 1995. This allowed us to use bird community data from both years in our models, thereby increasing the sample size and improving the power of the analyses. Lack of a significant difference in these parameters is not surprising; significant changes would not be expected over such a short period as long as major habitat changes have not occurred. The interim flow guidelines governing Glen Canyon Dam operations during this period limited the likelihood and occurrence of major short-term habitat changes (such as those that occurred during the 1983 high-water releases from Glen Canyon Dam).

There was a strong positive correlation among the three response variables (richness, abundance, and SDI) and 14 of the 17 covariates. Eleven covariates were considered good predictors (Table 9). Covariates associated with large vegetation structures (e.g. tree area and volume, NHWZ area) and tamarisk species (e.g., tamarisk area and volume) were the best predictors for each of the three bird community response variables. Additionally, total area and vegetated area also had good potential as predictors for abundance of breeding pairs. Thus, the overall pattern is that as patch size increased, the number of breeding bird species and individuals increased. This is one of the most universal and generally noted relationships between habitat and bird community patterns (Wiens 1989a; Block and Brennan 1993); it was qualitatively described for riparian patches in the Grand Canyon by Willson and Carothers (1979). In our study, SDI also increased with patch size.

Our best predictive models worked better for bird abundance than for species richness or diversity (Table 9). This is interesting in that many bird community studies find better correlations between habitat parameters and species richness than with bird abundance or density (Wiens 1989a). This is generally assumed to occur because the number of bird species is often more accurately determined than the actual number of breeding pairs. In our study, the use of species-specific decision rules, seldom used in other studies, may have produced relatively accurate estimations of the actual number of breeding pairs, thereby increasing the fit of our models relative to habitat parameters.

For all bird community variables, the lowest average values were found in river segment B. This is not surprising, given that patches in segment B were generally small and isolated, and often bordered bare rock rather than upland vegetation. The highest bird abundance, richness, and diversity values were generally found in river segment C, where patches are largest and most contiguous, and often bordered by extensive upland vegetation. This region of the canyon is also adjacent to the species-rich Sonoran desert biome.

The area and volume of tree habitat (defined as woody vegetation 2 m or higher) was a much better predictor of bird community response variables than were shrub (woody vegetation > 2 m high) and herb habitat (nonwoody vegetation, usually ground cover). This suggests that larger woody vegetation plays a more important role in structuring the Colorado River riparian bird community than do shrubs and herbs. The older, dense tamarisk and mesquite vegetation that comprises most of the tree-structured habitat is heavily used by breeding birds, and provides more nesting substrate, song perches, and environmental buffering than does shrub habitat. Mills et al. (1991) found a similar pattern in studies in Arizona and New Mexico, reporting that total vegetation volume (measured using a different methodology) was strongly correlated with breeding bird density, both within and between habitat types.

Of the major dominant plant species in the riparian habitat along the Colorado River, only mesquite and tamarisk func-

Table 9. The best models (based on correlation coefficients and significant tests) in terms of individual covariates for predicting the response variables ABUND, RICH, and SDI. Only the best model is presented for each covariate and community response variable pair, from those given in Tables 6, 7, and 8. The first line in each cell is the model correlation coefficient and P-value. The second line is the term in the model that had the highest P-value within each covariate–response variable pair. Comparisons within columns allow selection of the best covariates for predicting a given response variable. Models are only given for covariates considered to be good predictors, defined as $R^2 \geq 0.50$ for ABUND and RICH, and $R^2 \geq 0.4$ for SDI. Models are not given for SHRB, HERB, ARROW, BACH, GRASS, and WILLOW as no models were considered good predictors. Parameter estimates for these models are presented in Table 10.

Covariate	ABUND (no. breeding pairs)	RICH (no. breeding species)	SDI (Shannon Diversity Index)
AREA	$R^2 = 0.646$; $P = 0.015$ AREA & SEG & SEG*AREA	$R^2 = 0.515$; $P < 0.001$ AREA & SEG	$R^2 = 0.403$; $P < 0.001$ AREA & SEG
VEG	$R^2 = 0.658$; $P = 0.017$ VEG & SEG & SEG*VEG	$R^2 = 0.524$; $P < 0.001$ VEG & SEG	$R^2 = 0.407$; $P < 0.001$ VEG & SEG
TREE	$R^2 = 0.625$; $P < 0.001$ TREE & SEG	$R^2 = 0.533$; $P < 0.001$ TREE & SEG	$R^2 = 0.408$; $P < 0.001$ TREE & SEG
NHWZ	$R^2 = 0.639$; $P = 0.010$ NHWZ & SEG & SEG*NHWZ	$R^2 = 0.544$; $P < 0.001$ NHWZ & SEG	$R^2 = 0.465$; $P < 0.001$ NHWZ & SEG
OHWZ	$R^2 = 0.542$; $P < 0.001$ OHWZ & SEG	—	—
TREEVOL	$R^2 = 0.685$; $P = 0.003$ TREEVOL & SEG & SEG*TREEVOL	$R^2 = 0.611$; $P = 0.011$ TREEVOL & SEG & SEG*TREEVOL	$R^2 = 0.483$; $P = 0.011$ TREEVOL & SEG & SEG*TREEVOL
SHRBVOL	$R^2 = 0.617$; $P = 0.007$ SHRBVOL & SEG & SEG*SHRBVOL	$R^2 = 0.498$; $P < 0.005$ SHRBVOL & SEG	—
TAMVOL	$R^2 = 0.555$; $P < 0.001$ TAMVOL & SEG & SEG*TAMVOL	—	$R^2 = 0.532$; $P < 0.005$ TAMVOL & SEG& VOL & SEG
MESQVOL	$R^2 = 0.563$; $P = 0.011$ MESQVOL & SEG & SEG*MESQVOL	—	—
MESQ	$R^2 = 0.574$; $P = 0.015$ MESQ & SEG & SEG*MESQ	—	—
TAM	$R^2 = 0.647$; $P < 0.001$ TAM & SEG & SEG*TAM	$R^2 = 0.611$; $P < 0.001$ TAM & SEG	$R^2 = 0.545$; $P < 0.01$ TAM & SEG & SEG*TAM

Table 10. Parameter estimates for the best models (based on correlation coefficients and significant tests) in terms of individual covariates for predicting the response variables ABUND, RICH, and SDI (see Table 9). The factor SEG is presented in the equations by SEG = A or B, and is replaced by 1 for the respective river segment, and replaced by 0 (zero) otherwise. For river segment C, both SEG = A and SEG = B would be 0 (zero).

Covariate	ABUND = Number of Breeding Pairs
AREA	Log (ABUND) = 0.798 – 0.163(SEG=A) – 0.621(SEG=B) + 1.344E–05 (AREA) – 5.511E-07(AREA)(SEG=A) + 6.036E-05 (AREA)(SEG=B)
VEG	Log (ABUND) = 0.801 – 0.177(SEG=A) – 0.600(SEG=B) +1.335E-05(VEG) + 7.766E-07(VEG)(SEG=A) + 6.200E-05(VEG)(SEG=B)
TREE	Log (ABUND) = 0.839 – 0.183(SEG=A) – 0.325(SEG=B) + 1.976E-05(TREE)
NHWZ	Log (ABUND) = 0.797 – 0.175(SEG=A) – 0.54(SEG=B) + 3.594E-05(NHWZ) – 6.427E-06(NHWZ)(SEG=A) + 1.146E-04(NHWZ)(SEG=B)
OHWZ	Log (ABUND) = 0.853 – 0.134(SEG=A) – 0.334(SEG=B) + 2.099E-05(OHWZ)
TREEVOL	Log (ABUND) = 0.822 – 0.157(SEG=A) – 0.574(SEG=B) + 9.570E-06(TREEVOL) – 2.106E-06(TREEVOL)(SEG=A) + 7.073E-05(TREEVOL)(SEG=B)
SHRBVOL	Log (ABUND) = 0.809 – 0.156(SEG=A) – 0.614(SEG=B) + 2.103E-05(SHRBVOL) + 5.345E-06(SHRBVOL)(SEG=A) + 1.456E-04(SHRBVOL)(SEG=B)
TAMVOL	Log (ABUND) = 0.832 - 9.637E-02(SEG=A) – 0.433(SEG=B) + 3.253E-05(TAMVOL) – 2.387E-05(TAMVOL)(SEG=A) + 1.372E-04(TAMVOL)(SEG=B)
MESQVOL	Log (ABUND) = 0.884 – 0.136(SEG=A) – 0.257(SEG=B) + 7.945E-06(MESQVOL) + 4.131E-06(MESQVOL)(SEG=A) – 2.530E-04(MESQVOL)(SEG=B)
MESQ	Log (ABUND) = 0.871 – 0.139(SEG=A) – 0.245(SEG=B) + 1.987E-05(MESQ) + 2.409E-06(MESQ)(SEG=A) – 2.648E-04(MESQ)(SEG=B)
TAM	Log (ABUND) = 0.761 – 0.131(SEG=A) – 0.467(SEG=B) + 8.445E-05(TAM) – 4.968E-05(TAM)(SEG=A) + 7.119E-05(TAM)(SEG=B)
	RICH – number of breeding species
AREA	RICH = 5.794 – 2.210(SEG=A) – 3.083(SEG=B) + 1.322E-04(AREA)
VEG	RICH = 5.741 – 2.123(SEG=A) – 2.996(SEG=B) + 1.368E-04(VEG)
TREE	RICH = 6.205 – 2.303(SEG=A) – 3.156(SEG=B) + 1.942E-04(TREE)
NHWZ	RICH = 5.849 – 2.732(SEG=A) – 3.139(SEG=B) + 3.440E-04(NHWZ)
TREEVOL	RICH = 5.721 – 1.522(SEG=A) – 4.849(SEG=B) + 1.094E-04(TREEVOL) – 4.385E-05(TREEVOL)(SEG=A) + 5.611E-04(TREEVOL)(SEG=B)
SHRBVOL	RICH = 5.610 – 1.561(SEG=A) – 2.736(SEG=B) + 2.360E-04(SHRBVOL)
TAM	RICH = 5.035 – 1.501(SEG=A) – 3.793(SEG=B) + 9.601E-04(TAM) – 6.022E-04(TAM)(SEG=A) + 3.462E-04(TAM)(SEG=B)
	SDI = Shannon Diversity Index
AREA	SDI2 = 2.914 – 1.041(SEG=A) – 1.778(SEG=B) + 4.742E-05(AREA)
VEG	SDI2 = 2.899 – 1.010(SEG=A) – 1.749(SEG=B) + 4.887E-05(VEG)
TREE	SDI2 = 3.070 – 1.073(SEG=A) – 1.810(SEG=B) + 6.879E-05(TREE)
NHWZ	SDI2 = 2.857 – 1.255(SEG=A) – 1.745(SEG=B) + 1.358E-04(NHWZ)
TREEVOL	SDI2 = 3.009 – 1.051(SEG=A) – 3.134(SEG=B) + 3.337E-05(TREEVOL) – 4.823E-06(TREEVOL)(SEG=A) + 3.685E-04(TREEVOL)(SEG=B)
TAMVOL	SDI2 = 2.828 – 0.725(SEG=A) – 2.357(SEG=B) + 1.506E-04(TAMVOL) – 1.094E-04(TAMVOL)(SEG=A) + 8.768E-04(TAMVOL)(SEG=B)
TAM	SDI2 = 2.694 – 1.081(SEG=A) – 2.6(SEG=B) + 3.275E-04(TAM) – 1.634E-04(TAM x SEG=A) + 4.569E-04(TAM x SEG=B)

tioned as good predictors of bird community parameters. Areas of arrowweed, baccharis, and grass vegetation types contributed less predictive value, as might be expected based on general observations of lower relative use by birds. We were surprised that area of willow was not a stronger predictor, given the generally higher value of native riparian habitats as compared to introduced habitats (Wiens 1989a; Morrison et al. 1992). This may be because NHWZ vegetation in our study sites was overwhelmingly dominated by tamarisk, which may have overshadowed any bird community patterns relative to willow.

Several of our models accounted for approximately two-thirds of the variation in bird communities along the canyon. We think that this is a remarkably strong model response given that we addressed primarily vegetation-related habitat components, and did not measure a number of other potential influencing factors such as predators, competitors, and environmental characteristics that also function to shape the breeding bird community.

The remaining bird community variation not accounted for in our models is likely influenced by a number of regional and local factors. For example, the Colorado River through the Grand Canyon stretches for more than 360 km (280 river miles) and across an elevation gradient of 540 m. This entails changes in environment (e.g., temperature) and geomorphology that in turn influence vegetation differences (both patch-scale and floristically) such that patches in the upper canyon are often very different from similar-sized patches in the lower end. Also, the adjacent upland habitat is very different in different portions of the canyon, and avian assemblages in riparian tracts and adjoining uplands are not independent (Knopf and Sansom 1994).

Factors such as species range limits and recent range expansions by some species also affect the bird communities of an area. For example, the northern range limit of the Brown-crested Flycatcher (*Myiarchus tyrannulus*) occurs along the lower section of the Grand Canyon (Felley and Sogge, this volume), and this flycatcher has not moved upstream as a breeding species (and may never do so) even as major habitat changes have occurred. Conversely, Brown and Johnson (1988) noted that some species (Bell's Vireo [(*Vireo bellii*] and Hooded Oriole [*Icterus cucullatus*]) are moving into and up the Grand Canyon in recent years, and the Summer Tanager (*Piranga rubra*) is recorded in ever-increasing numbers. It is unknown whether these species will eventually breed as far upstream as Lees Ferry, but this illustrates the ongoing temporal and spatial dynamics that can be expected to continue within the canyon.

Brown-headed Cowbird (*Molothrus ater*) nest parasitism greatly affects some species along the river in the Grand Canyon (Brown 1994; Sogge et al. 1997). Cowbird parasitism has the potential of affecting host distribution and abundance in some systems (Lowther 1993; Robinson et al. 1995), and it may be playing a significant role in determining abundance and breeding distribution of some of the commonly parasitized hosts in the Grand Canyon.

Most of the riparian breeding birds along the river are neotropical migrants, and many of them spend half or more of the year migrating and wintering outside of the canyon. There has been increasing attention and research (e.g., Keast and Morton 1980; Martin and Finch 1995; Rappole 1995) directed toward understanding the importance of nonbreeding habitats to avian survival. Although much remains to be known, it has become clear that factors such as the quality and quantity of migratory stopover areas and wintering habitat can have major effects on the overwinter survival of birds. Therefore, factors outside the Grand Canyon affect the survival of locally breeding neotropical birds, and can add temporal and spatial variability to habitat use patterns that cannot be accounted for in models.

Our study looked at overall bird community patterns, and not individual bird species. However, species-specific life history traits can be expected to affect at least some community-wide patterns. For example, the presence of marshy vegetation is relatively unimportant to the overall breeding bird community because most species currently present do not require it or nest in it, but it is an absolute determinant of Common Yellowthroat (*Geothlypis trichas*) presence/absence. Large birds such as Brown-crested Flycatchers and Summer Tanagers may require larger patches and the presence of at least some very large trees in order to breed in a patch, even though most of the other, smaller bird species do not. Thus, individual life-history traits play a role in habitat use in the Grand Canyon.

It is also critical to remember that the results and models of this study are specific for the portion of the river corridor that we sampled, from Lees Ferry downstream to Diamond Creek. If our sampling had extended downstream below Separation Canyon to the Lake Mead interface, different and/or additional patterns would likely have emerged. Much of the habitat below Separation Canyon differs greatly from that upstream, being wider and more contiguous, with more marsh areas, and dominated by native willows, many of which are very tall and dense. As a result, the riparian bird community below Separation differs in having higher abundance of breeding Brown-crested Flycatchers, Song Sparrows (*Melospiza melodia*), Common Yellowthroats, Blue Grosbeaks (*Guiraca caerulea*), and Summer Tanagers. Also, some birds such as the Yellow-billed Cuckoo (*Coccyzus americanus*), Red-winged Blackbird (*Agelaius phoeniceus*), and Black-crowned Night Heron (*Nycticorax nycticorax*) breed below Separation but not above. These differences in bird community and habitat would have affected the details of our models, although the broad patterns of more species and individuals with increasing patch size would probably have been the same. This emphasizes the point made by Wiens (1989a), Hansen and Urban (1992), and Morrison et al. (1992) that care must be used in extrapolating habitat models from one area to another, even in what might be considered a single riparian system along the Colorado River.

Relationship to Flow-Induced Vegetation Changes

The general pattern that emerged is one whereby increasing patch size leads to increased bird abundance, richness, and diversity. This is true whether the vegetation responsible for the increase is NHWZ (dominated by tamarisk) or OHWZ (dominated by mesquite). Thus, the floristic details of habitat change are generally less important than changes in size, structure, and/or life form.

Flow patterns that result in smaller, more isolated patches can be expected to decrease bird numbers, richness, and diversity. If flow patterns create larger and more contiguous patches of habitat, regardless of the tree or shrub species, bird abundance and richness will increase in each patch. However, patch size in the Grand Canyon is strongly affected by local topography and geomorphology (Turner and Karpiscak 1980), and it may not be possible under any management strategy to greatly increase the size of many existing patches. Instead, future vegetation changes may take the form of loss of habitat (e.g., the gradual loss of mesquite in the OHWZ), increased number of patches (of a variety of sizes), and/or shifts in dominant species composition (e.g., from tamarisk to willow). Loss of mesquite vegetation will likely decrease bird abundance, as it was a significant factor in the habitat models. Increases in the number of patches will increase the overall number of birds and bird species in the canyon, and the exact size, width, and structure will determine which species will colonize and breed in the new patches. Patch size and structure can also affect predation, brood parasitism, and other habitat factors. Based on our models, floristic changes such as from tamarisk shrub/tree to willow shrub/tree are not likely to greatly affect the bird abundance and species richness, although it may result in some changes in the actual species composition at the patch level.

Implications to Future Monitoring

We found a strong positive correlation among most of the pairs of covariates. Some of these correlations are simply due to one covariate being a component of another covariate (e.g., tamarisk is the major component of NHWZ, and mesquite of OHWZ). Other correlations reflect the relatively simple vegetation community structure of the riparian corridor, which is dominated by only a few types of trees and shrubs. It also reflects the fact that patch size in this riparian system is strongly affected by geomorphology, and many areas that are topographically and geologically suitable for development of large patches of NHWZ are also suitable for large areas of OHWZ.

This correlation among covariates allows us to simplify our habitat model by excluding most covariates and focusing on those that are responsible for the most variation in bird abundance and richness. For a given level of predictive ability, estimation and measurement of fewer variables decreases research time and study costs. Given the choice between model parameters that can be measured relatively quickly, as opposed to those taking much more time, the manager and researcher may be best served by choosing the former. In our case, useful predictive covariates such as the total patch area and total vegetated area, as well as the area of NHWZ, OHWZ, mesquite, and tamarisk can be readily estimated on aerial photographs and rapidly ground-truthed. Tree volume produced the best predictive model, but the additional field effort needed to gather data to estimate volume (as opposed to area) may not be worthwhile given that several other area covariates produce very good models with nearly as high correlation coefficients. The use of remote sensing techniques holds great promise in simplifying habitat measurements, but may not work well in all systems or for all habitat questions (Sader et al. 1991).

Diversity indices have been widely used and provide interesting information; however, most diversity indices have severe analytical and statistical drawbacks (Wiens 1989a). For example, use of different indices on the same data sets may produce different patterns, and each diversity index is influenced somewhat differently by different aspects of the bird community. SDI, for example, is particularly sensitive to the presence of rare species. Although diversity indices can be expected to change with changes in the bird community, diversity index patterns can only be interpreted by looking at changes in each individual species and determining how these changes affect the diversity values (Wiens 1989a; Morrison et al. 1992). Such analyses are beyond the scope of our study, but we present SDI values because they are based on the most widely used index and may be useful for future comparative purposes.

The part of the river in which a patch was located (e.g., model parameter SEGMENT) clearly affected the bird community–habitat patterns that we noted. During subsequent research along the Colorado River corridor in the Grand Canyon, Spence (2004) also found a strong influence of patch location on bird habitat and communities. These results clearly indicate that the riparian bird community is not uniform throughout the length of the river corridor. In addition, future environmental changes are not likely to be equal or concurrent in all portions of the more than 350 km river corridor. Therefore, the canyon's bird species and communities will be subject to differing conditions, and may respond differently, depending on their location. Future bird monitoring programs should take these factors into account by (1) deciding which segments/river reaches are of interest and value for monitoring (e.g., all vs. a subset), (2) placing survey sites accordingly (e.g., throughout the river corridor or in a subset), and (3) assuring sufficient sample size in each segment to detect statistically meaningful trends (see discussion in Spence 2004). In addition, bird abundance and richness patterns may have to be analyzed and interpreted separately for different segments of the river to avoid missing spatially differing trends. Finally, it may be useful to develop and test novel ways to analyze bird survey data from a geographic distribution perspective.

ACKNOWLEDGMENTS

We thank the staffs of Grand Canyon National Park and Glen Canyon National Recreation Area for their support of this project, and the U.S. Bureau of Reclamation Glen Canyon Environmental Studies program (currently the U.S. Geological Survey Grand Canyon Monitoring and Research Center) for funding and substantial technical and logistical support. Salary support for Sogge and Wotawa was provided by the U.S. Geological Survey. Patricia Hodgetts and Helen Yard conducted hundreds of bird surveys over the course of this project; their efforts and contribution are greatly appreciated. Support from Christine Karas (Bureau of Reclamation) was crucial to this project. Paul Deshler provided critical GIS support, and Dan Spotsky (National Park Service) provided insight on spatial data issues. We are grateful for volunteer field assistance from John Grahame, Aurora Hindman, Bill Howe, Matthew Johnson, Albert Miranda, and Lia Spiegel. We appreciate the thorough and insightful reviews by Bryan Brown, John Spence, and an anonymous reviewer; their comments greatly strengthened this manuscript.

LITERATURE CITED

Anderson, L. S., and G. A. Ruffner. 1988. The effects of recent flooding on riparian plant establishment in Grand Canyon. U.S. Department of Commerce. National Technical Information Service. NTIS PB88-183504/AS.

Block, W. M., and L. A. Brennan. 1993. The habitat concept in ornithology. In Current Ornithology, edited by D. M. Power, no. 11, pp. 35–91.

Bonham, C. D. 1989. Measurements for Terrestrial Vegetation. J. Wiley & Sons, New York.

Brown, B. T. 1988a. Breeding ecology of a willow flycatcher population in Grand Canyon, Arizona. Western Birds 19: 25–33.

Brown, B. T. 1988b. Monitoring bird population densities along the Colorado River in Grand Canyon: 1987 breeding season. U.S. Department of Commerce. National Technical Information Service. NTIS P89-103311.

Brown, B. T. 1994. Rates of brood-parasitism by brown-headed cowbirds on riparian passerines in Arizona. Journal of Field Ornithology 65: 160–168.

Brown, B. T., and R. R. Johnson. 1985. Glen Canyon Dam, fluctuating water levels, and riparian breeding birds: The need for a management compromise on the Colorado River in Grand Canyon. USDA Forest Service General Technical Report RM-120: 76–80.

Brown, B. T., and R. R. Johnson. 1987. Fluctuating flows from Glen Canyon Dam and their effect on breeding birds of the Colorado River. U.S. Department of Commerce. National Technical Information Service. NTIS PB88-183512/AS.

Brown, B. T., and R. R. Johnson. 1988. Utilization of the Colorado River in Grand Canyon by migratory passeriformes as a corridor through arid lands. In The Proceedings of the Second Conference of Scientific Research in the National Parks. Volume 12: Terrestrial Biology: Zoology, pp. 85–96. National Park Service and American Institute of Biological Science.

Brown, B. T., and M. W. Trosset. 1989. Nesting-habitat relationships of riparian birds along the Colorado River in Grand Canyon, Arizona. Southwestern Naturalist 34: 260–270.

Brown, B. T., S. W. Carothers, and R. R. Johnson. 1987. Grand Canyon Birds. University of Arizona Press, Tucson.

Carothers, S. W., and B. T. Brown. 1991. The Colorado River through the Grand Canyon. University of Arizona Press, Tucson.

Collier, M., R. H. Webb, and J. C. Schmidt. 1996. Dams and rivers; primer on the downstream effects of dams. U.S. Geological Survey Circular 1126. USGS, Tucson, AZ.

Dieni, J. S. , and S. L. Jones. 2002. A field test of the area search method for measuring breeding bird populations. Journal of Field Ornithology 73: 253–257.

Felley, D. L., and M. K. Sogge. 1997. Comparison of techniques for monitoring riparian birds in Grand Canyon National Park. In Proceedings of the Third Biennial Conference of Research on the Colorado Plateau, edited by C. van Riper III and E. T. Deshler, pp. 73–83. National Park Service Transactions and Proceedings Series NPS/NRNAU/NRTP-97/12.

Hansen, A. J., and D. L. Urban. 1992. Avian response to landscape pattern: The role of species' life histories. Landscape Ecology 7: 163–180.

Keast, A., and E. S. Morton (Eds.). 1980. Migrant Birds in the Neotropics: Ecology, Behavior, Distribution, and Conservation. Smithsonian Institution Press, Washington, D.C.

Knopf, F. L., and F. B. Samson. 1994. Scale perspectives on avian diversity in western riparian ecosystems. Conservation Biology 8: 669–676.

Knopf, F. L., R. R. Johnson, T. Rich, F. B. Samson, and R. C. Szaro. 1988. Conservation of riparian ecosystems in the United States. Wilson Bulletin 100: 272–294.

Lowther, P. E. 1993. Brown-headed cowbird (*Molothrus ater*). In The Birds of North America, No. 47, edited by A. Poole and F. Gills. The Academy of Natural Sciences (Philadelphia) and The American Ornithologist's Union (Washington D.C.).

MacArthur, R. H., and J. W. MacArthur. 1961. On bird species diversity. Ecology 42: 594–598.

Martin, T. E., and D. M. Finch (Eds). 1995. Ecology and Management of Neotropical Migratory Birds: A synthesis and review of critical issues. Oxford University Press, New York.

Mills, G. S, J. B. Dunning, Jr., and J. M. Bates. 1991. The relationship between breeding bird density and vegetation volume. Wilson Bulletin 103: 468–479.

Morrison, M. L. 1986. Bird populations as indicators of environmental change. In Current Ornithology, edited by R. F. Johnston, no. 3, pp. 429–451.

Morrison, M. L., B. G. Marcot, and R. W. Mannon. 1992. Wildlife-Habitat Relationships; Concepts and applications. University of Wisconsin Press, Madison.

Neter, J., W. Wasserman, and M. H. Kutner. 1990. Applied Linear Statistical Models. Richard D. Irwin, Inc. Homewood, Illinois.

Norvell, R. E., F. P. Howe, and J. R. Parrish. 2003. A seven-year comparison of relative-abundance and distance-sampling methods. Auk 120: 1013–1028.

Ralph, C. J., G. F. Geupel, P. Pyle, T. E. Martin, and D. F. DeSante. 1993. Handbook of field methods for monitoring landbirds. General Technical Report PSW-GTR-144. USDA Forest Service.

Rappole, J. H. 1995. The Ecology of Migrant Birds: A Neotropical Perspective. Smithsonian Institution Press, Washington, D.C.

Remsen, J. V. Jr. 1994. Use and misuse of bird lists in community ecology and conservation. Auk 111: 225–227.

Robinson, S. K., S. I. Rothstein, M. C. Brittingham, L. J. Petit, and J. A. Gryzbowski. 1995. Ecology and behavior of cowbirds and their impact on host populations. In Ecology and Management of Neotropical Migratory Birds: A Synthesis and Review of Critical Issues, edited by T. E. Martin and D. M. Finch, pp. 428–460. Oxford University Press, New York.

Sader, S. A., G. V. N. Powell, and J. H. Rappole. 1991. Migratory bird habitat monitoring through remote sensing. International Journal of Remote Sensing 12: 363–372.

Schmidt, J. C., and J. B. Graf. 1988. Aggradation and degradation of alluvial sand deposits, 1965 to 1986, Colorado River, Grand Canyon National Park, Arizona. NTIS No. PB88-195458/AS.

Sogge, M. K., D. Felley, and M. Wotawa. 1998. Riparian bird community ecology in the Grand Canyon – Final report. U.S. Geological Survey Colorado Plateau Field Station report. Flagstaff, Arizona.

Sogge, M. K., T. E. Tibbitts, and J. A. Petterson. 1997. Status and ecology of the southwestern willow flycatcher in the Grand Canyon. Western Birds 28: 142–157.

Sokal, R. R., and F. L. Rolf. 1995. Biometry, 3rd ed. W. H. Freeman, New York.

Spence, J. R. 2004. The riparian and aquatic bird communities along the Colorado River from Glen Canyon Dam to Lake Mead, 1996–2002. Final report to the USGS Grand Canyon Monitoring and Research Center, Flagstaff, Arizona.

Stevens, L. 1983. The Colorado River in Grand Canyon: A Guide. Red Lake Books, Flagstaff, Arizona.

Stevens, L. E., and T. E. Ayers. 1994. The effects of interim flows from Glen Canyon Dam on riparian vegetation along the Colorado River in Grand Canyon National Park, Arizona: Draft 1994 Annual Technical and Administrative Report. U.S. Department of Interior, NBS, CPRS.

Turner, R. M., and M. M Karpiscak. 1980. Recent vegetation changes along the Colorado River between Glen Canyon Dam and Lake Mead, Arizona. U.S. Department of Interior. USGS Professional Paper 1132.

Verner, J. 1985. Assessment of counting techniques. Current Ornithology 2: 247–302.

Vickery, P. D., M. L. Hunter Jr., and J. V. Wells. 1992. Is density a misleading indicator of breeding success? Auk 109: 706–710.

Wiens, J. 1989a. The Ecology of Bird Communities: Volume 1—Foundations and Patterns. Cambridge University Press, Cambridge.

Wiens, J. 1989b. The Ecology of Bird Communities: Volume 2—Processes and Variations. Cambridge University Press, Cambridge.

Willson, M. F., and S. W. Carothers. 1979. Avifauna of habitat islands in the Grand Canyon. Southwestern Naturalist 24: 563–576.

Wilson, R. R., D. J. Twedt, and A. B. Elliott. 2000. Comparison of line transect and point counts for monitoring spring migration in forested wetlands. Journal of Field Ornithology 71: 345–355.

ANNOTATED SPECIES LIST OF BIRDS OF THE COLORADO RIVER CORRIDOR IN THE GRAND CANYON: 1993–1995

David L. Felley and Mark K. Sogge

The birds listed here were detected during our 1993–1995 studies of the riparian avian community of Colorado River, between Lees Ferry and Diamond Creek. We have included all records of birds that we saw or heard, including non-riparian species such as waterfowl, waders and shorebirds, raptors, and upland-associated songbirds. Other reports and publications also contain more detailed discussions of riparian breeding bird habitat associations and banding results (Sogge et al. 1998; Sogge et al. this volume), survey methods (Felley and Sogge 1997), and avian diet (Yard et al. 2004).

This is not intended to be a cumulative list of all bird species ever detected along the Colorado River corridor, as was provided by Brown et al. (1984) and Brown et al. (1987). Although not "comprehensive" from that standpoint, the information that we present adds substantially to previously published information, and further refines the status, distribution, and seasonality of many species. Furthermore, our data provide a "snapshot in time" of the diversity, distribution, and abundance of birds in the canyon during the period of our survey; such information can facilitate valuable comparisons with future avian research and monitoring data.

STUDY SITES & SURVEY TRIP DATES

We established study sites in a representative sample of riparian habitats along the Colorado River corridor between Lees Ferry and Diamond Creek. Vegetation communities change from Great Basin desert-scrub to Sonoran desert-scrub along the elevational gradient of the river corridor in the canyon (Turner and Karpiscak 1980). Also, characteristics of riparian vegetation relate to the geomorphology of the 11 reaches of the river (as defined by Schmidt and Graf 1988; Figure 1). For these reasons, we surveyed sites in each reach, within the constraints of river-trip logistics. We adjusted the number of survey sites according to the abundance of riparian vegetation in a reach. For example, we spent more time in the Lower Canyon (which supports extensive riparian habitat) than in the Upper Granite Gorge (with few and smaller habitat patches). At the intermediate scale, within each reach, we selected the habitat patches to survey in a non-random fashion, in order to obtain a broad distribution of spatial, geomorphological, and vegetative characteristics. Details of location and habitat characteristics for all study sites are presented in Sogge et al. (1998).

We conducted survey trips in May, June, July, and September of 1993; January, March, April, May, June, July, and September of 1994; and February, March, April, May, and June of 1995. Though the focus of this project was to survey riparian breeding birds, trips in September allowed us to document use of the river corridor by fall migrants, and January and February trips allowed us to add to the small amount of information on winter bird use in the canyon.

Birds included in this list were detected during formal surveys (point-counts, walk-

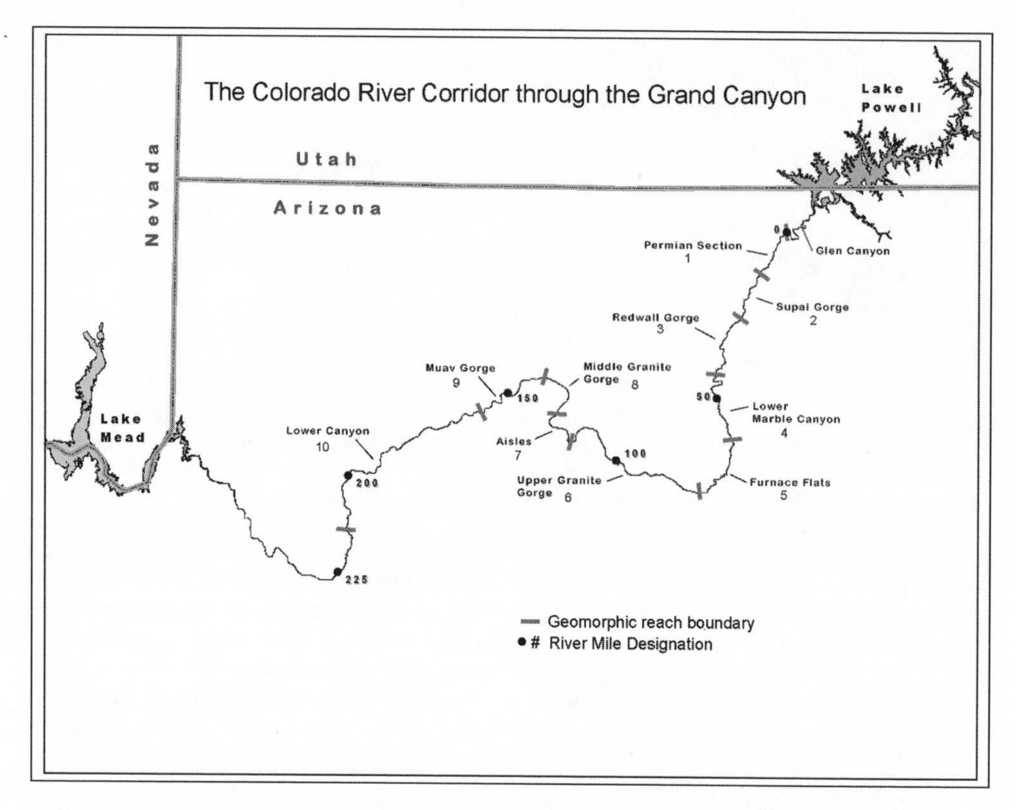

Figure 1. The Colorado River corridor in Arizona, between Lake Powell and Lake Mead. River miles are indicated by the filled circles and associated numbers. The geomorphic reaches referred to in the text are indicated by reach names and numbers (1–10); shaded lines that bisect the river represent geomorphic reach boundaries.

ing counts, and floating counts) and mist-netting (see Felley and Sogge 1997, Sogge et al. 1998 for details), as well as through incidental observations. Avian taxonomy and the order of the species list follows the American Ornithologists Union (AOU) Checklist of North American Birds (1998), updated through the July 2002 AOU Checklist Supplement (AOU 2002). Scientific names of all species referred to in the text are presented in the annotated species accounts. Locations of study sites and detections are given in river miles (RM) below Lees Ferry (Stevens 1983) followed by the designation right bank (R) or left bank (L) as viewed facing downstream.

WINTER BIRDS

Although our two winter visits amounted to only 23 days, our observations are discussed in detail because of the paucity of information on wintering birds in the canyon. The 17 January to 1 February 1994 trip included surveys from Lees Ferry to Parashant Wash, and mist-netting/banding at major sites: RM 1.0R, 46.7R, 171.0R, 198.0R, and RM 204.5R. Our 9–18 February 1995 trip entailed surveys from Lees Ferry to RM 214, but no banding.

Passeriformes, the perching or song birds, account for the largest number of species wintering in riparian habitats in the canyon. During our January 1994 trip we recorded 13 species: Say's Phoebe, Bewick's Wren, Winter Wren, Common Raven, Bushtit, Mountain Chickadee, Ruby-crowned Kinglet, Western Bluebird, Hermit Thrush, Yellow-rumped Warbler, Rufous-crowned Sparrow, Dark-eyed Junco, and Song Sparrow. Canyon and Rock Wrens were present but they rarely used riparian habitats. In addition to passerines, we found Red-naped Sapsucker, Northern Flicker (red-shafted subspecies), and Gambel's Quail. In February of 1995 we detected 23 passerine species. In addition to those already listed were Phainopepla, American Pipit, Crissal Thrasher, American Robin, Townsend's Solitaire, Scrub Jay, Canyon Towhee, Lincoln's Sparrow, and White-crowned Sparrow. The Mountain Chickadee and Winter Wren were the only species present in January 1994 but absent in February 1995.

During our surveys, riparian birds were less common in January 1994 (74 individuals, 12 species) than in February 1995 (126 individuals, 17 species). Abundance and diversity may have been higher in 1995 simply because we surveyed more sites and went farther down the river (to RM 214), or because a milder winter in 1994–95 may have brought more wintering species into the canyon. Granivorous species were much more common in winter than in other seasons, and insectivorous species were correspondingly rare.

MIGRATING BIRDS

We found 53 species using riparian habitat only during their migration. These may be separated into short-distance and long-distance migrants. Short-distance migrants winter elsewhere in the state, or in northern Mexico. They typically arrive in the canyon earlier in the spring (as early as February for the Costa's Hummingbird) and many depart later in the fall, even in November. A few, such as the Phainopepla, may behave as winter resident or short-distance migrant in different years depending on food availability and winter weather. Short-distance migrants include the Mourning Dove, Common Poorwill, White-throated Swift, Costa's Hummingbird, Belted Kingfisher, Ash-throated Flycatcher, Say's and Black Phoebes, Western Scrub Jay; House, Winter, and Marsh wrens; Northern Mockingbird, most thrushes, Blue-gray Gnatcatcher, Phainopepla, Loggerhead Shrike, Lucy's Warbler, most sparrows, Lesser Goldfinch, and House Finch.

Long-distance migrants winter from southern Mexico into South America. Most of these species pass through the canyon in late April and May, with fall migration starting in late July or August. In the canyon, this group includes the Vaux's Swift, Black-chinned and Broad-tailed Hummingbirds, Western and Cassin's Kingbirds, *Empidonax* flycatchers, swallows, vireos, most warblers, blackbirds and orioles, tanagers, and grosbeaks. Species such as the Lazuli Bunting fall between the categories of short-distance and long-distance migrants (Phillips et al. 1964).

Several avian species are represented in the canyon by both migrant and breeding populations, which presents a problem for long-term monitoring programs (Sogge et al., this volume) and for determining which species are actually breeding in the canyon. Repeated surveys of the same sites, and examination of the breeding condition of captured birds, allowed us to identify migratory peaks of abundance and to frequently differentiate breeders from migrants. It was clear from our findings that May counts of species such as Common Yellowthroat, Yellow-breasted Chat, and especially Yellow Warbler are inflated by migrants. Future surveys of riparian breeders should take this into account when timing surveys and developing population models (Sogge et al., this volume).

Mist nets provided an invaluable tool for distinguishing migrants from breeders, and for detecting furtive species. Unlike breeders, many migrant species make no sounds, or only quiet and nondescript call notes, and stay low in thick brush-heavy cover. This makes them difficult to find and identify, while making mist-netting more effective. Eight species of migrants were observed only through mist net capture: Dusky and Hammond's Flycatchers, Plumbeous and

Cassin's Vireos, Virginia's Warbler, Northern Waterthrush, Black-throated Blue Warbler, Blackpoll Warbler, and Golden-crowned Sparrow.

BREEDING BIRDS

The breeding bird community in the canyon is fairly typical of riparian areas in the Southwest. We found 30 species breeding in riparian habitat in our study sites; only five were year-round residents of the canyon: Bewick's Wren (but see species account below), Song Sparrow, Verdin, Gambel's Quail, and Wild Turkey. These last four species occur as localized populations of relatively few individuals.

Hence, the breeding community is dominated by species that migrate south during the non-breeding season. The breeding season starts first in the lower canyon as early as February, with the arrival of the Costa's Hummingbird, and is in full swing throughout the canyon by March with the arrival of short-distance migrants, most notably the Lucy's Warbler. Ash-throated Flycatchers, Yellow Warblers, Common Yellowthroats, and Yellow-breasted Chats start arriving in April, and most of the rest of the breeders arrive in May. A few species, the Blue Grosbeak in particular, arrive as late as June in some locations. Blue Grosbeaks are still observed in undiminished numbers in July, when the abundance of most other breeding species declines dramatically.

The distribution of breeding species varied spatially in the canyon. Some species are more upland associated, or able to use very small patches of habitat for breeding. These species, including the Lucy's Warbler, Blue-gray Gnatcatcher, and Ash-throated Flycatcher, bred in every reach surveyed and were the only riparian species that we found in many stretches of the sparsely vegetated Upper Granite Gorge. Other species appeared to be tied to very specific habitats. The Common Yellowthroat, for example, only bred where there were patches of emergent vegetation (esp. *Phragmites* and *Typha* spp.) suitable for nesting. Most of the obligate riparian species in the canyon were found only in the more heavily vegetated reaches.

Broader scale biogeographic factors were also important in determining bird distribution throughout the canyon. Species most closely associated with the Sonoran Desert were found only in the Lower Canyon. These include the Gambel's Quail, Ladder-backed Woodpecker, Brown-crested Flycatcher, Phainopepla, and Verdin. The presence of these Sonoran Desert species gave the Lower Canyon the highest breeding bird diversity of any reach.

ANNOTATED SPECIES ACCOUNTS

The following list includes all bird species that we observed during fieldwork on the Grand Canyon Avian Community Monitoring project, conducted between May 1993 and June 1995, between Lees Ferry and Diamond Creek (see details in Sogge et al. 1998). Observations are from walk-through and point count surveys ("Surveys"), banding operations ("Banding"), and general observations as recorded in field notes ("Incidental"). Note that banding was conducted only at RM 1.0R, 46.7R, 171.0R, 198.0R, and 204.5R. Unless otherwise noted observers consisted of one or more of the regular field crew—David Felley, Patti Hodgetts, Mark Sogge, and Helen Yard. Where appropriate, observers have been identified by initials as follows: David Felley–DF, Mark Sogge–MS.

Because this project focused on birds of Colorado River riparian habitats, groups not frequenting these habitats are not as well represented. These include waterfowl, raptors, aerial foragers such as swifts and swallows, and more upland-associated species. For a discussion of wintering Bald Eagles, see van Riper and Sogge (2004).

Definitions of abundance and residency status are taken from Brown et al. (1984) and are consistent with Brown et al. (1987). However, these categories are used to refer to the river corridor only rather than the larger geographic scope of these references. Species for which we confirmed breeding in the riparian habitat of the Colorado River are marked with an asterisk (*). Observations are listed by RM and date, generally in upriver to downriver order. Unless noted otherwise, each listed observation is of a single bird.

Definitions of abundance and residency status are provided below.

Relative Abundance

Common—Easily found in the proper habitat and season and always to be seen.

Fairly Common—Found in low numbers or scattered throughout the proper habitat during the right season; seldom or infrequently seen on any given day.

Uncommon—May or may not be found with difficulty in the proper habitat and right season.

Rare—Not to be expected, occurrence unpredictable.

Residency Status

Permanent resident—Species that remain within the region throughout the year.

Summer resident—Species that remain within the region during summer, usually nesting.

Summer visitor—Occurring but not nesting in an area during summer.

Winter resident—Species that reside within the region during winter.

Winter visitor—Species occurring in the region during winter on a temporary basis.

Transient—Species that pass through the area, usually during spring & fall migration.

Casual—A species out of its normal range, but to be expected occasionally.

Accidental—A species far from its normal range and not to be expected again.

Loons and Grebes

Pacific Loon (*Gavia pacifica*): Accidental.
Incidental: Lees Ferry, 1/7–8, and 3/12/94 (MS). This is the first record of this loon within the study area, although Spence and Bobowski (2003) have subsequently reported it as a rare migrant and possible winter resident upriver on Lake Powell.

Common Loon (*Gavia immer*): Casual.
Incidental: One individual in basic plumage; first seen at RM 72, then several more times before last sighting at RM 88 (at Phantom Ranch), 3/19/95.

Pied-billed Grebe (*Podilymbus podiceps*): Rare transient.
Incidental: Lees Ferry 3/13/94 and 4/13/95.

Eared Grebe (*Podiceps nigricollis*): Rare transient.
Incidental: Between Lees Ferry and RM 10, 4/13/94; just above Crystal Rapid (RM 98), 4/18/95.

Western Grebe (*Aechmophorus occidentalis*): Rare transient.
Incidental: RM 93.3, 4/18/95; RM 165, 5/20/95.

Cormorants

Double-crested Cormorant (*Phalacrocorax auritus*): Uncommon winter and summer visitor.
Incidental: Between Lees Ferry and RM 41, 2/14/93; RM 123, 4/19/95; RM 199.5, 6/10/94 (2 adults); RM 214.0, 6/24/94; RM 10, 6/14/95 (second-year plumage).

Herons and Egrets

Great Blue Heron (*Ardea herodias*): Fairly common summer and winter visitor.
Surveys: Eight – January through July. Most frequently in Reaches 1, 2, and 4.
Incidental: RM 1.0R–RM 1.6R, 6/2/94 and 7/12/94; "one every mile or so" between Phantom Ranch (RM 88) and RM 114, 2/12/95 (DF field notes); earliest on 1/20/94 below Lees Ferry; latest near RM 190, 9/22/94.

Great Egret (*Ardea alba*): Rare transient.
Incidental: RM 194, 1/30/94 (3).

Snowy Egret (*Egretta thula*): Uncommon transient.
Surveys: RM 214.0L, 4/26/95.
Incidental: RM 1.0R, 9/12/94; Reach 1, 5/7/94; RM 72, 4/17/95; above RM 88 and at RM 108, 4/18/95; RM 122.8, 4/20/94 (6); RM 190, 9/22/94.

Cattle Egret (*Bubulcus ibis*): Casual visitor.
Incidental: RM 93.3, 4/18/95.

Green Heron (*Butorides virescens*): Rare summer visitor.
Surveys: RM 1.6R, 5/7/94; RM 204.5R, 4/27/94.

Black-crowned Night-Heron (*Nycticorax nycticorax*): Rare summer visitor.
Surveys: RM 5.2R, 4/14/95 (sub-adult).
Incidental: Lees Ferry, 6/1/94 (2 subadults); RM 168.8, 5/14/94 (sub-adult).

White-faced Ibis (*Plegadis chihi*): Uncommon transient.
Incidental: RM 71 (Cardenas Marsh), 4/17/95; RM 167, 7/21/94 (3); RM 198.0R, 4/22/95.

New World Vultures

Turkey Vulture (*Cathartes aura*): Common transient.
Surveys: RM 174.2L, 5/14/94; RM 174.7, 4/20/95 (2); RM 174.4, 5/21/95 (3).
Incidental: Numerous groups. First spring migrants were 3/30/94 and 3/23/95. Latest sightings 9/22/94.

Ducks, Geese, and Swans

Note: Due to differences in water turbidity and depth, the abundance and distribution of waterfowl varies greatly upriver and downriver of Lees Ferry, and on nearby Lake Powell. Therefore, the status described for each species below—derived primarily from below Lees Ferry—may differ from that described by Spence (2004) and Spence and Bobowski (2003). Also, waterfowl numbers were often tallied during downriver transit, so some totals represent relatively long stretches of river.

Canada Goose (*Branta canadensis*): Uncommon winter visitor.
Surveys: RM 1.6R, 1/20/94 (17); RM 1.6R, 4/13/94.
Incidental: 35 between RM 52 and RM 88, 1/22/94; RM 52, 2/11/95 (20–25); RM 171.0R, 1/26/94 (10–15).

Wood Duck (*Aix sponsa*): Casual transient.
Incidental: Above RM 31.6 (South Canyon), 5/15/95 (female).

Gadwall (*Anas strepera*): Common winter resident.
Incidental: Commonly seen at Lees Ferry in winter. Also between Lees Ferry and RM 40, 2/14/93.

American Wigeon (*Anas americana*): Common winter resident.
Incidental: Common at Lees Ferry, 2/9/95; 76 between Lees Ferry and RM 52, 2/19/94; 10 between RM 5.2 and RM 46.7, 3/16/95; 4 between RM 52 and RM 88, 1/22/94; between RM 88 and RM 123, 9/19/94.

* Mallard (*Anas platyrynchos*): Fairly common summer and winter resident.
Surveys: Numerous, mostly in Reach 4. January (3 males), February (26, incl. 23 males), March (two pairs), May (5), June (3). In Reach 1–April (2) and June.
Incidental: Female incubating 10 eggs at Lees Ferry, 6/1/94; 30 between RM 5.2 and RM 46.7, 2/16/95; ducklings (7) at RM 43, 5/15/95; duckling at RM 46.7, 5/15/95; female flushed from grassy uplands (probably off a nest) at RM 48.5L, 4/17/94; nine ducklings in eddy below RM 49.1R, 5/16/95.
Note: This was the only species of duck that we found breeding in the canyon. Broods most often are seen in April, May, and June between RM 31.6 (South Canyon) and RM 52 (Nankoweap Creek). Although most common in winter and summer, some seen in every month of the year.

Blue-winged Teal (*Anas discors*): Uncommon transient.
Incidental: RM 66, 4/17/94 (2); RM 70, 5/11/94 (ca. 90); 21 between RM 123 and RM 167, 9/19/94; six between RM 171 and RM 176, 9/20/94; RM 190, 9/22/94 (18).

Cinnamon Teal (*Anas cyanoptera*): Uncommon transient.
Incidental: RM 1.6R, 5/15/95; six between Lees Ferry and RM 54, 2/19/94; three between RM 10 and RM 46, 4/13/94; RM 55,

4/17/94 (5); between RM 97 and RM 123, 5/12/94; RM 107, 4/18/95 (pair).

Northern Shoveler (*Anas clypeata*): Rare transient.
Incidental: One between Lees Ferry and RM 54, 2/19/94; two between Lees Ferry and RM 47, 4/19/94; one between RM 5 and RM 46.7, 4/14/95; 21 between RM 86 and RM 123, 9/19/94.

Northern Pintail (*Anas acuta*): Uncommon transient and winter visitor.
Surveys: RM 1.6R, 2/9/95 (10).
Incidental: 21 between Lees Ferry and RM 54, 2/19/94; 20 attacked by a Peregrine Falcon at RM 69, 9/17/94.

Green-winged Teal (*Anas crecca*): Fairly common transient and winter visitor.
Incidental: Numerous sightings. Detections in January (22), February (35), March (30), May (5), September (54), ranging from Lees Ferry to RM 190.

Canvasback (*Aythya valisineria*): Rare transient.
Incidental: "Occasionally seen" at Lees Ferry (DF field notes).

Redhead (*Aythya americana*): Uncommon winter visitor and transient.
Incidental: Lees Ferry, 6/4/93; 20 between Lees Ferry and RM 52, 2/19/94; RM 1.4, 5/12/95 (3).

Ring-necked Duck (*Aythya collaris*): Uncommon winter visitor and transient.
Incidental: 120 between Lees Ferry and RM 52, 2/19/94; two between RM 52 and RM 88, 2/23/93; RM 171.1R, 9/22/94 (2).

Lesser Scaup (*Aythya affinis*): Fairly common transient.
Incidental: A few between RM 1.0 and 46.7, 1/20/94; five between RM 52 and RM 88, 1/22/94; four between RM 52 and RM 88, 2/23/93; one taken by Peregrine Falcon at RM 60, 3/20/95.

Surf Scoter (*Melanitta perspicillata*): Casual.
Incidental: Lees Ferry, 1/18/94 and 4/11/94.
Notes: Also at Lees Ferry, 7/21/92 (prestudy detection by MS). Considered by Spence and Bobowski (2003) as rare migrant on Lake Powell.

Bufflehead (*Bucephala albeola*): Fairly common winter resident and transient.
Surveys: RM 1.0R, 3/13/94 (3); RM 1.6R, 3/16/94; RM 74.4L, 4/18/95 (2).
Incidental: 190 between Lees Ferry and RM 52, 2/19/94; "numerous" between RM 1.0 and 46.7, 3/16/94; "several" between RM 1.0 and RM 46.7, 1/20/94; 5 at RM 7, 2/9/95; five between RM 10.0 and RM 46.7, 4/13/94; 8 between RM 52 and RM 88, 1/22/94.

Common Goldeneye (*Bucephala clangula*): Common winter resident.
Surveys: 13 in January 1994, 17 in February 1995, and 9 April 1995.
Incidental: Many detections of numerous individuals. For example, 650 between RM 1.0 and RM 40, 2/14/93; 163 between RM 5.0 and 46.7, 3/16/95. Though most common in Reach 1, also seen in Reaches 4, 5, and 10.

Hooded Merganser (*Lophodytes cucullatus*): Rare winter visitor and transient.
Incidental: Lees Ferry, 1/17/94 (2 males); Lees Ferry, 2/14/93 (2 males, 3 females); just above Lees Ferry, 3/13/94 (4 females).

Common Merganser (*Mergus merganser*): Fairly common winter resident and transient, uncommon summer resident.
Surveys: RM 49.2L, 2/11/95 (3).
Incidental: Detected on 8 occasions, totaling 72 individuals, between Lees Ferry and RM 87; most before RM 52.

Ruddy Duck (*Oxyura jamaicensis*): Rare transient.
Incidental: Lees Ferry, 1/18/94 (5); Lees Ferry, 3/12/94.

Kites, Eagles, Hawks, and Falcons

Osprey (*Pandion haliaetus*): Uncommon transient.
Incidental: Lees Ferry, 4/14/94 and 5/15/95; above RM 98 (Crystal Rapid), 4/18/95; RM 74.4, 9/18/94; RM 196, 9/24/95; RM 204, 9/25/94.

Bald Eagle (*Haliaeetus leucocephalus*): Fairly common winter resident.
Surveys: RM 50.0R, 2/11/95 (adult).
Incidental: Two between RM 1.0 and RM 46.7, 3/16/94; five to eight between RM 17 and RM 46.7, 1/20/94; RM 52, 1/22/94 (2); five between RM 32 and RM 34, 2/9/95; Nankoweap Creek (RM 52), 2/11/95 (2).
Note: See van Riper and Sogge (2004) for additional details on Bald Eagles in the Colorado River corridor.

Northern Harrier (*Circus cyaneus*): Rare transient.
Incidental: Lees Ferry, 1/6/95; over uplands at RM 1.0, 1/18/94; RM 218, 7/27/95.

Sharp-shinned Hawk (*Accipiter striatus*): Uncommon transient.
Surveys: RM 46.7R, 3/17/95; RM 198.0R, 3/30/94 (sub-adult); RM 205.8R, 9/28/94 (sub-adult).
Incidental: Lees Ferry, 1/6/95 (adult and sub-adult); RM 15, 1/20/94; RM 50, 9/17/94; RM 14.5, 3/16/94; RM 198.0R, 3/24/95 and 9/23/94 (sub-adult); RM 204, 9/28/94.

Cooper's Hawk (*Accipiter cooperii*): Rare transient.
Surveys: RM 204.5R, 9/27/94.
Incidental: RM 1.0R, 8/12–13/94 and 10/13/94; RM 198.0R, 9/23/94 (sub-adult).

Northern Goshawk (*Accipiter gentilis*): Rare transient.
Incidental: RM 52, 4/17/94. Not previously reported along river by Brown et al. (1987).

* Red-tailed Hawk (*Buteo jamaicensis*): Uncommon year-round visitor.
Surveys: March (3), April (1), May (2), July (2).
Incidental: High above river at RM 46.7, 3/17/95; RM 80, 9/19/94 (eating a rock squirrel); RM 125, 3/21/95.

Golden Eagle (*Aquila chrysaetos*): Uncommon transient and rare summer visitor.
Incidental: One between Lees Ferry and RM 5.6, 4/13/94; high above river at RM 46.7, 3/17/95 and RM 74, 2/10/95; RM 71, 4/17/94; RM 204.5, 6/26/95 (sub-adult).

American Kestrel (*Falco sparverius*): Uncommon transient.
Surveys: RM 50.0R, 7/18/94; RM 174.7R, 4/20/95; RM 200.4R, 9/25/04.
Incidental: Lees Ferry, 1/6/95; three between RM 1.0 and RM 46.7, 3/16/94; one between RM 10 and RM 46.7, 4/13/94; RM 49.2L, 7/10/93; RM 74.4, 9/18/94; one between RM 88 and RM 123, 9/19/94; ca. RM 200, 9/25/94.

* Peregrine Falcon (*Falco peregrinus*): Uncommon year-round resident, fairly common transient.
Surveys: 12 total – February (1), March (5), May (4), June (2).
Incidental: Numerous observations. Most in March and September, and in Reaches 1, 4, and 5. Breeding confirmed at RM 46.7R, 7/15/94 (pair with a first-year bird). A pair harassing a Red-tailed Hawk above the river at RM 125.5, 3/21/95—probably territorial.
Notes: Multiple observations of Peregrines foraging. Female captured a Brewer's Sparrow below RM 1.0R, 9/13/94; a pair, one carrying probable teal, at RM 13, 3/16/95; one killed a Lesser Scaup at RM 60, 3/20/95; one stooped on a flock of Northern Pintails at RM 69, 9/17/94; pair diving on White-throated Swifts and bats at RM 204.5R, 5/18/94; one killed a female teal at RM 209, 4/1/94.

Prairie Falcon (*Falco mexicanus*): Rare transient.
Incidental: Above river just below RM 180 (Lava Falls), 6/9/94.
Notes: Although described by Brown et al. (1987) as "frequently seen along river and on the rims," we detected only one.

Gallinaceous Birds

* Wild Turkey (*Meleagris gallopavo*): Uncommon localized summer resident.
Surveys: RM 46.7R, 7/15/94; RM 49.1R, 4/17/95; RM 50.0R, 6/5/94.
Incidental: We observed this species many times, always in Reach 4: tracks at RM 46.7R, 1/22/94; two flew across river at 46.7L, 5/8/94; RM 47, 6/9/93 (female); tracks in wet sand at RM 49.1R, 4/17/95; RM 49.2L, 5/10/94; one flew across river at RM 50.0R, 6/5/94; RM 50.0R, 9/17/94 (2); one male and three females at RM 52L, 4/17/94. Dave DeRoissers (National Park Service River Ranger) showed us Wild Turkey eggshells at RM 51.0L, 6/5/94, confirming breeding along the river.
Notes: Monson and Phillips 1981 (p. 30) has one record from "the bottom of the Grand Canyon, 9 Aug. 1970." Brown et al. (1987, p. 177) did not consider this species a permanent resident along the river, but stated, "a few are even reported from along the river." Turkeys are now seen fairly frequently by river guides, and have a self-sustaining, summer resident population along the river. Local populations probably originated from reintroductions on the Kaibab Plateau (Phillips et al. 1964, p. 30).

Gambel's Quail (*Callipepla gambelii*): Uncommon permanent resident.
Surveys: RM 198.0R, 1/30/94 (3); at RM 208.7R in 1994, May (1), June (3), and July (13).
Notes: The status of this species along the river corridor is unclear. Statewide, it is an "abundant resident in all areas where mesquite occurs, including the Grand Canyon" (Monson and Phillips 1981, p. 28). In the Grand Canyon, Brown et al. 1987 (p. 178) consider it "common and localized," but only upstream to around RM 249.

Rails and Coots

Virginia Rail (*Rallus limicola*): Rare transient.
Incidental: Lees Ferry, 5/5/94; carcass found on beach at RM 1.0R, 6/20/95; RM 59L, 9/17/94.

Sora (*Porzana carolina*): Rare transient.
Surveys: In a marsh 6.5 miles above Lees Ferry, 6/3/95; in arrowweed patch at RM 202.5R, 4/27/94.
Notes: There are very few records for the Sora in the Grand Canyon (Brown et al. 1987, p. 180).

American Coot (*Fulica americana*): Fairly common winter resident.
Surveys: RM 1.0R, 4/12/95.
Incidental: Lees Ferry, 1/7/94 (7), 1/18/94 (2), 2/9/95 (several), and 4/14/95.
Note: Commonly seen at and above Lees Ferry during winter; uncommon downstream (MS).

Shorebirds

Killdeer (*Charadrius vociferus*): Fairly common summer resident and uncommon winter visitor.
Surveys: 17 detections, February through May. All but two between RM 1.0 and RM 5.2; the exceptions at RM 108R, 3/22/93 and RM 213.6L, 3/28/95.
Incidental: Lees Ferry 4/14/95; between Lees Ferry and RM 10, 4/13/94; between RM 54 and RM 88, 1/22/94.

Black-necked Stilt (*Himantopus mexicanus*): Rare transient.
Incidental: RM 1.4, 4/13/94.

Lesser Yellowlegs (*Tringa flavipes*): Casual transient.
Incidental: RM 204.5R, 4/25/95.

* Spotted Sandpiper (*Actitus macularia*): Fairly common summer resident.
Surveys: 23 detections. March (1), May (14), June (2), July (3), and September (3). Primarily Reaches 1, 4, and 5, but also 7 and 10.
Incidental: RM 1.0R, 7/12/94; between Lees Ferry and RM 46.7, 9/14/94 (8); between RM 5.0 and 46.7, 5/15/95 (18); RM 52, 1/22/94.
Notes: An adult with two chicks at RM 1.6R, 7/12/94, was our only breeding record.

Wilson's Snipe (*Gallinago delicata*): Rare transient.
Incidental: Lees Ferry, 1/19/94 and 1/6/95; RM 122.8L, 3/23/94.

Gulls

Ring-billed Gull (*Larus delawarensis*): Uncommon transient.
Surveys: RM 171.1R, 5/14/94.
Incidental: Lees Ferry, 4/14/95 (20); RM 1.4, 5/12/95 (3 sub-adults); RM 46.7R, 4/14/94 (6); RM 46.7R, 4/15/94 (44); RM 52, 5/18/95 (sub-adult); above Little Colorado River (RM 62), 3/21/94 (3); RM 70, 3/18/95 (8); "small flock" between RM 97 and RM 123, 5/12/94.

California Gull (*Larus californicus*): Rare transient.
Incidental: One with group of Ring-billed Gulls between RM 97 and RM 123, 5/12/94; below RM 204.5, 3/28/95.

Pigeons and Doves

Rock Dove (*Columba livia*): Rare transient.
Incidental: Lees Ferry on 1/18/94, 3/13/94, and 4/12/94.

* Mourning Dove (*Zenaida macroura*): Common summer resident.
Surveys: 74 observed on 65 surveys, most common in May (25) and June (29). Most in Reaches 1 (33) and 10 (23); also in Reaches 4 (10), 7 (5), and 5 (3).
Incidental: Too numerous to list. Earliest and latest dates: 4/17 and 9/21.
Notes: We found four active nests between 5/18 and 7/6; Reaches 1, 7, and 10.

Roadrunners

Greater Roadrunner (*Geococcyx californianus*): Uncommon permanent resident.
Incidental: Lees Ferry on 5/8/94, 5/30/94, 7/11/94, and 10/12/94; RM 1.0R, April 1993; RM 185, 6/18/93; RM 216, June 1993.
Notes: This species may be expanding its range in the canyon, as the farthest upstream record in Brown et al. 1987 (p. 192) is at RM 52.

Owls

Great Horned Owl (*Bubo virginianus*): Uncommon permanent resident.
Incidental: RM 5.2R, 5/13/95; RM 43, 4/15/95 and 5/15/95; RM 46.7R in June 1993, 3/19/94, 4/15/94, 3/14/95 (calling); mouth of Mohawk Canyon, RM 171.5L, 1/28/94 (calling); one between RM 171.0 and RM 198.0, 9/22/94.

Spotted Owl (*Strix occidentalis*): Rare permanent resident.
Incidental: One was heard calling at narrowest section of Spring Canyon, a tributary at RM 204.5R, several miles from the river, 4/24/95 and 6/24/95. This is a very rare record below the canyon's rim.

Goatsuckers

Common Poorwill (*Phalaenoptilus nuttallii*): Uncommon summer resident.
Incidental: One between RM 1.0R and RM 46.7R, 5/15/95; RM 47, 6/8/93 (calling); RM 47R, 4/15/95 (calling); remains found at RM 206R, 5/20/94; flushed from dry wash above RM 224L, 3/28/95.

Swifts

Vaux's Swift (*Chaetura vauxi*): Rare transient.
Surveys: RM 1.0R, 9/14/94 (2).
Incidental: RM 122.8L, 9/19/94 (5).

White-throated Swift (*Aeronautes saxatalis*): Common summer resident.
Surveys: 152 on 12 surveys. Date range 3/17 to 9/17. Most commonly seen in Reach 4.

Hummingbirds

* Black-chinned Hummingbird (*Archilochus alexandri*): Common summer resident.
Surveys: 284 individuals on 264 surveys. March (51), April (90), May (80), June (46), July (15). We found them breeding in every survey reach and at most survey sites. Earliest dates were 3/19/95 and 3/20/94, both at RM 46.7R. Latest date was 9/7/93 at RM 171.0R.
Notes: Brown (1992) reported all nests in tamarisk, and none in old high water zone vegetation. Three of the 13 nests we found

(23%) were in hackberry trees in the old high water zone, and only five (38%) were in tamarisk. However, these differences may be due to different nest-search methodologies.

* Costa's Hummingbird (*Calypte costae*): Fairly common summer resident.
Surveys: 34 detections – February (1), March (15), April (16), May (2). Earliest date was 2/17/95 at RM 206.5L, and latest was 5/22/95 at RM 197.6L.
Incidental: RM 46.7R, 3/17/95 (male); RM 108R, 3/22/94 and 4/19/94 (male displaying/singing); RM 171.0R, 3/27/94 (male); RM 179, 3/28/94 (male singing); RM 198.0R, 3/29/94; RM 204.5R, 3/25–27/95 and 4/24–26/95; RM 217L, 3/28/95; RM 224L, 4/26/95.
Notes: This species is "the dry desert hummingbird *par excellence*" (Phillips et al. 1964, p. 62) and is among the first spring migrants, 17 February not being exceptionally early. It is a Sonoran Desert species, and 31 of 34 survey records are from the lower canyon (Reaches 10 and 11). By late May, they have all left the canyon (Brown et al. 1987, p. 201), apparently for the coast of California and Baja (Phillips et al. 1964, p. 62).

Broad-tailed Hummingbird (*Selasphorus platycercus*): Rare transient.
Incidental: RM 204.5, 4/27/94 (male).

Rufous Hummingbird (*Selasphorus rufus*): Uncommon fall transient.
Incidental: RM 1.0R, 7/26/94 (adult male); RM 1.0R, 8/11/94 (2); RM 49.0, 7/17/94.

Kingfishers

Belted Kingfisher (*Ceryle alcyon*): Fairly common transient.
Surveys: RM 204.5R, 4/24/95.
Incidental: Between RM 1.0 and RM 46.7, 4/14/95; four between Lees Ferry and RM 46.7, 9/14/94; RM 90, 3/19/95 1995 (female); "numerous" (all males) along river above RM 88 (Phantom Ranch), 9/19/94; five or six between RM 122.8 and RM 167.0, 9/20/94; RM 119.0, 7/12/93; several between RM 120 and RM 167, 4/19/95; RM 131.3, 4/20/94 (3).
Notes: Though they appeared more abundant in reaches with little vegetation, such as the Upper Granite Gorge and the Muav Gorge, this may be due to greater visibility/detectability.

Woodpeckers

Red-naped Sapsucker (*Sphyrapicus nuchalis*): Uncommon winter resident and transient.
Surveys: 11 detections; all but two between RM 168 and RM 214; exceptions at RM 1.0 and RM 46.7.
Banding: RM 198.0R, 1/31/94 (2) and 9/24/94 (2).
Incidental: RM 46.7R, 1/22/94; RM 87L, 9/18/94 (in tamarisk); RM 108R, 4/19/94; RM 171.0R, 1/27/94 (visiting tamarisk sap wells) and 9/21/94 (foraging in tamarisk).
Notes: This was the most common woodpecker we observed in the canyon, and the only one frequently seen on winter surveys (both in January 1994 and February 1995). According to Phillips et al. (1964), transients pass through Arizona as late as November, and as early as January, so the birds we saw may have been early migrants. However, Brown et al. (1987, p. 207) considered them as uncommon winter residents along the river corridor, and attendance by several at well-developed sap wells suggests residence. Red-naped Sapsuckers in the canyon make extensive use of the exotic tamarisk for drilling sap wells. We also found sap wells in mesquite.

Ladder-backed Woodpecker (*Picoides scalaris*): Uncommon summer resident.
Surveys: RM 198.0R, 5/15/94; 198.0R, 7/23/94; RM 204.1R, 6/24/95; RM 204.5R, 7/26/94; RM 204.5R, 9/26/94.
Incidental: RM 198.0R, 7/23/94; RM 200.5R, 7/25/94.
Notes: Brown et al. (1987, p. 207) considered it a rare permanent resident of the western part of the canyon. Our observations suggest that its occurrence in Colorado River riparian habitat is seasonal, during the hottest months. However, because breeding on the lower Colorado River occurs from February through June, predominantly in riparian habitats, with juveniles dispersing in late

June and July (Rosenberg et al. 1991, p. 225), some of the birds that we saw may have been post-breeding dispersers from more distant populations.

Northern (Red-shafted) Flicker (*Colaptes auratus cafer*): Uncommon winter visitor.
Surveys: 6 detections. February (1), September (5).
Banding: RM 46.7R, 9/16/94.
Incidental: 171.0L, 1/27–28/94; RM 198.0R, 1/30/94.

Tyrant Flycatchers

Olive-sided Flycatcher (*Contopus cooperi*): Rare transient.
Incidental: RM 204.5R, 5/19/94.

Western Wood-Pewee (*Contopus sordidulus*): Fairly common transient.
Surveys: RM 1.0R, 9/13/94; RM 5.1L and 5.2R, 6/7/93; RM 49.1R, 5/10/94; RM 49.1R, 5/17/95; RM 25.5R, 6/19/95; RM 204.1, 5/25/95; RM 204.5R, 5/19/94; RM 204.5R, 5/25/95.
Banding: RM 1.0R, 9/13/94; RM 198.0R, 5/17/94; RM 204.5R, 5/9/94 and 5/25/95.
Incidental: Lees Ferry, 6/9/95; RM 1.0R, 5/31/94 (2); RM 1.0R, 5/14/95; RM 1.0R, 9/24/94.

A note on the following species of the genus *Empidonax:* These species are very difficult to properly identify in the field, their songs being the only good diagnostic character in many cases. Unfortunately, they rarely sing during migration and only the Willow Flycatcher breeds in the canyon. Thus our banding results provide the best records to date for some species in the canyon, because diagnostic morphometric characters can only be measured in the hand.

* Willow Flycatcher (*Empidonax traillii*): Rare local summer resident.
Surveys: RM 1.0R (2) and RM 1.6R, 6/2/94; RM 46.7R, 7/10/94.
Banding: 7 birds. RM 1.0R, 8/12/94 and 10/13/94; RM 46.7R, 6/10/94 and 7/9/94; RM 198.0R, 4/26/94 and 4/23/95; RM 204.5R, 5/27/95.
Incidental: RM 1.0R, 6/2/94 (2) and 8/12/94; RM 46.7R, 7/9/93; RM 71, 5/11/94; RM 204.5R, 5/19/94.
Note: Our project specifically avoided conducting surveys at known Willow Flycatcher breeding patches. For more information on this species in the Grand Canyon, see Sogge et al. (1997).

Hammond's Flycatcher (*Empidonax hammondii*): Rare transient.
Banding: RM 198.0R, 4/27/94.
Notes: Only three confirmed previous records exist for the canyon (Brown et al. 1987, p. 210).

Gray Flycatcher (*Empidonax wrightii*): Uncommon transient.
Surveys: RM 1.0R, 5/7/95; RM 46.0L, 5/16/95; RM 204.5R, 4/27/94 (2); RM 208.7R, 4/27/95.
Banding: 6 birds. RM 46.7R, 5/17/95; RM 198.0R, 4/26–27/94 (2); RM 198.0R, 4/21/95; RM 204.5R, 4/24/95 (2).
Notes: This *Empidonax* flycatcher is the only one that flicks its tail downward; therefore, several were visually identified during surveys (above). Only one previous record from the river in the canyon (Brown et al. 1987, p. 211).

Dusky Flycatcher (*Empidonax oberholseri*): Uncommon transient.
Banding: 6 birds. RM 1.0R, 5/12/95; RM 46.7R, 5/8/94 and 5/17/95; RM 198.0R, 5/16/94; RM 204.5R, 5/19/94 and 5/25/95.
Notes: The only two previous records in the canyon are both from September (Brown et al. 1987, p. 210).

"Western" Flycatcher (*Empidonax difficilis* and *E. occidentalis*): Uncommon transient.
Surveys: RM 49.1R, 5/10/94.
Banding: RM 1.0R, 9/13/94; RM 204.5, 5/19–20/94 (2).
Notes: At the time of our project, the Pacific-slope (*Empidonax difficilis*) and Cordillerian (*Empidonax occidentalis*) flycatchers were considered as a single species—the Western Flycatcher. Unfortunately, even the morphometric data taken from banded birds is not

adequate to separate the birds we detected into either of the currently recognized species.

* Black Phoebe (*Sayornis nigricans*): Common summer and uncommon winter resident.
Surveys: 32 detections, from March through July. All March records from Reach 10; other months included Reaches 1, 4, 5, 7, and 10.
Incidental: Too numerous to list. Earliest date was 2/17/95 below RM 214; latest on 7/16/94 at RM 174.4R.
Notes: Though considered an uncommon and localized breeder in the Grand Canyon (Monson and Phillips 1981, p. 101), we found eight nests—in Reach 1 (1), Reach 4 (2), Reach 5 (2), Reach 7 (3), and Reach 10 (22)—plus several old nests in the Muav Gorge (Reach 9). Nests may be more common in the lower reaches due to the availability of fine sediments for nest building. Earliest nest was under construction at RM 110L on 5/21/95. The preferred nest site of this species is under a ledge over water (Phillips et al. 1964, p. 83), making this species potentially vulnerable to nest inundation due to fluctuating river flows in the canyon.

* Say's Phoebe (*Sayornis saya*): Common summer and uncommon winter resident.
Surveys: 159 detections. January (2), February (13), March (44), April (27), May (26), June (24), July (19), September (4). Found in all reaches surveyed.
Notes: This species is not tied to riparian habitat in the canyon, so our surveys may underestimate their numbers around our study sites. We found several nests. A pair was building a nest at RM 171.0R, 3/27/94; the nest contained two eggs on 4/23, but appeared abandoned. Another nest had two very young nestlings and one egg on 3/28/95.

This was the only flycatcher we found wintering in the canyon. It is listed by Monson and Phillips 1981 (p. 102) as "wintering sparingly north to the Navajo Indian Reservation and inside the Grand Canyon." In January 1994, we found only two birds during surveys (RM 1.6 and RM 168), but we found them "numerous below RM 110" (DF field notes). In February 1995, we began seeing them at RM 52, and surveys detected 12 individuals at 10 sites, all below RM 125. Phillips et al. (1964, p. 84) put the beginning of spring migration in early March. From our observations, either migration begins somewhat earlier in the canyon, or birds wintering lower in the canyon begin moving upstream in February. Higher counts in March may represent a wave of migrants moving through the canyon. On the lower Colorado River, numbers decline markedly in late April (Rosenberg et al. 1991, p. 234), later than expected if migrants in March are coming from the lower Colorado.

Vermilion Flycatcher (*Pyrocephalus rubinus*): Rare summer visitor.
Incidental: RM 198 (Parashant Wash), 7/22/94 (adult female).

* Ash-throated Flycatcher (*Myiarchus cinerascens*): Common summer resident.
Surveys: 235 detections. April (14), May (58), June (92), July (71). All April records are from Reach 10. In May, June, and July, they were found in all reaches surveyed.
Banding: 51 birds. April (2), May (12), June (18), July (19). Twenty-eight in breeding condition; April (2), May (7), June (13), July (6).
Notes: This was the seventh most abundant riparian breeder in the canyon, breeding (based on enlarged cloacal protuberance or brood patch) in approximately 60 percent of our study sites. Most common in Reach 10, breeding at 20 of 22 sites. Only one nest was found in a broken-topped hackberry snag at RM 197.6L. First found on 5/22/95, possibly in nest-building stage. On 6/22, the adults were feeding young.

* Brown-crested Flycatcher (*Myiarchus tyrannulus*): Rare, localized summer resident.
Surveys: 28 detections: 22 at RM 198.0R, remainder within 1 mile of this site. May (6), June (11), July (11).
Banding: 7 individuals, 3 with brood patches. All from RM 198.0R.
Incidental: One heard calling at RM 196.5R.

Notes: Brown-crested Flycatchers occurred in the canyon only as a very localized population. Banding data were valuable to distinguish this species from the smaller (but visually similar) Ash-throated. Previous to this, the Brown-crested's status in the canyon was unclear. Brown et al. 1987 (p. 213) stated, "there are no specimens or photographs to help establish the status of this bird in the region." Now its status as a canyon breeder, though very localized, is well established through morphometric data and photographs (on file at Grand Canyon National Park).

Cassin's Kingbird (*Tyrannus vociferans*): Uncommon transient.
Surveys: RM 5.1L, 6/7/93; RM 5.1L and 5.2R, 5/15/95 (probably the same bird); RM 174.7R, 4/20/95.
Incidental: RM 46.7R, 7/15/94.

Western Kingbird (*Tyrannus verticalis*): Rare summer resident.
Surveys: RM 5.1L and RM 5.2R on 6/2/94 and 7/14/94; RM 74.4R, 7/18/94.
Incidental: One in cottonwoods at Deer Creek (RM 136), 4/20/94; RM 198.0R, 4/22/95; RM 204.5R, 6/13/94.
Notes: The sightings at RM 5.1L and 5.2R may be misidentifications of the Cassin's Kingbirds seen at that same location in 1993 and 1995.

Shrikes

Loggerhead Shrike (*Lanius ludovicianus*): Uncommon summer visitor.
Surveys: RM 198.0R, 7/23/94 (2); RM 206.5 L, 9/28/94; RM 214.04, 9/28/94.
Banding: RM 198.0R, 6/12/94; RM 198.0R, 7/23/94.
Incidental: RM 97.4R, 9/19/94; several near RM 198.0, 7/24/94; RM 204.5R, 9/27/94.

Vireos

* Bell's Vireo (*Vireo bellii*): Common summer resident.
Surveys: 541 detections on 441 surveys; most in May and June. 537 in Reach 10, 4 in Reach 4.
Banding: 120 birds, March (1), April (8), May (33), June (62), July (13), September (3); 67 hatch-year birds, May (12), June (43), July (12).
Incidental: Lees Ferry, 6/9 and 6/20/95 were farthest upstream detections. Several other singing males in Reach 4: RM 48.5R, 5/15/95; RM 50.0L, 4/17/95; RM 52, 5/10/94; (RM 53), 4/17/95 ("several"). Earliest date, 3/25/95 at RM 198.0R; latest on 9/20/94, just above RM 157.
Notes: This was the second most common breeder on our surveys, surpassed only by Lucy's Warbler. Bell's Vireos bred at 41 percent of sites surveyed (n = 133) between 1993 and 1995, including all but two sites below RM 170. We found 21 nests of this species; substrates included tamarisk (57%), mesquite (29%), coyote willow (5%), seep willow (5%), and arrowweed (5%). The earliest nest was under construction at RM 198.0R on 3/25/95. Despite this species' well-known status as a host of the Brown-headed Cowbird (Phillips et al. 1964, p. 143), we never found Bell's Vireos attending cowbird eggs, nestlings, or fledglings. This species has a low parasitism rate compared to other species in the canyon (Brown 1994), possibly due to the cowbird's low abundance in the lower reaches of the canyon where the vireo is most common. It is possible that higher cowbird parasitism rates above Nankoweap Creek have hindered the upstream range expansion of Bell's Vireo in the canyon (see Brown et al. 1983).

Gray Vireo (*Vireo vicinior*): Rare transient.
Surveys: RM 206.5L, 4/26/95.
Banding: RM 1.0R, 5/12/95.

A note on Solitary, Plumbeous, and Cassin's Vireos: At the time of our project, the Solitary Vireo had not been split into the Plumbeous and Cassin's species. However, comparing the wing and tail measurements of birds we captured with data from Pyle (1997), three of our four banded birds were assignable to the newer taxonomy (below).

Solitary Vireo (formerly *Vireo solitarius*): Rare transient.
Banding: RM 1.0R, 4/5/94.

Incidental: RM 46.7R, 9/16/94; RM 88, 9/19/94.
Note: The individuals listed above could not be assigned to either the Plumbeous or Cassin's species.

Plumbeous Vireo (*Vireo plumbeus*): Rare transient.
Banding: RM 204.5R, 4/24/95.

Cassin's Vireo (*Vireo cassinii*): Rare transient.
Banding: RM 198.0R, 3/31/94; RM 198.0R, 9/23/94.

Warbling Vireo (*Vireo gilvus*): Uncommon transient.
Banding: RM 1.0R, 9/11–13/94 (2); RM 198.0R 9/25/94; RM 204.5R, 5/19–20/94 (2); RM 204.5R, 9/27/94.

Crows and Jays

Western Scrub Jay (*Aphelocoma californica*): Uncommon transient and winter visitor.
Surveys: RM 125.5R, 2/13/95; RM 174.4R, 7/22/94; RM 198.0R, 4/26/94; RM 204.5R, 4/27/94.
Incidental: RM 198.0R, 4/25/94.

Pinyon Jay (*Gymnorhinus cyanocephalus*): Uncommon visitor.
Incidental: RM 1.0R, 4/12/94 (number not noted); RM 119.6L, 2/13/95 (flock of ca. 50).

American Crow (*Corvus brachyrhynchos*): Casual transient.
Incidental: RM 46.7R, 3/17/95 (2); RM 46.7 R, 4/16/95 (13).

* Common Raven (*Corvus corax*): Common permanent resident.
Surveys: 80 detections; observed in all months of surveys.
Incidental: Too numerous to list. One killed a fledgling Black-throated Sparrow at RM 120, 7/20/94. Nest in high cliffs, so few nests found. One being built on cliff ca. 150 m above the river at RM 198.0L, 3/31/94; two nestlings visible 5/15/94. A pair building at same location 3/24/95.

Swallows

Tree Swallow (*Tachycineta bicolor*): Uncommon transient.
Incidental: Lees Ferry, 5/14/95 ("a few"); mouth of Little Colorado River (RM 62) and below, 4/17/95 ("numerous"); below RM 180 (Lava Falls), 4/20/95 ("numerous ").

Violet-green Swallow (*Tachcineta thalassina*): Common summer resident.
Surveys: 336 individuals on 75 surveys. March (4), April (18), May (64), June (103), July (117), September (30). Earliest date, 3/28/95 at RM 208.7R; latest date, 9/14/94 at RM 1.0R. Most abundant in upper reaches, but found in every reach surveyed.
Incidental: Lees Ferry, 7/10/94 ("numerous") and 4/14/95 ("over 300"); large concentrations around RM 32 and RM 38, 5/7/94.

Northern Rough-winged Swallow (*Stelgidopteryx serripennis*): Uncommon transient.
Surveys: RM 1.0R, 5/6/94
Incidental: Lees Ferry, 4/14/95 (ca. 5) and 5/14/95 ("numerous"); RM 1.0R, 9/14/94 (3); two between RM 47 and RM 72, 9/17/94; RM 171.0R, 3/27/94; below RM 209, 4/27/94; RM 210, 7/20/93.

Bank Swallow (*Riparia riparia*): Rare transient.
Incidental: Lees Ferry, 5/14/95 May 1995; below RM 209, 4/27/94.

* Cliff Swallow (*Petrochelidon pyrrhonota*): Uncommon transient and rare summer resident.
Incidental: Lees Ferry, 4/14/95 (5); two between Lees Ferry and RM 46.7, 9/14/94; mouth of Little Colorado River (RM 62) and below, 4/17/95 ("numerous"); below Lava Falls (RM 180), 4/20/95 ("numerous"); RM 209, 4/27/94.
Notes: This species appears to be an irregular nester along the river corridor. We noted an old nest at RM 3.5R, 2/9/95, and an adult nest-building on a cliff above the river at RM 2.0 R, 6/14/95. The last previous documented nesting of this species was in 1975 (Brown et al. 1987, p. 217).

Barn Swallow (*Hirundo rustica*): Rare transient.
Incidental: Lees Ferry, 4/14/94; below RM 209, 4/27/94.

Chickadees and Titmice

Mountain Chickadee (*Poecile gambeli*): Uncommon winter visitor and transient.
Surveys: RM 1.0R, 1/19/94; RM 50.0R, 1/23/94; RM 74.4R, 3/21/94; RM 119.5R, 1/25/94.
Banding: RM 1.0R, 1/19/94 (2); RM 46.7R, 1/21/94.
Notes: Our winter records expand on the Brown et al. (1987, p. 223) classification of this species as a rare transient to the river in spring and fall only.

* Verdin (*Auriparus flaviceps*): Rare and localized permanent resident.
Banding: RM 204.5R, 7/26/94 (hatch-year bird).
Incidental: A used nest, possibly from 1993 but in good condition, in a graythorn bush toward upland side of RM 204.5R, 6/13/94. Another used nest at RM 204.5R, 5/26/95.
Notes: These are the first breeding records above the head of Lake Mead, where the Verdin's status was previously uncertain (Brown et al. 1987, p. 224). Occasional sightings have been made as far upstream as RM 157 (Havasu Creek; Brown et al. 1987, p. 224). The observation of nests in 1994 and 1995 suggests that there may be a small resident population at Spring Canyon (RM 204.5). State-wide, the Verdin is considered a common resident of Sonoran Desert-scrub, "except the bottom of the Grand Canyon" (Monson and Phillips 1981, p. 122). The nearest breeding records are from the Virgin River (Monson and Phillips 1981, p. 122). Rosenberg et al. (1991, p. 253) suggested that populations are limited by the severity of winter weather, so this species may be at the edge of its geographic range in the lower canyon.

* Bushtit (*Psaltriparus minimus*): Common winter visitor and rare summer resident.
Surveys: 29 individuals on 7 surveys. Detections at Lees Ferry, 9/28/93 (2); RM 1.6R, 3/16/95 (2); flocks at RM 5.1, RM 5.2, and RM 46.7, 1/20–22/94; RM 204.5R, 9/26/94 (3).
Banding: 57 birds; 41 at RM 1.0R, 16 at RM 46.7R.
Incidental: Lees Ferry, 6/20/95; RM 5.6R, 2/9/95 ("flock"); RM 204.5R, 9/27/94 (10). A pair was building a nest in a tamarisk at Lees Ferry behind the Park Service trailer, 14 March 1995; apparently, it was successful (Grahame 1995).
Notes: According to Brown et al. (1987, p. 224.) this is one of the most abundant birds using the Colorado's riparian habitat during the winter. In January and February, we found this species to be common in the upper reaches of the canyon and rare elsewhere.

Wrens

Rock Wren (*Salpinctes obsoletus*): Common permanent resident.
Surveys: 164 detections. January (3), February (10), March (38), April (34), May (21), June (30), July (16), September (12).
Banding: RM 46.7R, 7/8/93; RM 198.0R, 9/23/94.
Notes: Seasonal differences in detections may have to do with detectability rather than abundance. This very vocal species is much more easily detected in the breeding season when males are almost constantly singing. Seasonal changes in habitat use also appear to have affected abundance on our riparian surveys. Use of riparian habitat went from 0 percent in the winter to more than 50 percent of all observations in July. Despite these complicating factors, seasonal trends from our surveys are not consistent with Brown et al. (1987, p. 228), "winter population density of the rock wrens is somewhat higher than it is during the summer breeding season."

Canyon Wren (*Catherpes mexicanus*): Common permanent resident.
Surveys: 244 detections. January (7), February (12), March (29), April (27), May (27), June (69), July (47), September (26).
Banding: RM 1.0R, 9/28/93 and 9/12/94; RM 46.7R, 7/8/93; RM 50.0L, 7/18/94.

Notes: Like the Rock Wren, this species prefers rocky upland habitat and is highly vocal during the breeding season. It also makes higher seasonal use of riparian habitat during the summer: 16 percent of February observations vs. 51 percent of July observations were in riparian habitat.

* Bewick's Wren (*Thryomanes bewickii*): Common summer and fairly common winter resident.
Surveys: 403 individuals on 359 surveys. January (8), February (9), March (60), April (55), May (73), June (104), July (54), September (40). They were most abundant in Reaches 4 and 10, then Reaches 1 and 5.
Banding: 89 individuals. January (2), March (2), April (8), May (18), June (26), July (23), September (10). Peak of captures in summer includes hatch-year birds: May (13), June (13), July (20).
Notes: This was the fifth most abundant riparian breeder from survey data, breeding at 40 percent of sites surveyed (n = 133). Like the other wrens, detectability changes seasonally, with fall and winter surveys likely underestimating populations. This species is definitely a winter resident of riparian habitat in the canyon. Brown et al. (1987, p. 229) proposed that breeding populations in the canyon move out for the winter to be replaced by migrants from farther north. Our banding data suggest that this is generally true, though at least two banded birds were present and captured in both summer and winter.

House Wren (*Troglodytes aedon*): Uncommon transient.
Surveys: RM 1.0R, 5/14/95; RM 50.0R, 5/17/95; RM 76.5L, 4/18/95; RM 168.8R, 9/21/94; RM 171.0R, 7/22/94.
Banding: RM 1.0R, 5/14/95; RM 171.0R, 4/22/94; RM 198.0R, 3/31/94; RM 198.0R, 4/27/94.
Incidental: RM 198.0R, 4/21/95.

Winter Wren (*Troglodytes troglodytes*): Rare transient and winter visitor.
Surveys: RM 46.7R, 1/22/94; RM 46.7R, 3/17/95.

Marsh Wren (*Cistothorus palustris*): Fairly common transient.
Surveys: 19 detections. March (5), April (7), May (1), September (6).
Banding: 23 birds. RM 1.0R (16), RM 46.7R (1), RM 198.0R (3), RM 204.5R (3). March (4), April (11), May (2), September (4), October (2).
Incidental: RM 98 (Crystal Creek) in phragmites and cattails, 4/19/94.
Notes: Earliest date was 3/13, latest date was 10/13; both at RM 1.0R.

Dippers

American Dipper (*Cinclus mexicanus*): Uncommon winter resident.
Incidental: RM 1, RM 18.5, RM 27, RM 30, and RM 32—all on 2/9/95; five between RM 1.0R and RM 42, 2/14/93; one between RM 42 and RM 47, 2/15/93.
Note: The February 1993 sightings were by MS, during a separate avian research study.

Kinglets and Gnatcatchers

Golden-crowned Kinglet (*Regulus satrapa*): Rare winter visitor.
Incidental: RM 1.0R, 12/13–14/94; RM 1.0R, 1/9/95.

Ruby-crowned Kinglet (*Regulus calendula*): Common winter resident.
Surveys: 163 detections on 152 surveys. Highest counts in January (19), February (42), March (28), and April (26).
Banding: 66 birds. Most captures in April (23) and January (17), with 1–6 in all other months of surveys/banding. RM 1.0R (39), RM 46.7R (8), RM 171.0R (4), RM 198.0R (12), RM 204.5R (3).
Notes: This was both the most abundant and most widely distributed winter species observed. Found in all reaches surveyed on January and February trips, and on 61 percent of all winter surveys (n = 81). Phillips et al. (1964, p. 135) consider this species common from October to April in lower Sonoran riparian habitats in Arizona, but rare north of the Mogollon Plateau. Its occurrence in the canyon is in the northernmost extension of this habitat in Arizona, where it is one of the only foliage-gleaning insectivores win-

tering in the canyon, foraging extensively on tamarisk.

* Blue-gray Gnatcatcher (*Polioptila caerulea*): Common summer resident.
Surveys: 442 individuals on 382 surveys; found in all reaches surveyed. March (11, all in reach 10), April (107), May (149), June (135), July (35), September (5).
Banding: 41 individuals, including 13 hatch-year birds. RM 1.0R (8), RM 46.7R (4), RM 171.0R (9), RM 198.0R (13), RM 204.5R (7).
Notes: This was the fourth most abundant riparian breeder on our surveys, breeding at 57 percent of 133 sites surveyed. They were most abundant in Reaches 10 and 4. Earliest date was 3/23 at RM 46.7R, latest was 9/27 at RM 204.5R. This species is more upland-associated than many other riparian breeders. We found nests in tamarisk (7), acacia (3), and mesquite (4). Three mesquite nests were built inside dead mistletoe growth. We found adults feeding a cowbird nestling at RM 46.7R, and a cowbird fledgling at RM 49.2L; a pair chased a pair of adult cowbirds away from their nest at RM 1.0R, but the nest was later abandoned.

Thrushes

Western Bluebird (*Sialia mexicana*): Uncommon winter resident.
Surveys: RM 198.0R, 1/20/94 (15), 1/31/94 (4), and 2/16/95 (4); RM 204.5R, 2/17/95 (6) and 3/25/95.
Notes: We found large flocks feeding on mistletoe berries, and RM 198.0R and RM 204.5R were our only study sites with significant amounts of mistletoe. Therefore, this species may be more common than our surveys suggest, where mistletoe occurs in the lower canyon. According to Monson and Phillips (1981, p. 141), they "winter irregularly on the desert where mistletoe occurs" and Brown et al. (1987, p. 233) consider them "uncommon in early spring along the river."

Mountain Bluebird (*Sialia currucoides*): Rare winter visitor.
Incidental: Flock of 15 between Lees Ferry and RM 52, 2/19/94.

Townsend's Solitaire (*Myadestes townsendi*): Uncommon winter resident.
Surveys: 20 on 10 surveys; all in Reach 10 during 1995. February (4), March (14), April (2).
Incidental: RM 140, 3/20/95.
Notes: Like their relative, the Western Bluebird, they often occurred in flocks, feeding off mistletoe berries, but were also seen eating wolfberries (*Lyceum* sp.) at RM 171.0R. Unlike the bluebirds, they actively defended good food sources, keeping bluebirds and Phainopeplas away.

Hermit Thrush (*Catharus guttatus*): Rare transient and winter visitor.
Surveys: RM 122.8L, 4/19/95; RM 198.0R, 2/16/95; RM 205.8R, 5/20/94.
Banding: RM 1.0R on 11/15/95; RM 198.0R, 1/31/94 (2) and 3/23/95 (2).
Notes: There was only one previous winter record (Brown et al. 1987, p. 234).

American Robin (*Turdus migratorius*): Fairly common transient.
Surveys: 16 detections. February (7), March (2), April (4), May (3).
Banding: RM 198.0R, 3/94.
Incidental: RM 1.0R, 10/12/94 (5); RM 198.0 R, 3/24/95.
Notes: Detections in February may be early migrants, or wintering birds per Brown et al. (1987, p. 236).

Mockingbirds and Thrashers

Northern Mockingbird (*Mimus polyglottus*): Uncommon summer resident.
Surveys: 11 detections; 9 in Reach 10. April (2), May (4), June (2), July (3).
Banding: 9 birds. RM 1.0R (3), RM 198.0R (5), RM 204.5R (1). Eight in breeding condition (5 males with cloacal protuberances, 3 females with brood patches) between April and July.
Incidental: RM 55, 7/10/93; RM 122.8L, 6/8/94; RM 168.8R, 4/21/94; RM 171.0R, 4/20/95; RM 204.5R, 5/18/94 May 1994 (pair).
Notes: Although we found no nests, breeding condition of captured mockingbirds provided proof of nesting in the canyon

(though habitat unknown), something not previously documented (Brown et al. 1987, p. 236). Also, in 1994, a pair seen in both May and June at RM 204.5R.

Crissal Thrasher (*Toxostoma crissale*): Rare summer resident or visitor.
Surveys: RM 173.1R, 7/22/94; RM 198.0R, 2/17/95.
Incidental: RM 204.5R, 6/13/94; RM 204.5R, 7/26/94 (singing).
Notes: This is a secretive species of riparian thickets and there are few records from the Grand Canyon (Brown et al. 1987, p. 237). There are several September records from the Little Colorado River and Grand Canyon South Rim (Phillips et al. 1964, p. 124), which suggests that the birds we detected were migrants from a breeding population in northeastern Arizona.

Pipits and Wagtails

American Pipit (*Anthus rubescens*): Uncommon transient.
Surveys: RM 74.4R, 2/12/95.
Incidental: RM 1.0R, 12 October 1994 (ca. 150 hawking insects over the river); RM 6, 4/14/95; RM 30, 4/13/94.

Waxwings

Cedar Waxwing (*Bombycilla cedrorum*): Rare transient.
Surveys: RM 171.0R, 4/20/95.
Incidental: RM 198.0R, 5/22/95 (2); up Spring Canyon (RM 204.5), 6/23/95.

Silky Flycatchers

* Phainopepla (*Phainopepla nitens*): Uncommon summer resident.
Surveys: 18 detections; all at RM 204.1R and 204.5R. February (1), March (2), April (5), May (8), June (2).
Incidental: Lees Ferry, 5/31/94; RM 46.0L, 5/16/95 (male); RM 198.0R, 7/15/93; RM 209L, 4/1/94 (adults feeding fully fledged young); RM 224, 4/28/94 ("fairly abundant"); up Diamond Creek (RM 226), 1/31/94 ("several").
Notes: We found a nest in mesquite, with adults attending young, on 5/26/95 at RM 204.5R. The male fed one nestling a mistletoe berry.

Wood Warblers

Orange-crowned Warbler (*Vermivora celata*): Fairly common transient.
Surveys: 15 detections. May (2), September (13).
Banding: 19 birds. May (7), September (5), October (7). RM 1.0R (10), RM 46.7R (4), RM 198.0R (4), RM 204.5R (1).
Incidental: RM 198.0R, 4/22/95; RM 209.0, 5/93.

Nashville Warbler (*Vermivora ruficapilla*): Rare transient.
Surveys: RM 198.0R, 9/24/94 (the banded bird below).
Banding: RM 198.0R, 9/24/94 (hatch-year bird).
Notes: Only one previous record for this species along the river corridor (Brown et al. 1987, p. 244).

Virginia's Warbler (*Vermivora virginiae*): Uncommon transient.
Banding: 8 birds. RM 1.0R, 5/5–6/94 (2), 5/12–13/95 (2), and 8/11/94; RM 46.7R, 5/8/94 (2) and 5/18/95.
Notes: Our banding records add substantially to the status described in Brown et al. (1987, p. 245): "It is almost unknown as a transient below the rims, with only a few spring sightings from the river."

* Lucy's Warbler (*Vermivora luciae*): Common summer resident.
Surveys: 1779 detections on 1190 surveys. March (199), April (332), May (491), June (657), July (99), September (1). Earliest date, 3/17 at RM 46.7R; latest date, 9/2 at RM 49.1R.
Banding: 315 individuals. March (13), April (20), May (85), June (151), July (46). 159 of the 315 (50%) were hatch-year birds: May (46), June (77), July (36).
Notes: This was by far the most common and widespread riparian breeding species in the canyon, as was also found by Carothers and Johnson (1975). They bred in every reach surveyed and at 85 percent of the 113

sites surveyed. Despite this abundance, we found only nine nests, all in nestlings stage. Earliest nest was found 4/21 at RM 198.0R, latest nest fledged young on 6/17 at RM 46.7R. Though well known as the "mesquite warbler" (Bent 1953, p. 129), in the canyon it has adapted to the exotic tamarisk (Brown et al. 1987, p. 245). We found nests in tamarisk (4), mesquite (2), cavities in cliffs (2), and acacia (1). In tamarisk, the warblers nested in "cavities" created by hollowing out dense clumps of dead tamarisk needles piled on branches or adhering to the trunks.

* Yellow Warbler (*Dendroica petechia*): Fairly common summer resident.
Surveys: 229 detections on 218 surveys. April (17), May (109), June (61), July (20), September (22).
Banding: 118 birds. April (3), May (85), June (3), July (8), August (13), September (6).
Notes: This was the eighth most abundant breeding species. They bred at 32 percent of 133 sites surveyed; all in Reaches 1, 4, and 10. We found nests in May (1) and June (4); all were in tall tamarisks, in agreement with Brown et al. 1987 (p. 245). Based on banding data, only 4 of 48 captured females had brood patches, and only 16 of 44 males had cloacal protuberances. Therefore, most of the birds banded were probably migrants.

Black-throated Blue Warbler (*Dendroica caerulescens*): Accidental.
Banding: RM 1.0R, 10/12/94 (female).
Notes: This capture provided only the second of two records for the canyon (Brown et al. 1987, p. 246). Monson and Phillips (1981, p. 160) considered it a rare migrant, mostly occurring in the fall.

Yellow-rumped Warbler (*Dendroica coronata*): Uncommon winter resident, fairly common transient.
Surveys: 40 detections, all in Reaches 1, 4, and 10. January (2), February (4), March (13), April (9), May (5), September (7).
Banding: 9 birds. March (1), April (2), May (4), November (1), December (1).
Incidental: Lees Ferry, 1/6/95; two between RM 1.0R and RM 40, 2/14/93; RM 47.0R, 9/15/94 (2).
Notes: This is the only wood warbler we found wintering in the canyon, and it was not very abundant (two in January 1994 and four in February 1995). Across the continent this species winters farther north than all other *Dendroica* warblers. In Arizona, it is a common winter bird of the Lower Sonoran Zone and sycamore riparian habitats around the state. The western, Audubon's form predominates (Phillips et al. 1964, p. 151). Phillips et al. (1964, p. 152) considered the Myrtle form uncommon in winter and spring migration, and Brown et al. 1987 (p. 247) described it as accidental in the canyon during migration (only four observed, in April and May). However, two of the four birds we found in February 1995 were of the Myrtle race, as was one seen in March 1995. Based on this small sample of sightings, the Myrtle race may be more common than previously thought. However, our identifications of Myrtle were by the white throat patch of the male, a characteristic not recognized by Phillips et al. (1964, p. 152).

Black-throated Gray Warbler *(Dendroica nigrescens):* Rare transient.
Incidental: RM 46.7R, 5/9/94 (male).

Blackpoll Warbler *(Dendroica striata):* Accidental.
Banding: RM 204.5R, 9/26/94 (hatch-year).
Notes: This is a new record for the canyon. It is considered a rare visitor in the state, with few records (Monson and Phillips 1981, p. 165).

Northern Waterthrush *(Seiurus noveboracensis):* Rare transient.
Banding: RM 1.0R, 5/14/95; RM 198.0R, 4/27/94.

MacGillivray's Warbler *(Oporornis tolmiei):* Uncommon transient.
Surveys: Lees Ferry, 9/28/93; RM 1.0R, 9/13/94; RM 5.2R, 9/30/93; RM 198.0R, 9/23/94; RM 204.5R, 4/27/94.
Banding: 22 birds. May (17), September (5). RM 1.0R (8), RM 46.7R (2), RM 198.0R (4), RM 204.5R (8).

Incidental: RM 47.0R, 9/15/94; and several on Bright Angel Creek above Phantom Ranch (RM 88), 9/19/94.

* Common Yellowthroat *(Geothlypis trichas):* Common summer resident.
Surveys: 104 detections. March (1), April (9), May (40), June (34), July (12), September (8). Earliest date, 3/25 at RM 204.5R; latest date, 9/24 at RM 198.0R.
Banding: 45 birds. March (1), April (5), May (22), June (9), July (3), September (5). Nine were hatch-year birds.
Incidental: Male singing up Crystal Creek (RM 98) in phragmites and cattails, 4/19/94; male singing up Bright Angel Creek, 4/18/95.
Notes: This was the eleventh most abundant breeding species, breeding at 19 percent of 113 sites surveyed. High counts on surveys and from banding suggest that May is a migration peak. We found this species in phragmites and cattail (2), seep willows (2), and a clump of tall *Andropogon* grass (1). The two in phragmites and cattail were parasitized by Brown-headed Cowbirds, and one nest was consequently abandoned. This is one of the few terrestrial bird species in the canyon that may be directly affected by dam operation, through possible nest inundation (one we found was within 20,000 cfs inundation zone; another possibly was). However, yellowthroats probably benefited from recent dam operations because patches of marsh/emergent vegetation, the species' preferred nesting habitat, increased in number under interim flows. Common Yellowthroats appear able to nest in the smallest patch of cattails and phragmites (e.g., two pairs nesting at RM 46.7R in two patches of emergent vegetation amounting to less than 50 sq m).

Wilson's Warbler *(Wilsonia pusilla):* Common transient.
Surveys: Eight detections. May (4), September (4).
Banding: 45 birds. May (33), June (1), September (11). RM 1.0R (14), RM 46.7R (8), RM 171.0R (1), RM 198.0R (5), RM 204.5R (17).
Incidental: Lees Ferry, 5/6/94 ("several"); RM 136 (Deer Creek), 4/20/94.

* Yellow-breasted Chat *(Icteria virens):* Common summer resident.
Surveys: 149 detections on 132 surveys. April (2), May (53), June (78), July (15), September (1). Earliest date was 4/26/94 at RM 198.0R; latest was 9/18/94 at RM 74.4R.
Banding: 62 birds. April (3), May (27), June (19), July (9), August (2 hatch-year), September (2 hatch-year).
Notes: This was the ninth most abundant breeding species on our surveys, breeding at 29 percent of 133 sites surveyed. We found nests in tamarisk (2) and seep willow (1). This species has been indirectly affected by interim flows, as the dense riparian habitat it prefers (Brown et al. 1987, p. 252) has increased under interim flows.

Tanagers

Western Tanager (*Piranga ludoviciana*): Uncommon transient.
Surveys: RM 46.7R, 7/16/94; RM 47.5L and RM 50.0R, 5/17/95; RM 158.8L, 5/18/93; RM 204.5R, 5/18/94; RM 213.6L, 7/27/94.
Banding: 12 birds. May (8, all at RM 204.5R), July (2), September (2).
Incidental: RM 1.0R, 7/12/94; RM 1.0R, 5/14/95 (male); RM 47.0R, 9/15/94 (3); RM 204.5R, 5/18/94 ("numerous"); RM 204.5R, 7/26/94 (male and female); RM 213.6L, 7/27/94 (3).

* Summer Tanager (*Piranga rubra*): Rare localized summer resident, uncommon summer visitor.
Surveys: 22 detections; 20 in Reach 10, 2 in Reach 4. May (9), June (5), July (4), September (4).
Banding: Five banded, all at RM 198.0R. Male with cloacal protuberence, 6/10/94; female with brood patch 6/11/94; hatch-year bird, 7/19/93; adult female and hatch-year bird, 7/25/95.
Incidental: RM 46.7R, 5/16/95 (singing male) and 6/15/95 (adult and second-year male); adult male singing at RM 50.0L, 5/10/94 (singing male); RM 50.0L, 6/5/94 (male); RM 88 (Phantom Ranch), 6/6/94 and 7/19/94 (pair both dates); RM 136 (Deer

Creek), 5/13/94 (pair); RM 197.0L, 7/24/94 (male); RM 197.6L, 6/10/94 (pair, with female carrying nesting material); RM 198.0R, 5/15/94 (male singing); RM 198.0L, 7/23/94; RM 198.3R, 7/24/94 (banded male); RM 198.0L, 9/23/94 (banded male); RM 204.5R, 5/18/94 (male); RM 204.5R, 9/26/94.
Notes: Clearly, the pair banded at RM 198.0 R were breeding, probably successfully considering that two hatch-year birds were banded there. Though we found several males singing in Reach 4 during every year of the project, we saw no females or other evidence of breeding. This species has expanded its range upstream in the past 30 years (Brown et al. 1987, p. 252).

Sparrows and Towhees

Green-tailed Towhee (*Pipilo chlorurus*): Rare transient.
Surveys: RM 46.7R, 9/15/94 and 9/17/94; RM 49.1R, 9/17/94.
Banding: RM 1.0R, 4/11/94, 4/12/95, and 5/12/95.

Spotted Towhee (*Pipilo maculatus*): Uncommon transient.
Surveys: RM 46.7R, 3/20/94 and 5/16/95; RM 198.0R, 3/30/94.
Banding: RM 198.0R, 7/17/93.
Incidental: RM 198.0, 3/24/95 (2).

Canyon Towhee (*Pipilo fuscus*): Rare visitor.
Surveys: RM 74.4R, 2/12/95; RM 208.7R, 2/17/95.
Notes: The status of this species in the canyon is poorly known, with few records (Brown et al. 1987, p. 256).

* Rufous-crowned Sparrow (*Aimophila ruficeps*): Uncommon permanent resident.
Surveys: 21 detections; most (15) above RM 88 (Phantom Ranch). January (1), February (2), March (3), May (3), June (6), September (1).
Banding: 7 birds. RM 46.7R, 4/14/94 and 7/9/93; RM 50.0L, 7/18/94; RM 171.0R, 1/27/94; RM 198.0R, 6/21/95; RM 198.0R, 9/24/94 (2).
Incidental: RM 46.7R, 3/19/94 (singing).
Notes: This species made only occasional use of the riparian habitat, and so is more abundant within the canyon, but outside riparian, than appears here. A pair feeding a fledgling at RM46.0L, 6/15/95, was the only evidence of breeding in one of our study sites. They are more common along the river above Phantom Ranch (Brown et al. 1987, p. 257).

Chipping Sparrow (*Spizella passerina*): Uncommon transient.
Surveys: RM 1.6R, 4/13/95 (2); RM 5.6R, 4/13/94; RM 122.8L, 4/19/95; RM 204.5R, 9/26/94 (2).
Banding: RM 1.0R, 4/11/94 and 4/13/95.
Incidental: RM 76.0L, 9/18/94 ("several"; DF field notes).

Brewer's Sparrow (*Spizella breweri*): Uncommon transient.
Surveys: 10 detections; all in September. RM 1.0R (5), RM 198.0R (5).
Banding: 10 birds; one in August, others in September. RM 1.0R (5), RM 198.0R (5).
Incidental: Lees Ferry, 4/14/94 (2); RM 1.0R, 4/11/94 (2); RM 47.0R, 9/15/94; RM 52, 5/10/94; RM 204.5R, 4/26/95; on Bright Angel Creek above Phantom Ranch (RM 88), 9/19/94 ("several").

Black-chinned Sparrow (*Spizella atrogularis*): Rare transient.
Surveys: RM 206.5L, 3/26/95 (singing male).

Vesper Sparrow (*Pooecetes gramineus*): Rare transient.
Incidental: RM 209L, 4/1/94.

Lark Sparrow (*Chondestes grammacus*): Uncommon summer visitor.
Surveys: RM 204.5R, 5/18/94.
Banding: RM 46.7R, 7/16/94.
Incidental: RM 46.7R, 9/14/94; one between RM 46.7R and RM 73, 9/17/94; RM 65.5 (Lava Chuar), 7/19/94; RM 198.0R, 9/24/94; RM 209, 4/27/94.

* Black-throated Sparrow (*Amphispiza bilineata*): Common summer resident.
Surveys: 58 detections. March (6), April (13), May (12), June (14), July (13). Found in all

reaches surveyed but most common in Reaches 5, 7, 8, and 10.
Banding: 6 birds. RM 46.7R, 7/15–17/94 (2); RM 198.0R, 6/21/95; RM 198.0R, 7/18/93; RM 198.0R, 7/23/94; RM 204.5R, 5/27/94.
Incidental: RM 52, 3/20/94, 4/17/94 (pair), and 5/10/94 (pair); RM 88, 5/19/95 (3 drinking from river); fledgling eaten by raven at RM 120, 7/20/94.
Notes: Earliest date was 3/20/94 at RM 74.4R; latest was 7/27/94 at RM 206.5L. This species was much more associated with the uplands (30 of 51 observations in upland habitat), but commonly used riparian habitats during late June and July.

Sage Sparrow (*Amphispiza belli*): Rare transient.
Surveys: RM 214.0L, 9/28/94.
Banding: RM 1.0R, 11/15/94.

Savannah Sparrow (*Passerculus sandwichensis*): Rare transient.
Incidental: Lees Ferry, 4/14/95 (3); RM 52, 3/19/95.

* Song Sparrow *(Melospiza melodia):* Fairly common summer and winter resident.
Surveys: 61 detections on 53 surveys. January (1), February (12), March (13), April (8), May (15), June (9), September (3).
Banding: 29 birds. January (2), February (1), March (5), April (1), May (2), June (1), July (1), September (9), October (2), November (5).
Incidental: RM 46.7R, 6/15/95.
Notes: We observed adults feeding recently fledged young (tails < 20 mm) at RM 198.2L, 6/21/95. We also banded two hatch-year sparrows at RM 204.5R on 5/27/95, and one hatch-year at RM 198.0R on 6/22/95. Previously, there was only one confirmed breeding record from in the canyon (Brown et al. 1987, p. 262).

Lincoln's Sparrow (*Melospiza lincolnii*): Fairly common transient.
Surveys: 28 detections. February (1), March (10), April (6), May (2), September (9).
Banding: 29 birds. March (6), April (10), May (1), September (11), November (1).
Notes: The bird seen in February 1995 was at RM 206.5L. It may have been wintering there, as Brown et al. 1987 (p. 262) considered it possible that some birds winter in the canyon. Also, Root (1988) noted dense winter concentrations along the Colorado River near the borders of Nevada, Utah, and Arizona.

Swamp Sparrow (*Melospiza georgiana*): Casual visitor.
Incidental: Lees Ferry, 4/14/95 (MS).
Notes: There is only one previous record (a collection) from along the river in the canyon, in 1974 (Brown et al. 1987, p. 262).

White-crowned Sparrow (*Zonotrichia leucophrys*): Common winter visitor.
Surveys: 106 detections on 60 surveys. February (14), March (12), April (19), May (23), and September (38).
Banding: 61 birds. See notes below.
Incidental: Lees Ferry, 1/6/95 (6), 1/18/94 (17), and 4/12/94 (25–30); RM 47.0R, 9/14/94 (two in first basic plumage); RM 52, 3/20/94 (several in first basic plumage).
Notes: We observed two races—Z. *1. gambelii* (Nuttall) and Z. *1. oriantha* (Oberholser). Twelve Z. *1. oriantha* were banded, all in May at RM 1.0R. Phillips et al. (1964, p. 207) identified this race as transient in the state, found only in May and September. We also banded 49 Z. *1. gambelii*: January (2), February (1), March (3), April (10), May (8), September (12), October (4), November (4), December (5). Most (75%) were banded at RM 1.0R. This race is the abundant wintering race in Arizona (Phillips et al. 1964, p. 207), also wintering in the canyon.

Golden-crowned Sparrow (*Zonotrichia atricapilla*): Casual winter visitor.
Banding: RM 204.5R, 9/27/94 (hatch-year bird).
Notes: This is one of only a few records for the canyon.

Dark-eyed Junco (*Junco hyemalis*): Common winter resident.
Surveys: 122 detections on 50 surveys. January (20), February (49), March (46),

April (7). The Oregon race (*J. h. thurberi*) was most common.
Banding: 55 birds; 53 were of *thurberi* race. January (12), February (3), March (12), April (2), October (1), November (8), December (17). RM 1.0R (49), RM 46.7R (2), RM 198.0R (4).
Notes: The second most common wintering species, it uses all the habitats in the Grand Canyon, wandering widely throughout the day (Brown et al. 1987, p. 264). Seen more commonly in the upper reaches of the canyon, which lie in closer proximity to forests of the north rim where these birds commonly winter.

Grosbeaks and Allies

Black-headed Grosbeak (*Pheucticus melanocephalus*): Uncommon transient.
Surveys: RM 1.0R, 6/7/93 (2); RM 46.7R, 5/14/93; RM 50.0R, 5/15/93; RM 171.0R, 5/14/94; RM 198.0R 5/15/94; RM 198.0R, 9/11/93.
Banding: RM 1.0R, 7/13/94 and 5/12/95; RM 204.5R, 5/19/94.
Incidental: RM 198.0R, 5/15/94 (male singing); RM 204.5R, 5/18/94 (several).

* Blue Grosbeak (*Passerina caerulea*): Fairly common summer resident.
Surveys: 56 detections on 53 surveys, all in Reachs 1, 4, 5, 10. May (15), June (12), July (28), September (1).
Banding: RM 1.0R, 5/13/95, 6/1/94, and 7/12/94; RM 46.7R, 7/7/94.
Incidental: RM 1.0R, 5/30/94 (female) and 5/31/94 (pair); RM 46.7R, 5/9/94 (pair); RM 53, 9/17/94; RM 203.5R, 7/26/94; up Spring Canyon (RM 204.5), 5/19/94 (pair).
Notes: This species bred at 19 percent of 133 sites surveyed, and is one of the last species to arrive on the breeding grounds. Earliest date was 5/9 at RM 46.7R; latest was 9/17 at RM 47.5L. We found nests in May (1), June (2), and July (1), all in tamarisk. A June nest at RM 47.5L contained two cowbird eggs.

* Lazuli Bunting (*Passerina amoena*): Fairly common summer resident.
Surveys: 21 detections on 18 surveys. April (1), May (9), June (6), September (5).
Banding: RM 1.0R, 4/13/95; RM 1.0R, 5/13/95; RM 171.0R, 9/6/93 (hatch-year bird).
Incidental: At Lonely Dell Ranch (adjacent to Lees Ferry), 9/13/94 (6 hatch-years); Lees Ferry, 5/6/94 ("several"); RM 136 (Deer Creek), 5/13/94 (male singing in cottonwoods); on Deer Creek trail, 4/20/94 (male singing); RM 76.0L on 9/18/94 (2 hatch-years). Juveniles were generally abundant in September.
Notes: Earliest date was 4/13 at RM 1.0R (male); latest was 9/17 at RM 49.1R (hatch-year). We considered this species a breeder at one site in 1993, and two in 1995, including the female paired with the Indigo Bunting described below.

* Indigo Bunting (*Passerina cyanea*): Rare summer resident.
Surveys: RM 46.7R, 5/18/95 (male singing); RM 204.5R, 5/26/95 (2); RM 204.5R, 6/24/95 (banded male).
Banding: RM 204.5R, 5/27/95 and 6/26/95, both second-year males.
Incidental: RM 46.7R, 6/15/95.
Notes: One of the males at RM 204.5R was later seen paired with a female Lazuli Bunting. This was our only breeding record, though Indigo Buntings have bred in the canyon before (Brown et al. 1987, p. 255).

Blackbirds and Orioles

Red-winged Blackbird (*Agelaius phoeniceus*): Uncommon transient.
Surveys: RM 49.2L, 5/17/95 (male).
Banding: RM 198.0R, 4/22/95.
Incidental: Lees Ferry, 4/14/95 (11) and 5/6/94 (4 females); 14 females foraging in mesquites near RM 198, 4/22/95; 20 between RM 204.5R and RM 224, 9/28/94.

Western Meadowlark (*Sturnella neglecta*): Rare transient.
Incidental: RM 198.0R, 9/24/94.

Yellow-headed Blackbird (*Xanthocephalus xanthocephalus*): Rare transient.
Surveys: RM 5.1L, 9/14/94 (female).

Brewer's Blackbird (*Euphagus cyanocephalus*): Uncommon transient.

Surveys: RM 200.5R, 9/25/94 (16); RM 206.5 L, 9/28/94 (13).
Incidental: RM 50.0R, 9/17/94.

* Great-tailed Grackle (*Quiscalus mexicanus*): Uncommon summer resident.
Surveys: 89 detections on 41 surveys; 75 of these at Lees Ferry in 1993, the only year they nested during this study.
Banding: RM 204.5R, 5/18/94 (female).
Incidental: Lees Ferry, 5/13/95 and 6/1/94 (2 males); a female accompanied by a Rock Dove at Lees Ferry, 1/18, 3/13, and 4/12/94; RM 52, 5/10/94 (female).
Notes: This species has expanded its geographic range north during this century (first Arizona records from 1930s; Phillips et al. 1964, p. 172). It was never seen in the canyon before 1974 (Brown et al. 1987, p. 267). A colony bred at Lees Ferry in 1993, but not 1994 or 1995. Great-tailed Grackle frequently prey on eggs and nestlings of other birds. For example, in June 1993 at RM 75, a male and female drowned a fledgling Ash-throated Flycatcher, then carried it off (H. Yard and P. Hodgetts observation). Increasing populations of Great-tailed Grackles could affect riparian breeding bird populations in the canyon.

* Brown-headed Cowbird (*Molothrus ater*): Fairly common summer resident.
Surveys: 28 detections on 26 surveys. May (10), June (14), July (4). Reach 1 (13), Reach 4 (12), Reach 10 (3).
Banding: RM 1.0R, 5/30/94 (2 males); RM 198.0R, 6/19/93 (female; recaptured again here on 5/19/94).
Incidental: Lees Ferry, 5/6/94 ("several"); RM 1.0R, 5/30/94 (2 males, 1 female); RM 52, 5/10/94; RM 71, 5/18/95 ("abundant"; DF field notes).
Notes: Cowbirds are relatively late arrivals to the canyon; earliest date of 5/12/95 and latest of 7/20/93. We observed brood parasitism of Common Yellowthroat (RM 46.7R, 6/4/94; RM 198.0R, 7/23/94), Blue Grosbeak (RM 47.5L, 7/17/94), and Blue-gray Gnatcatcher (RM 46.7R, 6/15/95; RM 49.2L, 6/17/95). The Southwestern Willow Flycatcher is also parasitized in the canyon (Sogge et al. 1997). Despite greater diversity and abundance of breeding birds in Reach 10 (especially the Bell's Vireo, a well-documented host species), cowbirds were more abundant in Reaches 1 and 4. This may be because these narrow reaches (1 and 4) are closer to livestock concentrations on the Navajo Reservation and the federal rangelands of the Arizona Strip, where cowbirds prefer to forage. Phillips et al. (1964, p. 172) stated that cowbirds have been found to limit the geographic range of both Bell's Vireos and Yellow Warblers, and this mechanism may be limiting the upstream range expansion of the Bell's Vireo (Brown et al. 1983).

* Hooded Oriole (*Icterus cucullatus*): Uncommon summer resident.
Surveys: 17 detections. April (5), May (5), June (3), July (4).
Incidental: A male at RM 71 (Cardenas Marsh), 5/6/94 and 5/18/95; RM 168.8R, 4/21/94 (pair); an old nest collected at RM 168.8R in January 1994; RM 198.0R, 3/24/95 (male); pair at RM 204.5R, 6/14/94 and 4/25/95; pair at RM 214.0L, 6/15/94 and 6/26 /95 (nesting).
Notes: We considered them breeders at seven sites, all in Reach 10. Earliest date was 4/21/94 at RM 167.6R; latest was 7/22/94 at RM 168.8R. This species has only recently entered the canyon as a breeding species (Brown et al. 1987, p. 268).

* Bullock's Oriole (*Icterus bullockii*): Rare summer resident, fairly common transient.
Surveys: 13 detections. May (5), July (5), September (3).
Banding: Seven birds. RM 1.0R, 6/12/95; RM 198.0R, 7/19/93, 7/23/94 (2), and 9/9/93; RM 204.5R, 5/18/94 and 5/25/95.
Incidental: Lees Ferry, 6/1/94 (second-year male); RM 1.0R 7/12/94 and 8/11/94 (4); RM 1.6R, 7/12/94; RM 71, 7/11/93; RM 168.8R, 5/14/94; RM 171.0R, 5/21/95 (second year male); "fairly numerous" around RM 198.0R, especially females and hatch-year birds, 7/23/94 (DF field notes); RM 213.6L, 7/27/94 ("several").
Notes: Earliest date was 5/17/95 at RM 50.0

R; latest was 9/11/94 at RM 198.0R. Only the pair at RM 204.5R were confirmed breeders, the rest were assumed migrants.

Scott's Oriole (*Icterus parisorum*): Rare transient.
Incidental: RM 205.6R, 7/27/94 (adult male).

Finches

* House Finch (*Carpodacus mexicanus*): Common summer resident.
Surveys: 409 detections on 352 surveys; found in every survey reach except Reach 6. February (7), March (62), April (62), May (13), June (163), July (76), September (26).
Banding: 18 birds. April (1), May (4), June (2), July (11).
Notes: This was the third most abundant breeding species that we detected in the canyon. House Finches made only limited use of riparian vegetation, while ranging widely across the uplands. Earliest date was 2/6/95 at RM 1.6R; latest was 9/26/94 at RM 204.5R; however, has been recorded along the river in every month (Brown et al. 1987, p. 271). We found only two nests, both at RM 46.7R on 6/15/95.

* Lesser Goldfinch (*Carduelis psaltria*): Fairly common summer resident.
Surveys: 125 detections on 78 surveys. March (26), April (59), May (33), June (4), September (3).
Banding: RM 198.0R, 4/27/94 and 4/22/95 (2); RM 204.5R, 5/27/95.
Notes: Earliest date was 3/23/95 at RM 197.6L; latest was 9/28/94 at RM 208.7R. However, there are records along the river from every month (Brown et al. 1987, p. 272). We found a nest with four eggs in desert broom below RM 180, 5/14/94. We heard young birds begging for food at RM 204.5R, 5/25/95.

American Goldfinch (*Carduelis tristis*): Rare transient.
Incidental: One male in breeding plumage at the bottom of Deer Creek trail (RM 136), 4/20/94.

Old World Weaver Finches

House Sparrow (*Passer domesticus*): Rare transient.
Incidental: RM 46.7R, 4/14/94; RM 171.1R, 4/24/94 (pair); RM 171.0R, June 1993.

ACKNOWLEDGMENTS

We thank the management and staff of Grand Canyon National Park and the Glen Canyon National Recreation Area for their support of this project. Funding, technical assistance, and logistic support were provided by the Bureau of Reclamation's Glen Canyon Environmental Studies (GCES) program (now the U.S. Geological Survey's Grand Canyon Monitoring and Research Center). Patricia Hodgetts and Helen Yard conducted hundreds of surveys and operated mist-nets for hundreds of hours, often under difficult circumstances; we thank them for their knowledge, energy, skills, and dedication. We are grateful to Christine Karas (Bureau of Reclamation) and Margaret Rasmussen (USGS) for their valuable support throughout the study. Marlene Bennett and Don Henry (U.S. Fish and Wildlife Service), Paul Deshler (Northern Arizona University), and Dan Spotsky (Grand Canyon National Park) assisted with aspects of this research. We are grateful for volunteer field assistance from John Grahame, Aurora Hindman, Bill Howe, Matt Johnson, Bill Maynard, Albert Miranda, and Lia Spiegel. Our thanks to Thomas Gushue for creating the map in Figure 1. We truly appreciate the skilled and thoughtful reviews by Frank Brandt, William Howe, and R. Roy Johnson; their comments greatly improved the final result.

LITERATURE CITED

American Ornithologists' Union. 1998. Checklist of North American Birds, 7th edition. McLean, VA.

American Ornithologists' Union. 2002. 43rd Supplement to the Checklist of North American Birds. Auk 119: 897–906.

Bent, A. C. 1953. Life Histories of North American Wood Warblers, Volume 1. United States National Museum, Washington, DC.

Brown, B. T. 1992. Nesting chronology, density, and habitat use of Black-chinned Hummingbirds along the Colorado River, Arizona. Journal of Field Ornithology 63: 393–400.

Brown, B. T. 1994. Rates of brood parasitism by Brown-headed Cowbirds on riparian passerines in Arizona. Journal of Field Ornithology 65: 160–168.

Brown, B. T., S. W. Carothers, and R. R. Johnson. 1983. Breeding range expansion of Bell's Vireo in Grand Canyon, Arizona. Condor 85: 499–500.

Brown, B. T., S. W. Carothers, L. T. Haight, R. R. Johnson, and M. M Riffey. 1984. Birds of the Grand Grand Canyon Region: An Annotated Checklist, 2nd edition. Grand Canyon Historical Association Monograph 1, Grand Canyon, AZ.

Brown, B. T., S. W. Carothers, and R. R. Johnson. 1987. Grand Canyon Birds. University of Arizona Press, Tucson.

Carothers, S. W., and R. R. Johnson. 1975. Recent observations on the status and distribution of some birds of the Grand Canyon region. Plateau 47: 140–153.

Felley, D. L., and M. K. Sogge. 1997. Comparison of survey techniques for monitoring riparian birds along the Colorado River through Grand Canyon National Park. In Proceedings of the Third Biennial Conference on Research in Colorado Plateau National Parks, edited by C. van Riper III and E. T. Deshlor, pp. 73–83. October 1995. National Park Service Transactions and Proceedings Series NPS/NRN/NRTP-97/12.

Grahame, J. D. 1995. Breeding birds along the Colorado River through Glen Canyon: The 1995 report and an historical perspective. Final report. National Park Service, Resource Management Division. Glen Canyon National Resource Area. 52 pp.

Monson, G., and A. R. Phillips. 1981. Annotated Checklist of the Birds of Arizona. University of Arizona Press, Tucson.

Phillips, A. R., J. Marshall, and G. Monson. 1964. The Birds of Arizona. University of Arizona Press, Tucson.

Pyle, P. 1997. Identification Guide to North American Birds: Part I, Columbidae to Ploceidae. Slate Creek Press, Bolinas, CA.

Root, T. 1988. Atlas of Wintering North American Birds: An Analysis of Christmas Count Data. University of Chicago Press, Chicago and London.

Rosenberg, K. V., R. D. Ohmart, W. C. Hunter, and R. W. Anderson. 1991. Birds of the Lower Colorado River Valley. University of Arizona Press, Tucson.

Schmidt, J. C., and J. B. Graf. 1988. Aggradation and degradation of alluvial sand deposits, 1965 to 1986, Colorado River, Grand Canyon National Park, Arizona. NTIS publication PB88-195458/AS.

Sogge, M. K., T. J. Tibbitts, and J. Petterson. 1997. Status and breeding ecology of the Southwestern Willow Flycatcher in the Grand Canyon. Western Birds 28: 142–157.

Sogge, M. K., D. Felley, and M. Wotawa. 1998. Riparian bird community ecology in the Grand Canyon—Final report. USGS Colorado Plateau Field Station report to the National Park Service and Bureau of Reclamation. 260 pp.

Spence, J. R. 2004. The riparian and aquatic bird communities along the Colorado River from Glen Canyon Dam to Lake Mead, 1996–2002. Draft 2.0, Final report submitted to Grand Canyon Monitoring and Research Center, USGS, Flagstaff, Arizona.

Spence, J. R., and B. B. Bobowski. 2003. 1994–1997 water bird surveys of Lake Powell, a large, oligotrophic reservoir on the Colorado River, Utah and Arizona. Western Birds 34: 133–148.

Stevens, L. 1983. The Colorado River in Grand Canyon: A Guide. Red Lake Books, Flagstaff, Arizona.

Turner, R. M., and M. M. Karpiscak. 1980. Recent vegetation changes along the Colorado River between Glen Canyon Dam and Lake Mead, Arizona. U.S. Department of the Interior, USGS Professional Paper 1132.

van Riper, C. III, and M. K. Sogge. 2004. Bald Eagle abundance and relationship to prey base and human activity along the Colorado River in Grand Canyon National Park. In The Colorado Plateau: Cultural, Biological and Physical Research, edited by C. van Riper III and K. Cole, pp. 163–185. University of Arizona Press, Tucson.

Yard, H. K, C. van Riper III, B. T. Brown, and M. J. Kearsley. 2004. Diet of insectivorous birds along the Colorado River in Grand Canyon, Arizona. Condor 106: 106–115.

IMPLICATIONS OF MERRIAM'S TURKEY AGE, GENDER, CAUSE-SPECIFIC MORTALITY, AND REPRODUCTION ON POPULATION DEMOGRAPHICS BASED ON POPULATION MODELING

Brian F. Wakeling and Charles H. Lewis

Many demographic parameters influence turkey populations (Wakeling 1991; Roberts et al. 1995; Roberts and Porter 1996); these include age, gender, cause-specific mortality, and reproduction. In Arizona, turkey population size fluctuates substantially among years, as do the proportion of female turkeys that nest, the proportion of nests that successfully hatch, and age-specific survival rates for females (Wakeling 1991; Wakeling and Shaw 1994; Mollohan et al. 1995; Rumble et al. 2003). The objective of turkey management is generally to manage harvest or habitat to influence population size, yet imprecise knowledge of population demographics may limit the ability of management agencies to achieve the desired outcome. Population models can help biologists determine the relative importance of common demographic parameters that affect annual population change. Once identified, managers can concentrate on the determination of intrinsic or extrinsic factors responsible for the annual variation of the most meaningful demographic parameters. Our objective in this study was to improve our knowledge of population demographics by using deterministic and stochastic population modeling, based on published estimates of demographic parameters.

METHODS

We used a deterministic population model described by Wakeling (1991) to compare the effects of demographic parameters we obtained from the literature on turkey population trends. We used the model:

$$N = A + Y + J$$

where N is the August population estimate, A is the number of adults in the population, Y is the number of yearlings, and J is the number of juveniles (poults) in the population. To model adult numbers, we used the equation:

$$A = (A_{-1} \times S_A) + (Y_{-1} \times S_Y)$$

where A is the number of adults in this year's population, A_{-1} is the number of adults in last year's population, S_A is last year's adult annual probability of survival, Y_{-1} is the number of yearlings in last year's population, and S_Y is last year's yearling annual probability of survival. To model yearling numbers, we used the equation:

$$Y = J_{-1} \times S_J$$

where Y is the number of yearlings in this year's population, J_{-1} is the number of juveniles in last year's population, and S_J is the annual probability of survival for last year's juveniles. To model juvenile numbers, we used the equation:

$$J = (E_A \times A_{-1} \times N_{RA} \times N_{SA} \times S_J \times S_P) + (E_Y \times Y_{-1} \times N_{RY} \times N_{SY} \times S_J \times S_P)$$

where J is the number of juveniles in the population, E_A is the number of eggs laid by adults, N_{RA} is the proportion of adult females that initiate a nest, N_{SA} is the proportion of adult females that nest that hatch ≥ 1 egg, S_J is the probability of survival of juveniles from August of one year until August of the next, S_P is the probability of survival of young from hatching to August, E_Y is the number of eggs laid by yearlings, N_{RY} is the proportion of yearling females that initiate a nest, and N_{SY} is the proportion of yearling females that nest that hatch ≥ 1 egg.

Based on demographic parameters measured in Arizona, we modeled population trends using the aforementioned deterministic model. We based the model on mean demographic parameters measured in previous studies (Wakeling 1991; Mollohan et al. 1995). We compared an initial population estimate based on age structures observed in Arizona populations (500 adult female, 150 yearling female, 300 juvenile female, assuming sufficient males were available for breeding) with a population estimate from the model 10 years later. We then independently increased each demographic parameter by 10 percent while holding all others at their initial levels and deterministically modeled the effect on the turkey population 10 years later.

For this model, we used data from studies in Arizona. For each variable within the equation, we assumed that E_A and $E_Y = 4.50$; $N_{RA} = 0.70$; $N_{SA} = 0.90$; $N_{RY} = 0.00$; $S_P = 0.35$; $S_J = 0.49$; $S_Y = 0.70$; and $S_A = 0.66$.

We modified the deterministic model to allow for stochasticity. We allowed the demographic parameters to vary randomly within ranges observed during previous studies conducted in Arizona. These ranges were not always normally distributed, and random behavior within measured ranges seems to reflect natural variation observed in wild turkey populations (Vangilder 1992). We used 500 Monte Carlo simulations to detect the frequency with which the modeled population 10 years following initiation declined > 20%, declined ≤ 20%, increased ≤ 20%, or increased > 20%. The model was allowed to vary: N_{RA} = 0.35–0.65; N_{RY} = 0.00; N_{SA} = 0.60–0.90; S_P = 0.20–0.60; S_J = 0.35–0.85; S_Y = 0.52–0.87; and S_A = 0.50–0.85.

We also ran an additional 500 Monte Carlo simulations to model population numbers by allowing demographic parameters to vary within the range observed in studies of Merriam's turkeys from a productive, but non-native, habitat within the Black Hills of South Dakota (Rumble 1990; Rumble and Hodorff 1993). We modeled nesting parameters from this area because they differed substantially from those observed in Arizona. The modified parameters for this model varied: N_{RA} = 0.55–0.80; N_{SA} = 0.60–0.90; N_{RY} = 0.35–0.60; and N_{SY} = 0.60–0.90.

We modeled two additional hypothetical situations using 500 Monte Carlo simulations each. The first modification to the stochastic model was to allow yearling females to nest at rates equal to adults in Arizona and vary within the same range as adults. This modification was used to simulate the effect on the Arizona population if yearling females nested during their first spring. For this model, we modified parameters from our original stochastic model: N_{RY} = 0.35–0.65 and N_{SY} = 0.60–0.90.

The second modification altered the stochastic model by allowing yearling and adult female survival to improve by 0.20. This modification was used to simulate the possible non-compensatory effect the population might realize if predation were reduced. To model this possibility, we again modified parameters from our original stochastic model: S_Y = 0.70–1.00 and S_A = 0.70–1.00.

Ultimately, the assumptions of our stochastic models were (1) demographic parameters are not influenced by population size (i.e., no density-dependent survival or reproduction); (2) demographic parameters vary independently and randomly (i.e., no correlations exist among demographic parameters); and (3) age-specific differences exist among annual turkey survival and proportion of female turkeys that nest, but not for the proportion of nests that successfully hatch or clutch size.

RESULTS

Using mean values from previous studies in Arizona, the deterministic model resulted in a stable population size estimate among years. Adult survival had the greatest impact on the population size 10 years in the future, and the proportion of female yearlings that nest had the smallest effect (Table 1).

Stochastic models differed substantially in predictions of population change (Table 2). The population modeled for Arizona showed the greatest tendency for substantial population declines. Populations modeled using the proportion of female turkeys in the Black Hills of South Dakota that nest, and improved survival rates in Arizona had the greatest tendency to increase. Even the Arizona population modeled with the proportion of yearling female turkeys that nest equal to the proportion of adult female turkeys that nest showed a marked tendency to increase.

DISCUSSION

The sensitivity analysis showed that any demographic parameter altered by 10 percent could have a substantial effect on population size within 10 years. Turkey populations are highly dynamic and may fluctuate by as much as 50 percent between years (Mosby 1967), although a change of such magnitude is rare (Roberts and Porter 1996). Nonetheless, large variations among years are common in many demographic parameters in Arizona (Wakeling 1991; Mollohan et al. 1995; Wakeling and Rogers 1998). Still, because obtaining estimates of turkey population trends is logistically difficult, knowledge of the actual magnitude and frequency of annual fluctuations is limited.

Simulated demographic parameters had different effects on modeled population size. Simulated adult survival had the greatest effect, probably because this comprised the largest cohort in the model. Changes in simulated proportion of yearling female turkeys that nest had the smallest effect on the modeled population size. However, because Southwestern turkey populations typically experience no yearling nesting effect, the greatest potential for improvement exists in this demographic parameter. We are aware of only a single documented yearling nest within Arizona (Crites 1988), although Lockwood and Sutcliffe (1985) documented nesting by 8 percent of marked yearling females in New Mexico. Based on stochastic modeling, increases in the proportion of yearling female turkeys that nest can substantially increase turkey population densities.

Elsewhere in the species' range, yearling Merriam's turkey females are known to nest at considerably greater rates than that measured in the Southwest. Substantial nesting by yearling female turkeys has been noted in Colorado, Wyoming, South Dakota, and Montana (Hengel 1990; Rumble and Hodorff 1993; Thompson 1993; Hoffman et al. 1996), whereas other parts of Colorado, New Mexico, and Arizona have limited or no yearling nesting (Lockwood and Sutcliffe 1985; Hoffman 1990; Wakeling 1991). In habitats with yearling nesting, subadult females average about 0.5 kg heavier during midwinter than those from habitats with limited yearling nesting (Hoffman et al. 1996). The heavier turkeys generally had access to supplemental feeds during winter, whereas lighter birds did not (Hoffman et al. 1996).

In Arizona, access to supplemental feeds has been rare. Arizona lacks winter feed yards where livestock are supplementally fed near turkey range. Establishing supple-

Table 1. Effect on estimate of population size 10 years in future by altering deterministic model parameters by 10 percent.

Population Parameter	Percent Change to the Population Size 10 yrs in Future
Adult female survival	37.5
Yearling female survival	21.9
Juvenile female survival	23.1
Adult nesting proportion	25.3
Adult nesting success	26.3
Yearling nesting proportion[1]	11.1
Poult survival	26.3
Number of eggs	25.3

[1]Estimate derived by including a proportion of yearling female turkeys that nest at 0.5 within the deterministic equation, a parameter not included in the initial model.

Table 2. Stochastically modeled population performance 10 years in the future (percent of 500 simulations) relative to the beginning population size based on parameters measured in Arizona, in South Dakota, in Arizona with the proportion of yearling female turkeys that nest equal to those of adult female turkeys, and in Arizona with improved yearling and adult female survival.

Performance	Arizona	South Dakota	Arizona with Yearling Nesting	Arizona with Improved Survival
> 20% decline	63.0%	0.0%	7.0%	0.0%
≤ 20% decline	15.0%	0.0%	7.4%	0.0%
≤ 20% increase	11.0%	0.6%	5.8%	0.8%
> 20% increase	11.0%	99.4%	79.8%	99.2%

mental feeding areas for turkeys would be unlikely to be supported by management agencies because of the cost, influence on other wildlife species, and social implications. An area that has received limited research to date has been manipulating the availability of natural food sources to provide a better nutritional complement for turkeys.

In habitats where winter forage ability is limited in distribution, such as within livestock feed yards, anecdotal observations indicate that male turkeys can be socially dominant and may limit female turkey access to food (Harry Harju, Wyoming Game and Fish Department, personal communication). In Arizona, the fact that winter is the period of greatest mortality for turkeys has been tied to food availability (Wakeling 1991). In a situation where male turkeys have greater access to food resources during winter than do female turkeys, male turkey survival may be greater than female turkey survival. Supplemental feeding if provided in limited distribution may affect gender ratios, and male-biased gender ratios may have a negative impact on female survival.

All assumptions of our model may not be accurate. For example, demographic parameters may not be truly independent, as reproductive performance may be influenced by population density (Porter 1978), predation (Glidden 1977; Vander Haegen et al. 1988), weather (Beasom and Pattee 1980; Porter et al. 1983), habitat quality (Lockwood and Sutcliffe 1985; Wakeling et al. 1998), or disease (Rocke and Yuill 1987). Also, survival rates may vary dependently among age classes. Conversely, other assumptions seem accurate. For instance, demographic parameters seem random in distribution, rather than normally distributed (Vangilder 1992). We recognize that our models may not accurately reflect the true magnitude in population changes, but we believe that implications of changes to demographic parameters in our model and their effects on the population are correct.

MANAGEMENT IMPLICATIONS

Direct mortality of turkeys may be influenced through hunting regulations or predator management. Research has documented that female harvests of ≤ 10 percent of the female segment or ≤ 30 percent of the male segment do not limit population growth (Little et al. 1990; Healy and Powell 1999); harvest levels in Arizona are well below these thresholds (Wakeling 1991; Wakeling and Rogers 1998). In areas where gender ratios are male-biased, increased hunting of male turkeys may improve female survival. In management units where turkey populations are at low levels, closure of fall either-sex turkey hunts is currently practiced. Predator management can be difficult to achieve on a scale large enough to affect turkey survival and populations. Further, when populations of prey animals are at or near carrying capacity, predator management efforts are rarely successful in creating increases in prey, especially when site-specific predator management cannot be practiced (Ballard et al. 2001).

Perhaps the greatest potential for increasing Arizona turkey populations, or increasing the rate of recovery from declines, is improving the nutritional condition of yearling females. Heavier yearling females tend to have a greater likelihood to initiate a nest than do those that weigh about 0.5 kg less (Hoffman et al. 1996). An increased proportion of yearling female turkeys that nest seems to be correlated with an increased proportion of adult female turkeys that nest (Rumble et al. 2003), which probably has a nutritional nexus. Little information exists regarding optimal forage availability for turkey productivity. Yet, the information available suggests that mast crops are important to overwinter survival (Wakeling and Rogers 1996, 1998). Forest management practices that protect and improve mast yield from oak (*Quercus* spp.) and pine (*Pinus* spp.) species would favor turkey populations. Improving mast yield to sustain overwintering turkeys may be an area ripe for research and management.

ACKNOWLEDGMENTS

This research was funded by State Trust Wildlife Fund grants W-53 and W-78 through the Arizona Game and Fish Department. The reviews by C. Conway and two anonymous reviewers greatly benefited the manuscript. The National Wild Turkey Federation funded portions of the research on which this paper is based.

LITERATURE CITED

Ballard, W. B., D. L. Lutz, T. W. Keegan, L. H. Carpenter, and J. C. deVos, Jr. 2001. Deer-predator relationships: A review of recent North American studies with emphasis on mule and black-tailed deer. Wildlife Society Bulletin 29: 99–115.

Beasom, S. L., and O. H. Pattee. 1980. The effect of selected climatic variables on wild turkey productivity. Proceedings of the National Wild Turkey Symposium 4: 127–135.

Crites, M. J. 1988. Ecology of the Merriam's turkey in north-central Arizona. Master's thesis, University of Arizona, Tucson. 59 pp.

Glidden, J. W. 1977. Net productivity of a wild turkey population in southwestern New York. Transactions of the Northeast Fish and Wildlife Conference 34: 13–21.

Healy, W. M., and S. M. Powell. 1999. Wild turkey harvest management: Biology, strategies, and techniques. U.S. Fish and Wildlife Service Biological Technical Publication BTP-R5001-1999. 96 pp.

Hengel, D. A. 1990. Habitat use, diet and reproduction of Merriam's turkeys near Laramie Peak, Wyoming. Master's thesis, University of Wyoming, Laramie. 229 pp.

Hoffman, R. W. 1990. Chronology of gobbling and nesting activities of Merriam's wild turkeys. Proceedings of the National Wild Turkey Symposium 6: 25–31.

Hoffman, R. W., M. P. Luttrell, and W. R. Davidson. 1996. Reproductive performance of Merriam's wild turkeys with suspected *Mycoplasma* infection. Proceedings of the National Wild Turkey Symposium 7: 145–151.

Little, T. W., J. M. Kienzler, and G. A. Hanson. 1990. Effects of fall either-sex hunting on survival in an Iowa wild turkey population. Proceedings of the National Wild Turkey Symposium 6: 119–125.

Lockwood, D. R., and D. H. Sutcliffe. 1985. Distribution, mortality, and reproduction of Merriam's turkey in New Mexico. Proceedings of the National Wild Turkey Symposium 5: 309–316.

Mollohan, C. M., D. R. Patton, and B. F. Wakeling. 1995. Habitat selection and use by Merriam's turkey in northcentral Arizona. Arizona Game and Fish Department Technical Report 9, Phoenix. 46 pp.

Mosby, H. S. 1967. Population dynamics. In The Wild Turkey and Its Management, edited by O. H. Hewitt, pp. 113–136. The Wildlife Society, Washington, D. C.

Porter, W. F. 1978. The ecology and behavior of the wild turkey (*Meleagris gallopavo*) in southeastern Minnesota. Ph.D. dissertation, University of Minnesota, St. Paul. 122 pp.

Porter, W. F., G. C. Nelson, and K. Mattson. 1983. Effects of winter conditions on reproduction in a northern wild turkey population. Journal of Wildlife Management 47: 281–290.

Roberts, S. D., and W. F. Porter. 1996. Importance of demographic parameters to annual changes in wild turkey abundance. Proceedings of the National Wild Turkey Symposium 7: 15–20.

Roberts, S. D., J. M. Coffey, and W. F. Porter. 1995. Survival and reproduction of female wild turkeys in New York. Journal of Wildlife Management 59: 437–447.

Rocke, T. E., and T. M. Yuill. 1987. Microbial infections in a declining wild turkey population in Texas. Journal of Wildlife Management 51: 778–782.

Rumble, M. A. 1990. Ecology of Merriam's turkeys (*Meleagris gallopavo merriami*) in the Black Hills, South Dakota. Ph.D. dissertation, University of Wyoming, Laramie. 169 pp.

Rumble, M. A., and R. A. Hodorff. 1993. Nesting ecology of Merriam's turkeys in the Black Hills, South Dakota. Journal of Wildlife Management 57: 789–801.

Rumble, M. A., B. F. Wakeling, and L. D. Flake. 2003. Factors affecting survival and recruitment in female Merriam's turkeys. Intermountain Journal of Science 9: 26–37.

Thompson, W. T. 1993. Ecology of Merriam's turkeys in relation to burned and logged areas in southeastern Montana. Ph.D. dissertation, Montana State University, Bozeman. 195 pp.

Vander Haegen, W. M., E. E. Dodge, and M. W. Sayre. 1988. Factors affecting productivity in a northern wild turkey population. Journal of Wildlife Management 52: 127–133.

Vangilder, L. D. 1992. Population dynamics. In The Wild Turkey: Biology and Management, edited by J. G. Dickson, pp. 144–164. Stackpole Books, Harrisburg, PA.

Wakeling, B. F. 1991. Population and nesting characteristics of Merriam's turkey along the Mogollon Rim, Arizona. Arizona Game and Fish Department Technical Report 7, Phoenix. 48 pp.

Wakeling, B. F., and T. D. Rogers. 1996. Winter diet and habitat selection by Merriam's turkeys in north-central Arizona. Proceedings of the National Wild Turkey Symposium 7: 175–184.

Wakeling, B. F., and T. D. Rogers. 1998. Summer resource selection and yearlong survival of male Merriam's turkeys in north-central Arizona, with associated implications from demographic modeling. Arizona Game and Fish Department Technical Report 28, Phoenix. 50 pp.

Wakeling, B. F., and H. G. Shaw. 1994. Characteristics of managed forest habitat selected for nesting by Merriam's turkeys. In Sustainable Ecological Systems: Implementing an Ecological Approach to Land Management, technical coordinators W. W. Covington and L. F. DeBano, pp. 359–363. U.S. Forest Service General Technical Report RM 247.

Wakeling, B. F., S. S. Rosenstock, and H. G. Shaw. 1998. Forest stand characteristics of successful and unsuccessful Merriam's turkey nest sites in north-central Arizona. Southwestern Naturalist 43: 242–248.

MORPHOLOGIC CHARACTERISTICS OF A TRANSPLANTED POPULATION OF GOULD'S TURKEYS WITH COMPARISONS TO MERRIAM'S TURKEYS

Shelli Dubay, Brian Wakeling, and Tim Rogers

Gould's turkeys are believed to be the largest subspecies of turkey in North America (Schemnitz and Zeedyk 1992), but limited comparative data exist. Genetic data indicate that Gould's turkeys are the most genetically divergent subspecies of turkey in North America (Mock et al. 2001a). As a result, Gould's turkeys may be different genetically and morphologically, and potentially ecologically distinct as well. Our objective was to compare morphological traits from a reestablished population of Gould's turkeys in Arizona to Gould's turkeys from Mexico and to Merriam's turkeys in northern Arizona. Specifically, (1) we measured mean weight for Gould's turkeys in the Huachuca Mountains, Arizona and in Sonora, Mexico and for Merriam's turkeys (*M. g. merriami*) in northern Arizona; (2) we measured beard length in adult male Gould's turkeys in Arizona and in Sonora, Mexico; (3) we measured tarsometatarsus length for Gould's turkeys and Merriam's turkeys in Arizona; and (4) we compared morphological characteristics from Gould's turkeys captured in the Huachuca Mountains to published records for other turkey subspecies.

STUDY AREAS

Huachuca Mountains, Southeastern AZ

We studied Gould's turkeys in the Huachuca Mountains (289 sq km), primarily in the Coronado National Forest of southeastern Arizona (110°20'N, 31°25'W). Elevation varies from 1400 m in the surrounding foothills to over 2800 m at Miller and Carr Peaks. Average annual precipitation is 46 cm, with seasonal peaks in winter and late summer. At Canelo Hills, adjacent to the Huachuca Mountains, the average maximum temperature is 32.4° C in June and the average minimum is –3.3° C in January (Western Regional Climate Center records for Canelo, AZ).

Warm season perennial grasses, including grama (*Bouteloua* spp.) and three-awn (*Aristida* spp.), interspersed with honey mesquite (*Prosopis juliflora*), occupy elevations up to 1525 m. Madrean evergreen woodlands (Brown 1994) characterized by evergreen oaks (*Quercus emoryi, Q. oblongifolia, Q. arizonica*), junipers (*Juniperus deppeana, J. monosperma*), and pinyon pine (*Pinus discolor*) occur at moderate elevations of 1525 m to 2135 m. Montane riparian forests (Brown 1994) of Arizona sycamore (*Platanus wrightii*), Arizona ash (*Fraxinus velutina*), and Fremont cottonwood (*Populus fremontii*) predominate in canyon bottoms and along drainages. Madrean montane conifer forests (Brown 1994) with variable compositions of pine species (*P. ponderosa, P. engelmannii, P. leiophylla*), Gambel oak (*Q. gambelii*), and quaking aspen (*Populus tremuloides*) occupy interior portions of the range above 2135 m. Shrubs such as manzanita (*Arctostaphylos pungens*), catclaw (*Mimosa* spp.), and New-Mexican locust (*Robinia neomexicana*) occur throughout the lower foothills and mesic north-facing slopes, as well as in areas disturbed by fire and mining activities. The area includes private, state trust, military (Fort Huachuca), and U.S. Forest Service lands.

Mogollon Rim, Northern Arizona

We studied Merriam's turkeys on the Mogollon Rim approximately 65 km south of Winslow, Arizona on the Apache-Sitgreaves National Forest (Wakeling and Rogers 1995; 110°50'N, 34°30'W). Elevations range from 1700 m in the northern portion of the study area to 2430 m in the southern portion. Annual precipitation averages 47.2 cm, with peaks from January through March and from July through September, and temperature extremes range from –23 to 32° C (National Oceanic and Atmospheric Administration 1991).

Six vegetation cover types (mixed conifer, ponderosa pine–Gambel oak, ponderosa pine–pinyon-juniper, pinyon-juniper, aspen woodland, and forest meadow) are present in the study area (Laing et al. 1989). Mixed-conifer vegetation, which occurs above 2340 m, consists of Douglas-fir, white fir (*Abies concolor*), limber pine (*P. flexilis*), and Rocky Mountain maple (*Acer glabrum*). Ponderosa pine dominates west-facing slopes between 2340 and 1850 m, and pinyon-juniper habitat is found below 1850 m. Pinyon pine and alligator juniper are found at elevations below 2150 m. Gambel oak occurs regularly with ponderosa pine and in pinyon-juniper and mixed conifer habitats. Aspen and forest meadows are also found in mesic areas and at higher elevations.

Sonora, Mexico

We captured Gould's turkeys near Yecora, in Sonora, Mexico (28°22'N, 108°15'W). Topography of the area consists of forested mesas and steep canyons. Elevations range from 800 to 2200 m. Temperatures typically range between –3 and 38° C. The average annual precipitation is 80 cm (Haro Rodriguez 1993).

Vegetation includes several pine species, including Mexican pinyon (*P. cembroides),* Chihuahua (*P. leiophylla),* Mexican oocarpa (*P. oocarpa*), Durango (*P. durangensis*), and two-needle pinyon (*P. edulis*), and several oak species, including black oak (*Q. albocincta*), slender oak (*Q. graciliformis*), white oak (*Q. chihuahuensis*), Mexican blue oak (*Q oblongifolia*), emory oak (*Q. emoryi*), California scrub oak (*Q. dumosa*), and silverleaf oak (*Q. hypoleucoides*). Madrone (*Arbutus arizonica*) and manzanita (*Arctostaphylos pungens*) are common, as is beargrass (*Nolina microcarpa*). Herbaceous vegetation is also abundant. The land surrounding the town of Yecora consists largely of communal ranches and farms.

METHODS

Turkey Capture

Gould's turkeys in the Huachuca Mountains were captured in November through March of 1998–2002. Gould's turkeys in Sonora, Mexico were captured in January 1994 and January-February 1997. Merriam's turkeys were captured from December through March, beginning in January 1987 and ending in March 1997. We captured turkeys using rocket nets (Bailey et al. 1980) at sites baited with cracked corn, then released birds at the site. Baiting was initially done manually in the Huachuca Mountains, but was more effective after we switched to using a 115-L tripod game feeder (Moultrie Feeders, Alabaster, AL) in 2001. We baited Gould's turkeys in the Huachuca Mountains with approximately 900 g of cracked corn twice daily.

Trapped individuals were weighed and the general health condition of each bird was assessed. We measured tarsometatarsus length (Wakeling et al. 1997) with a ruler, and beard length was also measured for males. Tail feathers were counted on all Gould's turkeys trapped in the Huachuca Mountains beginning in 2000. If tail feathers were lost during capture, it was noted on the capture form.

Statistical Analyses

Due to small sample sizes, weights of subadult female Gould's turkeys in the Huachuca Mountains and Sonora, Mexico were not compared. An independent samples *t*-test was used to compare weights by sex and age class of Gould's turkeys captured in the Huachuca Mountains to those captured in Sonora (Zar 1999). A one-way ANOVA was used to compare weights of adult female and adult male Gould's turkeys from the Huachuca Mountains to Gould's turkeys from Mexico and to Merriam's turkeys from northern Arizona (Zar 1999). We used Tukey's multiple comparison tests to detect differences among areas. We compared mean tarsometatarsus length of adult female, juvenile male, and adult male Gould's turkeys captured in the Huachuca Mountains to those previously reported for Merriam's turkeys (Wakeling et al. 1997). Due to low sample sizes for juvenile females and adult males, only mean tarsometatarsus length of adult female and juvenile male Gould's turkeys from Sonora were compared to those from the Huachuca Mountains. We compared mean tarsometatarsus lengths for adult female, adult male, and juvenile male Gould's turkeys to values published by Wakeling et al. (1997). Mean number of tail feathers for Gould's turkeys from the Huachuca Mountains was calculated and compared to data previously reported for Gould's turkeys. Also, mean beard length of adult male Gould's turkeys captured in Sonora, adult male Gould's turkeys from the Huachuca Mountains, and adult male Merriam's turkeys from northern Arizona were compared using a one-way ANOVA and Tukey's multiple comparison tests. For all analyses, we considered $p < 0.05$ significant.

RESULTS

Weights of Turkeys

Gould's turkeys from the Huachuca Mountains were heavier than those from Mexico. Adult female Gould's turkeys captured in the Huachuca Mountains ($n = 22$, $\overline{X} = 5.1 + 0.2$ SE kg) were heavier than those captured in Mexico ($n = 32$, $\overline{X} = 4.8 + 0.1$ SE kg, $t = 2.1$, $P = 0.04$), subadult males captured in the Huachuca Mountains ($n = 17$, $\overline{X} = 6.6 + 0.3$ SE kg) were heavier than those from Sonora, Mexico ($n = 15$, $\overline{X} = 5.6 + 0.2$ SE kg, $t = 4.5$, $P < 0.01$), and adult males from the Huachuca Mountains ($n = 36$, $\overline{X} = 9.1 + 0.2$ SE kg) weighed more than those captured in Mexico ($n = 13$, $\overline{X} = 8.0 + 0.3$ SE kg, $t = 6.7$, $P < 0.01$).

Comparisons of all three populations of turkeys are shown in Table 1. Because subadult weights were not standardized by days post-hatching, we did not include them in analyses. An ANOVA revealed that adult males ($F = 41.4$, $P < 0.01$) and adult females ($F = 16.5$, $P < 0.01$) differed in weight among study areas.

When we compared weights of Gould's turkeys captured in the Huachuca Mountains to weights of other subspecies, we found that subadult female Gould's turkeys from the Huachuca Mountains were larger than reported subadult females of other subspecies (Table 2). Furthermore, they were larger than subadult females in other populations of Gould's turkeys. Adult females from the Huachuca Mountains were heavier than those from Sonora and from Chihuahua, Mexico, but similar in weight to Gould's turkeys from the Peloncillo Mountains, New Mexico (Table 2). Nonetheless, adult female Gould's turkeys from New Mexico and from Arizona were heavier than other subspecies compared.

Subadult male Gould's turkeys from the Huachuca Mountains, Arizona were heavier than other Gould's turkey subadults (Table 2). Subadult male Gould's turkeys from Sonora and subadult male Merriam's turkeys from northern Arizona were similar in weight to subadults of other subspecies. Adult male Gould's turkeys from the Huachuca Mountains were similar in size to adult males from New Mexico and from Chihuahua, Mexico, but these populations were heavier than adult males from Sonora. Some populations of Gould's turkeys seem to have adult males that are heavier than males from other subspecies.

Tarsometatarsus Length

Tarsometatarsus length of Gould's turkeys from the Huachuca Mountains did not differ from measurements taken from Gould's turkeys from Sonora, Mexico. Average tarsometatarsus length of adult females from the Huachuca Mountains ($n = 14$, $\overline{X} = 14.0 + 0.3$ SE cm) did not differ from the average length of the tarsometatarsus for adult

Table 1. Comparison of adult turkey weights from three sites in western North America.

Gender	Site and Subspecies	N	Mean Wt (kg)	+ SE in kg	Multiple Comparisons[1]		
					Huachuca	Sonora	Arizona
Female	Huachuca Gould's	22	5.1	0.2	—		X
	Sonora Gould's	32	4.8	0.1	—	—	X
	Arizona Merriam's	314	4.6	0.3	—	—	—
Male	Huachuca Gould's	36	9.1	0.2	—	X	X
	Sonora Gould's	13	8.0	0.3	—	—	—
	Arizona Merriam's	58	8.1	0.1	—	—	—

[1]X = significance at $p < 0.05$.

females in Sonora (n = 11, $\overline{X}$ = 13.8 + 0.1 SE cm, t = 0.6, P = 0.57). Average tarsometatarsus length for adult male Gould's turkeys from the Huachuca Mountains (n = 11) was 17.1 + 0.1 SE cm, and because measurements were not obtained from adult male Gould's turkeys in Sonora, we could not compare between populations. For juvenile males, the average tarsometatarsus length for 13 Gould's turkeys from the Huachuca Mountains was 16.7 + 0.2 SE cm, which did not differ from the average from 9 Gould's turkeys in Mexico (t = 0.6, P = 0.55).

Wakeling et al. (1997) determined tarsometatarsus lengths for all age and sex classes of Merriam's turkeys in north-central Arizona; subadult females had lengths less than 12.5 cm, adult females had lengths between 12.6 and 13.7 cm, subadult males had lengths from 13.8 to 15.6 cm, and adult males had lengths greater than 15.7 cm. The average tarsometatarsus length of adult male Gould's turkeys from the Huachuca Mountains, 17.1 + 0.1 cm, is longer than the minimum 15.7 cm reported by Wakeling et al. (1997). Juvenile male Gould's turkeys from the Huachuca Mountains and from Sonora had longer tarsometatarsus lengths ($\overline{X}$ = 16.7 + 0.2 SE and 16.6 + 0.2 SE) than those reported for Merriam's turkeys. Adult females from the Huachuca Mountains and adult females from Sonora had average tarsometatarsus lengths ($\overline{X}$ = 14.0 + 0.3 SE and 13.8 + 0.1 SE) that were slightly longer than those reported for Merriam's turkeys in northern Arizona.

Tail Feathers

Gould's turkeys in the Huachuca Mountains, Arizona had an average of 17 tail feathers (n = 39, range = 12–20).

Beard Length

Beard length of adult male turkeys differed among the turkey populations (F = 5.23, P < 0.01). Gould's turkeys captured in the Huachuca Mountains had an average beard length of 23.7 cm (range = 15.2–30.0 cm), beard length of Gould's turkeys in Sonora averaged 21.2 cm (range = 15.2–27.9 cm), and Merriam's turkeys from northern Arizona had an average beard length of 20.8 cm (range = 11.4–25.4 cm). Multiple comparisons revealed that Gould's turkeys from the Huachuca Mountains had longer beards than Merriam's turkeys in northern Arizona (P = 0.01), and all other comparisons were insignificant.

DISCUSSION

Gould's turkeys are thought to be the heaviest and largest turkey subspecies (Schemnitz and Zeedyk 1992), and it seems that some populations of Gould's turkeys are larger than other subspecies. Gould's turkeys are the most genetically distinct subspecies of turkeys, although they most closely resemble Merriam's turkeys (Mock et al. 2001a). Part of this genetic differentiation may manifest itself in that Gould's turkeys are overall heavier and taller than Merriam's turkeys. Average weight for adult Gould's

Table 2. Turkey weights (mean and range) from published reports by age, sex, and subspecies.

Age	Sex	Weight (kg)		Subspecies	Location	Reference
		Mean	Range			
Subadult	F	3.5	2.7–4.3	Gould's	Chihuahua, Mexico	Schemnitz and Zeedyk 1992
Subadult	F	4.1	only 1 animal	Gould's	New Mexico	Schemnitz and Zeedyk 1992
Subadult	F	4.5	3.6–4.9	Gould's	Southeastern Arizona	Current data
Subadult	F	3.7	3.2–3.9	Gould's	Sonora, Mexico	Current data
Subadult	F	4.0	not given	Merriam's	Nebraska	Compiled in Lewis 1967
Subadult	F	4.0	not given	Merriam's	Colorado	Compiled in Lewis 1967
Subadult	F	3.4	not given	Rio Grande	Texas	Compiled in Lewis 1967
Subadult	F	3.4	not given	Florida	Florida	Compiled in Lewis 1967
Adult	F	4.3	4.4–4.4	Gould's	Chihuahua, Mexico	Schemnitz and Zeedyk 1992
Adult	F	5.4	5.0–5.9	Gould's	New Mexico	Schemnitz and Zeedyk 1992
Adult	F	5.3	4.3–6.8	Gould's	New Mexico	York and Schemnitz 2003
Adult	F	5.1	4.3–5.9	Gould's	Southeastern Arizona	Current data
Adult	F	4.8	4.2–5.5	Gould's	Sonora, Mexico	Current data
Adult	F	5.2	not given	Gould's	New Mexico	York 1991
Adult	F	4.6	3.2–5.9	Merriam's	Northern Arizona	Current data
Adult	F	4.3	not given	Merriam's	Nebraska	Compiled in Lewis 1967
Adult	F	4.8	not given	Merriam's	Colorado	Compiled in Lewis 1967
Adult	F	3.9	not given	Rio Grande	Texas	Compiled in Lewis 1967
Adult	F	3.8	not given	Florida	Florida	Compiled in Lewis 1967
Subadult	M	5.6	5.4–5.9	Gould's	New Mexico	Schemnitz and Zeedyk 1992
Subadult	M	5.2	4.1–6.2	Gould's	Chihuahua, Mexico	Schemnitz and Zeedyk 1992
Subadult	M	6.6	5.5–7.3	Gould's	Southeastern Arizona	Current data
Subadult	M	5.6	5.5–6.5	Gould's	Sonora, Mexico	Current data
Subadult	M	5.5	3.5–8.0	Merriam's	Northern Arizona	Current data
Subadult	M	5.7	not given	Merriam's	Nebraska	Compiled in Lewis 1967
Subadult	M	5.3	not given	Merriam's	Colorado	Compiled in Lewis 1967
Subadult	M	5.9	not given	Rio Grande	Texas	Compiled in Lewis 1967
Subadult	M	4.7	not given	Florida	Florida	Compiled in Lewis 1967
Adult	M	7.1	6.8–7.5	Gould's	Chihuahua, Mexico	Schorger 1966
Adult	M	10.0	7.9–2.5	Gould's	Chihuahua, Mexico	Schemnitz and Zeedyk 1992
Adult	M	9.8	9.5–10.0	Gould's	New Mexico	Schemnitz and Zeedyk 1992
Adult	M	9.3	8.7–10.2	Gould's	New Mexico	Zornes 1993
Adult	M	9.1	8.2–10.5	Gould's	Southeastern Arizona	Current data
Adult	M	8.0	6.9–8.9	Gould's	Sonora, Mexico	Current data
Adult	M	8.2	6.9–9.5	Merriam's	Northern Arizona	Current data
Adult	M	8.1	not given	Merriam's	Nebraska	Compiled in Lewis 1967
Adult	M	8.3	not given	Merriam's	Colorado	Compiled in Lewis 1967
Adult	M	8.0	not given	Rio Grande	Texas	Compiled in Lewis 1967
Adult	M	6.6	not given	Florida	Florida	Compiled in Lewis 1967

turkeys differs with population (Table 2). Certain populations of Gould's turkeys are heavier than other subspecies of turkeys in North America, but some Gould's turkeys from Mexico are comparable in weight to other subspecies. Perhaps Gould's turkeys in the Huachuca Mountains are heavier due to higher quality habitat; the Arizona Game and Fish Department has been baiting Gould's turkeys in the Huachuca Mountains with cracked corn to trap these birds for another project. It is possible that consistent baiting has caused birds to be artificially heavier, but that is unlikely because baiting occurred only during the last 10 months of the study and weights were taken over several years.

A second possibility is that larger birds were used as transplant stock. Only 21 total birds served as a founder population, and if these individuals were heavier, the founder effect could explain the larger size. Mock et al. (2001b) found that Gould's turkeys from the Huachuca Mountains of Arizona had less genetic diversity than Merriam's turkeys from northern Arizona and Gould's turkeys from Mexico. As a result, this population could be genetically different from other Gould's turkey populations.

Tarsometatarsus length has also been used to age Merriam's turkeys (Wakeling et al. 1997). It seems that Gould's turkeys from the Huachuca Mountains had longer tarsometatarsus lengths than Merriam's turkeys in northern Arizona. Data from weights and tarsometatarsus lengths suggest that Gould's turkeys from Arizona might be larger overall than other subspecies of turkeys. A larger body frame would be consistent with greater weight.

The beard lengths of adult male Merriam's turkeys were shorter than those from Gould's turkeys from the Huachuca Mountains. Adult male Merriam's turkeys are harvested regularly in northern Arizona, whereas the first hunting season on Gould's turkeys in the Huachuca Mountains was implemented in 2002. Beards grow throughout the life of a turkey, and they only begin to wear when they reach the ground (Pelham and Dickson 1992). Therefore, it is possible that Gould's turkeys from the Huachuca Mountains are on average longer-lived than Merriam's turkeys from northern Arizona and therefore would be expected to have longer beards. However, the survivorship for birds in the Huachuca Mountains (0.640; Wakeling et al. 2001) is similar to the survivorship of Merriam's turkeys in northern Arizona (0.681 in Wakeling 1991 and 0.605–1.000 in Wakeling and Rogers 1998) so it seems unlikely that hunting pressure alone accounts for the longer beard length. Instead, perhaps overall taller turkeys, as evidenced by tarsometatarsus lengths, have longer beards. Gould's turkeys seem to have overall greater height and weight, and taller turkeys could have longer beards due to distance to the ground (Pelham and Dickson 1992).

It has been reported that Gould's turkeys have 20 to 22 tail feathers (Williams 2002). This suggests little variability in the population, which is not the case in our study. We found that the number of tail feathers differed substantially, ranging from 12 to 20 for subadult and adult males and females captured in the Huachuca Mountains. Since the average number of tail feathers was only 17, many birds had fewer than 20 tail feathers. Perhaps this is due to molt patterns, but in our experience turkeys often had fewer than 20–22 tail feathers. Therefore, number of tail feathers seems to be an unreliable trait for identification of Gould's turkeys.

Since an active restoration program for wild turkeys was initiated in the early 1900s, approximately 146,000 turkeys have been released into 7300 sites in the United States (Tapley et al. 2001). Many of these reestablishment events have placed non-native subspecies into an area. For example, Rio Grande turkeys (*M. g. intermedia*), which historically occupied Texas, Oklahoma, Kansas, and Mexico (Mock et al. 2001a), have been transplanted into California, Washington, Oregon, and Utah (Tapley et al. 2001). Such cross-transplantation and subsequent breeding of native subspecies with non-native transplants could cause loss of genotypes unique to particular subspecies. Given that the Gould's turkey subspecies was dramatically different genetically than other turkey subspecies (Mock et al. 2001a), particular care should be taken to preserve genotypes of Gould's turkeys in native range. Morphologic and genetic evidence from Gould's turkeys suggests that native subspecies should be considered for use in reestablishment events.

ACKNOWLEDGMENTS

M. Zornes, J. deVos, and R. Ockenfels reviewed earlier versions of the manuscript. R. Ockenfels and J. deVos also provided administrative support. Funding was provided by Arizona State Trust Grant W-78-R and by the National Wild Turkey Federation.

LITERATURE CITED

Bailey, W., D. Dennett, H. Gore, J. Pack, R. Simpson, and G. Wright. 1980. Basic considerations and general recommendations for trapping the wild turkey. Proceedings of the National Wild Turkey Symposium 4: 10–23.

Brown, D. E. 1994. Biotic communities: Southwestern United States and Northwestern Mexico. University of Utah Press, Salt Lake City.

Haro Rodriguez, J. M. E. 1993. Distribucion y abundancia del guajolote silvestre (*Meleagris gallopavo*) en el estado de Sonora. Centro Ecologico de Sonora, Hermosillo.

Laing, L., N. Ambos, T. Subirge, C. McDonald, C. Nelson, and W. Robbie. 1989. Terrestrial ecosystem survey of the Apache-Sitgreaves National Forests. U.S. Forest Service Report, Washington, D.C. 453 pp.

Lewis, J. C. 1967. Physical characteristics and physiology. In The Wild Turkey and its Management, edited by O. H. Hewitt, pp. 45–72. The Wildlife Society, Washington D.C.

Mock, K. E., T. C. Theimer, D. L. Greenburg, and P. Keim. 2001a. Conservation of genetic diversity within and among subspecies of wild turkey. Proceedings of the National Wild Turkey Symposium 8: 35–42.

Mock, K. E., T. C. Theimer, B. F. Wakeling, O. E. Rhodes, Jr., D. L. Greenberg, and P. Keim. 2001b. Verifying the origins of a reintroduced population of Gould's wild turkey. Journal of Wildlife Management 65: 871–879.

National Oceanic and Atmospheric Administration. 1991. Arizona climatological data. Volume 95, National Climatic Center, Asheville NC.

Pelham, P. K., and J. G. Dickson. 1992. Physical characteristics. In The Wild Turkey, Biology and Management, edited by J. G. Dickson, pp. 32–45. Stackpole Books, Harrisburg PA.

Schemnitz, S. D., and W. D. Zeedyk. 1992. Gould's turkey. The Wild Turkey, Biology and Management, edited by J. G. Dickson, pp. 350–360. Stackpole Books, Harrisburg PA.

Schorger, A. W. 1966. The Wild Turkey: Its History and Domestication. University of Oklahoma Press, Norman.

Tapley, J. L., R. K. Abernethy and J. E. Kennamer. 2001. Status and distribution of the wild turkey in 1999. Proceedings of the National Wild Turkey Symposium 8: 15–22.

Wakeling, B. F. 1991. Population nesting characteristics of Merriam's turkey along the Mogollon Rim, Arizona. Arizona Game and Fish Department Technical Report 7, Phoenix. 48 pp.

Wakeling, B. F., and T. D. Rogers. 1995. Winter habitat relationships of Merriam's turkeys along the Mogollon Rim, Arizona. Arizona Game and Fish Department Technical Report 16, Phoenix. 41 pp.

Wakeling, B. F., and T. D. Rogers. 1998. Summer resource selection and yearlong survival of male Merriam's turkeys in north-central Arizona, with associated implications from demographic modeling. Arizona Game and Fish Department Technical Report 28, Phoenix. 50 pp.

Wakeling, B. F., F. E. Phillips, and R. Engel-Wilson. 1997. Age and gender differences in Merriam's turkey tarsometatarsus measurements. Wildlife Society Bulletin 25: 706–708.

Wakeling, B. F., S. R. Boe, M. M. Koloszar, and T. D. Rogers. 2001. Gould's turkey survival and habitat selection modeling in southeastern Arizona. Proceedings of the National Wild Turkey Symposium 8: 101–108.

Williams, L. E. 2002. El cocoon; The real treasure of the Sierra Madre. Turkey and Turkey Hunting February: 36–39.

York, D. L. 1991. Habitat use, diet, movements, and home range of Gould's turkey in the Peloncillo Mountains, New Mexico. Master's thesis, New Mexico State University, Las Cruces. 104 pp.

York, D. L., and S. D. Schemnitz. 2003. Home range, habitat use, and diet of Gould's turkeys, Peloncillo Mountains, New Mexico. The Southwestern Naturalist 48: 231–240.

Zar 1999. Biostatistical Analysis, 4th ed. Prentice-Hall, Upper Saddle River NJ.

Zornes, M. L. 1993. Ecology and habitat evaluation of Gould's wild turkeys in the Peloncillo Mountains, New Mexico. Master's thesis, New Mexico State University, Las Cruces. 117 pp.

VERTEBRATES OF MONTEZUMA CASTLE NATIONAL MONUMENT: PRESENT STATUS AND HISTORICAL CHANGES

Charles A. Drost

An integrated, broad-based inventory of the flora and fauna of Montezuma Castle National Monument in central Arizona was undertaken between 1991 and 1994 to provide information on current status, trends, and potential management concerns of the natural resources of the monument. Components of the inventory included terrestrial vegetation (Rowlands 1999), aquatic invertebrates (Blinn et al. 1996), terrestrial invertebrates (Price and Fondriest 1998), fish (Montgomery et al. 1995), amphibians and reptiles (Drost and Nowak 1998), mammals (Drost and Ellison 1999), and birds (Sogge and Johnson 1998), as well as a series of historic photos documenting changes in the area (Richmond 1995). The work was a collaborative effort by researchers from Northern Arizona University and the Colorado Plateau Research Station in Flagstaff, Arizona, and was supported by funding from the National Park Service. Final project reports that provide detailed results are available from the USGS Colorado Plateau Research Station.

Montezuma Castle National Monument is a unit of the National Park Service in central Arizona, established to protect Sinagua culture cliff dwellings—notably the five-story, 20-room structure known as Montezuma Castle, perched imposingly on a limestone cliff face overlooking lower Beaver Creek. The monument consists of two separate sites—Montezuma Castle (the "Castle unit") and Montezuma Well—8 km apart in the Verde River valley between Flagstaff and Phoenix. The Montezuma Well section protects additional Sinagua and Hohokam sites, as well as a large, spring-fed limestone sink (from which the site receives its name) that has no known parallel anywhere in the world in its depth, its highly carbonated waters, and its uniquely adapted endemic invertebrate community.

Montezuma Castle is a relatively small national monument: the Castle section encompassees about 2.5 sq km, and the Well section is just over 1 sq km. The area is mid-elevation desert, ranging from 955 to 1090 m; it supports scattered juniper at its higher elevations and mesquite, acacia, and creosotebush at its lower elevations. There is a diverse cottonwood riparian association along Beaver Creek in its course through the monument.

As with many "cultural parks," Montezuma Castle National Monument has had relatively little research on plant and animal communities and other natural resources. However, the monument protects valuable samples of regional biological communities. In many respects, the same features that drew early humans to the area also provide for rich biological communities, like the extensive riparian habitat at Montezuma Castle, and these biological communities gain added importance by virtue of their long-term protection as monuments and parks. The poor state of knowledge of the biological resources at the monument was one of the driving forces behind this inventory project.

A second motivation for the study was the rapid growth in the Verde Valley area surrounding Montezuma Castle. Like most areas with ample water in the arid Southwest, the Verde Valley region is experiencing rapid urban, suburban, and recreational development. National Park Service management at Montezuma Castle anticipates steadily increasing pressure from such development on the monument's natural resources, both from habitat loss and fragmentation and also from a decline in riparian and aquatic communities due to water withdrawals and general water table drawdown (National Park Service 1989).

The four objectives of the inventory study were to conduct a general floral and faunal inventory; evaluate past changes in the plant and animal community as well as current threats; establish a baseline and methods for monitoring the natural resources of the monument; and provide information on natural resource management concerns and needs. This chapter gives an overview of the inventory results for vertebrate species, provides updates and additions from more recent work, and highlights known changes in the vertebrate communities and natural environment of the Montezuma Castle area and the surrounding Verde Valley.

STUDY DESIGN AND METHODS

For each component of the inventory study, project investigators conducted extensive literature and museum surveys to assemble existing information on the fauna at Montezuma Castle. This included computerized literature searches of library databases, searches of major museum collections, and interviews with biologists and other individuals knowledgeable about the area.

Fish

Field sampling of the fish fauna was conducted at three sites within the monument: Wet Beaver Creek just downstream of the outlet of Montezuma Well, Davis Hole within the Castle unit and the portions of Beaver Creek immediately downstream of Davis Hole, and Beaver Creek below (downstream of) the Castle unit visitor area and just upstream of the southern boundary fence at the Castle unit. The sites were sampled during fall, winter, and spring, using a variety of methods to capture the full range of species and size/age classes present in the creek. The methods included electrofishing, stationary trammel nets (in the deep water of Davis Hole), and seining.

Amphibians and Reptiles

Amphibians and reptiles have a wide range of activity patterns, from active, diurnal, fast-moving species, to nocturnal, slow-moving species that spend most of their lives under cover and are rarely seen. For this reason, a similarly wide range of sampling techniques was used for this group. Sampling was conducted throughout the year during the first year of the survey, but this was reduced to spring, summer, and fall during the following years because there were almost no amphibians and reptiles active during the winter months.

Methods used included (1) visual searches of aquatic habitats, primarily for breeding amphibians; (2) pitfall traps and artificial cover boards for a variety of ground-dwelling species (primarily lizards and toads); (3) funnel traps, which also trap ground-dwelling animals, particularly those living in fairly thick cover; (4) time-constrained visual encounter searches, which target animals active out in the open or those that can be found hiding under logs, boulders, and other cover; and (5) night searches, which sample snakes, lizards, and frogs abroad at night. Time-constrained searches consisted of walking surveys through pre-defined habitat areas for one hour, while visually searching the ground surface and turning and searching through vegetation, logs, and other cover that amphibians and reptiles might use as retreats. Night searches included driving along roads in and near the Castle and Well after dark, looking for animals on the road in the headlights, and also searching on foot with spotlights along the trails at both the Castle and Well.

Mammals

Various techniques were used to survey mammals. Transects or grids of small box traps (H. B. Sherman Traps) and pitfall traps were used for small mammals (primarily rodents); large, baited live traps (Havahart) and automatic camera systems were used for medium-sized, mostly nocturnal mammals; direct visual counts were taken of medium-sized and large mammals in the early morning and evening; and mist nets and searches of potential roost areas were used for bats. Particular attention was given to small mammals (rodents, shrews, and bats) because they were relatively poorly known. Incidental observations (both direct visual observations and presence of sign such as tracks and scat) were also recorded.

Given the small size of Montezuma Castle National Monument, most field sampling methods were intended to be extensive over the entire monument, stratified by major habitat type. This was true both for visual surveys conducted on foot and for trap-based methods. For example, pitfall and small mammal traps were set out in long transects across the entire length of major habitat strata, such as the mesquite grassland terrace south of Beaver Creek in the southern part of the Castle unit.

Birds

Birds were surveyed during the breeding season using timed, variable circular plot stations established in six broad habitat types within Montezuma Castle National Monument. Point counts were conducted once or twice per month through the breeding season (April through July) from 1991 to 1994. Surveys during the non-breeding season for migrants and winter birds consisted of general, walking surveys during the period from August through March.

RESULTS AND DISCUSSION

The 2–4 years of sampling at Montezuma Castle provided a relatively thorough inventory of all of the groups discussed herein (Table 1). Intensive sampling spanned 2 years, but additional data were collected for most groups, particularly as part of preliminary monitoring studies. The thoroughness of inventory is evidenced by the number of new species added to the known flora and fauna and by the sharply reduced number of "new" species recorded in successive years of the study.

Most areas of the western United States have relatively sparse detailed historical inventory information, making it difficult to evaluate long-term trends in plant and animal communities. However, the U.S. Biological Survey conducted a study of terrestrial vertebrates in and around Montezuma Castle National Monument in 1916 (Jackson and Taylor 1916), and a series of fish collections were made in Beaver Creek in the vicinity of the monument in the mid 1930s (see Montgomery et al. 1995). Comparison with these surveys reveals striking changes in the intervening years. For most vertebrate groups, losses of native species (population declines or extirpations) have surpassed population increases or new species arrivals. An exception is seen among the mammals, where ungulate populations are distinctly more diverse and numerous than they were in past times. The following sections note the main results of the study for each vertebrate class.

Known or probable causes for species loss from the Montezuma Castle area can be divided into three general categories: habitat alteration and loss combined with direct persecution (large carnivores, ungulates); alteration of aquatic habitat including introduction of non-native species (native fish, leopard frog); and loss or conversion of desert grassland habitat (probably hispid pocket mouse, Arizona cotton rat, and some grassland bird species). The greatest current natural resource threats at Montezuma Castle are changes in the local water regime and concomitant effects on riparian habitats, as well as potential isolation and fragmentation of natural habitats because of increased urban and suburban growth in the area. Fish and amphibians are of particular concern because of their direct dependence on aquatic habitats. Habitat fragmentation is most likely to affect birds and mammals, particularly large, wide-ranging species.

Fish

Because of the small size of the monument, the fish surveys provided only a narrow

Table 1. Vertebrate species at Montezuma Castle National Monument. Total numbers of species at the Montezuma Castle and Montezuma Well units combined, and new species documented by recent inventory. Total species does not include species no longer thought to be present at the monument and immediately surrounding area (cf. Appendix 1). The 32–33 listed for reptiles reflects uncertainty about the present occurrence of the Gila monster.

Vertebrates	Total Species	New Species
Fish	9	None
Amphibians	4	Red-spotted toad Bullfrog (non-native)
Reptiles	32–33	Spiny softshell turtle Glossy snake Lyre snake Mojave rattlesnake
Mammals	50	Desert shrew Fringed myotis Mexican woodrat
Birds	213	See Sogge et al. (in press) for details

window on the entire extent of Beaver Creek (the length of stream within the boundaries of the Castle unit is about 3.6 km, and in the Well unit about 1.7 km). Nonetheless, the data show dramatic changes in the fish community along this stretch of the creek (Table 2). As in much of the southwestern United States, over the last 70 years there has been a major shift in the community, from an assemblage of Southwest endemic species to a community dominated by introduced, non-native species. Sampling during this study documented the continued presence of four native fish species in low to moderate numbers, but the loss of at least two other natives (Montgomery et al. 1995). Spikedace (*Meda fulgida*) and speckled dace (*Rhinichthys osculus*) have not been recorded from the immediate vicinity of Montezuma Castle and Montezuma Well since the 1930s, and spikedace have been lost from much of the Verde River drainage (see Minckley 1973). The loach minnow (*Tiaroga cobitis*) formerly occurred very near the monument (Minckley 1973), though there are no definite records from within either the Castle or Well units.

Of three introduced fish species that were already present in the 1930s, two are still present and one has apparently been lost. In the meantime, three other non-natives have become solidly established. Records from the Arizona Game and Fish Department (Arizona Game and Fish 1993, in Montgomery et al. 1995) reflect that attempts have been made to stock other species in this section of Beaver Creek, including rainbow trout (*Oncorhynchus mykiss*), brook trout (*Salvelinus fontinalis*), and largemouth bass (*Micropterus salmoides*), but these species were not recorded in the early surveys or in recent surveys.

Amphibians and Reptiles

Amphibian diversity at Montezuma Castle is low, consisting of two toad species, one treefrog, and the introduced bullfrog. The native red-spotted toad (*Bufo punctatus*) had not been documented at the monument in earlier lists, but it is relatively rare and may simply have been overlooked. The bullfrog (*Rana catesbeiana*) is apparently a recent arrival (it had not been recorded at the monument prior to this study) and at present it is only found in low numbers. Although comprehensive historical survey data are not available for comparison, we do know from museum records that lowland leopard

Table 2. Changes in fish fauna of Beaver Creek in the vicinity of Montezuma Castle National Monument. The 1937 column aggregates the results of collections made along the creek in the 1930s. The 1996 column summarizes sampling results from this study. X indicates not found in this survey and believed to be extirpated from the Beaver Creek drainage; a dash indicates absent or not found (after Montgomery et al. 1995).

Fish Fauna	1937	1996
Native		
Desert sucker	Present	Common
Sonora sucker	Present	Uncommon
Roundtail chub	Present	Rare
Longfin dace	Present	Rare
Speckled dace	Present	X
Spikedace	Present	X
Non-Native		
Carp	Present	Common
Green sunfish	Present	Common
Black bullhead	Present	–
Yellow bullhead	–	Uncommon
Red shiner	–	Common
Smallmouth bass	–	Common

frogs (*R. yavapaiensis*) were historically recorded from very near the monument. This includes records as recent as 1969 (Arizona State University Museum No. 27640), but no leopard frogs were found within the monument or in nearby aquatic habitats. Leopard frogs have suffered serious declines throughout much of the Southwest (e.g. Sredl 1998).

The lizard fauna at Montezuma Castle is quite diverse (15 species), equaling or exceeding the number of species at much larger parks such as Grand Canyon (Miller et al. 1982) and Organ Pipe Cactus (Rosen and Lowe 1996). The snakes are equally diverse, with 16 species present. Our surveys documented four reptile species that had not previously been reported at Montezuma Castle National Monument (Table 1). Of these new additions, three (spiny softshell, lyre snake, and Mojave rattlesnake) are near the edge of their range, and all four are relatively rare or secretive. They could simply have been overlooked in previous studies. As with amphibians, there are no detailed comprehensive historical data available for comparison with our recent inventory results. Gila monsters have been documented at Montezuma Castle in the past, but they were not found in these surveys. This species can be difficult to detect, however, so it cannot be concluded that the species is now absent from the monument and vicinity.

Mammals

Our inventory sampling recorded three "new" mammal species (see Table 1), all of which may have simply been missed by earlier studies. For example, Hoffmeister (1986) noted that desert shrews were widespread and apparently common throughout much of Arizona, but were rarely seen without trapping methods targeted toward them. At the time of his writing, he could find no records at all for Yavapai County (where Montezuma Castle is located), in the middle of the species range. Montezuma Castle is well within the ranges of fringed myotis and Mexican woodrat, as well. Only one fringed myotis was captured during the mammal surveys, and an uncommon bat species could easily be missed without exhaustive surveys. The absence of Mexican woodrat from previous lists, in spite of relatively intensive small mammal trapping, is a little more surprising. This species may have increased in the local area, or it too may have been previously overlooked.

Like the fish community of Beaver Creek, the mammal fauna of the Montezuma Castle area has undergone major changes over the last 100 years, with the loss of a number of species and the colonization or return of other species. The primary reference point for evaluating changes in the mammal fauna of the Montezuma Castle area is a mammal inventory study conducted by the U.S. Biological Survey in July and August of 1916 (Jackson and Taylor 1916). Jackson and Taylor's work centered on the Camp Verde area, with two of their three sampling areas at or immediately adjacent to the Castle unit and the Well unit of Montezuma Castle National Monument. In addition to their own trapping, collecting, and general observations, Jackson and Taylor also interviewed local ranchers and other residents of the area about the contemporary mammal fauna, and these interviews provided significant additional insights into changes in the fauna up to that point in time.

Among small mammals, the Arizona cotton rat (*Sigmodon arizonae*) and hispid pocket mouse (*Chaetodipus hispidus*) have evidently disappeared from Montezuma Castle and the surrounding area since the time of Jackson and Taylor's surveys. The cotton rat was represented in this area by an unusually large form (referred to as *S. arizonae arizonae* by Hoffmeister 1986), which was apparently restricted to the Verde Valley area around Camp Verde. This unusual form may now be extirpated. In a similar manner, the hispid pocket mouse in the Camp Verde area was apparently a restricted population in this part of the Verde Valley, widely disjunct from the main range of the species in southeastern Arizona (cf. Hoffmeister 1986). This isolated population may also have been lost, though additional surveys in the area are needed.

Major losses in the carnivore fauna in the Verde Valley around Montezuma Castle appeared to be largely complete by the early 1900s, but there appears to have been a resurgence of some species since Jackson and Taylor's surveys (Table 3). Whether wolves occurred in the Verde Valley during historical time is not known; there are no specimens from the area (see Hoffmeister 1986). The nearest records are from the Fort Whipple area to the east (Coues, attributed by Davis 1982), and the Mogollon Rim area to the east (Hoffmeister 1986). One of the last jaguar (*Panthera onca*) records for Arizona north of the Mexican borderland area was from Bloody Basin, in the southern Verde Valley in 1939. The last record of grizzly (*Ursus arctos*) in the vicinity was from Brown Spring, 20 km south of Camp Verde, in 1904. Jackson and Taylor do not even list black bear (*Ursus americanus*) for the area, and their only note on mountain lion ("*Felis aztecus browni*" = *Puma concolor*) is from an interview with a rancher at Montezuma Well, who noted that he "killed a mountain lion across the creek (Wet Beaver Creek) some years ago." All of these species were targeted by a concerted predator control program during the late 1800s and early 1900s (e.g. Brown 1985; Brown and Davis 1995), and this probably accounts for the absence or near absence of all of these species by the time of Jackson and Taylor's 1916 surveys. It seems unlikely that any record or even suspicion of any of these species in the area would fail to be mentioned in the 1916 report, given the general public's concern with them, and their considerable scientific interest.

Table 3. Changes in the mammal fauna of the Montezuma Castle Area. Each column indicates species that were present during that time period. X indicates not found in this survey, and now extirpated from the Verde Valley; a dash indicates absent or not found.

Fauna	1910	1996
Carnivores		
Coyote	Present	Uncommon
Gray fox	Present	Uncommon
Black bear	–	Rare
River otter	X ?	Uncommon
Ocelot	–	X
Mountain lion	–	Rare
Bobcat	Present	Uncommon
Ungulates		
Collared peccary	–	Uncommon
Elk	–	Common
Mule deer	–	Common
White-tailed deer	–	Uncommon
Pronghorn	Rare	–

Medium-sized carnivores fared somewhat better. Jackson and Taylor had reports of bobcats ("*Lynx baileyi*" = *L. rufus*) at all of their survey areas. They characterized coyote ("*Canis mearnsi*" = *C. latrans*) as "apparently not abundant or even common about the Verde Valley." Though not reported by Jackson and Taylor, two of the only records for ocelot (*Leopardus pardalis*) in Arizona are from the Verde Valley area, from 1887 and 1932. There is some doubt whether the 1887 specimen was from the area or not (Hoffmeister 1986), but the 1932 record is of a specimen taken "near Camp Verde" by a predator control trapper (D. Brown, cited in Hoffmeister 1986). Hence, the ocelot apparently managed to persist largely unnoticed in the dense riparian community along the Verde River until relatively late.

Finally, Jackson and Taylor reported that they "saw no signs of otter anywhere in the Verde Valley" during their surveys, though a long-time resident of the area said that scattered numbers still occurred elsewhere on the Verde River. The local southwestern subspecies (*Lontra canadensis sonorae*) was evidently much reduced by this time. In the early 1980s river otters from Louisiana were introduced in the Verde Valley and apparently became successfully established (Raesley 2001; Savage and Klingel 2004). The Montezuma Castle mammal survey recorded only one sighting from Beaver Creek at Montezuma Castle, but the species is now commonly seen on parts of the Verde River.

In contrast to the overall trend among carnivores, ungulate populations at Montezuma Castle and the surrounding area are much more diverse now than they were when Jackson and Taylor (1916) made their report. This is due to the return of some species that had been largely eliminated, as well as to invasion by other species (both native and non-native). Jackson and Taylor (1916) reported that both mule deer (*Odocoileus hemionus*) and pronghorn (*Antilocapra americana*) were "formerly common" in the Montezuma Well and surrounding area. By 1916, however, mule deer were "now practically unknown west of the [Mogollon] rim, several miles to the east, and pronghorn were "now unknown in that locality [Montezuma Well]."

White-tailed deer (*Odocoileus virginianus*) were not mentioned as being present by Jackson and Taylor. Other sources suggest that white-tailed deer were rare or absent from northern Arizona, until relatively recently. Kellogg (1956) described the northern edge of the range of *O. virginianus couesi* in Arizona as "mountain regions ... south of the Mogollon Mesa in Arizona." Hoffmeister (1986) also noted that "whitetails have been rare ... until recently" in this northern part of their range. He cited several specimens from the highland areas of southern Coconino County, but all of these were from the 1960s and later, except one (Bill Williams Mountain area, 1884). This species is now fairly common in the riparian habitat of Montezuma Castle and Montezuma Well.

The native Arizona elk (*Cervus elaphus merriami*) did not range near the Verde Valley in historic time (Hoffmeister 1986), and was evidently driven to extinction in the early 1900s (Bailey 1931; Hoffmeister 1986). Elk from the Rocky Mountains were introduced in the Mogollon Rim area in 1913 (Hoffmeister 1986) and have spread widely throughout northern Arizona. They are now common winter visitors to Montezuma Castle.

The collared peccary (javelina, *Pecari tajacu*) is a southern species that has substantially expanded its distribution northward (Brown and Davis 1995). This species, also, was not mentioned by Jackson and Taylor, and even by 1956 (Knipe 1957) its known range limits were still to the south of Camp Verde and Montezuma Castle. Knipe (1957) did present tables showing a general increase in numbers in parts of the species' range, which has apparently translated into a significantly expanded distribution. The mammal survey recorded peccaries fairly commonly in the riparian habitats of Montezuma Castle and Well, and the species' range now extends north to Flagstaff and to the south rim of the Grand Canyon (R. V. Ward, Grand Canyon National Park, personal communication).

Birds

Bird populations at Montezuma Castle and the surrounding Verde Valley have been particularly dynamic, with loss of some species, arrival of other species not previously known from the area, and conspicuous changes in abundance in many others. Historic surveys were not as intensive as recent inventory studies and other bird research (e.g., Johnson and Sogge 1995; Johnson and van Riper 2004; Sogge et al., in press) Jackson and Taylor (1916) surveyed much more of the Verde Valley area (including Montezuma Castle and Montezuma Well) but were not as thorough; their surveys covered

Table 4. Changes in the bird fauna of Montezuma Castle National Monument and the surrounding Verde Valley over a 90-year span. Listing is based on comparison of recent inventory results with historic surveys and species list compilations (see text for details). Increase includes species that were previously absent but have recently colonized the area. X denotes regional extirpation of a species.

Increase	Decline
Wood Duck	Snowy Egret
Blue-winged Teal	Green Heron
Bald Eagle	Black-crowned Night Heron
Common Black Hawk	Wood Stork (X)
Rock Pigeon	Swainson's Hawk
Brown-crested Flycatcher	Peregrine Falcon
Common Raven	Ring-necked Pheasant
Juniper Titmouse	White-winged Dove
Verdin	Thick-billed Parrot (X)
Blue-gray Gnatcatcher	Barn Owl
European Starling	Vermilion Flycatcher
Indigo Bunting	Purple Martin
Great-tailed Grackle	Bank Swallow
Brown-headed Cowbird	Loggerhead Shrike
	Bell's Vireo
	Grasshopper Sparrow
	Western Meadowlark
	House Sparrow

a shorter time span and they divided their time between birds and mammals. Jackson (1941) and Sutton (1954) accumulated records over several years, including intensive observations (monthly surveys by Jackson), for the smaller area of Montezuma Castle National Monument and the immediate surrounding area. Taken together, however, these sources represent a substantial baseline on the bird fauna of the area over the last 100 years.

Assessing changes can be difficult without standardized survey methods, particularly for rare or cryptic species (e.g. owls). For this reason, I concentrate here on the records where the evidence for change is strongest (Table 4): species that were formerly present but are now unequivocally absent from Montezuma Castle and the Verde Valley, species that were not known formerly (and are not likely to have been overlooked) but occur regularly now, and species that have shown conspicuous, well-documented changes in abundance. There are several other cases of apparent changes in distribution and/or abundance, but documentation is limited or there are other problems in interpretation. One such species is the Blue-throated Hummingbird. Jackson (1941) reported regular occurrence and nesting of this conspicuous and distinctive species at Montezuma Castle in the 1940s. However, the occurrence is far north of the species' current distributional limit in southeastern Arizona; without additional documentation, the record must remain hypothetical. Such "ghosts" of changes in the bird community of the Verde Valley may never be revealed for certain.

Species that have declined include Snowy Egret, Green Heron, and Black-crowned Night Heron, which Jackson and Taylor (1916) recorded regularly; there are no recent records for the Egret or the Night Heron, and Green Heron is now a rare fall visitor. White-winged Dove was recorded as "abundant" by Jackson and Taylor, but there are no recent records of this species in the vicinity of Montezuma Castle. Purple Martin and Bank Swallow were both recorded from Beaver Creek and Montezuma Well by Jackson and Taylor, with Bank Swallow described as "abundant." There is one record of Purple Martin in the area in the last 25 years, but no records for Bank Swallow. Bell's Vireo was described by Jackson and Taylor as "abundant" in mesquite at all of their camps. The species was not listed at all by Jackson (1941), but is now listed as uncommon during the summer. Jackson and Taylor found the Grasshopper Sparrow to be "quite common" in alfalfa fields in the vicinity of Camp Verde. The species is now absent from the Verde Valley and is found only in southeastern Arizona (Monson and Phillips 1981). Some changes, including the Western Meadowlark and the House Sparrow, may be strictly local. The Meadowlark was formerly a common breeding resident in the pasture and fields at Montezuma Well (Sutton 1954). These areas have grown up and become more brushy, and the Meadowlark now occurs only as an uncommon winter visitor, though it remains common in nearby agricultural and suburban areas.

No studies have specifically addressed causes of bird population changes in the Verde Valley area, but some broad patterns are evident. Many of the increasing species have exhibited regional increases in the Southwest (Common Black Hawk, Brown-crested Flycatcher, Great-tailed Grackle, and Brown-headed Cowbird; DeSante and George 1994; Johnson 1994). Likewise, many increasing species have undoubtedly benefited from urbanization and agricultural practices (Rock Pigeon, Common Raven, European Starling, Great-tailed Grackle, and Brown-headed Cowbird). Others have benefited from specific protection and habitat recovery (e.g., Wood Duck, Bald Eagle).

Many of the species that have declined or have been extirpated also show broad regional declining trends. Wood Stork and Thick-billed Parrot have been extirpated from the Southwest, while other species like the Bank Swallow, Loggerhead Shrike, Bell's Vireo, and Grasshopper Sparrow have generally declined in numbers in the region (DeSante and George 1994). Loss of the Peregrine Falcon from much of its range was attributed to reproductive failure associated with DDT (Hickey 1969). Other declines are probably associated with degradation or loss of riparian habitat (e.g. Bell's Vireo, Vermilion Flycatcher; see Ohmart 1994) or grassland habitat (Grasshopper Sparrow). The causes of other bird species declines are unclear.

ACKNOWLEDGMENTS

This study would not have been successful without the substantial help and involvement of National Park Service staff at Montezuma Castle National Monument and the Southern Arizona Group Office. Special thanks go to former Superintendent Glen Henderson, Chief Ranger Steve Sandell, and current Superintendent Kathy Davis. The study designs, field data collection, analyses, and reports upon which this manuscript is based were carried out by many knowledgeable and dedicated individuals at Northern Arizona University and the USGS Southwest Biological Science Center. These individuals include Dr. Linn Montgomery, Bill Leibfried, and Gloria Hardwick (fish); Erika M. Nowak (amphibians and reptiles); Laura Ellison (mammals); and Mark Sogge, Charles van Riper III, and Matt Johnson (birds). This work was supported in part by the USGS Southwest Biological Science Center.

LITERATURE CITED

Arizona Game and Fish Department. 1993. AGFD Fisheries branch stocking records. Phoenix.

American Ornithologists Union. 2005. The AOU Check-list of North American Birds. 7th ed. (incorporating 42–45th supplements). http://www.aou.org/checklist/index.php3.

Bailey, V. 1931. Mammals of New Mexico. North American Fauna 53. U.S. Department of Agriculture, Bureau of Biological Survey.

Blinn, D., and G. E. Oberlin. 1996. Aquatic biota: Invertebrates and algae. Montezuma Castle National Monument Integrated Inventory and Monitoring Studies. USDI National Park Service, Colorado Plateau Research Station, Flagstaff AZ.

Brown, D. E. 1985. The grizzly in the Southwest. University of Oklahoma Press, Norman and London.

Brown, D. E., and R. Davis. 1995. One hundred years of vicissitude: Terrestrial bird and mammal distribution changes in the American Southwest, 1890–1990. In Biodiversity and Management of the Madrean Archipelago: The Sky Islands of Southwestern United States and Northwestern Mexico, technical coordinators L. F. DeBano, P. F. Ffolliott, A. Ortega-Rubio, G. J. Gottfried, R. H. Hamre, and C. B. Edminster, pp. 231–244. Gen. Tech. Rep. RM-GTR-264, USDA Forest Service, Rocky Mountain Forest and Range Experiment Station, 669 pp.

Davis, G. P. 1982. Man and Wildlife in Arizona: The American Exploration Period, 1824–1865, edited by N. B. Carmony and D. E. Brown. Arizona Game and Fish Department in cooperation with the Arizona Cooperative Wildlife Research Unit. Phoenix.

DeSante, D. F., and T. L. George. 1994. Population trends in the landbirds of western North America. Studies in Avian Biology 15: 173–190.

Drost, C. A., and L. E. Ellison. 1998. Mammals. Montezuma Castle National Monument Integrated Inventory and Monitoring Studies. USDI National Park Service, Colorado Plateau Research Station, Flagstaff AZ.

Drost, C. A., and E. Nowak. 1998. Amphibians and Reptiles. Montezuma Castle National Monument Integrated Inventory and Monitoring Studies. USDI National Park Service, Colorado Plateau Research Station, Flagstaff AZ.

Hickey, J. J., editor. 1969. Peregrine falcon populations: Their biology and decline. University of Wisconsin Press, Madison.

Hoffmeister, D. F. 1986. Mammals of Arizona. University of Arizona Press, Tucson.

Jackson, B. 1941. Birds of Montezuma Castle. Southwestern Monuments Association; Southwestern Monuments Special Report (No. 28). (Request from Montezuma Castle National Monument, P.O. Box 219, Camp Verde, AZ 86322.)

Johnson, M. J., and M. K. Sogge. 1995. A Checklist of Birds of Tuzigoot National Monument and Vicinity. Southwest Parks and Monuments Association.

Johnson, M. J., and C. van Riper III. 2004. Brown-headed Cowbird parasitism of the Black-throated sparrow in central Arizona. Journal of Field Ornithology 75: 303–311.

Johnson, N. K. 1994. Pioneering and natural expansion of breeding distributions in western North American birds. Studies in Avian Biology 15: 27–44.

Knipe, T. 1957. The Javelina in Arizona: A Research and Management Study. Arizona Game and Fish Department, Phoenix.

Kellogg, R. 1956. What and where are the whitetails? In The Deer of North America, edited by W. P. Taylor, pp. 31–55. The Wildlife Management Institute, Washington D.C.

Miller, D., R. Young, T. Gatlin, and J. Richardson. 1982. Amphibians and Reptiles of the Grand Canyon. Grand Canyon Natural History Association.

Minckley, W. L . 1973. Fishes of Arizona. Arizona Game and Fish Department, Phoenix.

Monson, G., and A. R. Phillips. 1981. Annotated Checklist of the Birds of Arizona. 2nd ed. University of Arizona Press, Tucson.

Montgomery, W. L., W. C. Leibfried, and G. C. Hardwick. 1995. Beaver Creek Aquatic Study: Fish and Herpetofauna. Montezuma Castle National Monument Integrated Inventory and Monitoring Studies. USDI NPS, Colorado Plateau Research Station, Flagstaff AZ.

National Park Service. 1989. Montezuma Castle National Monument Resources Management Plan.

Ohmart, R. D. 1994. The effects of human-induced changes on the avifauna of western riparian habitats. Studies in Avian Biology 15: 273–285.

Price, P. W., and S. M. Fondriest. 1998. Terrestrial Invertebrates. Montezuma Castle National Monument Integrated Inventory and Monitoring Studies. USDI National Park Service, Colorado Plateau Research Station, Flagstaff AZ.

Raesley, E. J. 2001. Progress and status of river otter reintroduction projects in the United States. Wildlife Society Bulletin 29 (3): 856–862.

Richmond, A. J. 1995. Historic Photograph Survey: Montezuma Castle/Well National Monument, 1876–1990. Technical Report NPS/NAU MOCA/NRTR-95/08. USDI National Park Service, Colorado Plateau Research Station, Flagstaff AZ.

Rosen, P. C., and C. H. Lowe. 1996. Ecology of the Amphibians and Reptiles at Organ Pipe Cactus National Monument, Arizona. Technical Report 53. National Biological Service, Cooperative Park Studies Unit. USGS Sonoran Desert Field Station, University of Arizona, Tucson.

Rowlands, P. 1999. Vegetation survey of Montezuma Castle National Monument. Montezuma Castle National Monument Integrated Inventory and Monitoring Studies. USDI National Park Service, Colorado Plateau Research Station, Flagstaff AZ.

Savage, M., and J. Klingel. 2004. The case for river otter restoration in New Mexico. A Report to the River Otter Working Group. http://www.amigosbravos.org/projects/riverotter /FinalOtterReport.htm.

Sogge, M. K., and M. J. Johnson. 1998. Bird community ecology at Montezuma Castle National Park: Avian inventory 1991–1994. Unpublished report. USGS Colorado Plateau Field Station (P.O. Box 5614, Flagstaff, AZ 86011).

Sogge, M. K., M. J. Johnson, and C. van Riper III. in press. The birds of Montezuma Castle National Monument, AZ. Western Birds.

Sutton, M. 1954. Bird Survey of the Verde Valley. Plateau, Museum of Northern Arizona 27 (2): 1–9.

Sredl, M. J. 1998. Arizona leopard frogs: Balanced on the brink? In Status and Trends of the Nation's Biological Resources, vol. 2, edited by M. J. Mac, P. A. Opler, C. E. Puckett Haecker, and P. D. Doran, pp. 573–574. U.S. Geological Survey, Reston VA.

H. H. T. Jackson, and W. P. Taylor. 1916. Biological Survey Reports, Verde Valley, 1916. U.S. Department of the Interior, Washington D.C.

Appendix A. Vertebrate fauna of Montezuma Castle National Monument, based on inventory studies conducted in 1992–1996, with recent updates. Scientific names of fish, amphibians, reptiles, and mammals follow the federal interagency Integrated Taxonomic Information System (ITIS, http://www.itis.usda.gov). Scientific and common names of birds follow AOU (2005). Synonyms or name changes are listed in brackets. R = rare, only known from a few records; U = uncommon, occurring in relatively low numbers, or scattered distribution; C = common, present in relatively high numbers, generally widespread; A = abundant, a common species that is quite numerous; X = extirpated, a species that occurred in the area at one time but is now absent; I = introduced, species not originally native to the area, introduced by humans. MOCA = Montezuma Castle unit and MOWE = Montezuma Well unit.

FISH

Suckers (Family Catostomidae)	
Sonora Sucker (*Catostomus insignis*)	U (C at MOWE)
Desert Sucker (*Catostomus clarkii*) [also called *Pantosteus clarki*]	C
Sunfishes (Family Centrarchidae)	
Green Sunfish (*Lepomis cyanellus*)	C (I)
Smallmouth Bass (*Micropterus dolomieu*)	C/A (I)
Spotted Bass (*Micropterus punctulatus*)	X (I)
Largemouth Bass (*Micropterus salmoides*)	(I)
Minnows (Family Cyprinidae)	
Longfin Dace (*Agosia chrysogaster*)	R
Red Shiner (*Cyprinella lutrensis*)	C/A (I)
Common Carp (*Cyprinus carpio*)	C (I)
Roundtail Chub (*Gila robusta*)	R
Spikedace (*Meda fulgida*)	X
Speckled Dace (*Rhinichthys osculus*)	X
Catfishes (Family Ictaluridae)	
Black Bullhead (*Ameiurus melas*)	X (I)
Yellow Bullhead (*Ameiurus natalis*)	U (I)
Channel Catfish (*Ictalurus punctatus*)	X (I)

AMPHIBIANS

True Toads (Family Bufonidae)	
Red-spotted Toad (*Bufo punctatus*)	R
Woodhouse's Toad (*Bufo woodhousii*)	C
Treefrogs (Family Hylidae)	
Canyon Treefrog (*Hyla arenicolor*)	U
True Frogs (Family Ranidae)	
Bullfrog (*Rana catesbeiana*)	R (I)

REPTILES

Pond Turtles, Terrapins and Box Turtles (Family Emydidae)	
Common Slider (*Trachemys scripta*)	U (I)
Mud and Musk Turtles (Family Kinosternidae)	
Sonoran Mud Turtle (*Kinosternon sonoriense*)	A
Softshell Turtles (Family Trionychidae)	
Spiny Softshell Turtle (*Apalone spinifera*) [formerly called *Trionyx spiniferus*]	R
Alligator and Glass Lizards (Family Anguidae)	
Arizona Alligator Lizard (*Elgaria kingii*) [formerly *Gerrhonotus kingii*]	U
Geckos (Family Gekkonidae)	
Western Banded Gecko (*Coleonyx variegatus*)	R
Beaded Lizards (Family Helodermatidae)	
Gila Monster (*Heloderma suspectum*)	X ?
Iguanid Lizards (Family Iguanidae)	
Greater Earless Lizard (*Cophosaurus texanus*)	U
Collared Lizard (*Crotaphytus collaris*)	U
Greater Short-horned Lizard (*Phrynosoma hernandesi*) [formerly included with *Phrynosoma douglasii*]	R
Clark's Spiny Lizard (*Sceloporus clarkii*)	U
Desert Spiny Lizard (*Sceloporus magister*)	C
Eastern Fence Lizard (*Sceloporus undulatus*)	U
Tree Lizard (*Urosaurus ornatus*)	C/A
Side-blotched Lizard (*Uta stansburiana*)	A

Appendix A (continued)

Whiptails and relatives (Family Teiidae)	
Gila Spotted Whiptail (*Cnemidophorus flagellicaudus*) [formerly included with *Cnemidophorus exsanguis*]	U
Western Whiptail (*Cnemidophorus tigris*)	A
Desert Grassland Whiptail (*Cnemidophorus uniparens*)	C
Colubrid Snakes (Family Colubridae)	
Glossy Snake (*Arizona elegans*)	R
Ring-necked Snake (*Diadophis punctatus*)	U
Night Snake (*Hypsiglena torquata*)	R
Common Kingsnake (*Lampropeltis getula*)	R
Coachwhip (*Masticophis flagellum*)	R
Striped Whipsnake (*Masticophis taeniatus*)	U
Gopher Snake (*Pituophis catenifer*) [formerly included with *Pituophis melanoleucus*]	U
Long-nosed Snake (*Rhinocheilus lecontei*)	R
Western Patch-nosed Snake (*Salvadora hexalepis*)	U
Ground Snake (*Sonora semiannulata*)	U
Black-necked Garter Snake (*Thamnophis cyrtopsis*)	U
Western Lyre Snake (*Trimorphodon biscutatus*)	R
Coral Snakes and relatives (Family Elapidae)	
Sonoran Coral Snake (*Micruroides euryxanthus*)	R
Vipers (Family Viperidae)	
Western Diamond-back Rattlesnake (*Crotalus atrox*)	U
Black-tailed Rattlesnake (*Crotalus molossus*)	U
Mojave Rattlesnake (*Crotalus scutulatus*)	R

MAMMALS

Insectivores	
Shrews (Family Soricidae)	
Desert Shrew (*Notiosorex crawfordi*)	R
Bats	
Free-tailed Bats (Family Molossidae)	
Brazilian Free-tailed Bat (*Tadarida brasiliensis*)	U
Evening Bats (Family Vespertilionidae)	
Pallid Bat (*Antrozous pallidus*)	U
Big Brown Bat (*Eptesicus fuscus*)	U
Red Bat (*Lasiurus borealis*)	R
California Myotis (*Myotis californicus*)	U
Small-footed Myotis (*Myotis leibii*) [formerly called the western small-footed Myotis, *Myotis ciliolabrum*]	R
Little Brown Myotis (*Myotis lucifugus*) [our form has been called the Arizona Myotis, *Myotis occultus*]	R
Fringed Myotis (*Myotis thysanodes*)	R
Cave Myotis (*Myotis velifer*)	U
Yuma Myotis (*Myotis yumanensis*)	C
Western Pipistrelle (*Pipistrellus hesperus*)	C
Townsend's Big-eared Bat (*Plecotus townsendii*) [also called *Corynorhinus townsendii*]	U
Rabbits, Hares, and Pikas	
Rabbits and Hares (Family Leporidae)	
Black-tailed Jack Rabbit (*Lepus californicus*)	C
Desert Cottontail (*Sylvilagus audubonii*)	C
Rodents	
Beavers (Family Castoridae)	
Beaver (*Castor canadensis*)	U
New World Rats and Mice (Family Cricetidae)	
White-throated Woodrat (*Neotoma albigula*)	C
Mexican Woodrat (*Neotoma mexicana*)	R
Stephens' Woodrat (*Neotoma stephensi*)	U
Muskrat (*Ondatra zibethicus*)	U
Northern Grasshopper Mouse (*Onychomys leucogaster*)	U
Brush Mouse (*Peromyscus boylii*)	A
Cactus Mouse (*Peromyscus eremicus*)	A
White-footed Mouse (*Peromyscus leucopus*)	R

Appendix A (continued)

Deer Mouse (*Peromyscus maniculatus*)	U
Western Harvest Mouse (*Reithrodontomys megalotis*)	U
Arizona Cotton Rat (*Sigmodon arizonae*)	X
Porcupines (Family Erethizontidae)	
Porcupine (*Erethizon dorsatum*)	R
Pocket Gophers (Family Geomyidae)	
Botta's Pocket Gopher (*Thomomys bottae*)	C
Kangaroo Rats, Pocket Mice and relatives (Family Heteromyidae)	
Ord's Kangaroo Rat (*Dipodomys ordii*)	U
Hispid Pocket Mouse (*Chaetodipus hispidus*)	X
Rock Pocket Mouse (*Chaetodipus intermedius*)	U
Squirrels and relatives (Family Sciuridae)	
Harris' Antelope Squirrel (*Ammospermophilus harrisii*)	U
Arizona Gray Squirrel (*Sciurus arizonensis*)	R
Rock Ground Squirrel (*Spermophilus variegatus*)	C
Cliff Chipmunk (*Tamias dorsalis*)	U
Carnivores	
Dogs, Wolves and Foxes (Family Canidae)	
Coyote (*Canis latrans*)	U
Gray Fox (*Urocyon cinereoargenteus*)	U
Cats (Family Felidae)	
Mountain Lion (*Puma concolor*) [formerly called *Felis concolor*]	R
Ocelot (*Leopardus pardalis*)	X
Bobcat (*Lynx rufus*) [formerly called *Felis rufus*]	U
Weasels, Skunks and relatives (Family Mustelidae)	
Hog-nosed Skunk (*Conepatus mesoleucus*)	U
Striped Skunk (*Mephitis mephitis*)	U
Spotted Skunk (*Spilogale putorius*) [The western spotted skunk is sometimes treated as a separate species (*Spilogale gracilis*)]	R
River Otter (*Lontra canadensis*) [formerly called *Lutra canadensis*]	R
Badger (*Taxidea taxus*)	R
Raccoons and relatives (Family Procyonidae)	
Ringtail (*Bassariscus astutus*)	U
Raccoon (*Procyon lotor*)	C
Bears (Family Ursidae)	
Black Bear (*Ursus americanus*)	R
Ungulates	
Pronghorns (Family Antilocapridae)	
Pronghorn (*Antilocapra americana*)	X
Deer, Elk and relatives (Family Cervidae)	
American Elk (*Cervus elaphus*)	C (I)
Mule Deer (*Odocoileus hemionus*)	C
White-tailed Deer (*Odocoileus virginianus*)	U
Peccaries and Pigs (Family Tayassuidae)	
Collared Peccary (*Pecari tajacu*) [formerly called *Tayassu tajacu*]	U

BIRDS

Loons (Family Gaviidae)
Pacific Loon (*Gavia pacifica*)
Grebes (Family Podicipedidae)
Pied-billed Grebe (*Podilymbus podiceps*)
Herons and Bitterns (Family Ardeidae)
Great Blue Heron (*Ardea herodias*)
Snowy Egret (*Egretta thula*)
Green Heron (*Butorides virescens*)
Black-crowned Night-heron (*Nycticorax nycticorax*)
Ibises and Spoonbills (Family Threskiornithidae)
White-faced Ibis (*Plegadis chihi*)
Storks (Family Ciconiidae)
Wood Stork (*Mycteria americana*)
Ducks, Geese and Swans (Family Anatidae)

Appendix A (continued)

Greater White-fronted Goose (*Anser albifrons*)
Snow Goose (Chen caerulescens)
Canada Goose (*Branta canadensis*)
Wood Duck (*Aix sponsa*)
Green-winged Teal (*Anas crecca*)
Mallard (*Anas platyrhynchos*)
Northern Pintail (*Anas acuta*)
Blue-winged Teal (*Anas discors*)
Cinnamon Teal (*Anas cyanoptera*)
Northern Shoveler (*Anas clypeata*)
Gadwall (*Anas strepera*)
Eurasian Wigeon (*Anas penelope*)
American Wigeon (*Anas americana*)
Canvasback (*Aythya valisineria*)
Redhead (*Aythya americana*)
Ring-necked Duck (*Aythya collaris*)
Lesser Scaup (*Aythya affinis*)
Common Goldeneye (*Bucephala clangula*)
Bufflehead (*Bucephala albeola*)
Hooded Merganser (*Lophodytes cucullatus*)
Common Merganser (*Mergus merganser*)
Ruddy Duck (*Oxyura jamaicensis*)
Pheasants, Grouse, and allies (Family Phasianidae)
Ring-necked Pheasant (*Phasianus colchicus*)
Quail (Family Odontophoridae)
Gambel's Quail (*Callipepla gambelii*)
New World Vultures (Family Cathartidae)
Turkey Vulture (*Cathartes aura*)
Hawks and Eagles (Family Accipitridae)
Osprey (*Pandion haliaetus*)
Bald Eagle (*Haliaeetus leucocephalus*)
Northern Harrier (*Circus cyaneus*)
Sharp-shinned Hawk (*Accipiter striatus*)
Cooper's Hawk (*Accipiter cooperii*)
Northern Goshawk (*Accipiter gentilis*)
Common Black-hawk (*Buteogallus anthracinus*)
Harris' Hawk (*Parabuteo unicinctus*)
Swainson's Hawk (*Buteo swainsoni*)
Red-tailed Hawk (*Buteo jamaicensis*)
Ferruginous Hawk (*Buteo regalis*)
Rough-legged Hawk (*Buteo lagopus*)
Golden Eagle (*Aquila chrysaetos*)
Falcons and Caracaras (Family Falconidae)
American Kestrel (*Falco sparverius*)
Merlin (*Falco columbarius*)
Peregrine Falcon (*Falco peregrinus*)
Rails and Coots (Family Rallidae)
Virginia Rail (*Rallus limicola*)
Sora (*Porzana carolina*)
Common Moorhen (*Gallinula chloropus*)
American Coot (*Fulica americana*)
Plovers (Family Charadriidae)
Black-bellied Plover (*Pluvialis squatarola*)
Killdeer (*Charadrius vociferus*)
Sandpipers and Phalaropes (Family Scolopacidae)
Willet (*Catoptrophorus semipalmatus*)
Spotted Sandpiper (*Actitis macularius*)
Baird's Sandpiper (*Calidris bairdii*)
Wilson's Snipe (*Gallinago delicata*)
Wilson's Phalarope (*Phalaropus tricolor*)
Pigeons and Doves (Family Columbidae)
Rock Pigeon (*Columba livia*)
White-winged Dove (*Zenaida asiatica*)
Mourning Dove (*Zenaida macroura*)
Common Ground-dove (*Columbina passerina*)

Appendix A (continued)

Parrots (Family Psittacidae)
Thick-billed Parrot (*Rhynchopsitta pachyrhyncha)*
Cuckoos and Anis (Family Cuculidae)
Yellow-billed Cuckoo (*Coccyzus americanus*)
Greater Roadrunner (*Geococcyx californianus*)
Barn Owls (Family Tytonidae)
Barn Owl (*Tyto alba*)
True Owls (Family Strigidae)
Western Screech-owl (*Megascops kennicottii*)
Great Horned Owl (*Bubo virginianus*)
Northern Pygmy-owl (*Glaucidium gnoma*)
Elf Owl (*Micrathene whitneyi*)
Short-eared Owl (*Asio flammeus*)
Nightjars (Family Caprimulgidae)
Lesser Nighthawk (*Chordeiles acutipennis*)
Common Nighthawk (*Chordeiles minor*)
Common Poorwill (*Phalaenoptilus nuttallii*)
Swifts (Family Apodidae)
Vaux's Swift (*Chaetura vauxi*)
White-throated Swift (*Aeronautes saxatalis*)
Hummingbirds (Family Trochilidae)
Blue-throated Hummingbird (*Lampornis clemenciae*)
Black-chinned Hummingbird (*Archilochus alexandri*)
Anna's Hummingbird (*Calypte anna*)
Costa's Hummingbird (*Calypte costae*)
Broad-tailed Hummingbird (*Selasphorus platycercus*)
Rufous Hummingbird (*Selasphorus rufus*)
Kingfishers (Family Alcedinidae)
Belted Kingfisher (*Ceryle alcyon*)
Woodpecker (Family Picidae)
Lewis' Woodpecker (*Melanerpes lewis*)
Red-headed Woodpecker (*Melanerpes erythrocephalus*)
Gila Woodpecker (*Melanerpes uropygialis*)
Red-naped Sapsucker (*Sphyrapicus nuchalis*)
Williamson's Sapsucker (*Sphyrapicus thyroideus*)
Ladder-backed Woodpecker (*Picoides scalaris*)
Northern Flicker (*Colaptes auratus*) [formerly called Red-shafted Flicker, *C. cafer*]
Flycatchers (Family Tyrannidae)
Olive-sided Flycatcher (*Contopus cooperi*)
Western Wood-pewee (*Contopus sordidulus*)
Willow Flycatcher (*Empidonax traillii*)
Gray Flycatcher (*Empidonax wrightii*)
Cordilleran Flycatcher (*Empidonax occidentalis*) [formerly lumped with Pacific-slope Flycatcher as Western Flycatcher, *E. difficilis*; Pacific-slope Flycatcher may occur on migration]
Black Phoebe (*Sayornis nigricans*)
Say's Phoebe (*Sayornis saya*)
Vermilion Flycatcher (*Pyrocephalus rubinus*)
Ash-throated Flycatcher (*Myiarchus cinerascens*)
Brown-crested Flycatcher (*Myiarchus tyrannulus*)
Cassin's Kingbird (*Tyrannus vociferans*)
Western Kingbird (*Tyrannus verticalis*)
Larks (Family Alaudidae)
Horned Lark (*Eremophila alpestris*)
Swallows (Family Hirundinidae)
Purple Martin (*Progne subis*)
Violet-green Swallow (*Tachycineta thalassina*)
Northern Rough-winged Swallow (*Stelgidopteryx serripennis*)
Bank Swallow (*Riparia riparia*)
Cliff Swallow (*Hirundo pyrrhonota*)
Jays, Crows and Ravens (Family Corvidae)
Steller's Jay (*Cyanocitta stelleri*)
Western Scrub-Jay (*Aphelocoma californica*)
Mexican Jay (*Aphelocoma ultramarina*)
Pinyon Jay (*Gymnorhinus cyanocephalus*)
American Crow (*Corvus brachyrhynchos*)

Appendix A (continued)

Common Raven (*Corvus corax*)
Chickadees and Titmice (Family Paridae)
Mountain Chickadee (*Poecile gambeli*)
Bridled Titmouse (*Baeolophus wollweberi*)
Juniper Titmouse (*Baeolophus* ridgwayi)
Verdin and allies (Family Remizidae)
Verdin (*Auriparus flaviceps*)
Long-tailed Tits (Family Aegithalidae)
Bushtit (*Psaltriparus minimus*)
Nuthatches (Family Sittidae)
White-breasted Nuthatch (*Sitta carolinensis*)
Pygmy Nuthatch (*Sitta pygmaea*)
Creepers (Family Certhiidae)
Brown Creeper (*Certhia americana*)
Wrens (Family Troglodytidae)
Cactus Wren (*Campylorhynchus brunneicapillus*)
Rock Wren (*Salpinctes obsoletus*)
Canyon Wren (*Catherpes mexicanus*)
Bewick's Wren (*Thryomanes bewickii*)
House Wren (*Troglodytes aedon*)
Marsh Wren (*Cistothorus palustris*)
Kinglets (Family Regulidae)
Ruby-crowned Kinglet (*Regulus calendula*)
Old World Warblers and Gnatcatchers (Family Sylviidae)
Blue-gray Gnatcatcher (*Polioptila caerulea*)
Thrushes (Family Turdidae)
Western Bluebird (*Sialia mexicana*)
Mountain Bluebird (*Sialia currucoides*)
Townsend's Solitaire (*Myadestes townsendi*)
Hermit Thrush (*Catharus guttatus*)
American Robin (*Turdus migratorius*)
Mockingbirds and Thrashers (Family Mimidae)
Northern Mockingbird (*Mimus polyglottos*)
Sage Thrasher (*Oreoscoptes montanus*)
Brown Thrasher (*Toxostoma rufum*)
Crissal Thrasher (*Toxostoma crissale*)
Pipits and Wagtails (Family Motacillidae)
American Pipit (*Anthus rubescens*)
Waxwings (Family Bombycillidae)
Cedar Waxwing (*Bombycilla cedrorum*)
Silky Flycatchers (Family Ptilogonatidae)
Phainopepla (*Phainopepla nitens*)
Shrikes (Family Laniidae)
Loggerhead Shrike (*Lanius ludovicianus*)
Starlings (Family Sturnidae)
European Starling (*Sturnus vulgaris*)
Vireos (Family Vireonidae)
Bell's Vireo (*Vireo bellii*)
Gray Vireo (*Vireo vicinior*)
Plumbeous Vireo (*Vireo plumbeus*) [formerly grouped with Solitary Vireo, *V. solitarius*; very similar Cassin's Vireo may occur on migration]
Hutton's Vireo (*Vireo huttoni*)
Warbling Vireo (*Vireo gilvus*)
Wood Warblers (Family Parulidae)
Golden-winged Warbler (*Vermivora chrysoptera*)
Orange-crowned Warbler (*Vermivora celata*)
Nashville Warbler (*Vermivora ruficapilla*)
Virginia's Warbler (*Vermivora virginiae*)
Lucy's Warbler (*Vermivora luciae*)
Yellow Warbler (*Dendroica petechia*)
Black-throated Blue Warbler (*Dendroica caerulescens*)
Yellow-rumped Warbler (*Dendroica coronata*) [formerly called Audubon's Warbler, *D. auduboni*]
Black-throated Gray Warbler (*Dendroica nigrescens*)
Townsend's Warbler (*Dendroica townsendi*)
Grace's Warbler (*Dendroica graciae*)

Appendix A (continued)

Northern Waterthrush (*Seiurus noveboracensis*)
Macgillivray's Warbler (*Oporornis tolmiei*)
Common Yellowthroat (*Geothlypis trichas*)
Hooded Warbler (*Wilsonia citrina*)
Wilson's Warbler (*Wilsonia pusilla*)
Yellow-breasted Chat (*Icteria virens*)
Tanagers (Family Thraupidae)
Hepatic Tanager (*Piranga flava*)
Summer Tanager (*Piranga rubra*)
Western Tanager (*Piranga ludoviciana*)
Sparrows, Towhees, and allies (Family Emberizidae)
Green-tailed Towhee (*Pipilo chlorurus*)
Spotted Towhee (*Pipilo maculatus*) [formerly called Rufous-sided Towhee, *P. erythrophthalmus*]
Canyon Towhee (*Pipilo fuscus*)
Abert's Towhee (*Pipilo aberti*)
Rufous-crowned Sparrow (*Aimophila ruficeps*)
American Tree Sparrow (*Spizella arborea*)
Chipping Sparrow (*Spizella passerina*)
Brewer's Sparrow (*Spizella breweri*)
Black-chinned Sparrow (*Spizella atrogularis*)
Vesper Sparrow (*Pooecetes gramineus*)
Lark Sparrow (*Chondestes grammacus*)
Black-throated Sparrow (*Amphispiza bilineata*)
Sage Sparrow (*Amphispiza belli*)
Song Sparrow (*Melospiza melodia*)
Lincoln's Sparrow (*Melospiza lincolnii*)
White-throated Sparrow (*Zonotrichia albicollis*)
White-crowned Sparrow (*Zonotrichia leucophrys*)
Dark-eyed Junco (*Junco hyemalis*)
Cardinals and Grosbeaks (Family Cardinalidae)
Northern Cardinal (*Cardinalis cardinalis*)
Black-headed Grosbeak (*Pheucticus melanocephalus*)
Blue Grosbeak (*Guiraca caerulea*)
Lazuli Bunting (*Passerina amoena*)
Indigo Bunting (*Passerina cyanea*)
Blackbirds, Orioles, and allies (Family Icteridae)
Red-winged Blackbird (*Agelaius phoeniceus*)
Western Meadowlark (*Sturnella neglecta*)
Yellow-headed Blackbird (*Xanthocephalus xanthocephalus*)
Brewer's Blackbird (*Euphagus cyanocephalus*)
Great-tailed Grackle (*Quiscalus mexicanus*)
Brown-headed Cowbird (*Molothrus ater*)
Hooded Oriole (*Icterus cucullatus*)
Bullock's Oriole (*Icterus bullockii*) [segregated from Northern Oriole, *I. galbula*]
Scott's Oriole (*Icterus parisorum*)
Finches and allies (Family Fringillidae)
Purple Finch (*Carpodacus purpureus*)
Cassin's Finch (*Carpodacus cassinii*)
House Finch (*Carpodacus mexicanus*)
Red Crossbill (*Loxia curvirostra*)
Pine Siskin (*Carduelis pinus*)
Lesser Goldfinch (*Carduelis psaltria*)
American Goldfinch (*Carduelis tristis*)
Evening Grosbeak (*Coccothraustes vespertinus*)
Old World Sparrows (Family Passeridae)
House Sparrow (*Passer domesticus*)

MOVEMENT PATTERNS AND NATURAL HISTORY OF WESTERN DIAMOND-BACKED RATTLESNAKES AT TUZIGOOT NATIONAL MONUMENT, ARIZONA

Erika M. Nowak

The 1916 National Park Service Organic Act states, in part, that the purpose of national parks is to "conserve the scenery and natural and historic objects and the wildlife therein and to provide for the enjoyment of the same in such manner and by such means as will leave them unimpaired for the enjoyment of future generations" (National Park Service 1980). Park wildlife that is potentially hazardous to visitors and staff, such as rattlesnakes, should be managed only after "scientific research and planning that ... will protect the resources within parks in an effective and ecologically sound manner" (National Park Service 1991).

Tuzigoot National Monument, which was established solely to protect hilltop Sinagua culture sites, is located on approximately 43 acres in the Verde Valley of north-central Arizona. At least two species of rattlesnakes are found there: western diamond-backed rattlesnakes (*Crotalus atrox*) and black-tailed rattlesnakes (*C. molossus*). Although no rattlesnake studies had been completed at Tuzigoot prior to this work, *C. atrox* have been studied in detail in the eastern portion of the Verde Valley at Montezuma Castle National Monument by Nowak and van Riper (1999), and in central and southern Arizona by Taylor et al. (2004, 2005), Repp (1998), Hare and McNally (1997), Beaupre (1995), Beck (1995), and Schuett and Repp (unpublished data).

Several factors suggest that the rattlesnakes at Tuzigoot may behave differently or use available habitat differently when compared to these previous studies. Tuzigoot is a very small monument that does not contain a high diversity of habitats or physical features. There are no riparian habitats or extensive limestone cliff outcrops within the monument, although these features do exist within 0.5 km of its boundary. Just east of Tuzigoot lies the large, spring-fed Tavasci Marsh; no field research on Southwestern rattlesnakes has yet occurred in proximity to marsh systems. The original marsh, which was re-created by the Arizona Game and Fish Department, is currently expanding, with some help from beavers (*Castor canadensis*). The marsh is within the normal range of rattlesnakes using the monument, and it may influence their habitat use patterns. The small size of the monument may cause migratory wildlife such as rattlesnakes to be particularly vulnerable to encroaching urban development. For example, construction of a residential golf course community by the Phelps Dodge Corporation is ongoing on the west and south boundaries of the monument. Adjacent development may have immediate or long-term effects on rattlesnakes (e.g. Nowak et al. 2002; Goode et al. 2003), but without baseline data on the habitat use and resource needs of rattlesnakes at Tuzigoot, any impacts may go undetected.

To help develop scientifically based management strategies for Tuzigoot rattlesnakes, with the help of Manuel Santana-Bendix, I undertook a 3-year telemetry study of the natural history of resident *Crotalus atrox*. My study had the following objectives: (1) record life history traits and behavioral observations; (2) determine seasonal activity, movement patterns, and habitat use; (3) determine activity range size; (4) compare life history traits, movement, and activity patterns between sexes; and (5) make recommendations for future rattlesnake management at the monument.

METHODS

Study Site

Tuzigoot National Monument is located in the Verde Valley of north-central Arizona at approximately 3400 ft (1036 m) in elevation (Figure 1); it consists of less than 43 acres (17 ha). The monument is bordered on all sides by private property, currently owned by Phelps Dodge Corporation. Forest Service land lies to the north of the private property, and Arizona State Park land lies to the east and south. Upland vegetative communities in the area are dominated by the Creosotebush-Crucifixionthorn Series of the Arizona Upland Division of the Sonoran Desertscrub, but vegetation representative of Chihuahuan, Mojave, and Great Basin Desertscrubs is also present (Turner and Brown 1994). Riparian vegetation outside the monument adjacent to the Verde River is characteristic of Sonoran Riparian Deciduous Forest Scrubland, and is characteristic of Sonoran Interior Marshland in Tavasci Marsh (Minckley and Brown 1994). The region also contains many porous limestone outcroppings and cliffs.

Capture and Telemetry Methods

Between March 1995 and December 2003, a total of 54 *Crotalus atrox* were encountered opportunistically at Tuzigoot by myself, technicians, and park staff. We restrained all snakes in plastic tubes, weighed them to the nearest 0.5 gram, measured them to the nearest 0.5 cm (snout-vent length, SVL), and sexed them. We painted them on the basal three rattles with individual color combinations using semi-permanent Testor's model paint, and then released them near their capture sites.

For permanent identification of individuals, we injected a uniquely coded passive integrated transponder (PIT) tag into the coelomic (gut) cavity of each snake starting in 1998. We followed the methods of Fagerstone and Johns (1987). Every time a rattlesnake was recaptured, researchers or park staff scanned it to verify its identity.

We implanted seven adult *Crotalus atrox* (five males and two non-pregnant females) with radio transmitters between 1995 and 1997. Each snake was large enough so that a transmitter would be less than 5 percent of its body weight. We used 11–13 g temperature-sensing transmitters (Telonics Telemetry-Electronics Consultants, Mesa, Arizona; and Holohil Systems, Ltd., Ontario, Canada); transmitting life was approximately 12 months for Telonics and 24 months for Holohil models. We anesthetized rattlesnakes in a specialized aquarium setup using gaseous isoflurane at a local veterinary hospital, and we administered anesthetic directly by tracheal entubation during surgery. Veterinarians followed the sterile surgical implantation procedures of Hardy and Greene (1999b) to implant a radio transmitter in the coelomic cavity of each rattlesnake. After surgery, we gave each snake an injection of saline equal to at least 5 percent of its body weight to help prevent dehydration, and several animals with obvious wounds or infections were given a small dose of Amikacin antibiotic. We held all rattlesnakes in a quiet heated room for at least 6 hours to ensure adequate recovery from anesthesia, and provided them with water ad libitum. After recovery, we returned all snakes to their original capture site and recorded their positions once every 7–12 days. We located snakes with a Telonics model TR-4 receiver and H-style directional antenna. When a rattlesnake was found, we recorded the time and date, air and substrate temperature, body temperature, microhabitat association, and behavior. We used a Trimble Navigation (Sunnyvale, California) Geo-Explorer global positioning system (GPS) unit to record positions of the snakes in universal trans-mercators (UTMs; NAD 83 Datum).

Data Analyses

I analyzed all data using the statistical computer program SPSS v. 7.0 (1996). Signifi-

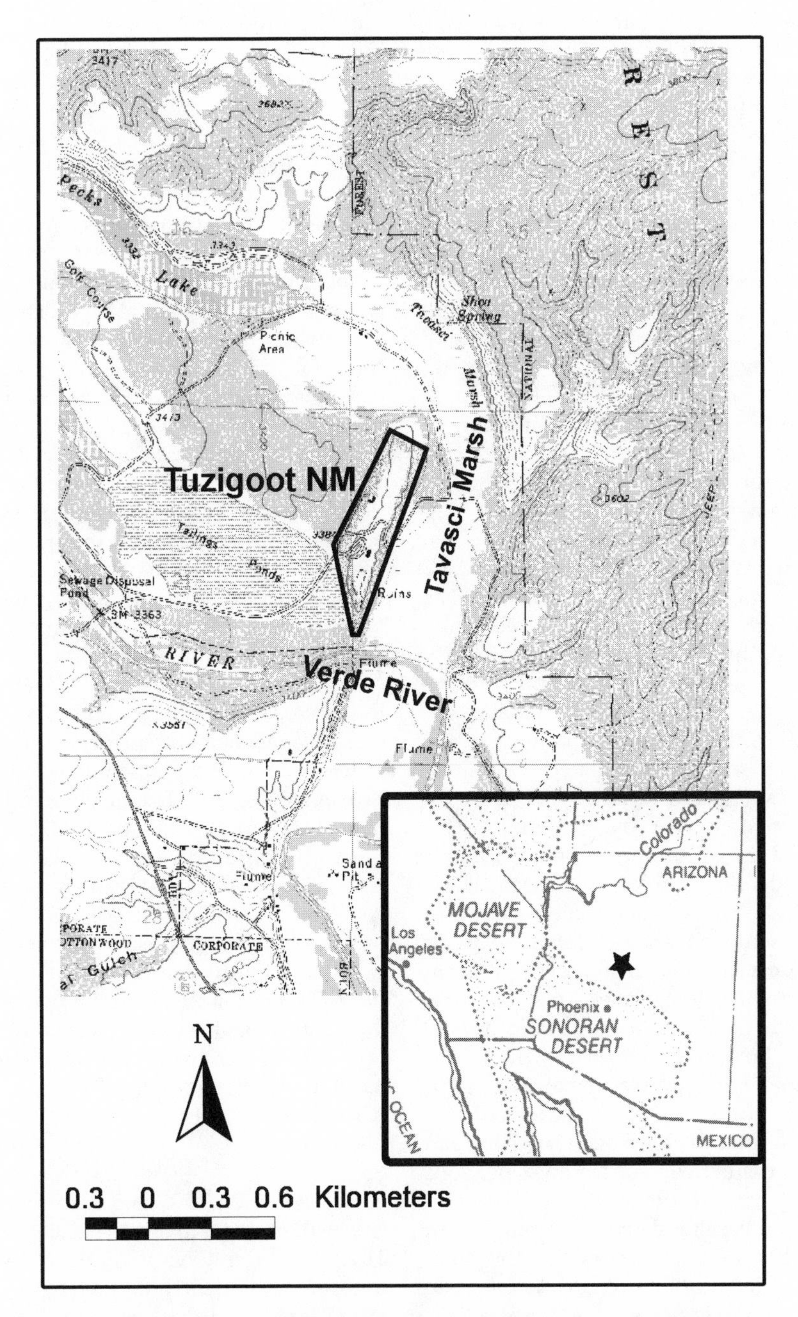

Figure 1. Location of natural history study of *C. atrox* at Tuzigoot National Monument between 1995 and 1997, showing monument location and key natural features.

cance was determined at the $p \leq 0.05$ level. Means are reported followed by ± one standard deviation. Due to small sample sizes, I analyzed all data using non-parametric statistical tests. Between-group comparisons were made with the Mann-Whitney-U-Wilcoxon two-sample test for two-sample data and with the Kruskal-Wallis one-way ANOVA test for multi-sample data (Sokal and Rohlf 1981). Frequency data were analyzed using the chi-square (χ^2) test with the Pearson estimator of χ^2 (Sokal and Rohlf 1981). To develop standardized body condition indices for the rattlesnakes, I calculated the residuals of the linear regression of weight to SVL. These residuals were then compared for males and females.

Rattlesnake locations, movement patterns, and activity ranges were mapped using Trimble Pathfinder, ARC-Info, and ARC View v. 3.3 (2002) mapping programs. The original GPS points were corrected using Trimble Pathfinder to give more precise coordinates for each location, and were reprojected into NAD 27 Datum for mapping.

To assess the movement patterns of the telemetered snakes, I calculated the following indices: (1) total distance traveled over the course of a year by each individual snake; and (2) total number of new locations used divided by total locations used. These parameters were calculated solely as indices of movement and are not intended to be precise descriptions of actual movement patterns made by the snakes. To standardize these data, I defined a movement as any distance between successive locations greater than 6 m (6 m was the average error for most corrected GPS locations). To determine total distance traveled per year, the distances between successive locations were summed. To determine frequency of movement, I divided the total number of movements of each snake by the total number of its locations. To determine total number of new locations used, I sorted the UTM coordinates for each snake and counted the number of locations greater than 6 m apart. To estimate the activity range size for each snake, I used the minimum convex polygon method (White and Garrott 1990) in the computer program Telem (K. McKelvey, personal communication). Because the size of the activity range of any animal is largely dependent on the number of its locations (Reinert 1992), I standardized the data by only including snakes for which there were at least 4 months of data (at least 15 locations).

To quantify habitat use, I recorded the number of locations of each snake in one of seven pre-defined habitat types: (1) areas on and adjacent to the monument not used by humans, generally scattered shrub and grassland habitats; (2) disturbed areas on the monument within 50 m of the visitor center area and/or ruins complex trails; (3) Tavasci Marsh, east and south of the monument; (4) dense, brushy herbaceous and grassy areas adjacent to the marsh; (5) mesquite bosque east and west of the marsh; (6) canyons and washes east of the marsh; and (7) the Verde River and its present floodplain, south and east of the monument. To determine the relative proportion of each habitat available to the snakes, I calculated the area (in hectares) of each habitat type over all combined ranges of the Tuzigoot snakes. I drew approximate polygons for each vegetation type over the Clarkdale SE (1992) USGS Digital Ortho-topographic Quarter Quadrangle (DOQQ) and calculated their areas using the ArcView extension XTools (9/15/2003). Frequency of habitat use by the snakes was calculated by dividing the number of locations of snakes in each habitat (weighted by the relative habitat area) by the total number of locations for the snakes. Frequency of habitat use versus availability was compared using a χ^2 test.

RESULTS

Between March 1995 and December 2003, we encountered a total of 54 *Crotalus atrox* at Tuzigoot. From 1995 to 1997, snakes were not yet permanently tagged with passive integrated transponders, and their paint wore off after about a year, so I do not know the exact number of individuals detected. Captured snakes included 3 unsexed adults, 2 unsexed immature snakes, 1 unaged female, 1 unaged male, 5 immature females,

16 adult females, 5 immature males, and 21 adult males. We encountered 11 adults after the initial capture, usually just once, but a few were encountered two or more times. Male #42 was originally telemetered in 1995, and was recaptured in 1998 and 1999. Between his first and last captures he grew approximately 4 cm in length and added 181 g in weight. A third, non-telemetered male was captured first in 2001 and again in 2003 (not measured).

Most of the rattlesnakes captured were originally encountered by visitors or staff personnel on the trails in and around the main ruins complex, or on adjacent walls. We sighted rattlesnakes in every month of the year except December. We encountered more rattlesnakes (20) in the fall (October and November) than during any other time of year.

We obtained a total of 234 locations for seven telemetered rattlesnakes between 1995 and 1997. The average number of locations of each individual rattlesnake was highly variable (31.6 ± 24.9), and varied between 4 (#27, captured in the early fall of 1997) and 78 (#15, followed continuously since April 1995).

Life History Traits

Although most Tuzigoot *Crotalus atrox* males were longer and heavier than females, this was not the case for all adults (Figure 2). Adults in this population were snakes reaching an SVL of at least 60 cm, based on the approximate size at which courtship was first observed in females. The average SVL for adult male *C. atrox* was 94.3 ± 14.0 cm (n = 30), and adult females averaged 86.4 ± 14.3 cm (n = 19). Adult males averaged 618.0 ± 320.6 g in mass (n = 25; not all snakes were weighed), and adult females averaged 540.8 ± 302.8 g (n = 19). When the body condition indices (residuals of the linear regression of mass on snout-vent length) were compared for all *C. atrox* captured in 1995–1997, there was no significant difference between the sexes (Z = –0.704, p = 0.482).

A volunteer observed two males (one telemetered) fighting near Tavasci Marsh with a female coiled nearby on September 2, 1999 (M. Spille, personal communication). No other reproductive activities were documented. Two male rattlesnakes found during the 1995–1997 research had noticeable amounts of seminal fluid in their hemipenes (when probed to determine sex) during the summer monsoon period. I encountered one neonate rattlesnake outside the visitor center building in October 1995 (see below), suggesting that birth occurs late in the summer or early fall. No evidence of spring mating aggregation at any hibernation site was observed.

Female #26 was found dead during the study in early November 1996. This snake was very thin when captured in mid July, moved less than 500 m from her original capture point, and was active late into the fall (as were two other telemetered rattlesnakes). I found her transmitter with pieces of skin and vertebrae just outside the monument fence. Perhaps she had some type of disease (thus being thin), and either succumbed to overnight freezing temperatures and was scavenged after death, or was killed and eaten by a non-human predator.

On October 21, 1995, I released adult male #33 outside the visitor center, next to a rock wall. The rattlesnake immediately began crawling along the wall toward the north. At the same time, I noticed that a neonate western diamondback, approximately 25 cm in total length with one segment on its rattle, was crawling along the wall toward the adult. When the rattlesnakes were approximately half a meter apart, the neonate stopped and became very alert. It stiffened, and curled its body up and off the ground into a "bridged" position identical to that used by snakes to thwart potential snake predators such as kingsnakes (Chiszar et al. 1992). It held this position as the adult continued to crawl toward the neonate. As the adult passed by the neonate, the small snake struck at the adult two times in rapid succession. At the second strike, the adult retracted its head and forebody for a few seconds, then resumed his path of travel. The neonate struck a third time, but the adult did not pause again. As the adult kept moving, the neonate pulled itself into a half

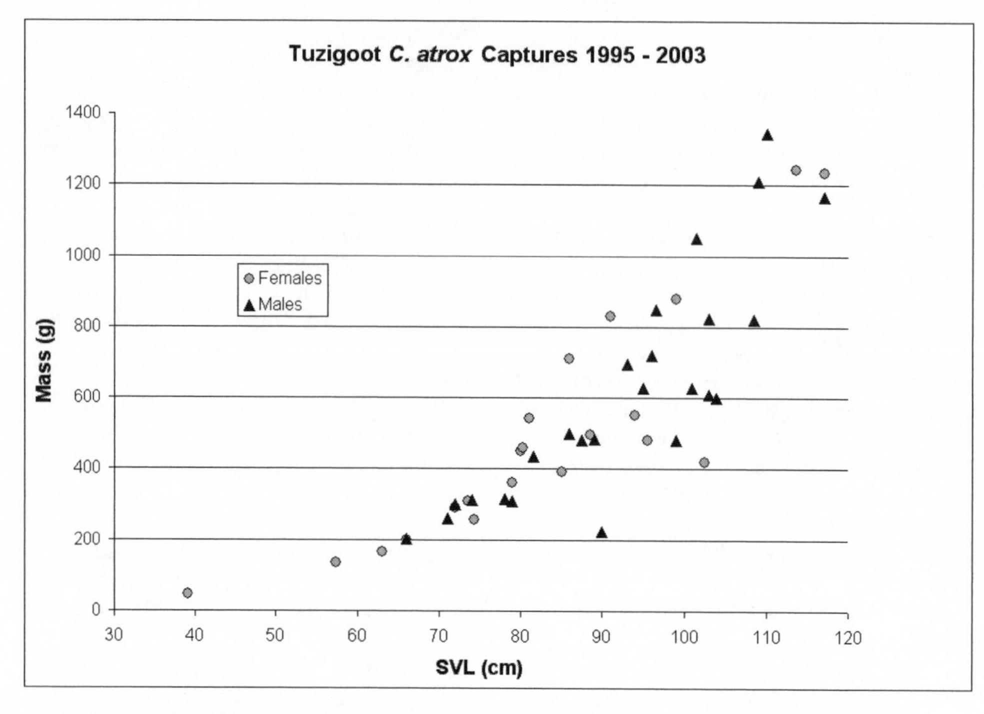

Figure 2. Weight (mass) in grams and snout-vent length in centimeters of initial captures of 49 western diamond-backed rattlesnakes (*C. atrox*) at Tuzigoot National Monument between April 1995 and December 2003.

coil and watched him for a few seconds, then resumed its southward path of travel along the wall. Neither snake was observed to tongue flick during this encounter.

Seasonal Activity

In 1995 and 1997, telemetered rattlesnakes at Tuzigoot hibernated from approximately mid-October to mid March. In 1996, most snakes entered hibernation later, in early November. Rattlesnakes were occasionally found basking at the entrance to their den sites, usually in early November and early March, but they did not appear to move between hibernation sites after initially entering the dens. Rattlesnakes were encountered on the main ruins visitor trail in late November and in January, during the hibernation period. This suggests that they were hibernating within the ruins or in other adjacent man-made structures (e.g., walls and buildings). In fact, two telemetered rattlesnakes did hibernate in such structures: #15 hibernated in the old pump house or an adjacent wall in 1995 and 1996, and #27 hibernated in ruins below the main visitor area in 1997. Several rattlesnakes seen on the trail near the opening of a drainage pipe under the ruins indicate a likely point of access into the main ruins complex for other snakes. Other hibernation sites included burrow and tunnel systems under large limestone boulders on the southwest-facing slope below the ruins complex (four snakes), and small caves in sandstone outcrop systems in two washes approximately 0.5 km east of the monument (two snakes, including #15 in 1997). Hibernation sites in the rock outcrops contained woodrat (*Neotoma* sp.) scat, but no hibernation site contained obvious *Neotoma* middens. Most telemetered rattlesnakes at Tuzigoot showed strong fidelity to hibernation sites, returning to the same den sites year after year. One exception to this pattern was #15, which apparently hibernated in a new location in 1997. However, he had used this same site as an active season retreat during late September in 1995 and 1996.

On average, telemetered rattlesnakes at Tuzigoot dispersed from their hibernation sites by late March and were active until mid October. The only year with detailed information on spring emergence patterns was 1997, when rattlesnakes dispersed from their hibernation sites initially in mid March, and then most spent several weeks inactive under boulders or in *Neotoma* middens. Male #42 dispersed approximately 300 m from his hibernation site over about two weeks, then returned to the den with the onset of a cold, wet storm in early April. He remained at the den for at least one more week before dispersing.

Activity Range

All telemetered snakes had activity ranges (estimated by the minimum convex polygon method) of less than 1 km in length (Figure 3a, b). Average activity range size for *Crotalus atrox* at Tuzigoot over two to three seasons was 12.5 ± 11.9 ha, ranging from 30.4 ha (#15, followed for three full seasons) to 0.20 ha (#26, followed for 4 months during the active season). The activity ranges for Tuzigoot rattlesnakes between March 1995 and December 1997 are shown in Figure 3a (females) and Figure 3b (males). Locations for animals with too few locations to estimate an activity range are also shown.

There was no significant difference in average activity range size between years ($\chi^2 = 2.244$, df = 2, p = 0.326). Average range size was 18.6 ± 16.7 ha in 1995 (n = 2); 6.6 ± 7.1 ha in 1996 (n = 4); and 6.0 ± 3.7 ha in 1997 (n = 4). Most individuals' ranges overlapped a great deal between years of the study, and ranges overlapped between different snakes as well. An example of overlap between years for one individual can be seen in the annual ranges of male #15 from 1995 to 1997 (Figure 4).

Movement Patterns

The average number of days between locations (7 in 1995 and 12 in 1996 and 1997) was too great to permit detailed discussion of movement patterns, but a general pattern is detectable. I estimated the total distance moved by telemetered rattlesnakes between successive locations each year over the course of the study. There was no statistical significance in total distance moved by the snakes between years ($\chi^2 = 1.091$, df = 2, p = 0.580). In 1995, *Crotalus atrox* (n = 2) moved an average distance of at least 2.9 ± 1.6 km. In 1996, they moved an average distance of at least 1.9 ± 1.4 km (n = 4), and in 1997 they moved an average distance of at least 1.8 ± 1.1 km (n = 4). When the average total distance moved by males and females was compared across years, there was also no significant difference (Z = –0.426, p = 0.670). Males moved an average distance of at least 2.1 ± 1.1 km per year, and females moved an average of at least 1.9 ± 1.6 km per year.

The proportion of new locations used by all *Crotalus atrox* was 74 percent (i.e., 26% of the sites had been used by one or more snakes previously). The proportion of new locations used by individual rattlesnakes during the active seasons over the course of the study varied between 57 and 82 percent (i.e., 43–18% of the sites used by each rattlesnake had been previously used by that snake, in either the same or a previous year). There was no significant difference among years when the proportion of new sites used was compared ($\chi^2 = 17.875$, df = 18, p = 0.464). The proportion of new sites used in 1995 was 66 percent (n = 2), 82 percent in 1996 (n = 4), and 68 percent in 1997 (n = 4). When the numbers of new locations used by males and females over the course of the study were compared, there was also no significant difference ($\chi^2 = 4.958$, df = 5, p = 0.421). Seventy-three percent of the locations that males used were new, and 66 percent of the locations that females used were new.

Habitat Use

I examined patterns of habitat use for six of the seven telemetered *Crotalus atrox* in 1995–

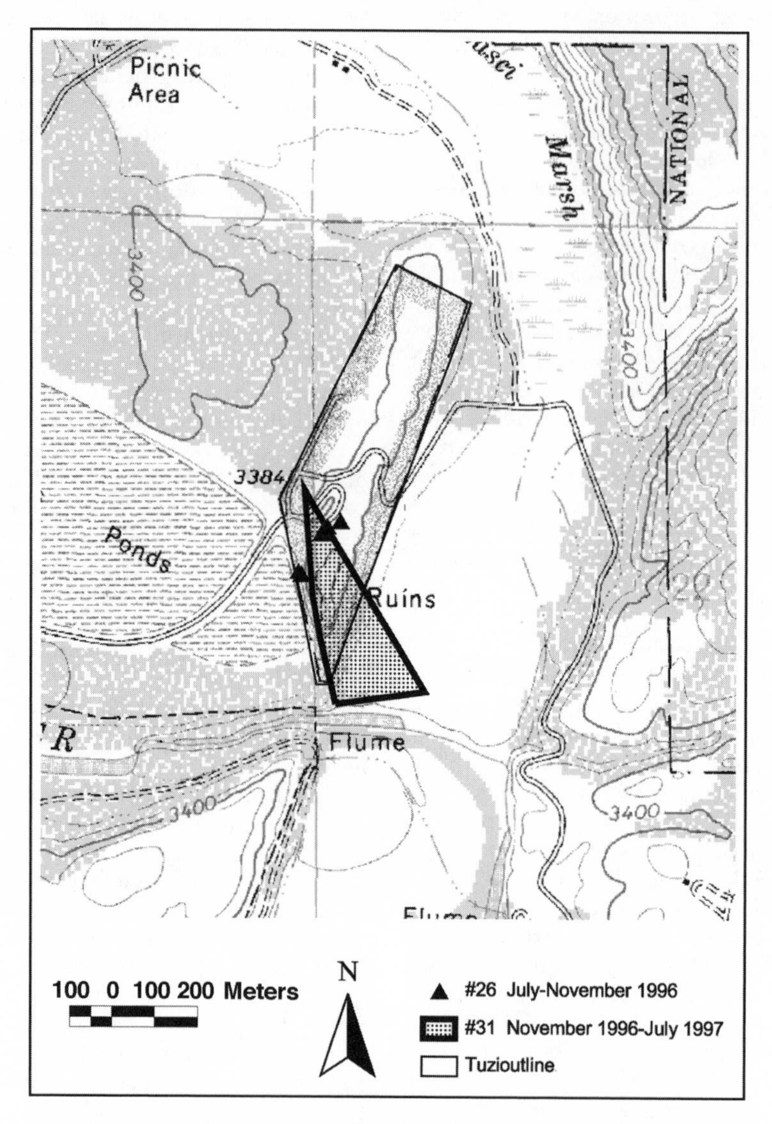

Figure 3a. Locations of activity ranges or locations for two telemetered adult female western diamond-backed rattlesnakes (*C. atrox*) at Tuzigoot National Monument. Monument boundaries are outlined in thin black lines in the center of the image. The shaded area indicates the activity range of #31 between November 1996 and July 1997; triangles indicate locations for #26 between July and November 1996.

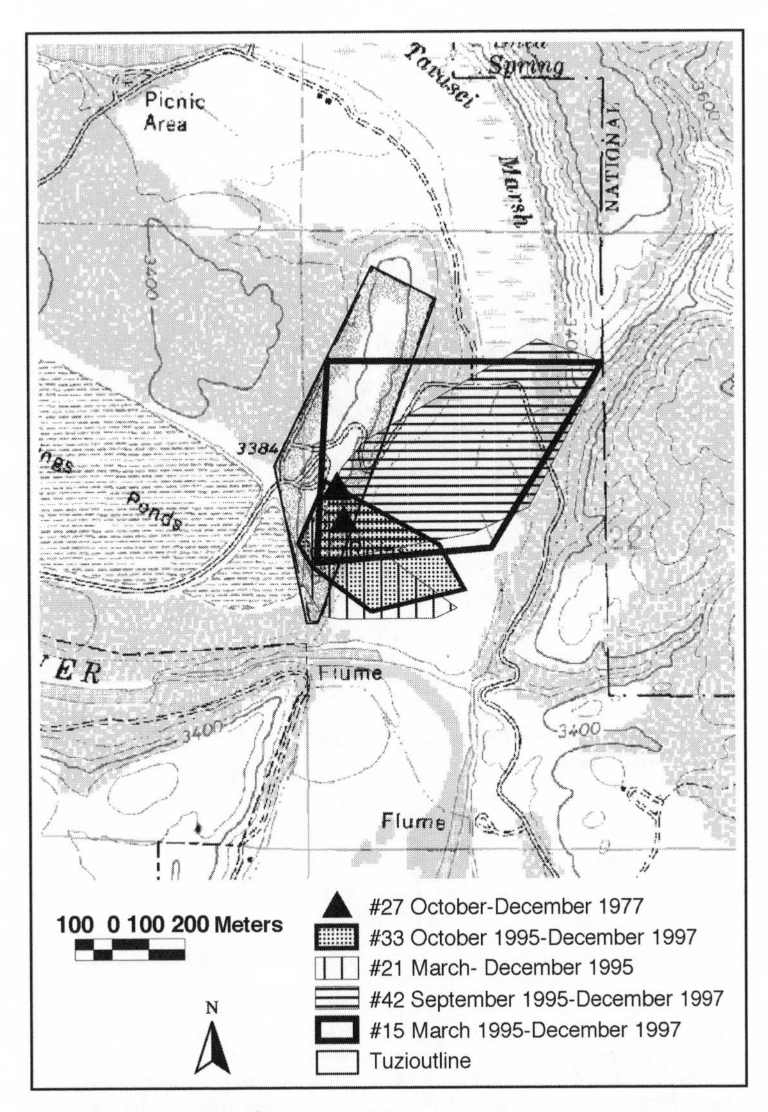

Figure 3b. Locations of activity ranges or locations for five telemetered adult male western diamond-backed rattlesnakes (*C. atrox*) at Tuzigoot National Monument. Monument boundaries are outlined in thin black lines in the center of the image. Triangles indicate locations for #27 in 1997. The stippled area indicates the activity range of #33 from 1995 to 1997; vertical lines indicate range for #21 in 1995; horizontal lines indicate range for #42 from 1995 to 1997; and the clear area bordered in thick black lines indicates the range of #15 from 1995 to July.

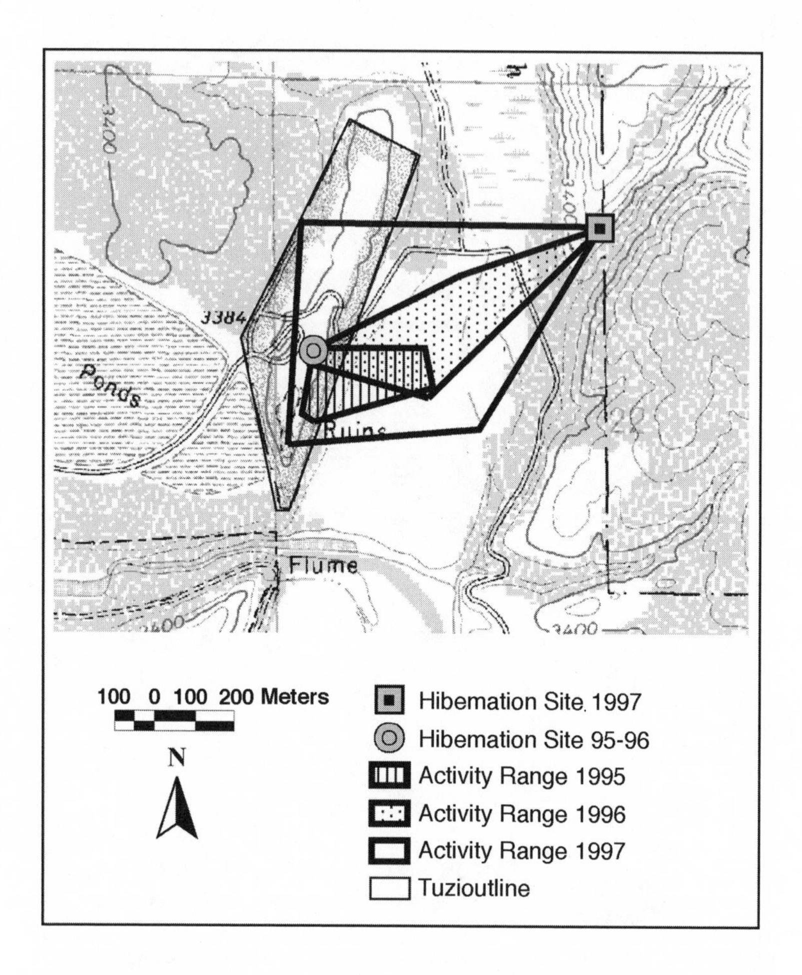

Figure 4. Locations of activity range and hibernation sites for #15, a telemetered adult male western diamond-backed rattlesnake (*C. atrox*) at Tuzigoot National Monument from 1995 to 1997. Monument boundaries are outlined in thin black lines in the center of the image. The clear area outlined in thick black indicates the activity range in 1995; vertical lines indicate 1996 range; and dotted area indicates 1997 range. A circle indicates the hibernation site in 1995 and 1996 and a square indicates the hibernation site in 1997.

1997. Proportions of each habitat type available over all the snake activity ranges are given in Figure 5a. There was no significant difference in the proportion of habitat used by the snakes compared to that available over their combined ranges ($\chi^2 = 6.000$, df = 6, p = 0.423; Figure 5b), suggesting that for this population there was no statistical evidence for habitat selection. However, there was a trend toward avoidance of mesquite bosque and increased use of open shrub-grassland habitats. As well, individual snakes appeared to have individual preferences for certain microhabitats (Table 1). Although all rattlesnakes were originally captured on the monument, five of the telemetered rattlesnakes spent large amounts of time outside its boundaries. Rattlesnakes routinely used U.S. Forest Service land, Arizona State Park land, and private Phelps Dodge Corporation land adjacent to the monument (see Figure 1 for orientation). Of 175 active locations for six telemetered rattlesnakes between April 1995 and November 1997, less than 50 percent were on monument property.

Although 17 of 19 rattlesnakes in 1995–1997 were captured within 50 m of the visitor center or on the main ruins complex trails, telemetered rattlesnakes did not appear to stay in these areas for extended periods of time. Approximately 36 percent of all locations of telemetered rattlesnakes occurred on the monument, but less than 10 percent were within 50 m of the visitor center or visitor trails.

It is also noteworthy that at least three telemetered rattlesnakes spent time either in water or at water's edge in Tavasci Marsh, an expanding perennial native marsh dominated by cattails and tules. Several times snakes were observed on islands of emergent vegetation surrounded by water.

Few patterns emerged regarding differences in habitat use during the active season between the sexes. Both sexes used all available habitats, with the exception of the mesquite bosque: males used this habitat significantly more often than females (Z = –2.087, p = 0.037; Figure 6). We located males more often in shrub/forb habitats, and females were observed more often in marsh and arroyo habitats.

DISCUSSION

Life History Traits

Western diamond-backed rattlesnakes at Tuzigoot do not share all of the life history traits previously reported from the species in other parts of its range. One probable reason is that this species is at the elevational limit of its range in Arizona in the Verde Valley (Lowe et al. 1986). Occurrence at this ecological limit may be correlated with the apparent small population size at Tuzigoot. Beaupre (1995) estimated a population density of 500–1000 *Crotalus atrox* per square kilometer at much lower elevations, near Phoenix. Although the scope of this study did not permit accurate estimates of population size, and sampling effort varied between years, only 54 *C. atrox* were observed at the monument between March 1995 and December 2003 (several were recaptured one or more times). The paucity of sightings may be due in part to reliance on procrypsis to avoid detection by predators, including humans (Duvall et al. 1985; Reinert and Zappalorti 1988b; May et al. 1996; Nowak and van Riper 1999), but it is not likely that large numbers of rattlesnakes would escape detection if they were present. During June 1994 to December 1996, Nowak and van Riper (1999) and the staff at Montezuma Castle NM detected 28 *C. atrox*. It is likely that relatively fewer rattlesnakes are encountered in the Verde Valley due to a shorter summer active period and more marginal habitat as compared with southern Arizona populations. Moreover, at Tuzigoot there is a lower diversity of habitat types, and Tuzigoot National Monument is small (less than 2 sq km).

Another difference in the *Crotalus atrox* population at Tuzigoot is its relative body size compared to other populations. Beaupre (1995) found that the largest snakes in a population near Phoenix were 105 cm in SVL and 750 g in weight; the largest examined at roundups in Oklahoma by Fitch and Pisani (1993) were 152 cm SVL and 2776 grams. A population of *C. atrox* in southern

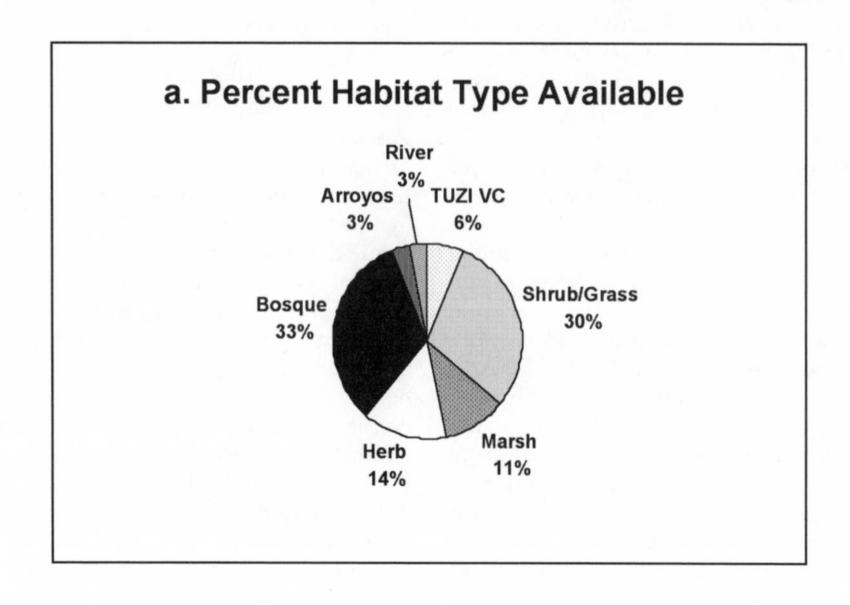

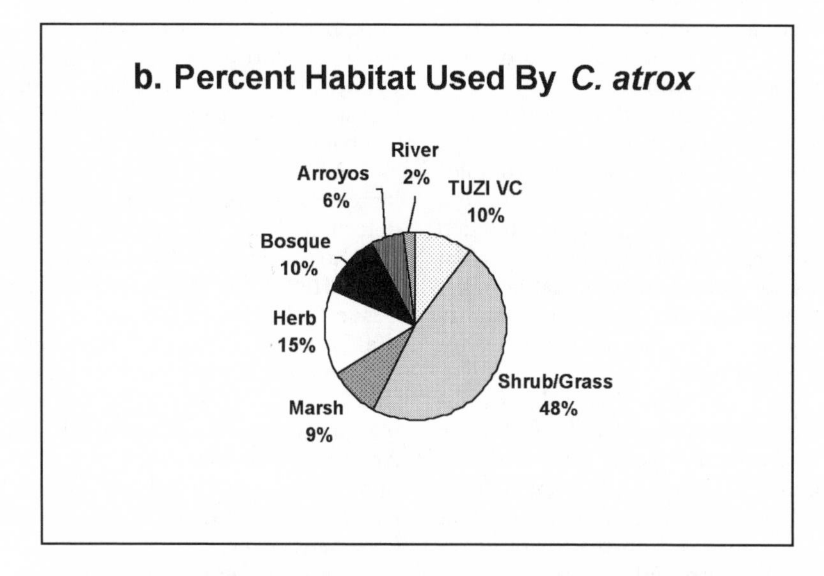

Figure 5. (a) Total proportion of each general habitat type available within the ranges of six western diamond-backed rattlesnakes (*Crotalus atrox*) around Tuzigoot National Monument from 1995 to 1997. (b) Total proportion of each general habitat type actually used by the six *C. atrox* during their active periods between March 1995 and October 1997. See Table 1 for habitat type descriptions.

Table 1. Proportion of available habitat and time spent during the active season in different habitats around Tuzigoot National Monument, Arizona, by six *Crotalus atrox* between April 1995 and November 1997. The rows indicate the proportion of locations for each snake in each habitat type. TUZI VC = near the monument visitor center area and trails; Shrub/Grass = on and adjacent to the monument, generally open shrub-grasslands; Marsh = Tavasci Marsh; Herb = dense herbaceous/grass areas adjacent to the marsh; Bosque = mesquite bosque; Arroyos = canyons and washes; River = the Verde River and its present floodplain.

Snake No.	TUZI VC	Shrub/ Grass	Marsh	Herb	Bosque	Arroyos	River
15	.10	.34	.10	.38	.05	.03	0
21	.05	.55	0	0	.27	0	.13
26	.20	.80	0	0	0	0	0
31	0	.78	0	0	0	0	0
33	.08	.36	0	.24	.32	0	0
42	.07	.07	.38	.14	0	.34	0

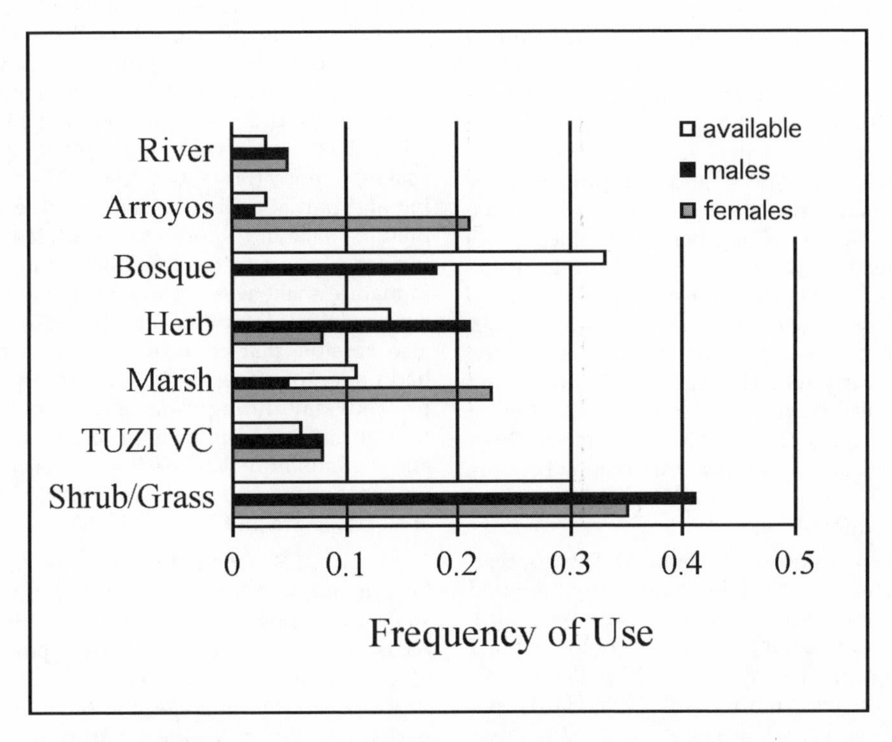

Figure 6. Frequency of use of habitat types by four male and two female western diamond-backed rattlesnakes (*C. atrox*) at Tuzigoot National Monument during their active periods between March 1995 and October 1997. The proportion of each habitat type available in their combined ranges is also given; see Table 1 for habitat type descriptions.

Arizona in the Owlhead Buttes were similar in length to the Tuzigoot population (males averaged 93.5 cm and females averaged 82 cm), but their weights were much smaller (males averaged 515 g and females averaged 356 g; Taylor et al. 2005). The population of *C. atrox* at Tuzigoot was intermediate in size between the above-mentioned studies, with the average male size 93 cm SVL and 634 g mass, and females averaging 83 cm SVL and 494 g. A few females at Tuzigoot were very large (the largest snake discovered, a female, was 113.5 cm in SVL and 1243 g in weight). This is probably not the norm for the population as a whole, as sex differences were not statistically significant, but it is an interesting phenomenon. A typical pattern for most rattlesnake species is that males are longer and heavier than females (King and Duvall 1990; Shine 1993; Fitch and Pisani 1993; Beaupre 1995).

As was the case for rattlesnakes studied by Nowak and van Riper (1999) at Montezuma Castle National Monument, I did not observe mating at hibernation sites in early spring such as that seen by Taylor et al. (2005), Repp (1998), and Beaupre (1995) for southern Arizona rattlesnakes. I cannot rule out a spring mating period in which males instead search for females away from the dens (e.g. Hardy and Greene 1999a; Nowak and van Riper 1999); however, we have observed a mating season during the summer and early fall. This second mating period has also been documented for *C. atrox* at Montezuma Castle by Nowak and van Riper (1999), and in southern Arizona by Beaupre (1995), Taylor et al. (2005), and Schuett and Repp (unpublished data).

Few studies of the natural history of free-ranging rattlesnakes have documented interactions between immature and adult rattlesnakes, other than those of mother and offspring (Duvall et al. 1985; Price 1988; Martin 1992; Butler et al. 1995; Hardy and Greene 1999a). Most intraspecific interactions recorded between adult snakes and possibly unrelated immature snakes are those of young snakes following the adults using scent trailing or perhaps vision, presumably to locate suitable hibernation sites (Brown and Maclean 1983; Graves et al. 1986; Reinert and Zappalorti 1988a). To my knowledge, there have been no observations of free-ranging immature rattlesnakes reacting defensively toward adults of the same species. "Body bridging" is a stereotypical response in pitvipers elicited by the presence, skin secretions, or even models (D. Duvall, personal communication) of snake-eating (ophiophagous) snakes such as kingsnakes (*Lampropeltis* spp.). Bridging involves stiffening of the body and curving the mid-body up off the substrate (Chiszar et al. 1992). There appear to be no other observations of the bridging response being elicited in a rattlesnake by a conspecific, but Weldon et al. (1992) noted that elicitation of bridging varies between individual rattlesnakes, and is further modified by prior experience or habituation to ophiophagous snake chemicals and odors. Cannibalism has been documented in captivity for adult *Crotalus atrox* (Klauber 1972) and immature sidewinders, *C. cerastes* (Brown and Lillywhite 1992), and in wild twin-spotted rattlesnakes, *C. pricei* (Prival et al. 2002). The striking and partial coiling behavior by the immature snake in response to the adult adds further support to the hypothesis that the immature snake was reacting as it would to a potential predator (Duvall et al. 1985). It is also possible that because the adult snake had undergone anesthesia and surgery the previous day during transmitter implantation, it did not emit an appropriate smell, and the immature was reacting primarily to the sensory anomaly (Halpern 1992; Weldon et al. 1992; Ford and Burghardt 1993).

Seasonal Activity

Hibernation and foraging periods of rattlesnakes at Tuzigoot were similar to those recorded for *C. atrox* in the eastern portion of the Verde Valley by Nowak and van Riper (1999). On average, the hibernation period for Verde Valley *C. atrox* started approximately 3–4 weeks earlier, and lasted 2–3 weeks longer than that seen in southern Arizona populations (Beaupre 1995; Repp 1998; Schuett and Repp, unpublished data).

The sighting of rattlesnakes in the monument ruins complex and on the visitor trails around the ruins in November and January is a significant finding. I believe that this observation implies that rattlesnakes are hibernating in or under the ruins and/or visitor trails (and, in at least one case, in or under a relatively recent structure). It is likely that rattlesnakes are accessing hibernation sites in the visitor use area through drainage pipes built into the ruins, or through rodent burrows which tunnel under man-made features. Rattlesnakes are seen when they briefly emerge for basking, likely to stimulate gonadal development, to help fight off diseases or parasites, or to increase digestion of food in the gut (Gregory 1984; Beck 1995; D. Duvall, personal communication). In this particular situation, when the rattlesnakes emerge to bask, they must venture out onto trails or onto ruin walls to find sunny locations, and are consequently seen by park staff or visitors.

Other observations of rattlesnakes hibernating in man-made structures include a *Crotalus molossus* at the Montezuma Well section of Montezuma Castle NM hibernating in a cliff dwelling, a *C. atrox* at Montezuma Castle NM proper hibernating in an abandoned root cellar-like excavation, a *C. atrox* at Tonto National Monument, Arizona, hibernating under a visitor trail (personal observation), and timber rattlesnakes (*C. horridus*) in New York State hibernating in pipes in a concrete retaining wall above a road (Stechert and Michell 1995). These observations suggest that man-made rock and concrete structures may provide thermal environments for hibernation that are similar in quality to natural sites, and that rattlesnake hibernation site selection is labile enough to allow the snakes to utilize non-natural structures.

The Tuzigoot population showed high fidelity to hibernation sites, a common pattern in many other rattlesnake species (Means 1977; Brown 1992; Sexton et al. 1992; Repp 1998; Hardy and Greene 1999a; M. Goode, personal communication). The location of several dens in large rock outcrops with south to southeast exposures is also typical of rattlesnakes (reviewed by Sexton et al. 1992). The use of burrows under limestone boulders as hibernation sites was not seen by Nowak and van Riper (1999) for *C. atrox* at Montezuma Castle NM in the eastern Verde Valley, but was noted for southern Arizona populations by Repp (1998).

The duration of the active period for rattlesnakes at Tuzigoot is generally 4–6 weeks shorter than that reported for rattlesnakes in southern Arizona, with the exception of southern *Crotalus molossus* populations (Beaupre 1995; Beck 1995; Repp 1998; Hardy and Greene 1999a). Short-distance dispersal movements from hibernation sites and basking in these locations for several weeks before permanent dispersal has been noted for other higher-elevation and/or higher-latitude populations of rattlesnakes, primarily *C. horridus* and prairie rattlesnakes, *C. viridis* (Duvall et al. 1985; Brown 1992; Martin 1992). This strategy would be useful in areas such as the Verde Valley, where sudden freezing temperatures associated with early spring storms may occur for several weeks after the typical spring emergence date, forcing snakes to either seek deep shelter sites or return to their hibernation sites.

Activity Range and Movement Patterns

In general, rattlesnakes do not have large activity or home ranges (Macartney et al. 1988), and this was also true of the Tuzigoot population, where rattlesnakes did not disperse more than 1 km in a straight line from their hibernacula. The average range size was 12.5 ha, which is comparable to *C. atrox* at Montezuma Castle NM (Nowak and van Riper 1999); the ranges there averaged 11 ha in 1995 and 17 ha in 1996. However, Beck (1995) found that the range size of *C. atrox* in southeastern Arizona averaged 5.4 ha over 6–15 month periods between 1988 and 1991. The great deal of overlap between years in the location and even shape of the activity ranges of Tuzigoot rattlesnakes is typical (Gregory et al. 1987; Reinert and Zappalorti 1988b; Secor 1994; Nowak and van Riper 1999). Adult rattlesnakes tend to remain in the same area year after year, provided that the area remains suitable for meeting hibernation and foraging needs.

Movement patterns of rattlesnakes at Tuzigoot are comparable to those noted for other *Crotalus atrox* populations. The total distance moved per year appears to be less than these other populations, although the relatively fewer locations of the Tuzigoot snakes per year likely accounts for some of this difference. Nowak and van Riper (1999) found that *C. atrox* moved an average of 1.7 km in 3 months during late summer at Montezuma Castle NM. Beck (1995) found that *C. atrox* in southeastern Arizona traveled an average of 12.9 km per year. In addition to small sample size, the proximity of the monument to riparian habitats and their possibly richer prey resources (e.g. Ellison and van Riper 1998) may be another explanation for the relatively shorter distance traveled by the Tuzigoot rattlesnakes every year.

The number of new locations used by *C. atrox* at Tuzigoot was the same as that used by rattlesnakes at Montezuma Castle NM in 1995 and 1996: an average 26 percent of sites were reused (Nowak and van Riper 1999). As in the Montezuma Castle NM study, different individual rattlesnakes used the same foraging or retreat sites at different times of the year (i.e., two rattlesnakes used the same *Neotoma* midden). Individual rattlesnakes in both studies also exhibited fidelity to general activity (foraging) areas as well as specific foraging sites year after year, returning repeatedly to the same preferred locations. Site fidelity during the active season has also been found for *C. atrox* by Beaupre (1995), for *C. horridus* by Sealy (1997), and for other pitvipers, including European adders (*Vipera berus*; Viitanen 1967). This phenomenon is usually explained by optimal sites being refound using non-spatially mapped olfactory, chemosensory, visual, thermal, or other environmental cues (Landreth 1973; Newcomer et al. 1974; Graves et al. 1986; Lawson 1991; Halpern 1992; Sexton et al. 1992; Macmillan 1995). Alternatively, Reinert (1993) has hypothesized that rattlesnakes learn preferred foraging regions or habitats while young and then return to these favored sites or to similar habitats throughout their adult lives. The precise homing behaviors exhibited by translocated *C. atrox* (Nowak et al. 2002) lend weight to this alternative hypothesis.

Habitat Use

Reinert (1993) made a distinction between habitat *correlation*, which he defined as "a relationship between the spatial distribution of an organism and specific environmental factors" and habitat *selection*, which "additionally seeks to determine the proximal environmental cues used by animals in the process of occupying specific habitats, and the underlying ultimate factors responsible for such habitat properties." Reinert further defined habitat *utilization* as the spatial pattern of habitat correlation for any species; this definition most closely matches my term *habitat use* (proportion of use of different habitats by the rattlesnakes around Tuzigoot).

Overall, the rattlesnakes used habitats in proportion to availability, but there was a trend toward avoidance of mesquite bosque and increased use of open shrub-grassland habitats. This may be related to availability of favorable thermal microclimates in the two types of habitats. Reinert (1993) has noted the importance of open habitats in supplying warm microclimates for basking and foraging by rattlesnakes. The mesquite bosque habitat at Tuzigoot has a dense annual grass (*Bromus* spp.) understory with a closed canopy of mesquite trees (*Prosopis velutina*), and thus may be less suitable than the more open over- and understory of the shrub-grassland habitats.

Overall, the habitats used for summer foraging by the Tuzigoot population seemed to be determined on the basis of individual snake preference. Some rattlesnakes were located primarily in riparian habitats during the active season, whereas others foraged in both upland and riparian habitats. At least one snake never left the monument (although she was only followed for 4 months).

There did not appear to be sex differences

in the types of habitats used; this observation has also been noted for male and nongravid female *Crotalus horridus* by Reinert and Zappalorti (1988b) and *C. viridis* by Duvall et al. (1985). As well, given apparent individual preferences, the presence or lack of differences between the sexes in habitat use may not be ecologically significant. Beck (1995) found that *C. atrox* used more creosotebush flat habitat during the summer foraging period than would be expected based on the amount of this habitat available (suggesting a preference for this habitat type in that population). In his study, *C. atrox* hibernated in rocky slope areas, and there was no riparian habitat available to the snakes.

A unique situation for Tuzigoot rattlesnakes during the active season was the availability of Tavasci Marsh, which contains a potentially rich prey base. A study of the comparative distribution and abundance of small mammals at nearby Montezuma Castle NM in 1993 and 1994 demonstrated that small mammals were more abundant in floodplain areas than in the surrounding uplands (Ellison and van Riper 1998). Although snakes did not spend disproportionate amounts of time in the marsh habitat, the location of several rattlesnakes within the marsh, perhaps while engaged in foraging, is unusual. Few other Southwestern rattlesnake populations have been documented using such a habitat; unpublished records include *Crotalus atrox* documented in a marsh in the San Bernadino National Wildlife Refuge (C. Schwalbe, personal communication). Viitanen (1967) noted the importance of wetlands for summer foraging in *V. berus*, as did Seigel (1986) for massasaguas (*Sistrurus catenatus*) in the southeastern United States. To determine a habitat selection relationship between the abundance of prey and the presence of rattlesnakes resulting in differential fitness, more information is needed on the diet and growth rates of this population of rattlesnakes. In particular, the unusually large females in the population suggest that energetic studies of diet and growth rates may be informative (see also Taylor et al. 2005).

MANAGEMENT IMPLICATIONS

One important finding of this study was that rattlesnakes at Tuzigoot showed high site fidelity, both to hibernacula and to active (foraging) locations. Snakes in this population did not disperse more than 1 km in a straight line from their hibernacula during the summer foraging months. In addition to riparian areas around the monument, rattlesnakes were both hibernating and foraging in the pueblo ruins and other man-made dwellings at Tuzigoot.

It is significant for management of rattlesnakes at Tuzigoot that we found more than half of the locations of telemetered rattlesnakes during the active seasons off the monument, on adjacent U.S. Forest Service, Arizona State Park, or private land. Given that snakes spent most of the summer off the monument, this begs the question of why rattlesnakes were encountered on the monument at all during the active season. One obvious explanation is migratory movement of the rattlesnakes to and from suitable hibernation sites within the visitor use areas. This phenomenon has also been documented in other national park areas. Nowak and van Riper (1999) found that nine *C. atrox* were hibernating in rock outcrops less than 50 m from the visitor trails, and some snakes dispersed across trails during migrations between foraging sites and den sites. Graham (1991) found that rattlesnakes sighted near the visitor center and staff housing areas at Natural Bridges National Monument in Utah were also primarily migrating through these areas to and from hibernacula near the visitor center. I have found similar patterns (unpublished data) for rattlesnakes at Tonto NM. The predictable presence of rattlesnakes during the mating season could also attract other rattlesnakes.

Another reason for the presence of rattlesnakes around human use areas in national parks during the summer active period may be foraging opportunities. Artificially increased prey abundance and the presence of artificial refuges (i.e., under buildings) were cited as potentially important influences of rattlesnake distribution at the Arizona Sonoran Desert Museum in southern Arizona by Perry-Richardson and Ivanyi (1992), and were documented for some rattlesnakes at Natural Bridges National Monument by Graham (1991). Similar patterns of increased prey abundance may exist for rattlesnakes in developed areas at Tonto National Monument (unpublished data), and rattlesnakes on golf courses in the Tucson area are known to use golf course structures as refugia (M. Goode, unpublished data). Increased abundance of prey in developed areas might be due to the presence of nonnative vegetation, which generally produces more biomass and seeds than native species, to garbage left by visitors or put out by staff, to bird feeding and watering, and to the presence of refugia in the form of wood, lumber, brush piles, thick vegetation, and loose soil for burrowing.

Regardless of where rattlesnakes are located on the monument, they are not likely to be seen by humans. Nowak and van Riper (1999) found that telemetered rattlesnake detectability was low for Montezuma Castle NM staff and visitors, even though snakes were occasionally located less than 5 m from the heavily used main visitor trail. In several other studies of free-ranging rattlesnakes, it has been demonstrated that rattlesnakes seldom reveal their presence, even if approached closely. The reason is thought to be that rattlesnakes use procrypsis as the primary mechanism to avoid detection by predators, including humans (Klauber 1972; Duvall et al. 1985; Reinert and Zappalorti 1988b; May et al. 1996; Parent and Weatherhead 2000). Additionally, rattlesnakes foraging frequently near or in concentrated human use areas may be habituated to the presence of people, and perhaps less likely to display defensive behaviors. I found that rattlesnakes at Tuzigoot tended to rely heavily on procrypsis, rarely rattled when approached, and did not strike unless handled. Thus, unless inadvertently stepped on or handled, rattlesnakes in national parks pose an overstated threat to public safety (see also Curry et al. 1989; Dart et al. 1992; Hare and McNally 1997), even though rattle snakes may be repeat "nuisances" over the course of their lives, given their site fidelity.

One potential solution to dealing with repeat nuisance rattlesnakes is long-distance translocation of the offending animal. However, based on the negative impacts seen by Nowak et al. (2002) and Reinert and Rupert (1999), long-distance translocation of problem rattlesnakes is not recommended. To discourage wandering behaviors, no snake should be relocated farther than 1 km (i.e., not moved out of its potential activity range; Nowak et al. 2002). I recommend short-distance relocation of less than 50 m for individual nuisance rattlesnakes into suitable habitat. In this scenario, snakes are simply removed from trails or from proximity to buildings into adequate vegetative cover, where they are likely to continue their travels without again encountering humans. Translocation should always occur away from heavily used roads into habitats similar to that from which the rattlesnake was displaced, in the direction of the snake's likely travel path if known. During migration periods snakes should be moved toward potential hibernacula after early October and away from these sites after early April. Basking rattlesnakes encountered around the ruins complex in the winter should not be relocated from this area, as they will be at risk of succumbing to freezing temperatures before they find suitable hibernation sites (Johnson 1996, unpublished Internet communication; Reinert and Rupert 1999). Rather, snakes encountered in the winter should be placed either in a part of the ruins inaccessible to tourists, or into a likely crevice, drainage pipe, or burrow. Interestingly, Sealy (1997, 2002) demonstrated that *C. horridus* in a North Carolina state park did not recur as nuisance animals in the same location once captured; however this was not the pattern for northern Pacific rattlesnakes (*C. oreganus*) in British Columbia (J. Brown, unpublished data).

A long-term solution to rattlesnake-human conflicts in developed areas could be to make microhabitats less attractive to potential rattlesnake prey and, therefore, less attractive to rattlesnakes during the foraging season. To decrease foraging of nuisance snakes around ruins and visitor buildings during the summer, garbage left by visitors should be controlled, as should bird feeding and dripping or standing water, all of which may be attractive to rodent and bird prey. Other tactics for controlling rodents around human dwellings could include continuous trapping and removal, stacking all wood off the ground, removing scrap material piles, removing and stopping watering of non-native vegetation, and sealing off rodent access to buildings.

Near suitable hibernacula at Tuzigoot, rattlesnakes could be physically rerouted around the ruins and trails by installing and maintaining snake-proof fencing, provided the snakes are not cut off from all potential den sites (limestone boulders) on the hilltop or forced onto roadways. People could also be rerouted from migratory paths of snakes by creating elevated boardwalks in place of the existing trails (the snakes might not be as likely to crawl onto such elevated surfaces due to their exposed nature). Since the end of this study, the openings of drainage pipes exiting the ruins have been covered with fine-gauge wire screening, in hopes of preventing snakes and rodents from using these pipes. However, snakes are still being seen on the trails near former locations, so rodent burrows may also be a point of access for the snakes. At least one interpretive sign has been posted at the beginning of the trail advising visitors that they may encounter rattlesnakes on trails in ruins.

Due to their dependence on a small national monument next to a restored marsh habitat, the rattlesnakes of Tuzigoot represent a unique population of snakes in Arizona. Loss of suitable habitat outside the monument, especially in the face of encroaching urban development, must be considered when developing management strategies for rattlesnakes that may intentionally venture outside the monument's boundaries. In addition to further research on the dietary ecology and energetics of this population, Tuzigoot's rattlesnakes should continue to be monitored to determine potential impacts from the adjacent Verde Valley Ranch golf course and residential development. Specific efforts should be made to reach residents of this development with interpretive presentations and printed material about rattlesnakes, to increase the survival rates of Tuzigoot rattlesnakes that venture onto private property. A service to remove nuisance rattlesnakes on private property to adjacent undeveloped park or marsh land could be coordinated among state and federal land managers and researchers.

ACKNOWLEDGMENTS

Many people and organizations contributed to the success of this project. This research would not have been possible without the dedication and inspiration of Manuel Santana-Bendix. Field technicians included Matthew Spille, Jon Bortle, and Justin Schofer. The study was funded by the Southern Arizona Group office of the National Park Service, Kathy Davis, and the Southwest Parks and Monuments Association. Permission to conduct the project on U.S. Forest Service land was granted by Ken Anderson and Tom Britt. Tim Graham and Canyonlands National Park Resource Management provided several radio transmitters. Bruce Weber, Mike Walker, and the staff of the Verde Veterinary Hospital (Cottonwood, Arizona) performed the implantation surgeries on the snakes. Many staff members of Tuzigoot National Monument contributed to this project, in particular Glen Henderson (former superintendent), John Reid, and Jose Castillo. Many individuals provided helpful comments or field observations, including S. Beaupre, D. Beck, J. Brown, C. Drost, D. Duvall, M. Goode, J. Grahame, H. Greene, D. Hardy, T. Hare, A. Holycross, T. Graham, B. Johnson, J. McNally, J. Oleile, C. Parent, J. Perry, B. Porchuk, H. Reinert, R. Repp, P. Rosen, J. Sealy, D. Swann, E. Taylor, C. van Riper III, and S. Walker. This paper benefited from thoughtful reviews by Matt Goode, Carl Lieb, and Cecil Schwalbe.

LITERATURE CITED

Beaupre, S. J. 1995. Sexual size dimorphism in the western diamondback rattlesnake (*Crotalus atrox*): Integrating natural history, behavior, and physiology. Sonoran Herpetologist 8: 112–121.

Beck, D. D. 1995. Ecology and energetics of three sympatric rattlesnake species in the Sonoran desert. Journal of Herpetology 29: 211–223.

Brown, T. S., and H. B. Lillywhite. 1992. Autecology of the Mojave Desert sidewinder, *Crotalus cerastes*, at Kelso Dunes, Mojave Desert, California, USA. In Biology of the Pitvipers, edited by J. A. Campbell and E. D. Brodie, pp. 279–308. Selva Press, Tyler TX.

Brown, W. S. 1992. Emergence, ingress, and seasonal captures at dens of northern timber rattlesnakes, *Crotalus horridus*. In Biology of the Pitvipers, edited by J. A. Campbell and E. D. Brodie, pp. 251–258. Selva Press, Tyler TX.

Brown, W. S., and F. M. Maclean. 1983. Conspecific scent trailing by newborn timber rattlesnakes, *Crotalus horridus*. Herpetologica 39: 430–436.

Butler, J. A., T. W. Hull, and R. Franz. 1995. Neonate aggregations and maternal attendance of young in the eastern diamondback rattlesnake, *Crotalus adamanteus*. Copeia 1995: 196–198.

Chiszar, D., J. Perelman, H. M. Smith, and D. Duvall. 1992. 'Shouldering' in prairie rattlesnakes: A new hypothesis. Bulletin of the Maryland Herpetological Society 28(3): 69–76.

Curry, S. C., D. Horning, P. Brady, R. Requa, D. B. Kunkel, and M. V. Vance. 1989. The legitimacy of rattlesnake bites in central Arizona. Annals of Emergency Medicine 18: 658–663.

Dart, R. C., J. T. McNally, D. W. Spaite, and R. Gustafson. 1992. The sequelae of pitviper poisoning in the United States. In Biology of the Pitvipers, edited by J. A. Campbell and E. D. Brodie, pp. 395–404. Selva Press, Tyler TX.

Duvall, D., M. B. King, and K. J. Gutzwiller. 1985. Behavioral ecology and ethology of the prairie rattlesnake. National Geographic Research (Winter 1985): 80–111.

Ellison, L. E., and C. van Riper III. 1998. A comparison of small-mammal communities in a desert riparian floodplain. Journal of Mammalogy 79(3): 972–985.

Fagerstone, K. A., and B. E. Johns. 1987. Transponders as permanent identification markers for domestic ferrets, black-footed ferrets, and other wildlife. Journal of Wildlife Management 51: 294–297.

Fitch, H. S., and G. R. Pisani. 1993. Life history traits of the western diamondback rattlesnake (*Crotalus atrox*) studied from roundup samples in Oklahoma. Occasional Papers of the Museum of Natural History No. 156, pp. 1–24. University of Kansas, Lawrence.

Ford, N. B., and G. M. Burghardt. 1993. Perceptual mechanisms and the behavioral ecology of snakes. In Snakes: Ecology and Behavior, edited by R. A. Seigel and J. T. Collins. McGraw Hill, New York.

Goode, M. J., J. Smith, M. Amarello, K. Setser, and N. Favour. 2003. Effects of urban development on herpetofauna. Final report in fulfillment of Grant U02009 to Nongame and Endangered Wildlife Program. Arizona Game and Fish Department, Phoenix.

Graham, T. 1991. Western rattlesnake ecology at Natural Bridges NM. Utah Park Science 11, Fall 1991.

Graves, B. M., D. Duvall, M. B. King, S. L. Lindstedt, and W. A. Gern. 1986. Initial den location by neonatal prairie rattlesnakes: Functions, causes, and natural history in chemical ecology. In Chemical Signals in Vertebrates 4, edited by D. Duvall, D. Muller-Schwarze, and R. M. Silverstein. Plenum Press, New York.

Gregory, P. T. 1984. Communal denning in snakes. In Vertebrate Ecology and Systematics—A Tribute to Henry S. Fitch, edited by R. A. Seigel, L. E. Hunt, J. L. Knight, L. Malaret, and N. L. Zuschlag, pp. 57–75. Museum of Natural History. University of Kansas, Lawrence.

Gregory, P. T., J. M. Macartney, and K. W. Larsen. 1987. Spatial patterns and movements. In Snakes: Ecology and Evolutionary Biology, edited by R. A. Seigel, J. T. Collins, and S. S. Novak. Macmillan, New York.

Halpern, M. 1992. Nasal chemical senses in reptiles: Structure and function. In Biology of the Reptilia Volume 18: Physiology E/Hormones, Brain, and Behavior, edited by C. Gans and D. Crews. University of Chicago Press, Chicago.

Hardy, D. L., Sr., and H. W. Greene. 1999a. Borderland blacktails: Radiotelemetry, natural history, and living with venomous snakes. In Toward Integrated Research, Land Management, and Ecosystem Protection in the Malpais Borderlands, edited by G. J. Gottfried, L. G. Eskew, C. G. Curtin, and C. B. Edminster, pp. 117–121. Conference Summary RMRS-P-10, USDA Forest Service, Rocky Mountain Research Station, Fort Collins CO.

Hardy, D. L., Sr., and H. W. Greene. 1999b. Surgery on rattlesnakes in the field for implantation of transmitters. Sonoran Herpetologist 12: 26–28.

Hare, T. A., and J. T. McNally. 1997. Evaluation of a rattlesnake relocation program in the Tucson, Arizona area. Sonoran Herpetologist 10: 26–31.

King, M. B., and D. Duvall. 1990. Prairie rattlesnake seasonal migrations: Episodes of movement, vernal foraging, and sex differences. Animal Behavior 39: 924–935.

Klauber, L. M. 1972. Rattlesnakes: Their Habits, Life Histories, and Influence on Mankind. 2nd ed. Vols. I and II. University of California Press, Berkeley.

Landreth, H. F. 1973. Orientation and behavior of the rattlesnake, *Crotalus atrox*. Copeia 1973: 26–31.

Lawson, P. A. 1991. Movement patterns and orientation mechanisms in garter snakes. Unpublished Ph.D. dissertation, University of Victoria, Canada.

Lowe, C. H., C. R. Schwalbe, and T. B. Johnson. 1986. The venomous reptiles of Arizona. Arizona Game and Fish Department, Phoenix.

Macartney, J. M., P. T. Gregory, and K. W. Larsen. 1988. A tabular survey of data on movements and home ranges of snakes. Journal of Herpetology 22: 61–73.

Macmillan, S. 1995. Restoration of an extirpated red-sided garter snake *(Thamnophis sirtalis parietalis)* population in the interlake region of Manitoba, Canada. Biological Conservation 72: 13–16.

Martin, W. H. 1992. Phenology of the timber rattlesnake (*Crotalus horridus*) in an unglaciated section of the Appalachian mountains. In Biology of the Pitvipers, edited by J. A. Campbell and E. D. Brodie, pp. 259–278. Selva Press, Tyler TX.

May, P. G., T. M. Farrell, S. T. Heulett, M. A. Pilgrim, L. A. Bishop, D. J. Spence, A. M. Rabatsaky, M. G. Campbell, A. D. Aycrigg, and W. E. Richardson II. 1996. Seasonal abundance and activity of a rattlesnake (*Sistrurus miliarius barbouri*) in central Florida. Copeia 1996: 389–401.

Means, D. B. 1977. Radio-tracking the eastern diamondback rattlesnake. National Geographic Society Research Reports (1977): 529–536.

Minckley, W. L., and D. E. Brown. 1994. Tropical-subtropical wetlands. In Biotic Communities of the Southwestern United States and Mexico, edited by D. E. Brown, pp. 268–287. University of Utah Press, Salt Lake City.

National Park Service. 1980. United States Codes: Department of the Interior, National Park Service Organic Act. Title 16, Subchapter 1. U.S. Government Printing Office, Washington D.C.

National Park Service. 1991. NPS-77: Natural Resources Management Guidelines. U.S. Government Printing Office, Washington D.C.

Newcomer, R. T., D. H. Taylor, and S. I. Guttman. 1974. Celestial orientation in two species of water snakes (*Natrix sipedon* and *Regina septemvittata*). Herpetologica 30: 194–200.

Nowak, E. M., and C. van Riper III. 1999. Effects and effectiveness of rattlesnake relocation at Montezuma Castle National Monument. USGS Technical Report USGSFRESC/COPL/1999/17. USGS Forest and Rangeland Ecosystem Science Center, Colorado Plateau Field Station, Flagstaff AZ.

Nowak, E. M., T. Hare, and J. McNally. 2002. Management of "nuisance" vipers: Effects of translocation on western diamondback rattlesnakes (*Crotalus atrox*). In Biology of the Vipers, edited by G. W. Schuett, M. Höggren, M. E. Douglas, and H. W. Greene, pp. 533–560. Eagle Mountain Publishing, Eagle Mountain UT.

Parent, C., and P. J. Weatherhead. 2000. Behavioral and life history responses of eastern massasauga rattlesnakes (*Sistrurus catenatus catenatus*) to human disturbance. Oecologia 125: 170–178.

Perry-Richardson, J., and C. Ivanyi. 1992. Preliminary analysis of a study on the free-ranging rattlesnakes on Arizona-Sonora Desert Museum grounds. AAZPA 1992 Regional Proceedings.

Price, A. H. 1988. Observations on maternal behavior and neonate aggregation in the western diamondback rattlesnake, *Crotalus atrox*. The Southwestern Naturalist 33: 370–373.

Prival, D. B., M. J. Goode, D. E. Swann, C. R. Schwalbe, and M. J. Schroff. 2002. Natural history of a northern population of twin-spotted rattlesnakes, *Crotalus pricei*. Journal of Herpetology 36 (4): 598–607.

Reinert, H. K. 1992. Radiotelemetric field studies of pitvipers: Data acquisition and analysis. In Biology of the Pitvipers, edited by J. A. Campbell and E. D. Brodie, pp. 185–198. Selva Press, Tyler TX.

Reinert, H. K. 1993. Habitat selection in snakes. In Snakes: Ecology and Behavior, edited by R. A. Seigel and J. T. Collins. Mc-Graw Hill, New York.

Reinert, H. K., and R. R. Rupert. 1999. Impacts of translocation on behavior and survival of timber rattlesnakes, *Crotalus horridus*. Journal of Herpetology 33: 45–61.

Reinert, H. K., and R. T. Zappalorti. 1988a. Field observations of the association of adult and neonatal timber rattlesnakes, *Crotalus horridus*, with possible evidence for conspecific trailing. Copeia 1988: 1057–1059.

Reinert, H. K., and R. T. Zappalorti. 1988b. Timber rattlesnakes (*Crotalus horridus*) of the Pine Barrens: Their movement patterns and habitat preference. Copeia 1988: 964–978.

Repp, R. 1998. Wintertime observations on five species of reptiles in the Tucson area: Sheltersite selections, fidelity to sheltersites, and notes on behavior. Bulletin of the Chicago Herpetological Society 33: 49–56.

Sealy, J. 1997. Short-distance translocations of timber rattlesnakes in a North Carolina state park: A successful conservation and management program. Sonoran Herpetologist 10: 94–99.

Sealy, J. B. 2002. Ecology and behavior of the timber rattlesnakes (*Crotalus horridus*) in the upper piedmont of North Carolina: Identified threats and conservation recommendations. In Biology of the Vipers, edited by G. W. Schuett, M. Höggren, M. E. Douglas, and H. W. Greene,, pp. 561–578. Eagle Mountain Publishing, Eagle Mountain UT.

Secor, S. 1994. Ecological significance of movements and activity range for the sidewinder, *Crotalus cerastes*. Copeia 1994: 631–645.

Sexton, O. J., P. Jacobson, and J. E. Bramble. 1992. Geographic variation in some activities associated with hibernation in neartic pitvipers. In Biology of the Pitvipers, edited by J. A. Campbell and E. D. Brodie, pp. 337–346. Selva Press, Tyler TX.

Shine, R. 1993. Sexual dimorphism in snakes. In Snakes: Ecology and Behavior, edited by R. A. Seigel and J. T. Collins. Mc-Graw Hill, New York.

Seigel, R. A. 1986. Ecology and conservation of an endangered rattlesnake, *Sistrurus catenatus*, in Missouri, USA. Biological Conservation 35: 333–346.

Sokal, R. R., and F. J. Rohlf. 1981. Biometry: The principles and practice of statistics in biological research. 2nd ed. W. H. Freeman, New York.

Stechert, R., and K. Michell. 1995. Current status, threats, and nuisance rattlesnake response programs in New York State. Paper given at the 38th Annual Meeting of the Society for Study of Amphibians and Reptiles, Appalachian State University, Boone NC.

Taylor, E. N., D. F. Denardo, and M. A. Malawy. 2004. A comparison between point- and semicontinuous sampling for assessing body temperature in a free-ranging ectotherm. Journal of Thermal Biology 29: 91–96.

Taylor, E. N., M. A. Malawy, D. M. Browning, S. V. Lemar, and D. F. DeNardo. 2005. Effects of food supplementation on the reproductive ecology of female western diamond-backed rattlesnakes (*Crotalus atrox*). Oecologia. In press.

Turner, R. M., and D. E. Brown. 1994. Tropical-subtropical desertland. In Biotic Communities of the Southwestern United States and Mexico, edited by D. E. Brown, pp. 180–221. University of Utah Press, Salt Lake City.

Viitanen, P. 1967. Hibernation and seasonal movements of the viper, *Vipera berus berus* (L.), in southern Finland. Annals of Zoology Fenn. (4) 1967: 473–546.

Weldon, P. J., R. Ortiz, and T. R. Sharp. 1992. The chemical ecology of crotaline snakes. In Biology of the Pitvipers, edited by J. A. Campbell and E. D. Brodie, pp. 309–320. Selva Press, Tyler TX.

White, G. C., and R. A. Garrott. 1990. Analysis of wildlife radio-tracking data. Academic Press, San Diego CA.

BEETLES OF SALT CREEK CANYON, CANYONLANDS NATIONAL PARK, UTAH

Louis L. Pech, Timothy B. Graham, Holly Demark, and Jennifer Mathis

In arid and semi-arid ecosystems, riparian zones possess greater biodiversity than surrounding uplands, with as much as 90 percent of species either found in or using riparian areas (Carothers 1977; Ohmart and Anderson 1982). Due to the availability of water, riparian zones are also the most productive regions in arid and semi-arid ecosystems (Lowe 1964; Lowe 1989). Increased structural complexity and increased productivity provide a greater diversity of niches to exploit (Crawford 1981; Crawford 1986; Whitford 1986).

Riparian zones in arid and semi-arid environments that support diverse biological communities are also subject to many human activities such as livestock grazing, water diversion, and channel alteration, as well as hiking, bicycling, and the use of off-road vehicles (Kondolf and Curry 1984; Mitchell and Woodward 1993; Green and Kauffman 1995; Briggs 1996). Despite the importance of these riparian zones, little is known about the effects of off-road vehicles (ORVs). It is possible that ORV effects are similar to trampling by livestock, which causes changes in the three-dimensional structure of habitats by crushing plants and compacting soil; community structure then changes over time due to the increased relative abundance of plants that are resistant or insensitive to disturbance (Kauffman et al. 1983; Kauffman and Krueger 1984; Fleischner 1994; Niwranski et al. 2002). Insect communities also change in response to these changes in habitat structure and the abundance and diversity of host plants (Capinera and Sechrist 1982; Jepson-Innes and Bock 1989; Quinn and Walgenbach 1990; Niwranski et al. 2002). Cessation of ORV use would presumably result in changes in the three-dimensional architecture and community structure of riparian zones; however, how a desert riparian insect community would change upon removal of ORVs is currently unknown.

Salt Creek Canyon in Canyonlands National Park (CANY) has for many years been subject to ORV use. Salt Creek flows from the Abajo Mountains into the Colorado River in the Needles District of CANY. It is especially significant because, other than the Colorado and Green Rivers, it is the largest perennial or semiperennial stream in CANY and one of the biologically most productive areas in the park. Because of its ecological significance to CANY, much of Salt Creek Canyon was closed to vehicles in 1998. As a result, the canyon is now divided into three regions: No Road (NR) where vehicle use ended in about 1964, Closed Road (CL) where vehicle use ended in 1998, and Road Open (RO) where limited vehicle use continues. These three regions provide an opportunity to examine the effects of ORVs on the community structure of a desert riparian ecosystem, and to evaluate how the community changes when ORV use ceases.

To this end, since April of 2000 Salt Creek Canyon has been the focus of a project surveying and monitoring invertebrates and amphibians in the canyon. The project's goal is to determine current conditions in each of the three sections of Salt Creek, including differences between sections, and to determine how conditions may be changing in the CL region now that vehicle use has ended (Graham 2002). In addition, the National Park Service began vegetation and stream channel monitoring in 1998. Combined, these two studies can provide the basic science needed to establish a monitoring program for the canyon.

Along with the need to monitor the ecological effects of human activities in national parks such as CANY there is also a need for descriptive work on the diversity and natural history of the flora and fauna of the Colorado Plateau (Stohlgren and Quinn 1992; St. Clair et al. 1994; Stohlgren et al. 1995). Although Mitchell and Woodward (1993) have examined some of the aquatic invertebrates of Salt Creek Canyon, there has been no extensive survey of the terrestrial insect community. Work in Salt Creek thus provides an opportunity to develop a description of its arthropod communities and to track changes over time, which will yield an extensive description of insect natural history and taxonomy in a desert riparian ecosystem. This project is large and ongoing; here we present a preliminary taxonomic survey and description of the beetle community in Salt Creek Canyon in 2000 as a prelude to more detailed studies on the ecological effects of ORVs.

MATERIALS AND METHODS

Salt Creek, in Canyonlands National Park, is located in southeastern Utah on the Colorado Plateau. The study area ranges in elevation from about 1510 to 1660 m above sea level. Temperatures range from an average low of –10.1° C in January to an average high of 34.9° C in July, with recorded extremes of –26.7° C and 41.7° C. Average annual precipitation is 216 mm with maximum precipitation typically occurring in August, April, and October (Ashcroft et al. 1992).

The study area (Figure 1) lies in the middle section of Salt Creek, which is a relatively low-grade stream beyond the footslope of the Abajos, and above Lower Jump, a 60 m fall. Salt Creek flows above ground intermittently through this section, with greater flows during winter and spring, diminishing to deep pools with long dry sections during summer, especially in periods of drought. The canyon's character changes somewhat about halfway through the study area, near CL5 (Figure 1). Below CL5 the canyon is fairly open, with broad alluvial benches between the rock cliffs and the stream channel; the cliffs do not exert as much influence on the riparian zone and stream channel below CL5. Above CL5 the canyon narrows, with narrower alluvial benches, and a greater proportion of the vegetation is associated with riparian ecosystems. The cliffs are closer to the stream channel and exclude the sun more than in the stream section below CL5. There are wetter and drier sections of stream channel both up and downstream from CL5, with comparable small-scale vegetation communities; similar sites in the three management zones were selected for study, based on the location of vegetation and stream channel morphology study sites established by the National Park Service. All samples collected in April have been processed, as well as those from RO3, CL1, and NR2 for June through October of 2000.

Trapping Methods

We used three trapping methods at each site (New 1995). Window traps for flying insects consisted of a 0.5 x 0.4 m sheet of Plexiglas suspended above a tub of soapy water, with the entire trap suspended in a tree or shrub at various heights and locations throughout the trapping site. Five windows were used at each site.

Bowl traps were used to collect invertebrates attracted to different colors: red, blue, yellow, and white. Each trap consisted of four Styrofoam bowls containing soapy water and the appropriate food coloring. Five traps were set in a transect across the riparian zone.

Pitfall traps were used to sample ground-active invertebrates. They were arranged in three rows, parallel to the creek and 2 m apart. Each row consisted of five pitfalls, spaced 2 m apart. Between each pitfall in a row, a plastic drift fence 10 cm high was erected to direct wandering invertebrates into a trap, which consisted of a plastic jar 12 cm in diameter and 16 cm deep buried flush

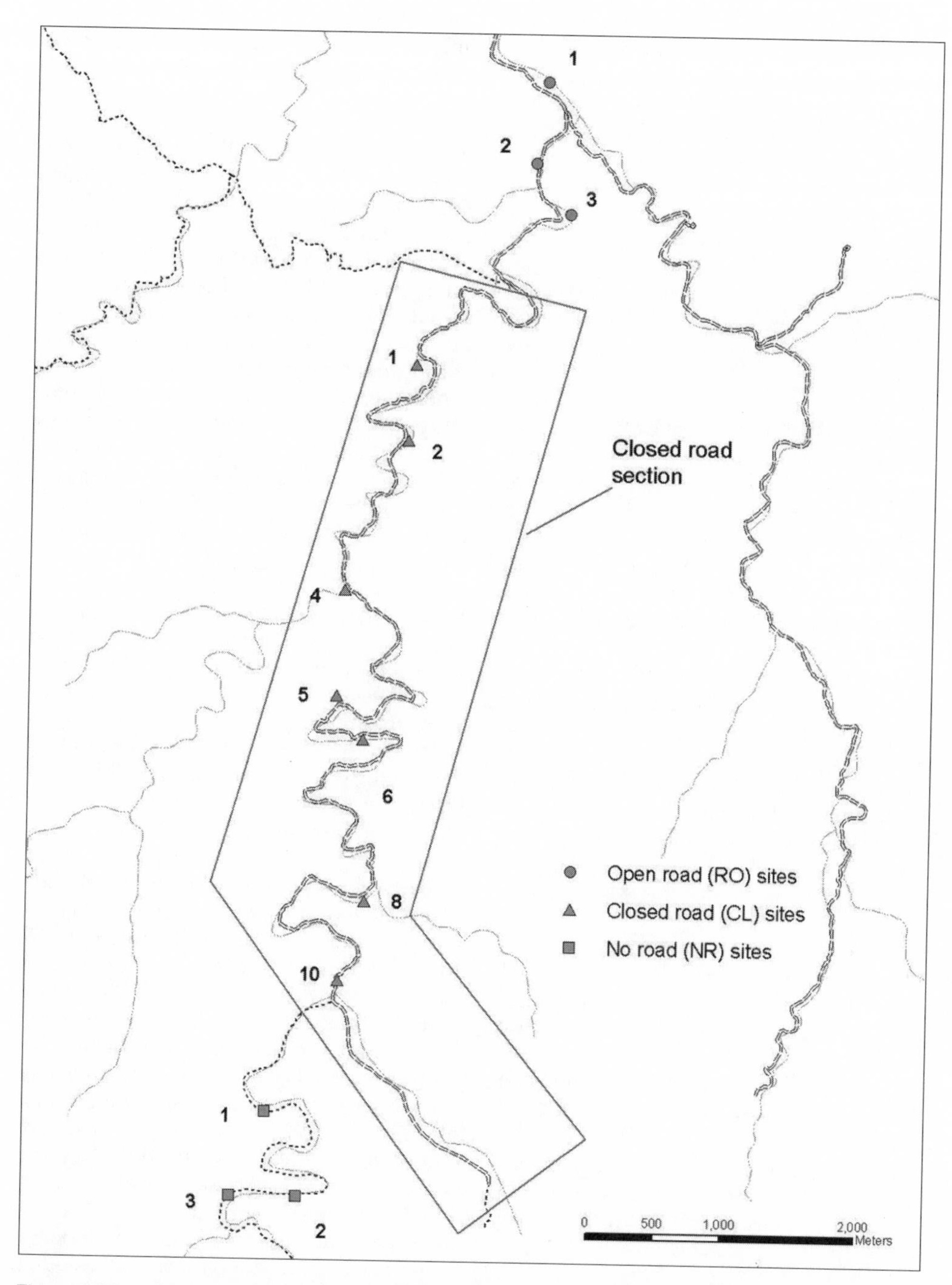

Figure 1. Map of a portion of Salt Creek, showing study site locations and closed portion of road. RO = Road Open (limited vehicle use continues), CL = Closed Road (vehicle use ended in 1998), and NR = No Road (vehicle use ended in 1964). Approximate section lengths RO 3.5 miles, CL 8.5 miles, NR 3.5 miles.

to the ground. Inside each jar was a steel can containing soapy water and a funnel that directed falling insects into the metal can. Funnels were 15 cm in maximum diameter and their upper surface was flush with the ground when inserted into the trap.

In total there were 40 traps at a site, kept open for 4 days and collected each day for a total of 160 samples per site per sampling period. The collection effort in 2000 consisted of 42 sampling periods, which resulted in 6720 samples. The beetles from these samples provide the data for this paper.

Specimen Collection

Trapping sites were open in April, June, July, August, September, and October of 2000. The number of sites sampled in a region for each month were as follows: April: NR = 3, CL = 3, RO = 3; June: NR = 2, CL = 6, RO = 2; July: NR = 0, CL = 6, RO = 2; August: NR = 0, CL = 1, RO = 2; September: NR = 2, CL = 4, RO = 2; and October: NR = 0, CL = 3, RO = 1. Window traps and bowl traps were processed by filtering the soapy water three times through a brine shrimp net, whereas pitfalls were processed by filtering the soapy water through a 1 mm mesh fishnet. Specimens were removed from fishnets with forceps and stored in film canisters containing 70% isopropyl alcohol.

Taxonomy

Samples were sorted to order using established taxonomic keys (Borror et al. 1989; Arnett et al. 2002). Coleoptera were identified to family, tribe, and genus using keys found in Arnett et al. (2002). *Anthrenus lepidus* was identified according to Beal (1998). Tenebrionids were identified to species according to Bell (1970) and Doyen (1984). Specimens that were not identified to a named genus or species were assigned to a designated morphospecies (Oliver and Beattie 1996).

Data Analysis

Community ordination was performed by principal components analysis (Palmer 2003) using SPSS 10.0. Only data from April 2000 were used for ordination because only specimens from April of 2000 have been completely processed.

RESULTS

Beetles collected during the summer of 2000 consisted of 963 specimens from 36 families. Eighty-five morphospecies have been recognized with 37 assigned to named genera (Table 1). The greatest diversity occurred in the Carabidae with 14 morphospecies, followed by the Tenebrionidae, Staphylinidae, and Coccinellidae with 8 morphospecies each. Most families (28 of 36) were represented by only one morphospecies. Some trapping sites were dominated by a few taxa. For example, 58 percent of the beetles at CL6 were from two genera—*Trichiorhyssemus* (Scarabaeidae) and *Melanactes* (Elateridae).

Notes on Select Taxa

We collected 26 *Anthrenus lepidus* (Dermestidae) in 2000; 81 percent (21 of 26) of these were found in the NR section. Only two were collected in the RO section and three were found in the CL section. *Anthrenus lepidus* consists of 13 forms or varieties that vary in the color and arrangement of dorsal setae (Beal 1998). The individuals in Salt Creek Canyon were of the Gold Streak variety.

Forty-two specimens from the Tenebrionidae were found in the 2000 collection. Forty (95%) of the tenebrionid specimens were found in the RO and CL sections of the canyon. *Trogloderus costatus* was the most abundant tenebrionid in our collection, found in both the RO and CL sections. The only tenebrionids found to date from the NR section were *Eleodes fusiformis*, which was also found in the other two sections, and *Araeoschizus*, which was also collected from the RO section. *Trogloderus costatus* and *Eleodes* spp. were represented by 10 specimens each, but the remaining tenebrionids were much rarer, with only one or two specimens per morphospecies.

The most abundant beetle collected in 2000 was *Anaspis*, the single genus found from the family Scraptiidae. *Anaspis* consti-

Table 1. Sampling data for study sites in Salt Creek Canyon.

	April	June	July	Aug	Sept	Oct	Moisture	Management
RO1	X	–	–	X	–	X	Wet	Open Road
RO2	X	X	X	–	X	–	Dry	Open Road
RO3	X	X	X	X	X	–	Wet	Open Road
CL1	–	X	X	X	X	X	Wet	Closed Road
CL2	X	X	X	–	X	–	Dry	Closed Road
CL4	X	X	X	–	–	X	Dry	Closed Road
CL5	–	–	X	–	–	–	Dry	Closed Road
CL6	X	X	X	–	–	X	Wet	Closed Road
CL8	–	X	X	–	X	–	Wet	Closed Road
CL10	–	X	–	–	X	–	Wet	Closed Road
NR1	X	X	–	–	X	–	Dry	No Road
NR2	X	X	–	–	X	–	Wet	No Road
NR3	X	–	–	–	–	–	Wet	No Road

tuted 17 percent of the beetle fauna in Salt Creek Canyon. However, it was completely absent in April of 2000, with 92 percent of the specimens examined so far being collected in June. *Anaspis* was generally distributed between the three canyon regions.

We have identified 134 Scarabaeoidea specimens. They were generally distributed throughout the canyon, but certain taxa showed preferences for specific sections. For example, 46 specimens of *Dichelonyx* (Scarabaeidae) were collected in 2000, mostly (89%) from the NR section. None were collected in April; all 46 specimens examined so far were collected in June. Four specimens of *Lichnanthe* (Glaphyridae) were collected, all in June and all from the NR section. In contrast, *Trichiorhyssemus* (Scarabaeidae) was absent from the NR section, but was collected from the RO and CL sections.

We have so far identified 58 specimens from six morphospecies of Chrysomelidae. However, only six specimens from two morphospecies in the subfamily Alticinae were collected in April of 2000. Chrysomelid abundance and richness were greater in July through September. The Alticinae, found in each region of the canyon, constituted 67 percent of the chrysomelids collected in 2000. The remaining specimens were from two morphospecies of Eumolpinae and two morphospecies of Chrysomelinae, all of which were absent from the NR section.

Eighty-two specimens from 14 morphospecies of Carabidae have been identified so far from the Salt Creek collection. As a group, carabids were found in approximately equal numbers in the three regions of the canyon. However, some genera were restricted to certain sections. For example, so far all five specimens of *Badister* were collected in the NR section, *Selenophorus* has been identified only from the RO and CL sections, and *Pterostichus* only from the CL and NR sections. In addition, distribution patterns changed throughout the summer. For example, *Amara* was present in the RO and CL sections in April, but absent from the NR section. However, four specimens of *Amara* were identified from the NR section in June and July.

Community Ordination

We used principal components analysis to compare the beetle assemblages at each trapping site. For the reasons described above, only samples collected in April of 2000 were used. The analysis extracted nine components. Components 1 (x-axis) and 2 (y-axis) explained 36 and 21 percent of the variance, respectively. A plot of the results is shown in Figure 2. Sites from different sections of the canyon showed considerable overlap along component 1. In particular,

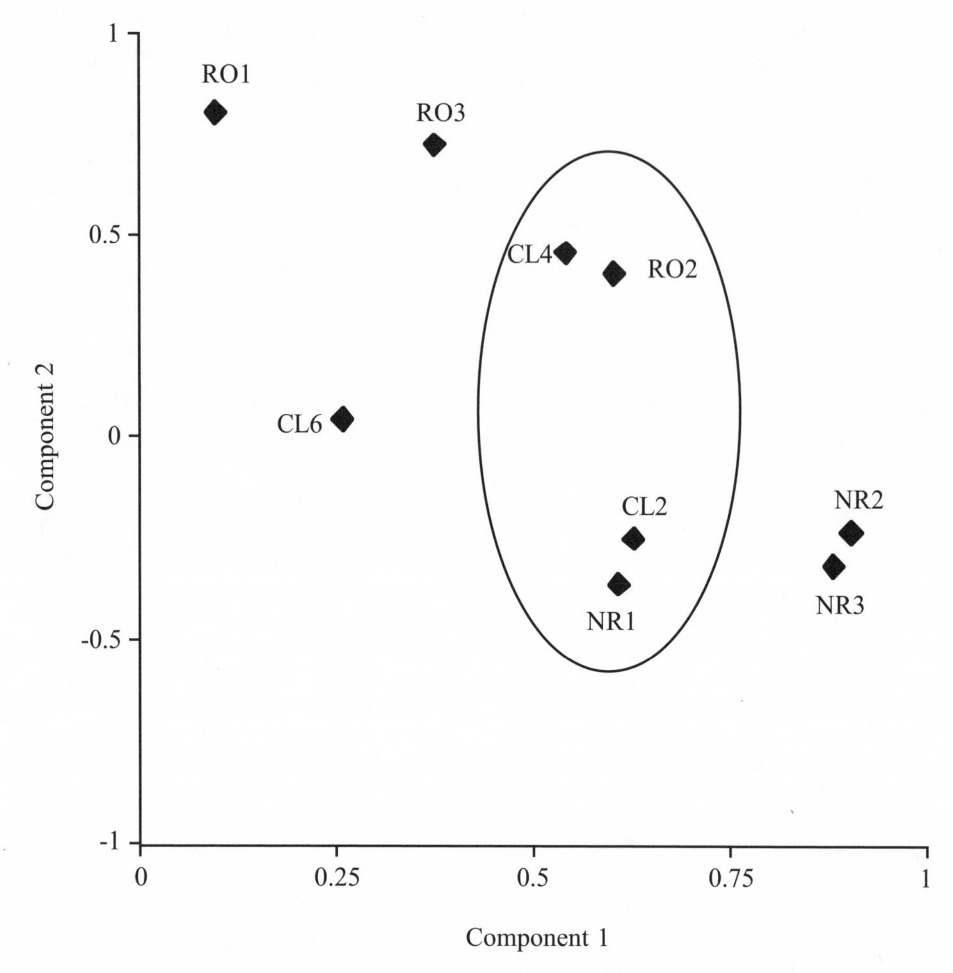

Figure 2. Principal components analysis was used to compare the beetle assemblages at each trapping site, using only samples collected in April of 2000. Sites enclosed in the oval represent upper riparian/alluvial bench transition habitat. Both components indicate that three pairs of sites support similar beetle assemblages: CL4 and RO2, CL2 and NR1, and NR2 and NR3.

sites NR1, RO2, CL2, and CL4 grouped together along this axis. These are all dry sites representing upper riparian/alluvial bench transition habitat. The wetter sites, NR2, NR3, RO1, RO3, and CL6, separated along both axes by canyon section, but there was some overlap between closed and open sites along axis 1. In contrast, the RO and NR sites were well separated with respect to component 2, and CL sites fell roughly in between the NR and RO sections, although the CL sites did not sort out in order up the canyon (e.g., CL2 is closest to RO3, but resembled NR1 more, whereas CL4, farther up the canyon, was closer to RO2 than NR1).

DISCUSSION

The ultimate goal of the Salt Creek Canyon study is to identify taxa with distributions that correlate with the different management zones of Salt Creek Canyon, and to identify which trapping methods are best for monitoring the indicator taxa, which includes determining when to monitor as well as which trapping technique(s) should be used (New 1995; Graham 2002). Beetles have been used as indicators in other systems (Epstein and Kulman 1990; French and Elliott 1999; Bess et al. 2002; Cartron et al. 2003), and thus were of interest in this study as well. Here, we have presented some of the characteristics of the beetle community in Salt Creek Canyon as a prelude to identifying potential indicator organisms.

To date, 85 beetle morphospecies from 36 families have been designated from our collection. In comparison, using pitfall traps, Bess et al. (2002) collected 57 beetle species from a riparian zone along the Rio Grande in New Mexico. Considering that at least 21 morphospecies from Salt Creek Canyon were collected exclusively or primarily with window or colored bowl traps, the number of ground-active beetle species in these two riparian zones may be quite similar (57 in Bess et al. 2002, 64 in this study). However, the species compositions appear to be quite different when two ground-active families, Carabidae and Tenebrionidae, are compared between the two sites. Only two carabid genera, *Amara* and *Harpalus*, and one tenebrionid species, *Eleodes fusiformis*, are shared between these two riparian zones (Bess et al. 2002; Cartron et al. 2003). In contrast, the carabid genera in Salt Creek Canyon are more similar to those described from old fields and upland and lowland forests in east-central Minnesota (Epstein and Kulman 1990). Of the 23 genera identified from Minnesota, 9 are also found in Salt Creek Canyon. Similarly, of the 46 carabid genera identified from riparian zones in Oklahoma (French and Elliott 1999), 7 were also found in Salt Creek Canyon. It is interesting that the number of carabid genera identified in studies from the Midwest is much higher than the number of carabid genera from Salt Creek Canyon (14), northern Arizona ponderosa pine forests (16; Villa-Castillo and Wagner 2002), and Rio Grande riparian zones (10; Cartron et al. 2003).

In contrast, beetle diversity in Salt Creek Canyon may be less than that from more arid environments. For example, roughly 50 percent of insect species inhabiting Sahara and Namib Desert sand dunes are beetles (Seeley 1991). Our preliminary data from Salt Creek Canyon indicate that beetles make up a smaller percentage of insect species, perhaps as low as 10 percent. One of the most abundant groups of insects in deserts and other arid environments are the Tenebrionidae (Crawford 1981). For example, of the 212 insect species described from the Sahara dunes, 27 percent were tenebrionids, and 34 percent of described insects from the Namib Desert dunes are also tenebrionids (Seeley 1991). More recently, Chikatunov et al. (1997) found 42 species of tenebrionids from 34 genera in one canyon near Mt. Carmel, Israel. In contrast, the eight species of tenebrionids constitute only about 9 percent of beetle species in Salt Creek Canyon.

The diversity of beetles in Salt Creek Canyon is similar to that of other taxa found on the Colorado Plateau and in the Great Basin. Nelson (1994) estimated that 14,000 to 26,000 insect species inhabit this region. Among these, the approximate number of species in select taxa includes 99 acridid grasshopper, 77 ant, 90 stonefly, 155 butterfly, and 160 robber fly species (Nelson 1994). Thus, the beetle diversity of Salt Creek Canyon approximates that of the acridids, ants, and stoneflies, but is considerably less than that of the Lepidoptera and Diptera.

Some taxa are limited to certain trapping sites and/or sections of the canyon. For example, 91 percent of *Anthrenus lepidus* were collected in the NR section. *Anthrenus lepidus* consists of 13 forms or varieties that vary in the color and arrangement of dorsal setae (Beal 1998). The individuals in Salt Creek Canyon are of the Gold Streak variety, which are distributed from California, Arizona, and New Mexico north to southern Oregon. Larval Gold Streak has been reared from the nests of packrat (*Neotoma*), phoebe (*Sayornis*), cliff swallow (*Petrochelidon pyrrhonata*), and magpie (*Pica pica*; Beal 1998). Adults have been collected from 13 species of plant including *Tamarix* (Beal 1998). Presumably, the restricted distribution of *A.*

lepidus in Salt Creek Canyon reflects a restricted distribution of suitable habitat for adults, larvae, or both. Similarly, of the 16 *Amara* (Carabidae) identified so far, nine were collected in the RO section. This is consistent with the known habitat choice of the tribe Zabrini, of which *Amara* is a member. Members of this tribe occupy disturbed sites that receive plenty of sunlight, and riparian members of the Zabrini prefer floodplains with dry sand (Arnett et al. 2002); the RO section of Salt Creek Canyon fits this description. However, later in the season four specimens of *Amara* were collected in the NR section, which is wetter and has more cover. It is possible that the *Amara* distribution shifts through the summer as sites become drier. Alternatively, the *Amara* collected in the NR and RO sections may be different species with very different habitat requirements. We are currently unable to distinguish between these possibilities, which emphasizes the need for long-term monitoring and increased taxonomic resolution.

Community ordination indicates that different trapping sites and sections of the canyon support different beetle communities that generally vary along some environmental gradient. There is a general pattern of position in the canyon from right to left along the first axis, and bottom to top on the second axis (see Figure 2), sorting sites from upstream to downstream. It is possible that the sites differ in their beetle fauna because the sites vary continuously up the canyon, and beetles are simply responding to changes in conditions and/or resources associated with geographic location. However, there are other factors involved in determining the ordination pattern. For example, CL6 is isolated at the left end of the first axis, with little influence in its position from the second axis; RO2 and CL4 are much closer to each other than either is to other sites in the same management zones. The same is true for NR1 and CL2. These latter four sites are the driest, with vegetation characteristic of dry alluvial benches with little riparian vegetation. It may be that these alluvial benches have more similar environmental conditions than the different riparian patches along Salt Creek. The riparian sites were widely separated in ordination space and were correlated with management zone and canyon character. Hydrology is likely different in these two reaches of Salt Creek because of differences in canyon structure, vegetation cover, and watershed area. Thus, the beetle communities may be structured, in part, by moisture levels or flooding regimes at a given trapping site.

These results may be particularly relevant in light of those presented recently by Cartron et al. (2003), who demonstrated that carabid abundance and diversity is higher at flood versus non-flood sites along the middle Rio Grande whereas other taxa were unaffected or only weakly affected by the flooding regime. Therefore, they suggested that carabids may function as sensitive indicators of the hydrogeological relationships between the Rio Grande and the surrounding riparian zone. Although our data are preliminary, they do suggest that the relationship between carabids and moisture levels or flooding may be more complex. In contrast to the situation described by Cartron et al. (2003), carabid abundance in April of 2000 was slightly higher at dry sites than wet sites, but our preliminary data from the remainder of the summer suggest that carabid abundance increases at wetter sites versus dry. This was due to the appearance of large numbers of *Pterostichus*, *Selenophorus*, and *Stereocerus*, which were represented by only one specimen each in April. In contrast, the Scarabaeoidea were most abundant at wetter sites throughout the summer. In particular, the most abundant scarabs, *Trichiorhyssemus* and *Dichelonyx*, were most common at CL6 and NR2, respectively. Thus, our results suggest that for some taxa, abundance at a site may be determined, at least in part, by moisture levels or flooding regimes.

Other factors may be more important for some taxa. For example, tenebrionids were all but absent from the No Road section, but were collected in approximately equal numbers from both wet and dry sites in the rest of the canyon. Our results are consistent with studies done on tenebrionids inhabiting shortgrass prairies (Stapp 1997; Crist, in press). In these studies, tenebrionids were captured more frequently in habitats with more shrub cover interrupted by bare ground. In contrast, fewer beetles were captured in low-lying areas with denser vegetation. The habitats in Salt Creek Canyon where tenebrionids are more frequently captured are similar in structure to those in prairies where tenebrionids are more abundant. The NR section, in contrast, has denser vegetation. In addition, Cartron et al. (2003) found no difference between flood and non-flood sites with respect to tenebrionid abundance and diversity. Therefore, it is perhaps unlikely that the low abundance of tenebrionids in the NR section is due to the higher moisture levels or different flooding regimes in that part of our study area, and is instead due to differences in spatial distribution of vegetation.

Specimen collection in Salt Creek Canyon has been ongoing since 2000. Future efforts will be directed at identifying the beetle specimens remaining in the collection to the lowest taxonomic level possible. In combination with ongoing studies at other sites on the Colorado Plateau and in the Great Basin, a fairly comprehensive catalog of beetle diversity will be established.

ACKNOWLEDGMENTS

This work was supported by grants from the USGS, NPS, and Earthwatch Institute to T. B. G. and a Carroll College Faculty Development Grant to L. L. P. We are especially grateful for the efforts of the many Earthwatch volunteers who have contributed their time and money in support of this project.

LITERATURE CITED

Ashcroft, G. L., D. T. Jensen, and J. L. Brown. 1992. Utah Climate. Utah Climate Center, Utah State University, Logan. 127 pp.

Arnett, R. H., M. C. Thomas, P. E. Skelley, and J. H. Frank. 2002. American Beetles. Vol. 2. CRC Press, Boca Raton FL.

Beal, R. S., Jr. 1998. Taxonomy and biology of nearctic species of *Anthrenus* (Coleoptera: Dermestidae). Transactions of the American Entomological Society 124: 271–332.

Bell, R. T. 1970. Identifying Tenebrionidae (darkling beetles). U.S. International Biological Program, Technical Report No. 58, Grassland Biome. Fort Collins CO.

Bess, E. C., R. R. Parmenter, S. McCoy, and M. C. Molles, Jr. 2002. Responses of a riparian forest-floor arthropod community to wildfire in the middle Rio Grande valley, New Mexico. Environmental Entomology 31: 774–784.

Borror, D. J., C. A. Triplehorn, and N. F. Johnson. 1989. An Introduction to the Study of Insects, 6th edition. Saunders College Publishing, Philadelphia.

Briggs, M. K. 1996. Riparian Ecosystem Recovery on Arid Lands: Strategies and References. University of Arizona Press, Tucson.

Capinera, J. L., and T. S. Sechrist. 1982. Grasshopper (Acrididae)-host plant associations: Response of grasshopper populations to cattle grazing intensity. The Canadian Entomologist 114: 1055–1062.

Carothers, C. 1977. Importance, preservation, and management of riparian habitats: An overview. Symposium on Importance, Preservation, and Management of the Riparian Habitat, Tucson, Arizona. USDA Forest Service, Rocky Mountain Forest and Range Experiment Station.

Cartron, J. E., M. C. Molles, Jr., J. F. Schuetz, C. S. Crawford, and C. N. Dahm. 2003. Ground arthropods as potential indicators of flooding regime in the riparian forest of the middle Rio Grande, New Mexico. Environmental Entomology 32: 1075–1084.

Chikatunov, V., M. Lillig, T. Pavlíek, L. Blaustein, and E. Nevo. 1997. Biodiversity of insects at a microsite, 'Evolution Canyon,' Nahal Oren, Mt. Carmel, Israel. Coleoptera: Tenebrionidae. Journal of Arid Environments 37: 367–377.

Crist, T. O. In press. Insect populations, community interactions, and ecosystem processes in shortgrass steppe. In Ecology of the Shortgrass-Steppe: Perspectives from Long-Term Research, edited by I. C. Burke and W. K. Lauenroth. Oxford University Press.

Crawford, C. S. 1981. Biology of Desert Invertebrates. Springer-Verlag, New York.

Crawford, C. S. 1986. The role of invertebrates in desert ecosystems. In Pattern and Process in Desert Ecosystems, edited by W. G. Whitford. University of New Mexico Press, Albuquerque.

Doyen, J. T. 1984. Systematics of *Eusattus* and *Conisattus* (Coleoptera; Tenebrionidae; Coniontini; Eusatti). Occasional Papers of the California Academy of Sciences No. 141, San Francisco.

Epstein, M. E., and H. M. Kulman. 1990. Habitat distribution and seasonal occurence of carabid beetles in east-central Minnesota. The American Midland Naturalist 123: 209–225.

Fleischner, T. L. 1994. Ecological costs of livestock grazing in western North America. Conservation Biology 8: 629–644.

French, B. W., and N. C. Elliott. 1999. Spatial and temporal distribution of ground beetle (Coleoptera: Carabidae) assemblages in riparian strips and adjacent wheat fields. Environmental Entomology 28: 597–607.

Graham, T. 2002. Expedition Briefing: Canyonland Creek Ecology. The Earthwatch Institute, Maynard MA.

Green, D. M., and J. B. Kauffman. 1995. Succession and livestock grazing in a northeastern Oregon riparian ecosystem. Journal of Range Management 48: 307–313.

Jepson-Innes, K., and C. E. Bock. 1989. Response of grasshoppers (Orthoptera: Acrididae) to livestock grazing in southeastern Arizona: Differences between seasons and subfamilies. Oecologia 78: 430–431.

Kauffman, J. B., and W. C. Krueger. 1984. Livestock impacts on riparian ecosystems and streamside management implications: A review. Journal of Range Management 37: 430–437.

Kauffman, J. B., W. C. Krueger, and M. Vavra. 1983. Effects of late season cattle grazing on riparian plant communities. Journal of Range Management 36: 685–691.

Kondolf, G. M., and R. R. Curry. 1984. The role of riparian vegetation in channel bank stability: Carmel River, California. In California Riparian System: Ecology, Conservation, and Productive Management. University of California Press, Berkeley.

Lowe, C. H. 1964. The Vertebrates of Arizona. University of Arizona Press, Tucson.

Lowe, C. H. 1989. The riparianness of desert herpetofauna. In California Riparian Systems: Protection, Management, and Restoration for the 1990's. USDA Forest Service, Pacific Southwest Forest and Experiment Station. Davis CA.

Mitchell, S., and B. Woodward. 1993. Man's effects on aquatic and riparian organisms in the canyons of Canyonlands and Arches National Parks and Natural Bridges National Monument. National Park Service.

Nelson, C. R. 1994. Insects of the Great Basin and Colorado Plateau. In Natural History of the Colorado Plateau and Great Basin, edited by K. T. Harper, L. L. St. Clair, K. H. Thorne, and W. M. Hess, pp. 211–237. University Press of Colorado, Boulder.

New, T. R. 1995. Invertebrate Surveys for Conservation. Oxford University Press, Oxford, UK.

Niwranski, K., P. G. Kevan, and A. Fjellberg. 2002. Effects of vehicle disturbance and soil compaction on arctic collembolan abundance and diversity on Igloolik Island, Nunavut, Canada. European Journal of Soil Biology 38: 193–196.

Ohmart, R. D., and B. W. Anderson. 1982. North American desert riparian ecosystems. Reference Handbook on the Deserts of North America, edited by G. L. Bender. Greenwood Press, Westport CT.

Oliver, I., and A. J. Beattie. (1996). Invertebrate morphospecies as surrogates for species – a case study. Conservation Biology 10: 99–109.

Palmer, M. 2003. See http://www.okstate.edu/artsci/botany/ordinate/ (Mike Palmer, Botany Department, Oklahoma State University, accessed 2003).

Quinn, M. A., and D. D. Walgenbach. 1990. Influence of grazing history on the community structure of grasshoppers of a mixed-grass prairie. Environmental Entomology 19: 1756–1766.

Seeley, M. K. 1991. Sand dune communities. In The Ecology of Desert Communities, edited by G. A. Polis. University of Arizona Press, Tucson.

St. Clair, L. L., K. T. Harper, K. H. Thorne, and W. M. Hess. 1994. Recommendations for future research. In Natural History of the Colorado Plateau and Great Basin, edited by K. T. Harper, L. L. St. Clair, K. H. Thorne and W. M. Hess, pp. 277–279. University Press of Colorado.

Stapp, P. 1997. Microhabitat use and community structure of darkling beetles (Coleoptera: Tenebrionidae) in shortgrass prairie: Effects of season, shrub cover and soil type. American Midlands Naturalist 137: 298–311.

Stohlgren, T. J., and J. F. Quinn. 1992. An assessment of biotic inventories in western U.S. national parks. Natural Areas Journal 12: 145–154.

Stohlgren, T. J., J. F. Quinn, M. Ruggiero, and G. S. Waggoner. 1995. Status of biotic inventories in U.S. national parks. Biological Conservation 71: 97–106.

Villa-Castillo, J., and M. R. Wagner. 2002. Ground beetle (Coleoptera: Carabidae) species assemblages as an indicator of forest condition in northern Arizona ponderosa pine forests. Environmental Entomology 31: 242–252.

Whitford, W. G. 1986. Pattern and Process in Desert Ecosystems. University of New Mexico Press, Alburquerque.

LANDSCAPE HABITAT SELECTION BY FEMALE MULE DEER IN A PARTIALLY RESTORED PONDEROSA PINE FOREST IN NORTHWEST ARIZONA

Stan C. Cunningham, Stephen S. Germaine, Heather L. Germaine, and Susan R. Boe

Southwestern U.S. ponderosa pine (*Pinus ponderosa*) forest communities evolved with episodic ground fires (> 5,000 ha) approximately every 2–10 years pre 1880 (Swetnam and Betancourt 1990). However, since the mid to late 1800s, aggressive wildfire suppression, livestock grazing, and even-aged timber management have tended to homogenize ponderosa pine forests (Allen et al. 2002). Current condition ponderosa pine forests are densely stocked (up to > 500 trees/ha), and understory grasses, forbs, and shrubs have declined (Covington and Moore 1994; Bogan et al. 1998). Large trees have decreased in number due to logging (Allen et al. 2002), meadow size has decreased due to small tree invasion (Swetnam et al. 1999), and floral and faunal biodiversity levels have decreased (Allen 1998). These conditions have increased the risk, size, and number of stand-replacing fires (Swetnam and Betancourt 1998).

Although dense stands of suppressed ponderosa pine trees have always been present, they are more prevalent today (Cooper 1960), and a broad scientific, social, and political consensus has recently emerged that restoration of ecological sustainability in southwestern pine forests is necessary and urgent (Covington and Moore 1994; Covington et al. 1997; Allen et al. 2002). Restoration treatments to reduce trees to historic densities (± 80 trees/ha; Covington et al. 1997) are now proposed for > 81,000 ha annually in Arizona and New Mexico (Anonymous 2001).

Reducing tree density in landscape-scale areas (> 50 ha) should influence mule deer (*Odocoileus hemionus*) populations. Reductions in herbaceous and shrub productivity in dense forests, and lower diversity and senescence of existing browse have contributed to mule deer declines in recent decades (Julander and Low 1976; Carlson et al. 1993; Carpenter 1998). Mule deer population fitness is believed to be superior in early successional stage habitat rather than in later stages, when most nutrients are tied up in woody material (Wallmo and Schoen 1981).

Post-fire benefits to mule deer have been documented in ponderosa pine (Hungerford 1970), pinyon pine (*P. edulis*)–juniper (*Juniperus osteosperma*; McCulloch 1969; Stager and Klebenow 1987), and pine-oak (*Quercus* spp.) habitats (Kie 1984). After a burn, forbs and green grass increase (Thill et al. 1990; Kucera and Mayer 1999), and rapidly growing young or resprouting browse is usually more nutritious than older browse. Thus restoration treatments that open tree canopies and use controlled burning should increase forage and benefit mule deer. However, removal of thermal and hiding cover could be negative; mule deer distribution and density in presettlement (pre-1870) ponderosa pine forests remain unknown (Wagner et al. 2000; Block et al. 2001).

The Mt. Trumbull Resource Conservation Area (RCA), located in the Grand Canyon–Parashant National Monument in northwest Arizona, contains 1699 ha of experimental ponderosa pine forest (Taylor 2000). Treatments were designed to result in open forests ($\bar{x}$ = 88 trees/ha) with an increased herbaceous layer capable of sustaining periodic cool fires. These treatments require decreased stocking rates in order to retain herbaceous layers and an herb-dominated understory. Vegetation diversity and treatments within the Mt. Trumbull RCA provided an opportunity to examine how they were used by mule deer. We investigated whether female mule deer selected, avoided, or used landscape-scale areas in response to forest restoration treatments. We hypothesized that as earlier successional stages came available, female mule deer would select treated areas more and use current condition forest less.

STUDY AREA

Mt. Trumbull RCA is an isolated, high-elevation forest (sky island) on the southern Uinkaret Plateau just north of the Grand Canyon in northwest Arizona. Study area elevations range from 2000 to 2280 m. Precipitation is bimodal, occurring mostly in winter and summer, averaging 32.9 cm per year. Restoration occurred in a forested valley between Mt. Trumbull and Mt. Logan (Figure 1). The five major vegetative communities within the study area were homogeneous ponderosa pine stands; mixed ponderosa pine–Gambel oak (*Q. gambelii*); mixed ponderosa–pinyon-juniper woodland; deciduous forest of Gambel oaks, New Mexican locust (*Robinia neomexicana*), and aspen (*Populus tremuloides*); and naturally occurring meadows.

Mixed ponderosa pine–deciduous forest was the most common type in the study area, averaging 53.0 percent of the total area. Mixed ponderosa-pinyon-juniper forest was second most abundant, averaging 27.0 percent, followed by homogeneous ponderosa pine (7.4%), meadows (6.4%), and deciduous forest (6.0%). By May 1998, 44 ha were completely treated, 124 ha were treated by May 1999, 227 ha were treated by May 2000, 260 ha by May 2001, and 431 ha by September of 2002. The proportion of transition forest available ranged from 4.0 percent in 1997 to 32.0 percent in 2002.

METHODS

Restoration Prescriptions

The Mt. Trumbull RCA was divided into 33 units of varying size ($\bar{x}$ = 58.8 ha), with approximately 1092 ha to receive treatment and 607 ha reserved as current condition controls ($\bar{x}$ = 666 trees/ha). Approximately 431 ha were fully treated by September 2002. In ponderosa pine, the prescription was to leave 1.5–3 trees within 18.3 m of each presettlement (1870) aged tree. In pinyon-juniper, 2 trees within a 4.6 m radius of each presettlement aged tree were left. Gambel oak and New Mexican locust were not thinned, and care was taken to avoid damaging oak trees because of their demonstrated value to wildlife (Reynolds et al. 1970; Rosenstock 1998). Commercial logging, followed by sub-commercial thinning, was used to remove unmarked trees. Logged sites were burned, then reseeded with a native herb mix dominated by grasses. In this manner, pre-treatment stand densities were reduced 87 percent to densities ($\bar{x}$ = 88/ha) more closely approximating pre-settlement conditions (Covington et al. 1997). Domestic livestock grazing did not occur during our study.

Treatments were conducted in stages: cruising and marking, thinning, burn preparation, burning, and seeding, with some units taking more than 3 years to complete. We classified treatment stages as current condition (trees may have been marked but no cutting started), transition (cut but slash not removed), or treated (all stages completed).

Capture and Telemetry

We captured and transmittered 34 female mule deer between May 1997 and September 2002. After initial capture in 1997, we tried to maintain a minimum of 20 transmittered females. We immobilized mule deer with a mixture of Telazol-Xylazine (1 mg : 2 mg) administered via Pneudart 3 cc darts fired from Capshur or Marlin/Pneudart rifles. Captured mule deer were fitted with Telonics radio collars and unique ear tags. We administered Yohimbine (3 mg) as an

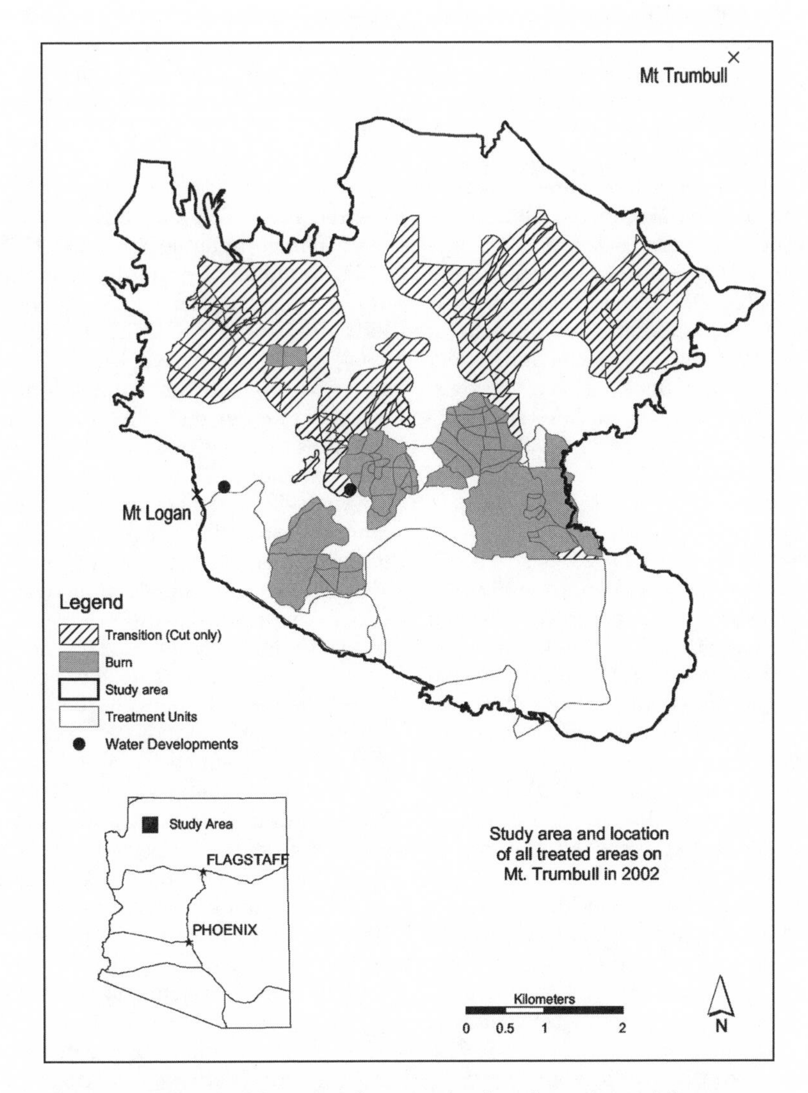

Figure 1. Mt. Trumbull Resource Conservation Area in northwest Arizona where transmittered female mule deer were located from 1997 through 2002. Units correspond to restoration treatment areas.

antagonist, and we visually monitored mule deer until they had safely recovered.

To locate transmittered mule deer, camouflaged observers quietly followed a signal on foot until the animal was visible. If the observer disturbed the animal prior to location, we aborted the observation. After the mule deer was observed, the site it was first located in was recorded using a Global Positioning System (GPS) receiver. Locations were collected between 0600 and 1800 hours MST between May 1 and September 15 from 1997 through 2002. Observation attempts were rotated incrementally among marked mule deer, with no individual mule deer observed more than once per day. For logistic reasons, we only searched for transmittered deer while they were on the study area. No area of the study site was more than 3 km from a treated area.

GIS Mapping and Data Analysis

Using a Geographic Information System (GIS) we created and field-verified a digital vegetation map of the five vegetation types and their treatment status annually. To determine if female mule deer locations were proportional to vegetation types and treatment stages available on an annual basis (Neu et al. 1974), we plotted all female mule deer locations collected on our current vegetation status map each year. We created a master annual minimum convex polygon (mcp) encompassing all locations, determined the area of each vegetation type and treatment stage inside the master mcp, and tallied frequency of locations. We compared percentage of locations in each vegetation type and treatment to the Bonferroni 90% simultaneous confidence interval (Byers et al. 1984) of expected proportions to determine whether selection (use > availability) or avoidance (use < availability) occurred. We calculated Jacob's *D* (Jacob 1974) values (–1.0 to 1.0) to examine the extent of selection or avoidance. Negative values indicate avoidance, positive values indicate selection, and the closer the value to 1.0 or –1.0 from 0, the stronger the relationship. Because fewer habitat categories increased test power (Alldredge and Ratti 1986), our second analysis was to pool all vegetation types and look at habitat use based on the three treatment stages. We also used the same analysis to look at frequency of female deer locations based on the amount of time since each treatment had been completed by yearly categories (0.5, 1.5, 2.5, etc.).

RESULTS

We recorded 1069 locations of 34 transmittered female mule deer from 1997 through 2002. Number of mule deer monitored in any year ranged from 9 to 24, and number of locations from 66 to 269. Number of locations per animal per year ranged from 1 to 41 ($\bar{x}$ = 18) with no animal contributing more than 8 percent of the total number of locations in any year except 1997, when one deer contributed 12.1 percent.

Analysis of female mule deer locations among restoration stages and vegetation type revealed some consistent trends (Table 1). The only areas female mule deer did not use as expected (avoided) were current condition forest types. Homogeneous ponderosa pine in 1998 and 2002, ponderosa-deciduous in 2001, ponderosa-pinyon-juniper in 1999 and 2002, and meadows located in current condition forests in 2002 were used less than available. Fully treated mixed ponderosa-deciduous forest was selected more than expected 3 of the 6 years (2000–2002). Transition areas were selected in mixed ponderosa-deciduous forest in 1998 and meadows in transition areas in 1999.

By pooling vegetation types and analyzing proportion of female mule deer locations with the availability of treatment stages only, selection patterns were clearer. Treated areas were selected more than expected 4 of 6 years, and transition areas were selected 3 of 5 years (Table 2). Current condition forests were used less than expected 4 of 6 years. The only year all restoration stages were used as expected was 2000. In 1997, treated areas were selected, but transition areas were not available.

With respect to age of treatment, female mule deer used transition and treated areas similarly (Figure 2). In transition areas, 80.2

Table 1. Jacob's *D* values of avoidance or selection of vegetation type and restoration stage by year of female mule deer on Mt. Trumbull, Arizona, 1997–2002. An x indicates that the proportion of locations compared to habitat availability was as expected as determined from 90% Bonferroni confidence intervals. A dash indicates that habitat was not available that year.

	1997	1998	1999	2000	2001	2002
Homogeneous ponderosa pine						
Current condition	x	–0.46	x	x	x	–0.35
Transition	–	–	–	x	x	x
Treated	–	–	–	–	x	x
Ponderosa-deciduous						
Current condition	x	x	x	x	–0.19	x
Transition	x	0.6	x	x	x	x
Treated	x	x	x	0.6	0.6	0.6
Ponderosa-pinyon-juniper						
Current condition	x	x	–0.29	x	x	–0.37
Transition	–	–	x	x	x	x
Treated	–	–	–	–	–	–
Deciduous						
Current condition	x	x	x	x	x	x
Transition	–	–	–	–	x	x
Treated	–	–	–	–	x	x
Meadow						
Current condition	x	x	x	x	x	–0.57
Transition	–	–	0.81	x	x	x
Treated	–	–	–	–	–	x

Table 2. Jacob's *D* values of avoidance or selection of vegetation type and restoration stage by year of female mule deer on Mt. Trumbull, Arizona, 1997–2002. An x indicates that the proportion of locations compared to habitat availability was as expected as determined from 90% Bonferroni confidence intervals. A dash indicates that habitat was not available that year.

Year	n	Current Condition	Transition	Treated
1997	66	x	–	0.55
1998	236	–0.63	0.60	0.58
1999	143	–0.34	x	0.53
2000	111	x	x	x
2001	269	–0.17	0.22	x
2002	244	–0.34	0.23	0.21

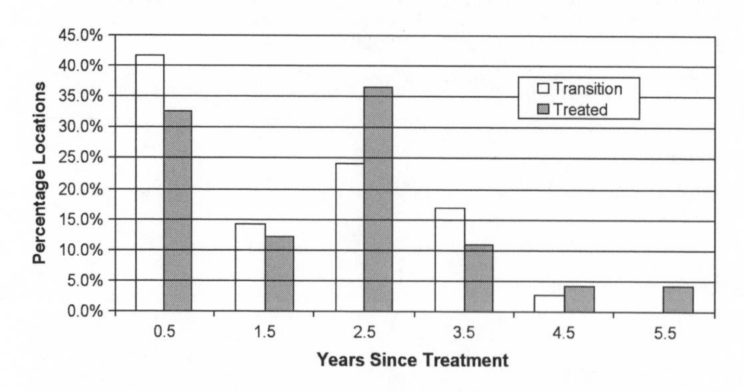

Figure 2. Percentage of female mule deer locations in transition or treated restoration units based on the time since the last stage was completed. Data were collected from 1997 though 2002 on the Mt. Trumbull Resource Conservation Area, northwest Arizona.

percent of locations occurred in treatments < 2.5 years old, and in fully treated areas, 81 percent of the locations occurred in areas treated < 2.5 year ago. With respect to proportion of female mule deer locations expected based on 90% Bonferroni confidence intervals, both transition and treated areas < 2.5 years of age were selected (D = 0.33 and 0.3 respectively), and treatments > 2.5 years old were avoided (D = –0.29 for transition, D = –0.3 for treated).

DISCUSSION

Female mule deer did not use treatment stages as available, selecting restoration treatment areas 4 of 6 years of study. Insufficient female mule deer location sampling in 1997 and 2000 (n = 66 and 111 respectively), coupled with the minimum number of treatments of 7 in 1997 to the maximum of 14 in 2002 probably increased our risk of type II error (failure to detect a change). We attribute selection of treated areas in 4 of 6 years to an increase in forage. Germaine et al. (2004) found that forage increased in treated forest in each of the years evaluated (1998–2000). Silvicultural prescriptions that open tree canopies usually increase forage abundance and diversity (Patten 1974; Wallmo and Schoen 1981; Masters et al. 1993).

Since more than 80 percent of the locations occurred in transition or treated areas less than 2.5 years old, forage and selection of restored or transition areas may diminish as revegetation occurs. The initial selection may be due to the controlled burns. Several studies have found increased use of burned areas in the first few years post-fire, but decreased use to pre-burn levels in subsequent years (Biswell 1961; Ashcraft 1979; Kie 1984; Klinger et al. 1989). Allen et al. (2002) pointed out the importance of periodic surface fires post-restoration to maintain the early and middle successional stages that mule deer evolved with.

Recent federal guidelines (Federal Register 2002) have emphasized the importance of clearing brush in forests to reduce fire danger; the guidelines however request categorical exclusion from the National Environmental Policy Act (NEPA) process in Southwestern forest thinning projects. We believe that female mule deer would not have selected the early successional stages had not oak and New Mexico locust patches been preserved. Germaine et al. (2004) found that mule deer preferred bedding in current condition forests from 1998 to 2000; we found that female mule deer avoided current condition forests in 1998 and 1999 when all locations, regardless of behavior, were combined. However, the majority of bedded mule deer seen in treated or transition areas were in Gambel oak motts or small patches of New Mexican locust that were not thinned or burned (Germaine et al. 2004). This underscores the importance of leaving patches of cover in restoration treatments. Patches of dense vegetation are important for bedsites, predator avoidance, and thermoregulation (Smith et al. 1986) and are likely critical for fawning (Huegel et al. 1986; Gerlach and Vaughan 1991; Fox and Krausman 1994). Female mule deer with fawns prefer to forage in areas that have adequate escape cover (King and Smith 1980). An increase in vulnerability of fawns in treated areas to predation could limit or even decrease mule deer populations (Ballard et al. 2001). We believe Gambel oak patches should be left in Arizona restoration projects, or if not available, other sources of ground cover should be made available for at least 5 years until succession restores the understory.

We feel comfortable stating that mule deer selected treated or transition areas when adjacent (< 3 km) to our study area, but we cannot infer that treated areas may change mule deer home ranges. Our methods were insufficient to determine if female mule deer changed their core use areas to use treated or transition areas, or even avoided more open areas during critical periods of short duration such as fawning. We suggest future research using GPS collars with multiple locations over a 24-hour period (diurnal and nocturnal) to study such questions.

ACKNOWLEDGMENTS

We thank the following individuals for help with capturing and observing deer, and collecting vegetation data for this study: D. Brown, D. Caputo, B. Croft, J. deVos, N. Dodd, M. Frieberg, T. Gehr, P. James, J. Johnson, G. Kneib, S. Knox, D. Rigo, I. Rodden, L. Sarten, R. Schweinsburg, and R. Vega. We thank the Arizona Strip District of the Bureau of Land Management (BLM) and the Ecological Restoration Institute at Northern Arizona University for administrative and coordination support. J. deVos, R. Schweinsburg, R. Ockenfels, and A. DiOrio improved earlier drafts of this paper. Funding was provided by the BLM Arizona Strip District and the Federal Aid in Wildlife Restoration Act Project W-78-R.

LITERATURE CITED

Alldredge, R. J., and J. T. Ratti. 1986. Comparison of some statistical techniques for analysis of resource selection. Journal of Wildlife Management 50: 157–165.

Allen, C. D. 1998. A ponderosa pine area reveals its secrets. In Status and Trends of the Nation's Biological Resources, edited by M. J. Mac, P. A. Opler, C. E. Puckett Haecker, and P. D. Doran, pp. 551–552. Two volumes. U.S. Department of Interior, U.S. Geological Survey, Reston VA.

Allen, C. D., M. Savage, D. A. Falk, K. F. Suckling, T. W. Swetnam, T. Schulke, P. B. Stacey, P. Morgan, M. Hoffman, and J. T. Klingel. 2002. Ecological restoration of southwestern ponderosa pine ecosystems: A broad perspective. Ecological Applications 12: 1418–1433.

Anonymous. 2001. Proposed 10 year federal/state/tribal/local partnership programs for restoration of Arizona/New Mexico urban interface and wildland forest areas. Prepared for the Western Governor's Association, Phoenix AZ.

Ashcraft, G. C. 1979. Effects of fire on deer in chaparral. California-Nevada Wildlife Transactions 1979: 177–189.

Ballard, W. B., D. Lutz, T. W. Keegan, L. H. Carpenter, and J. C. deVos, Jr. 2001. Deer-predator relationships: A review of recent North American studies with emphasis on mule and black-tailed deer. Wildlife Society Bulletin 29: 99–115.

Biswell, H. H. 1961. Manipulation of chamise brush for deer range improvement. California Fish and Game Quarterly 47: 125–144.

Block, W. M., A. B. Franklin, J. P. Ward, J. L. Ganey, and G. C. White. 2001. Design and implementation of monitoring studies to evaluate the success of ecological restoration on wildlife. Restoration Ecology 9: 293–303.

Bogan, M. A., C. D. Allen, E. H. Muldavin, S. P. Plantania, J. N. Stuart, G. H. Farley, P. Melhop, and J. Belnap. 1998. Southwest. In Status and Trends of the Nation's Biological Resources, edited by M. J. Mac, P. A. Opler, C. E. Puckett Haecker, and P. D. Doran, pp. 543–592. Two volumes. U.S. Department of Interior, U.S. Geological Survey, Reston VA.

Byers, C. R., Steinhorst, R. K., and P. R Krausman. 1984. Clarification of a technique for analysis of utilization-availability data. Journal of Wildlife Management 48: 1050–1053.

Carlson, P. C., G. W. Tanner, J. M. Wood, and S. R. Humphrey. 1993. Fire in key deer habitat improves browse, prevents succession, and preserves endemic herbs. Journal of Wildlife Management 57: 914–928.

Carpenter, L. H. 1998. Deer in the West. In Proceedings of the 1997 Deer/Elk Workshop, Rio Rico, AZ, edited by J. C. deVos, Jr., pp. 1–10. Arizona Game and Fish Department, Phoenix.

Cooper, C. F. 1960. Changes in vegetation, structure and growth of southwestern pine forests since European settlement. Ecological Monographs 30: 129–164.

Covington, W. W. and M. M. Moore. 1994. Southwestern ponderosa pine forest structure: Changes since Euro-American settlement. Journal of Forestry 92: 39–47.

Covington, W. W., P. Z. Fulé, M. M. Moore, S. C. Hart, T. E. Kolb, J. N. Mast, S. S. Sackett, and M. R. Wagner. 1997. Restoring ecosystem health in ponderosa pine forests of the Southwest. Journal of Forestry 95: 23–29.

Fox, K. B., and P. R. Krausman. 1994. Fawning habitat of desert mule deer. Southwestern Naturalist 39(3): 269–275.

Federal Register. 2002. National Environmental Policy Act documentation needed for fire management activities: Categorical exclusions. Vol. 67, No. 241. Washington D.C.

Gerlach, T. P., and M. R. Vaughan. 1991. Mule deer fawn bed site selection on the Pinyon Canyon maneuver site, Colorado. Southwestern Naturalist 36: 255–258.

Germaine, S. S., H. L. Germaine, and S. R. Boe. 2004. Characteristics of mule deer forage and day beds in current-condition and restoration-treated pine-oak forest. Wildlife Society Bulletin 32: 554–564.

Huegel, C. N., R. B. Dahlgren, and H. L. Gladfelter. 1986. Bedsite selection by white-tailed deer fawns in Iowa. Journal of Wildlife Management 50: 474–480.

Hungerford, C. R. 1970. Response of Kaibab mule deer to management of summer range. Journal of Wildlife Management 34: 852–862.

Jacob, J. 1974. Quantitative measure of food selection. Oecologia 14: 413–417.

Julander, O., and J. B. Low. 1976. A historic account and present status of mule deer in the West. In Mule Deer Decline in the West, edited by G. W. Workman and J. B. Lowe, pp. 3–20. Symposium Proceedings, Utah State University, Logan.

Kie, J. G. 1984. Deer habitat use after prescribed burning in northern California. Pacific Southwest Forest and Range Experiment Station. USDA Forest Service.

King, M. M., and H. D. Smith. 1980. Differential habitat utilization by the sexes of mule deer. Great Basin Naturalist 40: 273–281.

Klinger, R. C., M. J. Kutilek, and H. S. Shellhammer. 1989. Population responses of black-tailed deer to prescribed burning. Journal of Wildlife Management 53: 863–871.

Kucera, T. E., and K. E. Mayer. 1999. A sportsman's guide to improving deer habitat in California. California Department of Fish and Game, Sacramento.

Masters, R. E., R. L. Lochmiller, and E. M. Engle. 1993. Effects of timber harvest and prescribed fire on white-tailed deer forage production. Wildlife Society Bulletin 21: 401–411.

McCulloch, C. Y. 1969. Some effects of wildfire on deer habitat in pinyon-juniper woodland. Journal of Wildlife Management 33: 778–784.

Neu, C. W., C. R. Byers, and J. M. Peek. 1974. A technique for analysis of utilization-availability data. Journal of Wildlife Management 38: 541–545.

Patten, D. R. 1974. Patch cutting increases deer and elk use of a pine forest in Arizona. Journal of Forestry 72: 764–766.

Reynolds, H. G., W. P. Clary, and P. F. Ffolliott. 1970. Gambel oak for southwestern wildlife. Journal of Forestry 68: 545–547.

Rosenstock, S. S. 1998. Influence of Gambel oak on breeding birds in ponderosa pine forests of northern Arizona. Condor 100: 485–492.

Smith, H. D., M. C. Oveson, and C. L. Pritchett. 1986. Characteristics of mule deer beds. Great Basin Naturalist 46: 542–546.

Swetnam, T. W., and J. L. Betancourt. 1990. Fire–southern oscillation relations in the southwestern United States. Science 1017–1020.

Swetnam, T. W., and J. L. Betancourt. 1998. Mesoscale disturbance and ecological response to decadal climatic variability in the American Southwest. Journal of Climate 11: 3128–3147.

Swetnam, T. W., C. D. Allen, and J. L. Betancourt. 1999. Applied historical ecology: Using the past to manage for the future. Ecological Applications 9: 1189–1206.

Stager, D. W., and D. A. Klebenow. 1987. Mule deer response to wildfire in Great Basin pinyon-juniper woodland. U.S. Forest Service General Technical Report.

Taylor, G. J. 2000. Environmental assessment: Ponderosa pine ecosystem underburns. Bureau of Land Management, Arizona Strip Field Office, St. George UT.

Thill, R. E., H. F. Morris, Jr., and A. T. Harrel. 1990. Nutritional quality of deer diets from southern pine-hardwood forests. American Midland Naturalist 124: 413–417.

Wagner, M. R., W. M. Block, B. W. Geils, and K. F. Wenger. 2000. Restoration ecology: A new forest management paradigm, or another merit badge for foresters? Journal of Forestry 22–27.

Wallmo, O. C., and J. W. Schoen. 1981. Forest management for deer. In Mule and Black-Tailed Deer of North America, edited by O. C. Wallmo. University of Nebraska Press, Lincoln.

AN EVALUATION OF MULE DEER HARVEST ESTIMATES ON THE NORTH KAIBAB, ARIZONA

Amber A. Munig and Brian Wakeling

Harvest estimates are important because they provide data useful for managing hunted species. The Arizona Game and Fish Department (AGFD) manages Unit 12A for buck hunting opportunities that emphasize harvest of older age class animals, reduced hunter densities, and higher hunt success. Data collected from a mandatory hunter check station, a mail questionnaire, post-hunt population surveys, and forage monitoring are used when establishing hunts and permit levels. Knowledge of the bias and precision of harvest estimates is important because managers can then infer the degree of confidence from which they base decisions. The AGFD uses two techniques to estimate harvest: a mandatory check station for successful hunters in Unit 12A and a voluntary mail questionnaire that is used statewide, including Unit 12A. AGFD has used the mail questionnaire for harvest estimation for more than 40 years, but recent concerns regarding its accuracy and precision have arisen with the public because the two estimates differ. An evaluation of the mail questionnaire program has not been completed in about 10 years.

METHODS

We compared 13 years (1990–2002) of mule deer harvest estimates in Unit 12A obtained from voluntary mail questionnaires with those from a mandatory check station in the same area. Four hunts for antlered deer have been held annually in Unit 12A—in late October in Units 12AE (mean = 310 permits) and 12AW (mean = 897 permits), and in late November in Units 12AE (mean = 52 permits) and 12AW (mean = 140 permits). Other hunts have been held, such as general antlerless or muzzleloader antlered deer hunts, but we excluded these from analysis because they were not held each year within our study. Mail questionnaires are sent annually (an average of 2543 questionnaires) to all deer hunters in Unit 12A, with a response rate of about 52 percent. Total hunter effort and deer harvest are inferred from this sample. Data are not corrected for any type of response bias. The mandatory hunter check station was operated from 0800 to 2000 h each day of the hunt, and from 0800 to 1200 h the day after the hunt ended. Successful hunters were required to present their deer at the check station.

We tested for differences between harvest estimates using paired-sample t-tests (Zar 1996:150). We then evaluated linear relationships between harvest estimates using simple correlation. We tested for differences in slopes and elevations among the four hunt correlations by using multiple comparisons among correlation functions (Zar 1996:364).

RESULTS

We found differences between the mail questionnaires and the mandatory check station harvest estimates (t = 4.961, $P \leq 0.001$). The 13-year mean mule deer harvest as estimated from the mandatory check station was 1078 (SE = 234.17) deer. During the same period, the mean mail survey estimate was 1230 (SE = 262.97) deer. Because we detected no difference in slope or elevation of the correlation relationships in harvest estimates among the four hunts, we pooled annual harvest data. We found a significant ($P \leq 0.001$) linear relationship between the two estimates ($Y = 19.212 + 1.122 X$) that explained a substantial proportion of the variation ($r^2 = 0.998$; Figure 1).

DISCUSSION

The Arizona hunter questionnaire program was implemented in 1958, after it was compared to the hunter report card (attached to the tag itself) method that was used previously. The mail questionnaire was more effective in estimating harvest levels than the report card (Smith 1962); mail surveys lack some biases associated with other approaches (Hawn and Ryel 1969; Scheaffer et al. 1990; Steinert et al. 1994).

Research has shown that the use of mailed questionnaires could be used to estimate harvest levels and hunter effort, as well as to provide information on weapon type used, age class and gender of animal killed, unit hunted, and wounding rates. These have proved accurate enough to provide trend information to wildlife managers and administrators who then use the data to establish season dates, bag limits, and weapon types for upcoming hunts (Snyder 1963; MacDonald and Dillman 1968; Unsworth et al. 2002). Today AGFD has more than 45 years of comparable data on which to base management decisions. Deadline dates have shortened and hunt structures have become more complex, but basically the program remains the same today as in the 1960s and 1970s. This consistency is maintained so long-term trends may be analyzed and appropriate management decisions made.

Past analyses have shown that biases exist within our questionnaire data. Based on multiple-wave mailings (to increase return rates), check station data, and telephone interviews, it appears that hunter numbers, success, and harvest tend to be overestimated (Table 1) by the mail questionnaire. This occurs because successful hunters are more likely to return their questionnaires than are other people in the sample. In addition, permit holders who hunted but were not successful are more likely to respond than those who did not hunt at all. This generally results in harvest overestimates of about 10 percent. Check stations underestimate the harvest because even mandatory check stations do not always have complete compliance (Unsworth et al. 2002). Because our methods remain consistent, the biases also seem to be consistent and should therefore not compromise the comparability of data among years or areas. Fewer permits have been authorized recently, largely in response to declining deer herds. This results in smaller sample sizes, which may influence the bias and precision of harvest estimates.

Compared to the check station method, the mail questionnaire continues to show a slight overestimate of harvest and hunter participation. Currently, the cost savings of the voluntary mail questionnaire method on a statewide basis along with the historical data set justifies the continued use of the mail questionnaire program. Additional data are obtained from the questionnaire, such as total hunter days and participation. Testing of the mail questionnaire response bias has been proposed using a three-tiered mailing. The response bias test should be conducted for deer hunts statewide. The AGFD has also considered making the hunter questionnaire program mandatory, although this would require changes to existing rules governing hunting. Beyond harvest estimates, the check station allows for the collection of age structure, body mass, antler development, and body condition data that are not obtained through the mail questionnaire. Because Unit 12A is managed for high-quality harvest, which by nature is conservative in deer removal, this additional information is necessary. Because both mail questionnaire and check station harvest estimate methods provide consistently biased estimates, the final decision on which method or combination to use should be based on data needs, fiscal considerations, and public acceptance. During the years of our study, the AGFD estimated, based on population models, that 8000–12,000 mule deer occupied Unit 12A.

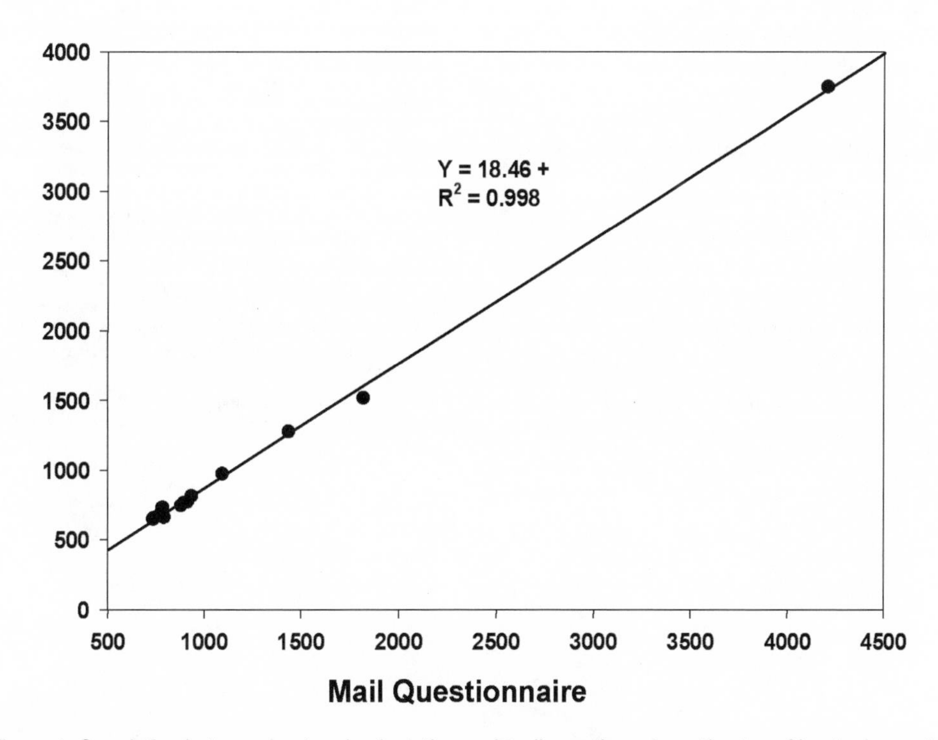

Figure 1. Correlation between hunter check station and mail questionnaire estimates of hunter harvest in Arizona, Unit 12A, 1990–2002.

Table 1. Summary of evaluations that have quantified the non-response error of questionnaire data.

Source	% Over Estimation of Hunters	% Over Estimation of Harvest
Arizona		
1961, deer field checks (Smith 1962)	–	8.2
1958–1963, multiple deer mail-outs with extrapolation (Snyder 1963)	–	12.6
1982, Region 3 javelina phone survey (AGFD unpublished data)	10.0	–
1983–1987, 3-Bar deer check station (AGFD unpublished data)	5.0	15.3
1983–1987, Tonto deer check station (AGFD unpublished data)	3.3	8.2
1984–1987, Kaibab deer check station (AGFD unpublished data) (assumes 100% check-out)	–	< 13.6
1990–1991, Unit 8 field checks, deer (AGFD unpublished data)	4.6	–
1991, Kaibab deer check station (AGFD unpublished data) (after correcting for non-checked deer)	4.4	5.7
1993, Region 5 phone survey, turkey (AGFD unpublished data)	11.4	–
New Mexico		
1960–1962, deer (Snyder 1963)	–	1.9
deer (MacDonald and Dillman 1968)	–	8.0

Differences between harvest estimates of more than 200 animals are likely to be inconsequential in this context.

ACKNOWLEDGMENTS

This analysis was undertaken and funded under the Federal Aid in Wildlife Restoration Project W-53-M of the Arizona Game and Fish Department. The manuscript benefited from reviews by Jim Unsworth and Dan Edge.

LITERATURE CITED

Hawn, L. J., and L. A. Ryel. 1969. Michigan deer harvest estimates: Sample surveys versus a complete count. Journal of Wildlife Management 33: 871–880.

MacDonald, D., and E. G. Dillman. 1968. Technique for estimating non-statistical bias in big game harvest surveys. Journal of Wildlife Management 32: 119–129.

Scheaffer, R. L., W. Mendenhall, and L. Ott. 1990. Elementary survey sampling. PWS-Kent, Boston MA.

Smith, R. H. 1962. Reporting bias in the deer hunt questionnaire. Arizona Game and Fish Department, Phoenix.

Snyder, W. A. 1963. Random card survey of big game license holders. New Mexico Game and Fish Department, Santa Fe.

Steinert, S. F., H. D. Riffel, and G. C. White. 1994. Comparisons of big game harvest estimates from check station and telephone survey. Journal of Wildlife Management 58: 335–340.

Unsworth, J. W., N. F. Johnson, L. J. Nelson, and H. M. Miyasaki. 2002. Estimating mule deer harvest in southwestern Idaho. Wildlife Society Bulletin 30: 487–491.

Zar, J. H. 1996. Biostatistical Analysis. 3rd ed. Simon and Schuster, Upper Saddle River NJ. 662 pp.

RECENT TRENDS IN NORTH AMERICAN MOUNTAIN LION POPULATIONS: A HYPOTHESIS

James C. deVos, Jr., and Ted McKinney

Mountain lion (*Puma concolor*) abundance has increased in recent decades throughout the western United States and Canada (Logan and Sweanor 2001; Riley and Malecki 2001). The mountain lion has also established or is reestablishing in areas of the Great Plains and eastern and midwestern North America (Riley and Malecki 2001; Tischendorf 2003). Resource managers consider the mountain lion to have been extirpated from these regions more than a century ago, but documented encounters between mountain lions and humans, including human deaths and injuries, have been increasing throughout western North America since the 1970s (Beier 1991; Danz 1999; Mattson et al. 2003).

Hopkins (2003) suggested the general notion that mountain lion abundance is increasing throughout the West over the last 2–3 decades, but the report lacks quantitative support. Recently, researchers in the northwestern United States have suggested that mountain lions might be declining in that region, despite an increase in reported human–mountain lion conflicts (Lambert et al. 2003). However, the consensus seems to be that mountain lion abundance in the West is increasing, a conclusion based largely on upward trends in annual hunter harvest (Logan and Sweanor 2001; Riley and Malecki 2001). Although Logan et al. (2003) suggested that recent mountain lion increases may be due to legal protection and increasing prey abundance during the 1970s to mid-1990s, the causal mechanisms of the upward trends in mountain lion abundance and distribution are poorly understood.

Sport hunting is a major cause of mountain lion deaths in most western states, and many wildlife management agencies use regulated sustained-yield harvest to manage mountain lion populations. Estimates of mountain lion population size and population dynamics relative to management prescriptions are generally derived from harvest estimates. Most states that allow legal hunting of mountain lions prohibit killing females with kittens, but this measure may be ineffective (Logan and Sweanor 2001). Mountain lion abundance has increased in many but not all states and provinces, whether mountain lion sport hunting is regulated, unregulated except for required possession of a hunting license as in Texas (Russ 1989), or prohibited as in California (Mansfield and Weaver 1989, Table 1).

We reviewed historic and recent trends of predator and ungulate population abundance and distribution, habitat loss and fragmentation, predator control, changes in vegetative cover, and the ecology of large carnivores in North America—mountain lions, grizzly bears (*Ursus arctos*), and gray wolves (*Canis lupus*)—in an effort to understand why the abundance and the distribution of mountain lions have increased in some areas of the western United States and Canada. We also reviewed information regarding bobcats (*Lynx rufus*) and coyotes (*Canis latrans*), particularly in relation to interference or exploitative competition between these smaller predators and the large carnivores. We hypothesize that increasing mountain lion abundance and distribution in North America is due to (1) the effects of extirpation of grizzly bears and wolves in reducing exploitive and interference competition between these predators and mountain lions, (2) increased absolute or relative abundance of ungulate prey available to mountain lions subsequent to extirpation of grizzly bears and wolves, and (3) changes in vegetation cover that has improved hunting efficiency for mountain lions.

PREDATOR ABUNDANCE AND DISTRIBUTION

Predator eradication programs and habitat loss and fragmentation coincided with the colonization of eastern North America by European human immigrants early in the 1600s, continuing into recent times and resulting in major changes in abundance and distribution of large carnivores. Grizzly bears and wolves were essentially eliminated in the continguous United States early in the twentieth century. Prior to European settlements in the eastern states, grizzly bears occupied most of the western half of North America. Early extinctions occurred particularly in the eastern and southwestern United States and Mexico, and grizzly bears today occupy only remote mountainous and coastal ranges in northwestern regions of North America. Survival of the grizzly bear in North America has been limited to spacious habitats insulated from excessive human-caused mortalities and protected by rugged physiography and inaccessibility. Human-caused mortalities may present a greater threat than habitat modification to the persistence of grizzly bear populations (Craighead and Mitchell 1982).

Similarly, the abundance and distribution of wolves began to decline coincident with the arrival of European immigrants. Wolves were extirpated in the eastern states between 1880 and 1914, and they were nearly gone from western states by 1944; however, the decline in range distribution from 1944 to 1974 was less than during the previous 60 years. The development of livestock agriculture, which is a crucial factor affecting wolf distribution, seems to have reached profitable limits by the 1940s, contributing less to the decline in wolves since then (Paradiso and Nowak 1982). Wolves now occur mainly in Alaska and Canada (Paradiso and Nowak 1982), but have increased in numbers (Ballard 1982) and successfully recolonized some areas of the western states in recent decades (Arjo et al. 2002).

Efforts to eradicate mountain lions and other predators in the United States extended from historic times into the mid or late twentieth century (Bekoff 1982; McCord and Cardoza 1982; Paradiso and Nowak 1982; McCulloch 1986; Cunningham et al. 2001). Mountain lions, once prevalent throughout the continental United States except Alaska, were nearly extirpated from the eastern half of the country by the late 1800s, and western populations were diminished by the early 1900s. Mountain lions are now restricted primarily to the western states and Canada, Mexico, and Central America (Nowak 1976; Dixon 1982; Culver et al. 2000). Mountain lions likely survived in western North America because of their solitary, cryptic nature and their tendency to inhabit remote, rugged terrain (Logan and Sweanor 2001).

Studies in North America indicate similar localized densities for mountain lions and wolves, and lower densities for grizzly bears; densities of the large carnivores may differ with habitat, season, and prey abundance. Estimated mountain lion densities have ranged from 0.7 to 5.6 per 100 sq km (Cunningham et al. 2001; Logan and Sweanor 2001). Wolf density estimates usually average fewer than 3.8 per 100 sq km, generally ranging from about 0.4 to 10 per 100 sq km (Paradiso and Nowak 1982; Messier 1991; Gasaway et al. 1992; Thurber et al. 1992; Vucetich et al. 2002). Estimated densities of grizzly bears generally have been 2 or fewer per 100 sq km, but have ranged from 0.4 to 8.1 per 100 sq km (Craighead and Mitchell 1982; Gasaway et al. 1992).

Habitat loss and fragmentation continue to be important factors affecting the abundance and distribution of large carnivores (Wilcox and Murphy 1985; Clark et al. 1996; Noss et al. 1996; Weaver et al. 1996; Sweanor et al. 2000). The variables that caused large carnivores to decline historically in North America, particularly killing and habitat destruction, continue to threaten grizzly bears, mountain lions, and wolves (Clark et al. 1996). Humans have undermined resiliency mechanisms and caused the decline of large carnivores by accelerating the rate and expanding the scope of environmental disturbance (Weaver et al. 1996). Habitat loss and fragmentation represent the single greatest threat to mountain lion conservation, but over-harvest remains a concern of resource managers (Logan and Sweanor 2001).

Although predator eradication also focused on bobcat and coyote populations, control efforts were unsuccessful, and habitat fragmentation and loss have had little impact on their abundance and distribution, which appear to have remained largely unchanged throughout North America following European colonization (Young 1958; Bekoff 1982; McCord and Cardoza 1982). Bobcats currently occupy most of the continental United States except Alaska and east-central regions, and occur from southern Canada through Mexico and Central America (McCord and Cardoza 1982). Coyotes are also distributed throughout most of North America, except for some northeastern parts of Canada (Bekoff 1982).

PREDATOR INTERSPECIFIC COMPETITION

Interspecific competition may be categorized as exploitation or interference. Exploitation competition is based on differential efficiency in obtaining the same food resources, whereas interference competition occurs when one species inhibits access of another to resources, via aggressive or avoidance behavior and interspecific killing. Overlap in resource use does not necessarily imply exploitation competition, but it may circumstantially indicate the potential for competition (Case and Gilpin 1974; Witmer and deCalesta 1986; Major and Sherburne 1987; Litvaitis and Harrison 1989; Thurber et al. 1992). Coexistence among sympatric carnivores may be enhanced by differences in body and prey sizes and spatio-temporal use of resources (i.e. resource specialization; Rosenzweig 1966; Gittleman 1985; Witmer and deCalesta 1986; Arjo et al. 2002).

The elimination of grizzly bears and wolves likely benefited mountain lion abundance and distribution in North America through reduced exploitation and interference competition. Murphy et al. (1999) suggested that the reduction, augmentation, or reintroduction of wolves or bears may influence mountain lions if there is significant dietary and spatial overlap. Predation, which is a function of relative abundances of predators and prey, is affected by interference among predators (Skalski and Gilliam 2001; Alonso et al. 2002); these concepts likely apply in multiple-predator and multiple-prey systems. For example, interference among mountain lions likely increases with their densities, reducing efficiencies of predation (Brown et al. 1999; Skalski and Gilliam 2001). Moreover, ungulate prey can control the risk of predation behaviorally by varying their vigilance according to mountain lion densities (Lima and Dill 1990; Brown et al. 1999). Many carnivores vary social organization in relation to prey availability (Pierce et al. 2000), and mutual interference between predators may produce nonuniform spatial distributions of predators and prey (Alonso et al. 2002). However, little is understood regarding how extirpations or recent trends to reintroduce or reestablish grizzly bears and wolves might affect the behavior, distribution, and population characteristics of mountain lions (Ruth et al. 2003a). Weaver et al. (1996) suggested that wolves are more resilient to human disturbance of habitat and populations than are mountain lions, and grizzly bears are the least resilient of these large North American carnivores. In contrast, Geist (1999) suggested that mountain lions are more adaptive to human presence than are grizzly bears and wolves.

Coyotes, bears, and wolves may displace mountain lions from their food caches, mountain lions and wolves may kill each other, and mountain lion–wolf interactions may favor wolves because of their pack behavior and the timidity of mountain lions in defending kills. Wolf predation also might alter the population characteristics, behavior, and distribution of prey, and reductions in prey abundance might adversely affect food acquisition by mountain lions (Murphy et al. 1999). Ballard (1982) observed wolf packs on 130 kills, of which 13 percent were contested by grizzly bears, and concluded that the predators sometimes interacted with each other, and both might scavenge the same kill. Wolves may kill coyotes, and interference competition between the species may reduce the abundance of coyotes in some areas, although differential use of food resources might enhance the coexistence of coyotes and wolves (Thurber et al. 1992; Arjo et al. 2002). Coyotes have established in some areas where wolves have been extirpated (Ballard et al. 1999). Limited studies (Lingle 2002) also suggest that increased coyote abundance might influence habitat use by mule deer (*Odocoileus hemionus*), possibly increasing prey availability in steeper, rougher terrain more suitable to mountain lions. Although interference competition between large and small carnivores may occur to some degree, the abundance and distribution of bobcats and coyotes likely have little effect on mountain lion abundance and distribution. Limited data suggest that coyote populations negatively influence bobcat abundance, and coyotes may not compete well with mountain lions and wolves (Young 1958; Bekoff 1982; Boyd and O'Gara 1985; Litvaitis and Harrison 1989; Koehler and Hornocker 1991; Murphy et al. 1999).

Canids and felids have evolved different hunting styles and mechanisms of social organization. Most felids are solitary hunters and feeders that rely on stealth and ambush to kill prey, whereas Canidae as a family are adapted primarily to cursorial hunting behavior based on cooperation within packs (Kleiman and Eisenberg 1973). Wolves tend to hunt in packs continually during the day, relying largely on scent or chance encounters to locate prey, and speed during chases to make the kill (Kleiman and Eisenberg 1973; Paradiso and Nowak 1982). In contrast, mountain lions are solitary hunters that concentrate their activities between dusk and dawn, rely on visual location of prey, and use cover and stealth to kill through ambush or brief attacks (Dixon 1982; Beier et al. 1995; Pierce et al. 2000). Grizzly bears may exhibit primarily crepuscular activities, scavenging winter-killed ungulates or preying on weakened moose (*Alces alces*) on winter ranges (Craighead and Mitchell 1982) and juvenile moose in the spring (White et al. 2001). Bobcats and coyotes incorporate variable hunting tactics, depending on prey size (McCord and Cardoza 1982; Pierce et al. 2000).

The coexistence of grizzly bears, mountain lions, and wolves in North America is apparent in their sympatry over some 100 centuries since the end of Pleistocene glaciations. We suggest that the coexistence of these species in the presence of relatively abundant ungulate prey and overlapping use of prey by each predator was enhanced largely via exploitation and interference competition. Although overlap is apparent in the use of ungulate prey by the large carnivores, interspecific avoidance behavior (particularly by mountain lions) and differences in hunting styles and diurnal activity patterns likely enhanced coexistence.

ABUNDANCE AND DISTRIBUTION OF UNGULATE PREY

Mule deer are currently distributed over most of western North America in the United States, Canada, and Mexico. Populations declined dramatically in most areas during or immediately following human settlement of their range, but by the 1950s, mule deer had become abundant throughout their range, followed later by declines in most of the West (Julander and Low 1976; Mackie et al. 1982; Ballard et al. 2001). In Arizona, mule deer abundance increased steadily during the 1970s to peak during 1985–1988, but has declined since then (Kamler et al. 2002); this characterizes the trend in many western states (Ballard et al.

2001). Regardless, the distribution and numbers of mule deer in North America today are higher than when herds were depleted by settlers in the early twentieth century (Denney 1976; Julander and Low 1976; Mackie et al. 1982). In contrast to mule deer, white-tailed deer (*O. virginianus*) are adaptable to human activities, and therefore evidenced smaller historical declines than mule deer, and their distribution and numbers have expanded in recent years throughout the contiguous United States except for the east-central regions, and in southern Canada and most of Mexico (Hessleton and Hessleton 1982).

Elk historically were distributed over most of temperate North America, except for the Great Basin and southeastern states. The abundance and distribution of the cervid declined severely by the early twentieth century, but numbers increased dramatically throughout the Rocky Mountains by about the 1930s, where elk are abundant today (Peek 1982). Elk have been successfully reintroduced into two eastern states, Michigan and Pennsylvania, and efforts are continuing to restore free-ranging populations in other eastern states (Larkin et al. 2003). Moose currently occupy extensive areas in the northern parts of the continent, and their numbers are stable or increasing in many parts of their range (Coady 1982).

Bighorn sheep (*Ovis canadensis*) abundance and distribution historically likely exceeded current levels, with abundance declining due to human impacts during the last half of the nineteenth century. The abundance and distribution of bighorn sheep increased during the second half of the twentieth century, largely through efforts to reintroduce them in areas where they had been extirpated (Krausman and Shackleton 2000). They now occur in patchy populations throughout western North America (Lawson and Johnson 1982; Valdez and Krausman 1999).

Collared peccary (*Pecari tajacu*) and domestic cattle are now both important dietary items for mountain lions. Day (1985) has cited lack of collared peccary remains from archeological digs in Arizona as evidence that they did not inhabit the American Southwest prior to European settlement. Based on the writings of early explorers of the Southwest, Day suggested that collared peccary likely began to colonize this region in approximately 1700. Since that time, collared peccary have continued to expand their range in Arizona (Knipe 1956; Day 1985), and today they occupy most of Arizona south of the Mogollon Rim. With European settlement, domestic cattle were grazed in most areas of Arizona and domestic cattle have proven to be a readily available supplement to mountain lion diets. In a study in central Arizona, Shaw (1977) determined that 37 percent of the mountain lion kills he investigated were domestic cattle and that 34 percent of the scats examined contained cattle remains.

Cunningham et al. (1995) examined the importance of domestic cattle as dietary components for mountain lions in southern Arizona and found that domestic cattle were the primary food item consumed by mountain lions. They found that in summer, domestic cattle comprised more than 51 percent of the biomass consumed, and in winter the mountain lion diet consisted of more than 44 percent cattle. Both of these values exceeded the biomass of deer consumed.

During recent times, mountain lion density appears to be positively related to the abundance of cervids in habitats with stalking cover. Wolf density is positively associated with ungulate biomass and may be higher where deer biomass is prevalent (Weaver et al. 1996). The abundance and distribution of mountain lions in some western states are also linked with the availability of ungulate prey (Shaw et al. 1988; Pierce et al. 2000; Riley and Malecki 2001). Laundré and Hernández (2000) reported that mountain lion numbers in Idaho and Utah increased from 1987 to 1996 due to high deer abundance, and then declined following high winter mortality of deer in 1994. Survival of young mountain lions was most dependent upon deer abundance. The relative density of wolves is also linked to differences in moose and white-tailed deer densities in some areas (Messier 1991; Gasaway et al. 1992).

Deer are the principal natural prey of mountain lions, but numerous studies indicate a variable prey base, including bighorn sheep, elk, collared peccary, pronghorn (*Antilocapra americana*), and domestic cattle (Iriarte et al. 1990; Ockenfels 1994; Cunningham et al. 1999; Kunkel and Pletscher 1999; Kamler et al. 2002). Recent upward trends in mountain lion abundance might reflect a shift to consumption of the increased alternate prey coincident with mule deer declines (Leopold and Krausman 1986; Branch et al. 1996; Kamler et al. 2002; Rominger et al. 2004).

PREDATOR DIETS AND PREDATION

Mountain lions and wolves are adapted to carnivorous diets, and both predators tend to hunt and kill large prey, although their diets also include small animals such as rabbits and rodents (Rosenzweig 1966; Holleman and Stephenson 1981; Dixon 1982; Paradiso and Nowak 1982; Kunkel et al. 1999). Predation by mountain lions and wolves can be significant mortality factors for ungulate populations (Bleich and Taylor 1998; Ballard et al. 2001; Jedrzejewski et al. 2002). Sympatric mountain lions, bears, and wolves prey on mule and white-tailed deer, elk, and moose (Kunkel and Pletscher 1999; Kunkel et al. 1999; Arjo et al. 2002), and sympatric grizzly bears and wolves in the absence of mountain lions prey on moose (Boutin 1992; Van Ballenberghe and Ballard 1994; White et al. 2001).

In comparison to mountain lions and wolves, grizzly bears are more omnivorous and have a diverse diet that includes primarily vegetation, but they consume animals ranging in size from ants and moths to deer, elk, livestock, and moose (Craighead and Mitchell 1982; Kunkel and Pletscher 1999; White et al. 2001). Relationships between grizzly bear abundance and habitat or key food resource variables have not been determined (Weaver et al. 1996), but combined predation by wolves and bears—grizzly bears and/or black bears (*Ursus americanus*)—can maintain moose populations within low-density equilibrium (Gasaway et al. 1992). Grizzly bear predation of ungulates tends to occur mainly during spring to early summer, and they commonly scavenge carcasses (Craighead and Mitchell 1982).

Coyotes and bobcats also prey on deer, other large game, livestock, and scavenge kills made by mountain lions and wolves, but their diets tend to consist primarily of smaller prey, such as rabbits and rodents (Young 1958; Rosenzweig 1966; Bekoff 1982; McCord and Cardoza 1982; Gittleman 1985; Koehler and Hornocker 1991; Ockenfels 1994; Pierce et al. 2000). Nonetheless, predation by coyotes may cause significant mortality in deer populations in some areas, whereas bobcats likely have little impact on populations of large prey (Ballard et al. 2001). Coyotes may scavenge wolf kills more often than they scavenge mountain lion kills (Arjo et al. 2002).

CHANGES IN LANDSCAPE-SCALE VEGETATIVE STRUCTURE

The vegetative structure in the southwestern United States was much different prior to European settlement than that found today. Swetnam and Betancourt (1990) found that the fire regime in southwestern coniferous forests was markedly different after 1900 than prior to that date. Before 1900, fires occurred in ponderosa pine (*Pinus ponderosa*) at 2–6 year intervals, but recently that interval has increased to more than 40 years, and the once open forests have become more densely vegetated with young trees. Pinyon (*Pinus* spp.) and juniper (*Juniperus* spp.) woodlands have also undergone marked change in density and distribution in recent times and now occupy large portions of the Southwest that once were grasslands (Miller 1921; Rogers et al. 1984; Cinnamon 1988). In a site north of Flagstaff, Arizona, Jameson (1962) reported that only a few scattered juniper trees occurred in 1907, but in 1960 there were 148 trees per ha, with some of the trees being up to 3 m tall. With the widespread introduction of domestic livestock and intense fire suppression efforts, woody invasion has occurred in virtually all vegetation communities in Arizona. We believe this increase in woody species has increased the ability of mountain lions to meet their dietary demands because they employ a stalk-pounce strategy, and the denser vegetation would limit the sight distance prey species rely on to avoid predation, a concept supported by Rominger et al. (2004).

HUMAN–MOUNTAIN LION ENCOUNTERS

Documented human encounters, injuries, and deaths involving mountain lions have increased in recent years throughout the West. The causes of this trend are poorly understood (Mattson et al. 2003). Subadult mountain lions, particularly males about 10–33 months of age, tend to disperse from birth areas to other locations, traveling through marginal or suitable habitats. Maximum dispersal distances of about 80 km and 215 km have been reported for females and males, respectively (Beier 1995; Sweanor et al. 2000; Logan and Sweanor 2001). Beier (1991) found that nearly 40 percent of mountain lions attacking humans were yearlings (12–23 months old), suggesting that they were dispersing animals or young animals beginning to hunt without maternal assistance. Dispersing mountain lions move from and into subpopulations via habitat corridors in fragmented landscapes, and encounters with human-related elements can increase mountain lion mortalities (Beier 1995), as well as the potential risk of dangerous encounters between humans and mountain lions (Logan et al. 2003). Survival in dispersing mountain lions is probably low (Beier 1995; Sweanor et al. 2000; Logan and Sweanor 2001), perhaps hampering colonization of viable populations in areas of the eastern-midwestern United States, where the predator was previously extirpated. Conversely, the presence of other mountain lions, particularly members of the opposite sex, might enhance successful colonization by individual dispersers in habitats where mountain lions were previously extirpated (Seidensticker et al. 1973).

Human activities may not lead to strong avoidance by mountain lions if repeated exposure to humans results in no negative consequences (Murphy et al. 1999). Sweanor et al. (2003) approached mountain lions unaccustomed to human activity and observed them at distances as little as 3 m for periods up to 2 hours on 262 occasions; mountain lions exhibited threat behavior in only 6 percent of encounters. Perhaps mountain lions are more tolerant of human presence than previously believed, and typically rare sightings may be related to the predators' cryptic behavior. Acquisition of prey near human activities may encourage mountain lions to habituate to humans. Habitat fragmentation and loss and habituation to human activities likely expose humans increasingly to both resident and dispersing, transient mountain lions.

The increase in human encounters, injuries, and deaths in recent years associated with mountain lions likely results from greater human encroachment of occupied mountain lion habitat through expanded human developments and increased recreational activities, as well as increased abundance of mountain lions (Murphy et al. 1999; Koehler and Nelson 2003; Mattson et al. 2003). Three factors likely explain much of the recently increased contact between humans and mountain lions (Danz 1999): increased abundance and distribution of the predator; encroachment of a growing human population onto marginal and suitable mountain lion habitat through construction and recreational activities; and possible adjustment or habituation of mountain lions to human presence.

SUMMARY

North American ecosystems were profoundly disturbed after Europeans arrived in the eastern United States in the early 1600s. Broad disturbances associated with human encroachments over North America have continued into the twenty-first century. Habitat fragmentation, loss, and modification as well as persecution continue to threaten grizzly bear, mountain lion, and wolf populations in some areas of North America. Interdependence among components and variability are fundamental properties of ecosystems, but only in recent years have scientists begun to study the interrelationships among large carnivores that might affect their behavior, distribution, and population characteristics.

We believe the coexistence of grizzly bears, wolves, and mountain lions in North America was likely enhanced through behavioral mechanisms. Although population level impacts are uncertain, several lines of evidence suggest that exploitation and interference competition occurred historically between sympatric mountain lions, wolves, and grizzly bears. Distributions and use of habitats by the large carnivores overlapped, as did their use of ungulate prey species, suggesting exploitation competition. Spatial displacements and direct mortalities of mountain lions, caused particularly by wolves, also suggest the importance of interference competition in community structure and function.

Predator eradication efforts began with the European occupation of eastern North America in the early 1600s, and extended into at least the mid twentieth century; grizzly bears and wolves were extirpated from most of the continent. Mountain lions, however, persisted in the western states and Canada, likely due at least in part to their cryptic behavior, greater tolerance of human disturbances, and ability to form inconspicuous suburban populations. Bobcats and coyotes also survived predator eradication efforts with little change in abundance and distribution, but it is unlikely that significant competition occurs between these predators and mountain lions.

Populations of wild ungulate prey, largely depleted in many areas of North America following European settlement and population expansion, have increased dramatically since the early twentieth century, with the end of extensive predator eradication campaigns. Persecution of predators alone does not provide causal explanation for these trends, but increased abundance and distribution of wild prey populations, coupled with the introduction of large numbers of domestic livestock (Rominger et al. 2004) likely provided greater absolute and relative availability of preferred and alternative prey for mountain lions. The pattern of greater ungulate prey availability to mountain lions in North America has persisted over recent decades, despite local or regional declines in some populations.

We hypothesize that reduced interference and exploitation competition among the large carnivores and greater availability of ungulate prey to mountain lions were important factors contributing to the apparent increases in abundance and possibly distribution of mountain lions in western North America during the past several decades. Landscape-scale invasion of woody plant species has also likely increased the ability of mountain lions to take advantage of the increased prey densities. Consistent with this notion, biologists have recently recognized the need for research to enhance our understanding of competition between large carnivores, and the necessity for conservation efforts to address the roles of habitat, prey availability, and human disturbance in interspecific predator relationships (Kunkel et al. 1999; Murphy et al. 1999; Logan and Sweanor 2001; Quigley et al. 2003; Ruth et al. 2003a, 2003b).

LITERATURE CITED

Alonso, D., F. Bartumeus, and J. Catalan. 2002. Mutual interference between predators can give rise to Turing spatial patterns. Ecology 83: 28–34.

Arjo, W. M., D. H. Pletscher, and R. R. Ream. 2002. Dietary overlap between wolves and coyotes in northwestern Montana. Journal of Mammalogy 83: 754–766.

Ballard, W. B. 1982. Gray wolf–brown bear relationships in the Nelchina Basin of south-central Alaska. In Wolves of the World—Perspectives of Behavior, Ecology, and Conservation, edited by F. H. Harrington and P. C. Pacquet, pp. 71–80. Noyes Publications, Park Ridge NJ.

Ballard, W. B., H. A. Whitlaw, S. J. Young, R. A. Jenkins, and G. J. Forbes. 1999. Predation and survival of white-tailed deer fawns in north-central New Brunswick. Journal of Wildlife Management 63: 574–579.

Ballard, W. B., D. Lutz, T. W. Keegan, L. H. Carpenter, and J. C. deVos, Jr. 2001. Deer-predator relationships: A review of recent North American studies with emphasis on mule and black-tailed deer. Wildlife Society Bulletin 29: 99–115.

Beier, P. 1991. Cougar attacks on humans in the United States and Canada. Wildlife Society Bulletin 19: 403–412.

Beier, P. 1995. Dispersal of juvenile cougars in fragmented habitat. Journal of Wildlife Management 59: 228–237.
Beier, P., D. Choate, and R. H. Barrett. 1995. Movement patterns of mountain lions during different behaviors. Journal of Mammalogy 76: 1056–1070.
Bekoff, M. 1982. Coyote. In Wild Mammals of North America, edited by J. A. Chapman and G. A. Feldhamer, pp. 447–459. Johns Hopkins University Press, Baltimore MD.
Bleich, V. C., and T. J. Taylor. 1998. Survivorship and cause-specific mortality in five populations of mule deer. Great Basin Naturalist 58: 265–272.
Boutin, S. 1992. Predation and moose population dynamics: A critique. Journal of Wildlife Management 56: 116–127.
Boyd, D., and B. O'Gara. 1985. Cougar predation on coyotes. Murrelet 66:17.
Branch, L. C., M. Pessing, and D. Villarreal. 1996. Response of pumas to a population decline of the plains vizcacha. Journal of Mammalogy 77: 1132–1140.
Brown, J. S., J. W. Laundré, and M. Guring. 1999. The ecology of fear: Optimal foraging, game theory, and trophic interactions. Journal of Mammalogy 80: 385-399.
Case, T. J., and M. R. Gilpin. 1974. Interference competition and niche theory. Proceedings of the National Academy of Science 71: 3073–3077.
Cinnamon, S. K. 1988. The plant community of Cedar Canyon, Wupatki National Monument, as influenced by prehistoric and historic environmental changes. Master's thesis, Northern Arizona University, Flagstaff.
Clark, T. W., A. P. Curlee, and R. P. Reading. 1996. Crafting effective solutions to the large carnivore conservation problem. Conservation Biology 10: 940–948.
Coady, J. W. 1982. Moose. In Wild Mammals of North America, edited by J. A. Chapman and G. A. Feldhamer, pp. 902–922. Johns Hopkins University Press, Baltimore MD.
Craighead, J. J., and J. A. Mitchell. 1982. In Wild Mammals of North America, edited by J. A. Chapman and G. A. Feldhamer, pp. 515–556. Johns Hopkins University Press, Baltimore MD.
Culver, M., W. E. Johnson, J. Pecon-Slattery, and S. J. O'Brien. 2000. Genomic ancestry of the American puma (*Puma concolor*). Journal of Heredity 91: 186–197.
Cunningham, S. C., L. A. Haynes, C. Gustuvson, and D. Haywood. 1995. Evaluation of the interaction between mountain lion and cattle in the Aravaipa-Klondyke area. Technical Report 17. Arizona Game and Fish Department, Phoenix.
Cunningham, S. C., C. R. Gustavson, and W. B. Ballard. 1999. Diet selection of mountain lions in southeastern Arizona. Journal of Range Management 52: 202–207.
Cunningham, S. C., W. B. Ballard, and H. A. Whitlow. 2001. Age structure, survival, and mortality of mountain lions in southeastern Arizona. Southwestern Naturalist 46: 76–80.
Danz, H. P. 1999. Cougar! Swallow Press/Ohio University Press, Athens.
Day, G. I. 1985. Collared peccary—research and management in Arizona. Arizona Game and Fish Department, Phoenix.
Denney, R. N. 1976. Regulations and mule deer harvest—political and biological management. In Mule Deer Decline in the West—A Symposium, edited by G. W. Workman and J. B. Low, pp. 87–92. Utah State University, Logan.
Dixon, K. R. 1982. Mountain lion. In Wild Mammals of North America, edited by J. A. Chapman and G. A. Feldhamer, pp. 711–727. Johns Hopkins University Press, Baltimore MD.
Gasaway, W. C., R. D. Boertje, D. V. Grangaard, D. G. Kelleyhouse, R. O. Stephenson, and D. G. Larsen. 1992. The role of predation in limiting moose at low densities in Alaska and Yukon and implications for conservation. Wildlife Monographs 120: 1–59.
Geist, V. 1999. Adaptive strategies in American mountain sheep. In Mountain Sheep of North America, edited by R. Valdez and P. R. Krausman, pp. 192–208. University of Arizona Press, Tucson.
Gittleman, J. L. 1985. Carnivore body size: Ecological and taxonomic correlates. Oecologia 67: 540–554.
Hessleton, W. T., and R. M. Hessleton 1982. White-tailed deer. In Wild Mammals of North America, edited by J. A. Chapman and G. A. Feldhamer, pp. 878–910. Johns Hopkins University Press, Baltimore MD.
Holleman, D. F., and R. O. Stephenson. 1981. Prey selection and consumption by Alaskan wolves in winter. Journal of Wildlife Management 45: 620–628.
Hopkins, R. A. 2003. Mystery, myth and legend: The politics of cougar management in the new millennium. Mountain Lion Workshop, Jackson WY (abstract) 7: 55.
Iriarte, J. A., W. L. Franklin, W. E. Johnson, and K. H. Redford. 1990. Biogeographic variation of food habits and body size of the American puma. Oecologia 85: 185–190.
Jedrzejewski, W., K. Schmidt, J. Theuerkauf, B. Kedrzejewska, N. Selva, K. Sub, and L. Szymura. 2002. Kill rates and predation by wolves on ungulate populations in Bia_owie_a primeval forest (Poland). Ecology 83: 1341–1356.
Julander, O., and J. B. Low. 1976. A historical account and present status of the mule deer in the West. In Mule Deer Decline in the West—A Symposium, edited by G. W. Workman and J. B. Low, pp. 3–19. Utah State University, Logan.
Kamler, J. F., R. M. Lee, J. C. deVos, Jr., W. B. Ballard, and H. A. Whitlaw. 2002. Survival and cougar predation of translocated bighorn sheep in Arizona. Journal of Wildlife Management 66: 1267–1272.
Kleiman, D. G., and J. F. Eisenberg. 1973. Comparisons of canid and felid social systems from an evolutionary perspective. Animal Behaviour 21: 637–659.
Knipe, T. 1956. Javelina in Arizona—a research and management study. Arizona Game and Fish Department, Phoenix.

Koehler, G. M., and M. G. Hornocker. 1991. Seasonal resource use among mountain lions, bobcats, and coyotes. Journal of Mammalogy 72: 391–396.

Koehler, G. M., and E. Nelson. 2003. Project CAT (cougars and teaching): Integrating science, schools and community development planning. Mountain Lion Workshop, Jackson WY (abstract) 7: 46.

Krausman, P. R., and D. M. Shackleton. 2000. Bighorn sheep. In Ecology and Management of Large Mammals in North America, edited by S. Demarais and P. R. Krausman, pp. 517–544. Prentice-Hall, Upper Saddle River NJ.

Kunkel, K., and D. H. Pletscher. 1999. Species-specific population dynamics of cervids in a multipredator ecosystem. Journal of Wildlife Management 63: 1082–1093.

Kunkel, K. E., T. K. Ruth, D. H. Pletscher, and M. G. Hornocker. 1999. Winter prey selection by wolves and cougars in and near Glacier National Park, Montana. Journal of Wildlife Management 63: 901–910.

Lambert, C., H. S. Robinson, H. Cruickshank, D. D. Katnik, and R. B. Wielgus. 2003. Cougar population dynamics in the northwest. Mountain Lion Workshop, Jackson WY (abstract) 7: 45.

Larkin, J. L., D. S. Maehr, J. J. Cox, D. C. Bolin, and M. W. Wichrowski. 2003. Demographic characteristics of a reintroduced elk population in Kentucky. Journal of Wildlife Management 67: 467–476.

Laundré, J. W., and L. Hernández. 2000. Long term population trends of mountain lions in southeastern Idaho and northwestern Utah. Mountain Lion Workshop, San Antonio, TX (abstract) 6: 24–25.

Lawson, B., and R. Johnson. 1982. Mountain sheep. In Wild Mammals of North America, edited by J. A. Chapman and G. A. Feldhamer, pp. 1036–1055. Johns Hopkins University Press, Baltimore MD.

Leopold, B. D., and P. R. Krausman. 1986. Diets of 3 predators in Big Bend National Park, Texas. Journal of Wildlife Management 50: 290–295.

Lima, S. L., and L. M. Dill. 1990. Behavioral decisions made under the risk of predation: A review and prospectus. Canadian Journal of Zoology 68: 619–640.

Lingle, S. 2002. Coyote predation and habitat segregation of white-tailed deer and mule deer. Ecology 83: 2037–2048.

Litvaitis, J. A., and D. J. Harrison. 1989. Bobcat-coyote niche relationships during a period of coyote population increase. Canadian Journal of Zoology 67: 1180–1188.

Logan, K. A., and L. L. Sweanor. 2001. Desert Puma. Island Press, Washington D.C.

Logan, K. A., L. L. Sweanor, and M. G. Hornocker. 2003. Reconciling science and politics in puma management in the West: New Mexico as a template. Mountain Lion Workshop, Jackson, WY (abstract) 7: 56.

Mackie, R. J., K. L. Hamlin, and D. F. Pac. 1982. Mule deer. In Wild Mammals of North America, edited by J. A. Chapman and G. A. Feldhamer, pp. 862–877. Johns Hopkins University Press, Baltimore MD.

Major, J. T., and J. A. Sherburne. 1987. Interspecific relationships of coyotes, bobcats, and red foxes in western Maine. Journal of Wildlife Management 51: 606–616.

Mansfield, T. M., and R. A. Weaver. 1989. The status of mountain lions in California. Proceedings of the Third Mountain Lion Workshop 3: 15–18. Prescott AZ.

Mattson, D. J., J. V. Hart, and P. Beier. 2003. A conceptual model and appraisal of existing research related to interactions between humans and pumas. Mountain Lion Workshop, Jackson WY (abstract) 7: 10.

McCord, C. M., and J. E. Cardoza. 1982. Bobcat and lynx. In Wild Mammals of North America, edited by J. A. Chapman and G. A. Feldhamer, pp. 728–766. Johns Hopkins University Press, Baltimore MD.

McCulloch, C. Y. 1986. A history of predator control and deer productivity in northern Arizona. Southwestern Naturalist 31: 215–220.

Messier, F. 1991. The significance of limiting and regulating factors on the demography of moose and white-tailed deer. Journal of Animal Ecology 60: 377–393.

Miller, F. H. 1921. Reclamation of grasslands by Utah juniper on the Tusuyan National Forest, Arizona. Journal of Forestry 19: 647–651.

Murphy, K. M., P. I. Ross, and M. G. Hornocker. 1999. The ecology of anthropogenic influences on cougars. In Carnivores in Ecosystems: The Yellowstone Experience, edited by T. W. Clark, A. P. Curlee, S. C. Minta, and P. M. Kareiva, pp. 77–101. Yale University Press, New Haven CT.

Noss, R. F., H. B. Quigley, M. G. Hornocker, T. Merrill, and P. C. Paquet. 1996. Conservation biology and carnivore conservation in the Rocky Mountains. Conservation Biology 10: 949–963.

Nowak, R. M. 1976. The cougar in the United States and Canada. U.S. Department of Interior Fish and Wildlife Service, Washington D.C., and New York Zoological Society, New York.

Ockenfels, R. A. 1994. Mountain lion predation on pronghorn in central Arizona. Southwestern Naturalist 39: 305–306.

Paradiso, J. L., and R. M. Nowak. 1982. Wolves. In Wild Mammals of North America, edited by J. A. Chapman and G. A. Feldhamer, pp. 460–474. Johns Hopkins University Press, Baltimore MD.

Peek, J. M. 1982. Elk. In Wild Mammals of North America, edited by J. A. Chapman and G. A. Feldhamer, pp. 851–861. Johns Hopkins University Press, Baltimore MD.

Pierce, B. M., V. C. Bleich, and R. T. Bowyer. 2000. Social organization of mountain lions: Does a land-tenure system regulate population size? Ecology 81: 1533–1543.

Quigley, H., and nine co-authors. 2003. Response by three large carnivores to recreational big game hunting along the Yellowstone National Park and Absaroka–Beartooth Wilderness boundary. Mountain Lion Workshop, Jackson, WY (abstract) 7: 40.

Riley, S. J., and R. A. Malecki. 2001. A landscape analysis of cougar distribution and abundance in Montana, USA. Environmental Management 28: 317–323.

Rogers, G. F., H. E. Malde, and R. M. Turner. 1984. Bibliography of repeat photography for evaluating landscape change. University of Utah Press, Salt Lake City.

Rominger, E. M., H. A. Whitlaw, D. L. Weybright, W. C. Dunn, and W. B. Ballard. 2004. The influence of mountain lion predation on bighorn sheep translocations. Journal of Wildlife Management 68 (4): 993–999.

Rosenzweig, M. L. 1966. Community structure in sympatric Carnivora. Journal of Mammalogy 47: 602–612.

Russ, W. B. 1989. Status of the mountain lion in Texas. Proceedings of the Third Mountain Lion Workshop 3: 30–31. Prescott AZ.

Ruth, T. K., P. C. Buotte, H. B. Quigley, and M. G. Hornocker. 2003a. Cougar ecology and cougar-wolf interactions in Yellowstone National Park: A guild approach to large carnivore conservation. Mountain Lion Workshop, Jackson, WY (abstract) 7: 27.

Ruth, T. K., P. C. Buotte, K. M. Murphy, M. G. Hornocker, and H. B. Quigley. 2003b. Cougar predation on prey in Yellowstone National Park: A preliminary comparison pre- and post-wolf reestablishment. Mountain Lion Workshop, Jackson, WY (abstract) 7: 33.

Seidensticker, J. C., M. G. Hornocker, W. V. Wiles, and J. P. Messick. 1973. Mountain lion social organization in the Idaho Primitive Area. Wildlife Monographs 35: 1–60.

Shaw, H. G. 1977. Impacts of mountain lion on mule deer and cattle in northwestern Arizona. In Proceedings of the 1975 Predator Symposium, edited by R. L. Phillips and C. Jonkel, pp. 306–318. University of Montana, Missoula.

Shaw, H. G., N. G. Woolsey, J. R. Wegge, and R. L. Day, Jr. 1988. Factors affecting mountain lion densities and cattle depredation in Arizona. Arizona Game and Fish Department Research Branch Final Report, Phoenix.

Skalski, G. T., and J. F. Gilliam. 2001. Functional responses with predator interference: Viable alternatives to the Holling type II model. Ecology 82: 3083–3092.

Sweanor, L. L., K. A. Logan, and M. G. Hornocker. 2000. Cougar dispersal patterns, metapopulation dynamics, and conservation. Conservation Biology 14: 798–808.

Sweanor, L. L., K. A. Logan, and M. G. Hornocker. 2003. Puma responses to close encounters with researchers. Mountain Lion Workshop, Jackson, WY (abstract) 7: 13.

Swetnam, T.W., and J. L. Betancourt. 1990. Fire–southern oscillation relationships in the southwestern United States. Science: 1017–1020.

Thurber, J. M., R. O. Peterson, J. D. Woolington, and J. A. Vucetich. 1992. Coyote coexistence with wolves on the Kenai Peninsula, Alaska. Canadian Journal of Zoology 70: 2494–2498.

Tischendorf, J. W. 2003. Cryptic cougars—perspectives on the puma in eastern, Midwestern, and Great Plains regions of North America. Mountain Lion Workshop, Jackson, WY (abstract) 7: 5.

Valdez, R., and P. R. Krausman. 1999. Description, distribution, and abundance of mountain sheep in North America. In Mountain Sheep of North America, edited by R. Valdez and P. R. Krausman, pp. 3–22. University of Arizona Press, Tucson.

Van Ballenberghe, V., and W. B. Ballard. 1994. Limitation and regulation of moose populations: The role of predation. Canadian Journal of Zoology 72: 2071–2077.

Vucetich, J. A., R. O. Peterson, and C. L. Schaefer. 2002. The effect of prey and predator densities on wolf predation. Ecology 83: 3003–3013.

Weaver, J. L., P. C. Paquet, and L. F. Ruggiero. 1996. Resilience and conservation of large carnivores in the Rocky Mountains. Conservation Biology 10: 964–976.

White, K. S., J. W. Testa, and J. Berger. 2001. Behavioral and ecologic effects of differential predation pressure on moose in Alaska. Journal of Mammalogy 82: 422–429.

Wilcox, B. A., and D. D. Murphy. 1985. Conservation strategy: The effects of fragmentation on extinction. American Naturalist 125: 879–887.

Witmer, G. W., and D. S. deCalesta. 1986. Resource use by unexploited sympatric bobcats and coyotes in Oregon. Canadian Journal of Zoology 64: 2333–2338.

Young, S. P. 1958. The Bobcat of North America. The Stackpole Company, Harrisburg PA, and the Wildlife Management Institute, Washington D.C.

PATTERNS OF CARNIVORE CO-OCCURRENCE ON THE NORTH RIM, GRAND CANYON NATIONAL PARK

Sarah E. Reed and Elaine F. Leslie

Extinctions of mammal species in U.S. national parks have led to increased concern about the fate of the remaining carnivores (Newmark 1995; Parks and Harcourt 2002). Predators affect trophic processes through predation (Estes and Palmisano 1974) and scavenger subsidies (Wilmers et al. 2003), and changes in species composition can have important effects on community structure and stability (Mittelbach et al. 1995; Crooks and Soulé 1999). Wide-ranging, low-density carnivores are particularly at risk for population declines due to edge effects and external threats posed by humans (Woodroffe and Ginsberg 1998).

Conserving multiple species across the landscape requires an understanding of spatial relationships among species as well as the habitat relationships of each species. Laboratory and field experiments have demonstrated local impacts of predation and competition on species distributions (e.g., Huffaker 1958; Connell 1961) and research on radio-collared animals has shown that mammalian carnivores avoid each other in space (Fedriani et al. 1999; Neale and Sacks 2001). However, few studies consider interaction processes in the design of non-invasive surveys and monitoring (e.g., Fedriani et al. 2000). Distribution models are typically based on site and landscape variables related to habitat quality (e.g., Virgos et al. 2002) and do not incorporate data on the distributions of co-occurring species. These sampling and modeling approaches implicitly assume that species distributions are not correlated in space.

Null model analyses of community structure and species co-occurrence are popular tools for investigating spatial relationships in species presence/absence data. Null model analysis has been proposed as a method for quantifying non-random co-occurrence patterns arising from competitive exclusion in island ecosystems (Diamond 1975; Connor and Simberloff 1979). The null model approach has been controversial, but many statistical concerns associated with early analyses have been addressed (Gotelli 2000). A recent meta-analysis of 96 data sets encompassing a variety of taxa and ecosystems showed non-random patterns of species co-occurrence according to several well-tested indices of community structure (Gotelli and McCabe 2002).

In this chapter, we describe a pilot study pairing multiple carnivore species surveys with analyses of community structure and co-occurrence. We conducted our research in the relatively homogeneous ponderosa pine forest of the North Rim of Grand Canyon National Park (GCNP), where we assumed that any co-occurrence patterns we observed would be primarily attributable to interspecific effects rather than habitat variation. Our objectives were to evaluate three non-invasive survey methods for the detection of mammalian carnivores and to assess analysis methods for testing patterns of community structure and species co-occurrence. We identify the combination of methods that were most effective for detecting the presence of all of our target species. We analyze data from our pilot season and discuss refining our approach for future research.

FIELD METHODS

The North Rim of the Grand Canyon, on the Kaibab Plateau, extends south-north from 36°7'S to 36°22'N and east-west from 112°6'E to 113°35'W. The North Rim is dominated by ponderosa pine and mixed conifer forests at an average elevation of 2438 m (range 2316–2682 m); it is the largest continuous unharvested and ungrazed forest ecosystem in Arizona (Fulé et al. 2002). Eleven species of terrestrial mammalian carnivores are believed to exist in GCNP: mountain lion (*Puma concolor*), black bear (*Ursus americanus*), coyote (*Canis latrans*), bobcat (*Lynx rufus*), gray fox (*Urocyon cinereoargenteus*), badger (*Taxidae taxus*), raccoon (*Procyon lotor*), striped skunk (*Mephitis mephitis*), spotted skunk (*Spilogale gracilis*), ringtail (*Bassariscus astutus*), and long-tailed weasel (*Mustela frenata*). Grizzly bears (*Ursus arctos*) and jaguars (*Panthera onca*) were extirpated from GCNP by the early twentieth century (Hoffmeister 1971), and the last known observation of gray wolves (*Canis lupus*) in the park occurred in 1935.

We selected 20 sites within the North Rim's 382 sq km of primary ponderosa pine forest (Figure 1) to sample carnivore species presence. We assumed that selecting sites within this forest type would minimize variation in habitat quality due to forest cover, elevation, and topography. Sites were 4 ha (200 m x 200 m) in area, and site center points were located a minimum of 4 km apart. We selected site center points iteratively. We randomly generated each point, buffered it at a distance of 4 km, and selected subsequent points outside the buffered area. The buffer area around each site is greater than home range sizes recorded for all target species except mountain lions (Crooks 2002), minimizing the chance that we would detect the same set of individuals at multiple sites. Sites were sampled three times each between June and August of 2003, with a sampling interval of 14–21 days.

During each sampling visit, we used three non-invasive methods to maximize the likelihood of detecting all target species. Our first method of transect searches involved observer(s) walking four 200 m transects through the site to record scat, tracks, burrows, and other target species sign. Transect lines were parallel, oriented north-south, and located 25 m, 75 m, 125 m, and 175 m from the western edge of the site. Scats, tracks, and active burrow entrances were measured and photographed, and scat samples were collected.

Our second method involved establishing a single north-south transect of five hair trap stations spaced 100 m apart through the center of each site. Hair trap stations included a visual attractant and a padded hair snag (Silver Cloud Associates, LLP, Libby MT) attached to a tree at a height of 40–50 cm. Stations were baited with a scented lure (1:1:8 ratio of propylene glycol, glycerine, and beaver castorium with several drops of catnip oil and puma or bobcat urine). This method was designed for surveys of mountain lions in GCNP, but has resulted in additional detections of bobcats, coyotes, and gray foxes in that study (E. Garding, GCNP, personal communication). Hair trap stations were checked for hair during each site visit, and re-baited and replaced if necessary.

Our third method involved placing a single remotely triggered camera at the center of the site. We used three different brands of passive infrared, 35 mm still camera systems (DeerCam, Park Falls WI; WildlifePro, Forestry Suppliers, Inc., Jackson MS; and VanCam Wildlife Detection System, Bakersfield CA). Cameras were mounted securely to trees at a height of 30–45 cm, parallel to the ground and facing a clearing. Camera stations were baited with a pork short rib, attached by fishing line to a stake approximately 5 m in front of the camera. We verified that motion detectors were triggered by movement at and in front of the bait. Cameras were set to trigger intervals of 1.5–3 minutes and night-only mode to minimize detections of birds and other non-target species. Film was collected during

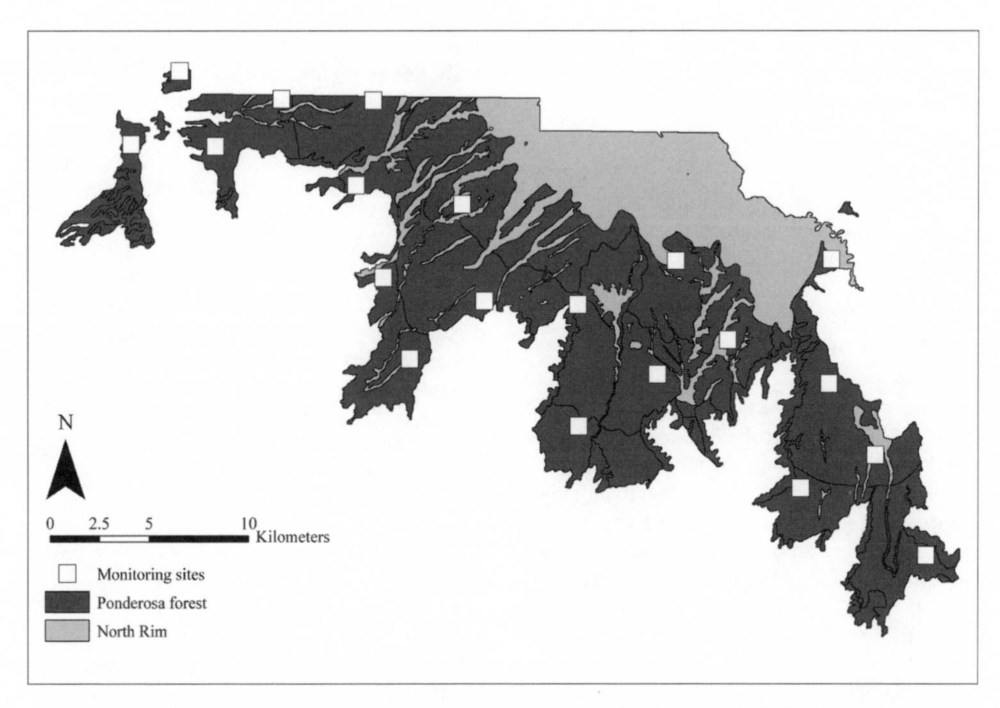

Figure 1. Locations of 20 sites sampled for occurrence of carnivores during 2003 in the ponderosa pine forest of the North Rim of Grand Canyon National Park.

each site visit and camera stations were rebaited if necessary.

ANALYSIS

Sign collected during searches was identified to species by size, shape, context, and content (Murie 1975; Halfpenny 1986). All identifications were verified by at least two observers. Sign that was too old or unclear for species identification was collected but not considered in the analysis. Verified species detections were organized in a presence/absence matrix with species as rows and sites as columns. Data collected during all three visits were compiled into a single matrix. A presence was noted when a species was detected during one or more visits to a site, and an absence was noted when a species was not detected in a site.

We analyzed the presence/absence matrix for patterns of community structure. We used the statistical software package EcoSim (Gotelli and Entsminger 2001) to generate random distribution patterns and to calculate an index of community structure. We ran 10,000 simulations using an algorithm that distributed the observed number of occurrences for each species with equal probability among the 20 sites. This approach assumed that habitat quality and sample effort were equal among sites. We compared the observed presence/absence matrix to the randomly generated distribution using the checkerboard (C-score) index (Stone and Roberts 1990). The C-score quantifies the average number of checkerboards (sites in which two species fail to co-occur) for all species pairs.

Let r_i and r_j equal the number of occurrences of species i and species j for all sites and S_{ij} equal the number of co-occurrences of those species. For P pairs of species:

$$C = \sum_i \sum_j \frac{(r_i - S_{ij})(r_j - S_{ij})}{P}. \qquad (1)$$

We selected the C-score index for this analysis because it measures patterns of interspecific exclusion without being overly sensitive to noise in the data and it has a low type II error rate when paired with equiprobable randomization (Gotelli 2000).

We also analyzed co-occurrence rates between pairs of species because pairs of species could be exclusively distributed even if the community, on average, were not. We limited this analysis to species present in two or more sites. For each pairwise comparison, we used Microsoft Excel to distribute the observed number of occurrences for each species with equal probability among the 20 sites. We ran 10,000 simulations and calculated a frequency distribution of co-occurrence rates. We compared the observed co-occurrence rate with the randomly generated distribution using the standard score (Z-score) index (Zar 1996). The Z-score quantifies the probability of observing a value less than or equal to the expected rate of co-occurrence.

We compared the relative effectiveness of the different monitoring methods for detecting the target species. We summarized numbers of detections by species and monitoring method. We identified which method produced the greatest number of detections, which method produced detections of the greatest number of species, which method produced the least number of detections, and which species were detected by only one of the methods.

RESULTS

Six carnivore species were detected in one or more sites—mountain lion, coyote, bobcat, gray fox, badger, and spotted skunk. We verified 40 of 57 total detections and compiled them in a presence/absence matrix for all sites (Table 1).

Table 1. Verified detections of six carnivore species in 20 North Rim sites. Although a species may have been detected more than once in a particular site, only species presence is indicated here.

Site	Mountain Lion	Coyote	Fox	Bobcat	Badger	Spotted Skunk	Total
A	0	0	0	0	1	0	1
B	0	0	0	1	0	0	1
C	0	0	0	1	1	0	2
D	1	0	0	1	1	0	3
E	0	0	0	0	0	0	0
F	0	0	0	1	0	0	1
G	0	0	0	1	0	0	1
H	0	0	1	0	0	0	1
I	0	1	0	0	1	0	2
J	0	1	0	0	0	0	1
K	0	1	0	0	0	0	1
L	0	0	1	0	0	0	1
M	0	1	0	0	0	0	1
N	0	0	0	0	0	0	0
O	0	1	0	0	0	0	1
P	0	1	0	0	0	0	1
Q	0	1	0	0	0	0	1
R	1	1	0	1	0	0	3
S	0	0	0	0	1	0	1
T	0	0	1	1	1	1	4
Total	2	8	3	7	6	1	27

Our two measures of community structure and species co-occurrence produced different results. In a competitively structured community, we would expect the C-score for the observed community to be significantly greater than the expected value generated by randomized simulations. We found that the C-score of the observed community was not significantly different (P = 0.38) from that expected by chance. If an individual pair of species were exclusively distributed, we would expect the observed rate of co-occurrence to be significantly less than the expected value generated by randomized simulations. Using the Z-score index of species co-occurrence, we found that three species pairs (coyotes and bobcats, coyotes and foxes, and coyotes and badgers) had rates of co-occurrence less than that expected by their detection rates (Table 2), and the rate of co-occurrence between bobcats and coyotes was significantly less than that expected by chance ($P < 0.05$).

The three primary survey methods—hair traps, camera traps, and transect searches—produced different numbers of carnivore species detections. Transect searches produced the greatest number of detections (n = 46) and were the most effective for detecting multiple carnivore species. Thirty of these detections were verified for five species, and scats and tracks provided the only verified detections of mountain lions and bobcats (Table 3). Cameras were particularly effective for documenting coyotes, indicated by instances where other methods failed to detect coyotes that were photographed. A camera also produced the sole detection of the rare spotted skunk (Figure 2). Hair traps were relatively ineffective, producing only one hair sample that was visually identified as coyote or gray fox.

DISCUSSION

We did not detect black bear, raccoon, ringtail, long-tailed weasel, or striped skunk, although these species are not known to be extirpated from GCNP. Black bears are believed to persist in very low densities on the North Rim. Raccoons are likely restricted to lower elevations and the South Rim of the park, whereas ringtails are primarily found in the developed areas and along the canyon rims. Striped skunks and long-tailed weasels are believed to be present in the North Rim forests, but we were unable to distinguish possible sign from that of other species, including spotted skunks and burrowing rodents. Whereas survey effort was not sufficiently intensive to determine the true absences of undetected species, further research regarding the fate of these species is warranted. Inventory data would benefit park management as well as regional assessments, which are limited by inconsistent records among parks (Parks and Harcourt 2002).

We were not surprised that we failed to detect significant spatial structure in the carnivore community. Community structure for such wide-ranging species would likely be more evident at broader scales than represented by our sampling sites. Null model methods are also better suited for investigating patterns of co-occurrence among interacting populations than among interacting individuals (Gotelli and McCabe 2002). It would be more appropriate to assess community structure using the North Rim and other natural islands (e.g. plateaus) or man-made habitat islands (e.g., protected areas) as sampling units when reliable inventory data become available at this scale. In addition, our approach did not account for temporal asynchrony. Species that co-occurred in our study sites could have been partitioning the habitat in time.

Our pairwise analysis of species co-occurrence did suggest that interspecific factors may affect individual species distributions. We found significant disassociation of coyotes and bobcats, and trends indicating limited co-occurrence between coyotes and foxes and coyotes and badgers. Our results are consistent with segregation of habitat use and limited home range overlaps among mid-level carnivore species observed by other researchers (Fedriani et al. 1999; Fedriani et al. 2000; Neale and Sacks 2001). For example, Neale and Sacks (2001) found that in undeveloped habitats only an average of 16 percent of a bobcat's home range

Table 2. Expected and observed rates of species co-occurrence among carnivores in North Rim sites. Expected rates were calculated as the mean of a distribution of co-occurrence rates generated by a Monte Carlo simulation (n = 10,000) that randomly distributed the observed numbers of coyote, bobcat, gray fox, and badger occurrences across the 20 sites.

Species Pair	Expected $(\bar{x} \pm s.d.)$	Observed	Z Score	*P*
Coyote/Bobcat	0.14 + 0.05	0.05	1.67	< 0.05
Coyote/Gray Fox	0.06 + 0.04	0	1.52	0.06
Coyote/Badger	0.12 + 0.05	0.05	1.38	0.08

Table 3. Comparison of total numbers of verified detections by species and monitoring method for 20 North Rim sites.

			Transect Search		
Species	Hair	Photo	Scat	Track	Burrow
Mountain Lion	0	0	3	0	0
Coyote	0	7	4	2	0
Fox	0	1	2	0	0
Bobcat	0	0	9	2	0
Badger	0	1	0	0	8
Spotted Skunk	0	1	0	0	0
Total	0	10	18	4	8

overlapped neighboring coyote home ranges. Given the limited scale of the sites (4 ha) relative to the range of bobcat home range sizes reported in the literature (e.g., 24–563 ha; Crooks 2002), it is reasonable to expect that species with segregated home ranges would rarely be detected in the same sites.

In this pilot study, the number of sites we sampled was relatively low (n = 20), our species were low-density and elusive, and sample effort was limited to one season. Significance levels of our results are highly sensitive to false absences. For example, if a bobcat had been detected in one more site where a coyote occurred, the difference between the expected and observed rates of co-occurrence would not have been significant. Sample effort and sample size need to be substantially increased on the North Rim and perhaps other study areas to confidently assess patterns of carnivore species co-occurrence.

The three monitoring methods employed in our research were not equally effective for detecting carnivores on the North Rim. Taken together, camera traps and transect searches produced detections of all six carnivore species. Transect searches yielded the majority (75%) of verified detections, whereas photographs provided unequivocal species identifications and detections of rarer species. For example, a camera produced the first detection of a spotted skunk on the North Rim since a museum sample was collected in 1963 (C. Hyde, GCNP, personal communication). We did not observe differences in performance between brands of cameras, but our sample effort was low and other researchers have noted differences

Figure 2. This remotely triggered photograph is the first confirmed detection of a spotted skunk on the North Rim of Grand Canyon National Park in 40 years.

in more extensive studies (Swann et al. 2004). Hair trapping was relatively inefficient, with 100 hair trap stations yielding only one sample of unclear origin. Although hair traps were ineffective in our effort to detect multiple carnivores, they may be useful for population and genetic studies of particular species. Our success may have been limited by our choices of hair snag design and scent lure, and other combinations of methods should be tested.

Our approach linked non-invasive field methods with null model analyses to incorporate consideration of interspecific effects into carnivore community surveys. This pilot study is part of a multi-year research project on carnivores in GCNP, and we plan to refine our approach in future years in light of our initial results. We plan to increase the number of sites we are sampling on the North Rim and expand our research to include a second study area on the South Rim. We plan to minimize effort expended on hair trapping and introduce covered track plates as an additional field method to improve our detection of smaller carnivore species (Zielinski and Kucera 1995). Because scats were an important source of detection data, we plan to develop laboratory analyses to increase the number of samples we are able to identify and to increase confidence in our species identifications. We will extract and amplify DNA from scat samples and use restriction fragment length polymorphisms (RFLPs) to distinguish among our target species (Paxinos et al. 1997; Farrell et al. 2000). Our preliminary results indicate limited co-occurrence among pairs of species in our study area, and we plan to focus future analyses on investigating these patterns and their consequences for species distribution models.

ACKNOWLEDGMENTS

The Grand Canyon National Park Foundation (GCNPF), Grand Canyon National Park, NSF Graduate Research Fellowship program, and the North Coast Integrated Hardwood Range Management Program (IHRMP) of the University of California provided funding and supplies for this research. We are grateful to the Eugene Polk Science Fellowship program and GCNP Science Center staff for field assistance and logistical support. A. Bidlack, M. Clark, E. Garding, and S. Palmer provided helpful comments on earlier drafts of this manuscript.

LITERATURE CITED

Connell, J. H. 1961. The influence of interspecific competition and other factors on the distribution of the barnacle *Chthamalus stellatus*. Ecology 42: 710–723.

Connor, E. F., and D. Simberloff. 1979. The assembly of species communities: Chance or competition? Ecology 60: 1132–1140.

Crooks, K. R. 2002. Relative sensitivities of mammalian carnivores to habitat fragmentation. Conservation Biology 16: 488–502.

Crooks, K. R., and M. E. Soulé. 1999. Mesopredator release and avifaunal extinctions in a fragmented system. Nature 400: 583–586.

Diamond, J. M. 1975. Assembly of species communities. In Ecology and Evolution of Communities, edited by M. L. Cody and J. M. Diamond, pp. 342–444. Harvard University Press, Cambridge MA.

Estes, J. A., and J. F. Palmisano. 1974. Sea otters: Their role in structuring nearshore communities. Science 185: 1058–1060.

Farrell, L. E., J. Roman, and M. E. Sunquist. 2000. Dietary separation of sympatric carnivores identified by molecular analysis of scats. Molecular Ecology 9: 1583–1590.

Fedriani, J. M., F. Palomares, and M. Delibes. 1999. Niche relations among three sympatric Mediterranean carnivores. Oecologia 121: 138–148.

Fedriani, J. M., T. K. Fuller, R. M. Sauvajot, and E. C. York. 2000. Competition and intraguild predation among three sympatric carnivores. Oecologia 125: 258–270.

Fulé, D. Z., W. W. Covington, M. M. Moore, T. A. Heinlein, and E. M. Waltz. 2002. Natural variability in forests of the Grand Canyon, USA. Journal of Biogeography 29: 31–47.

Gotelli, N. J. 2000. Null model analysis of species co-occurrence patterns. Ecology 81: 2606–2621.

Gotelli, N. J., and G. L. Entsminger. 2001. EcoSim: Null models software for ecology. Version 7.0. Acquired Intelligence Inc. & Kesey-Bear. http://homepages.together.net/~gentsmin/ecosim.htm.

Gotelli, N. J., and D. J. McCabe. 2002. Species co-occurrence: A meta-analysis of J. M. Diamond's assembly rules model. Ecology 83: 2091–2096.

Halfpenny, J. 1986. A Field Guide to Mammal Tracking in North America. Johnston Books, Boulder CO.

Hoffmeister, D. F. 1971. Mammals of Grand Canyon. University of Illinois Press, Urbana.

Huffaker, C. B. 1958. Experimental studies on predation: Dispersion factors and predator-prey oscillations. Hilgardia 27: 343–383.

Mittelbach, G. C., A. M. Turner, D. J. Hall, J. E. Rettig, and C. W. Osenberg. 1995. Perturbation and resilience: A long-term, whole-lake study of predator extinction and reintroduction. Ecology 76: 2347–2360.

Murie, O. J. 1975. A Field Guide to Animal Tracks. 2nd ed. Houghton Mifflin Company, Boston MA.

Neale, J. C. C., and B. N. Sacks. 2001. Resource utilization and interspecific relations of sympatric bobcats and coyotes. Oikos 94: 236–249.

Newmark, W. D. 1995. Extinction of mammal populations in western North American national parks. Conservation Biology 9: 512–526.

Parks, S. A., and A. H. Harcourt. 2002. Reserve size, local human density, and mammalian extinctions in U.S. protected areas. Conservation Biology 16: 800–808.

Paxinos, E., C. McIntosh, K. Ralls, and R. Fleischer. 1997. A noninvasive method for distinguishing among canid species: Amplification and enzyme restruction of DNA from dung. Molecular Ecology 6: 483–486.

Stone, L., and A. Roberts. 1990. The checkerboard score and species distributions. Oecologia 85: 74–79.

Swann, D. E., C. C. Hass, D. C. Dalton, and S. A. Wolf. 2004. Infrared-triggered cameras for detecting wildlife: An evaluation and review. Wildlife Society Bulletin 32: 351–356.

Virgos, E., J. L. Telleria, and T. Santos. 2002. A comparison on the response to forest fragmentation by medium-sized Iberian carnivores in central Spain. Biodiversity and Conservation 11: 1063–1079.

Wilmers, C. C., R. L. Crabtree, D. W. Smith, K. M. Murphy, and W. M. Getz. 2003. Trophic facilitation by introduced top predators: Gray wolf subsidies to scavengers in Yellowstone National Park. Journal of Animal Ecology 72: 909–916.

Woodroffe, R., and J. R. Ginsberg. 1998. Edge effects and the extinction of populations inside protected areas. Science 280: 2126–2128.

Zar, J. H. 1996. Biostatistical Analysis. 3rd ed. Prentice Hall, Upper Saddle River NJ.

Zielinski, W. J., and T. E. Kucera. 1995. American marten, fisher, lynx and wolverine: Survey methods for their detection. USDA Forest Service General Technical Report PSW GTR-157.

Cultural Resources

HELP FOR THE LOOTED ROCKSHELTERS OF THE COLORADO PLATEAU IN A NEW CENTURY OF ARCHAEOLOGY: NEW BASKETMAKER II RESEARCH ON THE GREAT COMB RIDGE

Francis E. Smiley and Michael R. Robins

The archaeology of southeastern Utah looms large in the history of archaeological investigations in the American Southwest because some of the earliest archaeological investigations on the Colorado Plateau took place in southeastern Utah and other areas of the Four Corners region before the turn of the twentieth century. The work focused on rockshelters, in many ways the most visible of archaeological sites. The shelters contained remarkably well preserved perishables such as baskets, sandals, atlatls, and corn. The excavators approached their work with insight, avarice, and unlimited energy, mowing down rockshelters like Dakota wheat.

Many excavators ripped and tore and sent the recovered materials into the void of the antiquities market. But responsible archaeologists, most notably Kidder and Guernsey (1919; Guernsey and Kidder 1921), contributed much to our understanding of the prehistory of the Southwest, and in national and university museums we have remarkable collections of virtually every aspect of the lives of the Basketmakers: the early Neolithic Anasazi culture, the material remains of which typically dominate rockshelter assemblages in the Four Corners area. We use the term "Neolithic" in the sense that Smiley (1997a) applied it to the beginnings of farming in the American Southwest. Just as the term has been in general use for many decades (e.g., Childe 1953) with reference to early farming societies in the Middle East, Africa, Europe, and the Far East, so is the term applicable to the societies of the Southwest.

A century later, the question remained: What, if anything, may be left of the remains of the Basketmakers? We present the results of ongoing archaeological research on the Colorado Plateau's earliest farmers, the Basketmaker II groups, who now appear to date as early as 4000 BP. We use the term "Basketmakers" to refer to the Basketmaker II peoples of the northern Southwest, not including later, ceramic-period Basketmaker III groups in the same regions.

A certain amount of confusion surrounds the temporal, material, and social organizational circumstances, as well as the origins of the earliest farmers and farming systems on the Colorado Plateau. The confusion—perhaps uncertainty might be a better term—devolves from the small store of chronometric data scattered over the geography of the northern Southwest. Across a truly grand geographic range, the early farmers originated the foundational Puebloan adaptation of dispersed settlements in small and relatively short-occupation encampments, hamlets, or villages. The early farmers began the "portable" village or at least the "portable settlement" pattern, settling for a decade or so, then moving ... always moving and settling (Smiley 1997a).

The "village-mobility" and small, dispersed settlement pattern persisted, with notable exceptions in the northern Southwest, from as early as 4000 BP until about 1000 or 900 BP—in other words, for two or three millennia. On one hand, material developments and social organizational changes clearly occurred during this two or three millennium period. Still, a remarkable feature of the American Neolithic adaptation is the persistence of small, dispersed settlements. Undoubtedly, reliance on farming in the grand region of the northern Southwest placed severe limits on population levels in antiquity, as it continues to do today. Probably, the levels of population permitted by early farming technologies did not greatly exceed those possible for the preceding hunter-gatherer societies.

Given the foregoing parameters, we want to examine the larger questions facing researchers on the American Neolithic in the Southwest. The large questions, while wide ranging and varied, tend first and foremost to have significant chronometric components, a situation that devolves from the fact that not knowing the absolute ages of changes, of events, or of processes makes developing estimates of the rates of change and the estimation of contemporaneity impossible.

A major process across the world has been the nearly universal human phenomenon of simply settling down. One of the major and measurable material consequences of the developing American Neolithic adaptation—and we are speaking here not only of the pre-ceramic early agricultural period, but also of the period from the agricultural transition that took place in the period from as early as 4000 BP all the way to Pueblo II–III times at approximately 900 BP—has been the development of sedentary settlement systems. Although clearly maize-dependent, the earliest farmers began the long process with a very different human-land relationship than the much more labor-intensive, landscape-modifying adaptations of the later sedentary Pueblo peoples.

Archaeologists understood decades ago (Childe 1953; Smith 1995) that, around the world, food production and sedentary adaptations go hand-in-hand in prehistory. What we have learned in more recent decades is that the relationship between sedentism and food production appears complex; one scenario does not fit all situations. Although correlated, the two faces of the Neolithic adaptation do not provide simple causal explanations for one another (e.g., Smith 1995). For "local" examples, consider the apparent sedentary adaptation of the early farming groups in, say, the Tucson Basin (Huckell 1995). Sedentism may even have preceded farming in that region owing to the abundance of "super-productive" native plants. Conversely, the Basketmaker II groups on the Colorado Plateau appear to have remained relatively mobile for one, even two millennia after they either moved onto the plateau from the south or adopted the materials and technology of farming as a diffusion phenomenon (Matson 1991; Smiley 2002).

The question of mobility looms large in the development of early farming societies because of the constraints that mobility places on both population growth and the extent to which groups can modify their environments to make shelter, storage, and the other accoutrements of settled village life. Mobility strategies, therefore, have important implications for the regional dispersal or concentration of groups. Regional population packing and settlement sizes, in turn, heavily condition the nature of social organization (Smiley 1997b). Social organization develops and changes in accord with local integrative mechanisms, means of conflict resolution, technological capacities, and the extent of pan-regional integration.

In our view, current major research into the Early American Neolithic in the Southwest has a great deal to do with changes in population mobility. As any farmer living in the past ten thousand years could attest, crops require attention and protection, which effectively tethers farming populations, or at least some portions of farming populations, to the land. The tethering effect of agriculture clearly added an interesting dimension to the economies of the earliest farmers.

Our observation (elaborated below) over more than 20 years of early farming research in the northern Southwest is that given the chancy nature of farming in antiquity, populations appear to have continued significant reliance on wild resources. Continued reliance on wild resources constitutes a "resilient" adaptation (sensu Yellen 1977; Holling 1973) and the scheduling of farming, foraging, and hunting activities must have been complicated. Moreover, the scheduling process and integration of strategies probably varied a great deal from year to year, just as primary climatic variables like precipitation, average temperature, and so on, exhibited major changes in annual and decadal frequency and amplitude.

Simply put, there remain a great many unanswered questions about the early plateau farmers, the Basketmaker II groups, of the northern Southwest. We can boil the questions down to these: (1) Where did the early Colorado Plateau farmers come from? A number of ideas have been advanced on this topic and research continues (Berry 1982; Wills 1988; Matson 1991; Smiley 2002; Smiley and Robins 2003). (2) Where did they go? Recent investigations on the Basketmaker II–III transition (Geib and Spurr 2000) have developed new evidence on that topic and promise to continue to do so. And finally we come to a basic question raised by the past decade of chronometric research revealing that the Basketmakers enjoyed a long cultural tradition lasting more than a millennium (Smiley 1984, 1985, 1994, 1997c, 1998, 2002): (3) What happened in the apparently long developmental period of the Early American Neolithic?

Chronometric data continue to accumulate that have significant implications for the antiquity and developmental trajectory of early plateau farmers. The question as to "what happened" devolves from the point in Basketmaker II research a decade or so ago when we discovered that corn agriculture dated to at least 3000 BP, about a millennium earlier than previously thought (Smiley et al. 1986; Wills 1988; Smiley 1994). Now, as Smiley (1994, 2002) has suggested, agriculture may have yet another millennium of time depth. Either way (one millennium of "new" time, or two), as far as we archaeologists were concerned, the bottom dropped out of the Basketmaker chronology. Suddenly, we had one—probably two—millennia of Basketmaker II history to dig up. And one—probably two—millennia of human social evolutionary development to discover.

Smiley has outlined the chronometric situation in previous work (1994, 2002). As Figure 1 shows, the antiquity of farming in the northern Southwest appears to be significantly deeper than previously thought (Smiley 1994, 1997c). The evidence from sites north of the San Juan River, and we are speaking of the Falls Creek Shelters on the Animas River north of Durango, Colorado and sites in the Butler Wash area of southeastern Utah, have produced radiocarbon dates on cultigens nearing 3000 BP. In addition, Butler Wash area dates from Bent Oak Shelter in Cottonwood Canyon (not to be confused with Cottonwood Wash, to the east) suggest even greater antiquity for early farming in southeastern Utah. Figure 2 shows the study area in southeastern Utah in which the sites discussed in this article lie.

In other previous works Smiley (1998, 2000) discussed the possibility that farming may date as early as 4000 BP in the northern Southwest. Smiley based this possibility on a calibrated radiocarbon date on corn from Three Fir Shelter on northern Black Mesa at about 3900 BP. A date on corn from Bat Cave at about 4100 BP (Wills 1988) lends additional support to the tentative, but interesting possibility that the Three Fir Shelter date is not simply a statistical outlier.

Recently, however, chronometric developments in the Tucson Basin have produced more support. Lascaux and Hesse (2001) report calibrated determinations on corn and associated wild seeds that, with the Three Fir Shelter and Bat Cave dates, fall within 150 years of one another, clustering at about 4000 BP, as Figure 1 shows. Because the Bat Cave and Three Fir Shelter assays had (previous to the dates revealed by Lascaux and Hesse) predated the oldest known cultigens in the southern Southwest, any previous attempts to argue the validity of the northern dates had been problematic (Smiley 1994). Now that farming appears to have begun by at least 4000 BP in the south,

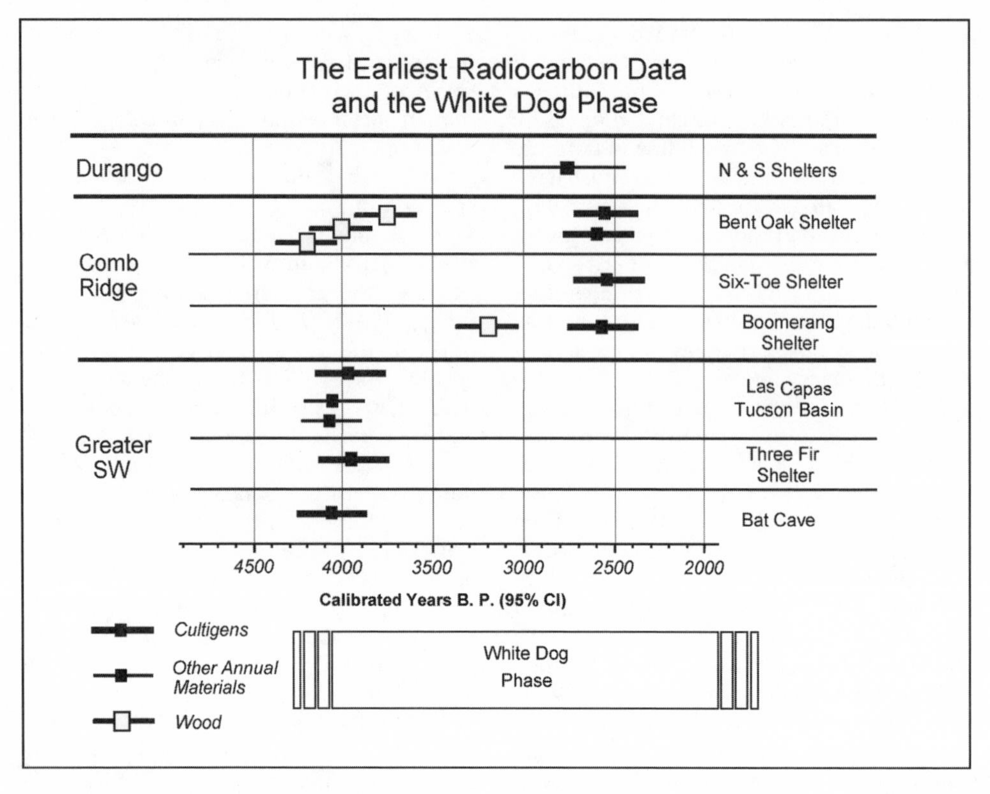

Figure 1. Graph showing selected radiocarbon dates on early Basketmaker II cultigens and other materials in the American Southwest.

the Three Fir Shelter and Bat Cave dates can conceivably accurately measure the age of agriculture in the north. Clearly, we cannot have early farming in the north before early farming in the south.

Accordingly, the narrow spread of earliest dates in the extreme southern and northern extents of the Southwest appears remarkable unless (1) the small sample is errant (an obvious possibility and a typical archaeological problem) or (2) the mechanism of transition across the huge geographic area from northern Mexico to southern Utah and Colorado was so rapid that the progress of farming into the Southwest cannot be regionally tracked by radiocarbon dating. The pesky error term of radiometric determinations would make impossible the geographic tracking of a process that required only a few centuries in time or less.

To examine the process of agricultural transition, Smiley (1997b) developed a computer simulation using the structural device of cellular automata, which are simple matrix elements with a limited set of decision rules designed to simulate regional population social organization and packing during the transition to food production in the Southwest. Of the two mechanisms suggested for the transition to food production across the Southwest—diffusion and migration (Matson 1991)—the simulation suggests that the diffusion process could readily provide the mechanism and the speed of transmission across the vast Southwest within the two-century period that the

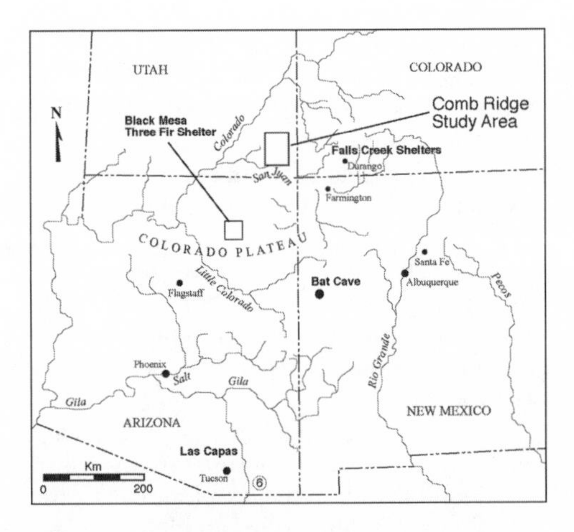

Figure 2. Map of the American Southwest showing locations of sites mentioned in the text. The Comb Ridge study area contains Fish Mouth, JJB, Six-toe, Boomerang, Bent Oak, and Davidson rockshelters.

current chronometric data suggest the transition may have required. The migration process remains a possibility, but in other world areas (e.g., western Europe) the process has been documented to be a great deal slower (Ammerman and Cavalli-Sforza 1973; Cavalli-Sforza 1983).

But now a couple of caveats are in order: Simulations behave as we program them to behave and archaeological samples are frequently suspect. Accordingly, the thought experiment of diffusion simulation by Smiley (1997b) and the inference drawn from the available chronometric data do not constitute strong evidence of a rapid diffusion of farming from south to north about 4000 years ago. They do, however, give a great deal of food for thought, which may be the most important result of computer simulation of cultural processes.

INFRA-TRIBAL ADAPTATIONS

Smiley has said elsewhere (2000, 2002) that perhaps the most interesting aspect of Basketmaker II peoples scattered across the northern Southwest is their apparent transitional status between the economic and social patterns and structures of band-level hunter-gatherers and the economic and social patterns of tribal-level sedentary farmers. The increasing social complexity attendant on the development of tribal organizational and economic forms correlates closely with the shift to farming and herding (Smith 1995).

Because the development of tribal social forms with the means for population integration, social cohesion, economic productivity, and much-reduced mobility options ranks as one of the most important in human prehistory, societies like the Basketmakers, who were clearly in transition, can provide important insights. So how or where do the Basketmakers figure in the human spectrum of social and economic forms? To what extent do our conceptions of the spectrum of human social forms require change, to accommodate the early farming societies of the northern Southwest? Maybe the quickest approach is to ask if the Basketmakers fit into the anthropologist's traditional (and useful) typology of human social organization: bands, tribes, chiefdoms, states (e.g., Flannery 1972). In other words, precisely who are these Early Neolithic peoples of the northern Southwest?

The early Basketmaker II groups were, in fact, unequivocally farmers. And as good farmers, they should be, if the traditional anthropological models of social and economic forms hold, essentially sedentary tribal groups. But as we discuss below, we are overwhelmed by explicit data that clearly indicate that they were not sedentary groups or even tribal groups. Moreover, by the traditional socio-economic models, farmers should be living in permanent dwellings and should be living in villages, but the Basketmakers clearly did neither of these things.

In some ways, the early farmers do appear to have been extensively adapted, exhibiting similar technological, decorative stylistic, and social organizational patterns over huge areas, a trait typical of extensively adapted hunter-gatherer, band-level societies. In other important respects, the early farmers appear more settled and territorial—after the fashion of most tribal societies.

The similarities that Basketmaker groups exhibit to "classical" anthropologically defined band-level hunter-gatherer societies (Service 1962) are significant. For example, we think we see Basketmaker societies operating at very low population levels, commensurate with the densities of hunter-gatherers—that is, small extended family groups of around 25 members.

Robins (1997) has suggested that some of the large-scale rock art sites may have had ceremonial functions within the macro-social sphere of classical hunter-gatherer societies, especially given that these large-scale rock art sites tend to occur in areas of potentially high agricultural productivity. Small, dispersed groups of people could have gathered at these locations for feasting, trade, social interaction, and the exchange of marriage partners. In addition, the Basketmakers continued to use the portable material culture of hunter-gatherers: basketry, far greater use of bifacial core technology than expedient core technology (e.g., Parry and Kelly 1987), and small manos. Basketmaker housing needs were met without the investment in stone axes and mauls, and their cooking did not require ceramic cooking pots. Like most hunter-gatherer groups, then, there is little evidence of the substantial permanent dwellings typical of even semi-sedentary people.

Finally, the Basketmakers appear to have maintained grand regional connections, communications with groups hundreds of kilometers away as evidenced by regional clines in rock art, artifact styles, and trade goods. Although some investigators have noted regional differences in rock art styles and other artifactual materials (Schaafsma 1980; Matson 1991; Robins and Hays-Gilpin 2000), the Basketmaker styles are still widespread within the western and eastern divisions set out by Matson. In fact, we expect band-level regional manifestations to show precisely that: clinal variation in material culture.

On the other hand, the Basketmakers across the northern Southwest exhibited a number of traits usually associated with sedentary, tribal farmers. In the first place, the Basketmakers were serious farmers, even from the earliest periods. By "serious farmers," we mean that all evidence points to a significant reliance on cultivated food. In the second place, and from our work at Boomerang Shelter, we can see the developmental trajectory of storage capacity and increased storage efficiency over the span of the early Basketmaker period called the White Dog phase and dated to the temporal range from possibly as early as 4000 BP to about 2000 BP.

Third, we think we can see a significant change in the human-land relationship mentioned earlier in the development of the entirely new (to the Colorado Plateau) practice of burying the dead in social context. The development of the practice sometime during the White Dog phase serves as an indicator of a profoundly changed ideology toward death and dying. Although chronometric data are scarce, Smiley's (2002) estimate is that the burial of the dead in social context did not begin until sometime around 2200–2000 BP. The Basketmakers appear to have been quite different from the preceding Archaic groups in that the Basketmakers tended to place the deceased in social context with elaborate funerary kit. Moreover, the practice likely served as a demonstration of ownership or at least stewardship of economically and spiritually valued places—that is, the rockshelters that protected the foodstuffs, persons, and dead of the Basketmaker societies.

In the fourth place, we have to consider that just as tribal societies tend to do, the Basketmakers began to remake the landscape. Although they operated on a very small scale, they altered their sheltered sites by constructing large numbers of storage facilities with the occasional small house, as at Bent Oak Shelter (Smiley and Robins 1997) and Three Fir Shelter (Smiley et al. 1986) for a couple of examples.

Fifth, the Basketmakers altered the natural environment, too, principally by introducing new plant species (cultigens) into the ecology of the Colorado Plateau. Such practices reverse the hunter-gatherer practice of *mapping onto* resources, of moving *to* the food and materials in the environment, of using the environment as a storehouse. The Basketmakers clearly began the process that we know to be so evolutionarily significant of mapping the resources onto their domain. In the sixth and final place, we begin to see clear indication of the sort of significant conflict well known in tribal societies, namely the violence evidenced in the burials excavated from Cave 7 not far from our Butler Wash research area (Hurst and Turner 1993).

So, the Basketmaker II groups across the White Dog phase, the early Neolithic of the northern Southwest, clearly appear transitional. They acted in many respects like tribal farmers and in many other important respects like mobile hunter-gatherer band societies. Clearly, the Basketmakers were what Smiley (2000) has termed infra-tribal cultures. As such, they incorporated new behaviors, technologies, ideologies, and social forms: farming, storage, violence and territoriality (Robins and Hays-Gilpin 2000; Robins 2002) on a level not previously evidenced, regional adaptations, and a very clearly altered human-land relationship.

And now we return to the point we made in our opening remarks. A great deal of confusion surrounds the temporal, material, and social organizational circumstances, as well as the origins of the earliest farmers on the Colorado Plateau. Such confusion should perhaps not be so surprising or irritating because, clearly, the Basketmakers simply do not fit the classical anthropological typology of human social forms. However we classify them, the Basketmakers constitute one of the most important cultural and scientific phenomena around. They provide rare, well-preserved instantiations of the band-to-tribe socio-political transition as well as the foraging-to-farming subsistence transition. While most archaeologists find it useful to classify societies into time-honored categories, many of us live to identify and study transitionals and, in the Basketmaker II peoples of the northern Southwest, we have world-class exemplars of a very archaeologically and ethnographically rare phenomenon.

RESEARCH ON THE GREAT COMB RIDGE

Motivated by the problems just cited, we have in recent years focused on a small corner of the American Southwest—the Comb Ridge just north of the San Juan River in southeastern Utah (see map). One of the major research domains outlined above is the attempt to discover what happened in the two-millennium temporal vacuum created by the new radiocarbon chronometry for the onset of the southwestern Neolithic. Since 1996 we have evaluated a number of Basketmaker II rockshelters ravaged by depressingly energetic and thorough looters.

But even depressingly energetic and thorough looters fail to carry away everything they unearth. The bombed-out surfaces of shelters in the canyons of Comb Ridge are often littered with bits of perishable materials from plant remains to corn to cordage to sandals. We have tried to turn the entropic activities of the looters to advantage. We have sampled the materials in the back dirt from several sites. Under the reasonable assumption that the materials scattered in the back dirt piles provide a fairly random sample of the chronometrically useful materials in the disturbed strata, we have assayed a number of samples and have derived useful chronologies of shelter use. Moreover, we have begun to amass enough chronometric data to begin to approach the ranges of occupations of the shelters. Using disturbed materials from bombed-out contexts appears to provide good evidence of the timing and lengths of shelter occupations. Figure 1 shows the earlier dates on the Comb Ridge shelters. The full tabulation of chronometric data can be found in Smiley 1997c.

A primary problem in looking for the "lost millennia" has been a lack of Basketmaker stratigraphy to reveal the events and processes of a period of more than 15 centuries of development. Most sheltered sites in the northern Southwest were excavated or looted before the advent of stratigraphically based work, and other sites consisted of many centuries of Basketmaker occupations compressed in 20 or so centimeters of "horizontal" stratigraphy, as Three Fir Shelter on Black Mesa demonstrates (Smiley et al. 1986). Either way, investigators cannot get a look at stratigraphic sequences that tell the story of cultural change.

For these southeastern Utah rockshelters, the assessment news is both good (much more matrix remains in situ than most thought) and bad (most sites look like the aftermath of B-52 strikes). We have learned that sometime between four and three thousand years ago, the first farmers in southeastern Utah began planting corn and squash north of the San Juan River. We have learned that Archaic occupation on the Comb began before 8300 BP and continued without evidence of interruption through the Basketmaker II White Dog phase and into the Christian Era. The Early Archaic date at about 8300 BP is an AMS determination on packing material from a sandal found in pothunter's spoil at Boomerang Shelter (Dale Davidson, personal communication 1996; Smiley 1997c). Although we have few data points, the data do not argue for a hiatus. Figure 3 is a photograph of the huge shelter from the high slickrock across the canyon. Photographing Boomerang Shelter is a little like trying to photograph a redwood tree: at distance the scale is meaningless and up close you only get a small portion.

We have learned that the Basketmakers have a *history*. The Basketmaker occupation is both long and developmental. While few, if any, sites in the northern Southwest evidence deeply stratified Basketmaker occupations, we do have such evidence in the Great Comb. At Boomerang Shelter we have a developmental trajectory for storage technology that begins with small, earthen cists at 2 m below the modern surface, moves to small rock slab cists higher up, then goes to larger stone cists, and finally to the familiar, large adobe-chinked cists near the modern surface (Figure 4). The sequence clearly indicates increasing storage requirements, which in turn suggests either demographic shifts (more people) or subsistence shifts (increasingly sedentary people), or both. The increase in storage registers in both the increase in average cist size over time and the fact that the frequency of cists does not diminish in this particularly heavily used portion of Boomerang Shelter.

Finally, we have learned that dating looters' debris from thoroughly scrambled sites can provide a good proxy for careful, random sampling; even looting can sometimes be turned to advantage (Smiley and Robins 1997). Figure 2 shows the location of the Great Comb Ridge and the study area in which lie the rockshelters—Bent Oak, Sixtoe, JJB, Boomerang, and Fishmouth—in the deeply incised canyons of the Comb. One additional site, Davidson Shelter, lies to the northwest on nearby Little Baullies Mesa.

One of the deeply stratified rockshelter sites in the Comb is Boomerang Shelter. The richness of cultural deposits at the site indicates its importance to the Basketmaker lifeway. The site is in many ways ideal. The north-facing exposure provides good shade from the summer heat, making the shelter especially attractive as an agricultural camp. Rich alluvial agricultural soils, fed by numerous springs, runoffs, and seeps abound in the immediate area.

Sadly, shelter accessibility and probable use as a cowboy camp in the late 1800s led to its rapid deterioration from casual looting and relic hunting. Fortunately, Boomerang Shelter was not subjected to the systematic broadside strip mining evident at more recently looted sites such as Atlatl Rock Cave (Geib et al. 1999). The evident fact that the earlier looters were less organized and not as thoroughly destructive as their criminal progeny has led to the preservation of small areas of intact stratigraphy.

What we are therefore delighted to report is that, bombed and blasted though most sites in the northern Southwest may appear, our research in these particular catastrophically disturbed rockshelters reveals that each site retains enormous scientific and cultural value. In some cases, such as at Bent Oak Shelter, deeply stratified Archaic-to-Pueblo sequences escaped looters' attention because the deposits were not dry and contained few

Figure 3. The interior of Boomerang Shelter, view to the south. Note crew members for shelter scale.

perishables. In other cases, at Boomerang Shelter for example, the looters appear only to have ripped out the upper meter of deposits.

In addition, scattered on the cratered surfaces of the Comb sites lie rich assemblages of perishable materials: corn cobs, cordage, basketry fragments, sandal fragments, faunal materials, remains of food processing, basket manufacture, coprolites, and on and on. This rich array of cultural material, while de-contextualized at the local site level, still allows comparative analyses at the inter-site level. With this rich archaeological resource we can begin to examine process and timing of the agricultural transition in southeastern Utah.

BOTANICAL DATA FROM TWO CONTRASTING SITES

Because we have tested and evaluated several rockshelters in the region, we are beginning to assemble comparative data to answer some of the questions we outlined earlier. Of all the sites investigated during our three field seasons of research, Boomerang Shelter and Davidson Shelter offer considerable potential for comparative analyses. Both sites contain predominantly Basketmaker II remains; however, Boomerang Shelter was occupied continuously from the onset of the White Dog Phase, whereas Davidson Shelter appears, on the strength of both radiocarbon data and shallow deposition, to have been occupied only toward the end of the Basketmaker II period (ca. AD 300; Barr 2001).

The sites differ in the patterns of local resource distribution; that is, the catchments differ significantly. Boomerang Shelter lies in a deep, riparian canyon environment at the edge of grassland and sparse juniper-savanna at an elevation of approximately 1400 m. The biotic community surrounding Boomerang, especially to the west along the eastern fringe of Cedar Mesa, is one of the

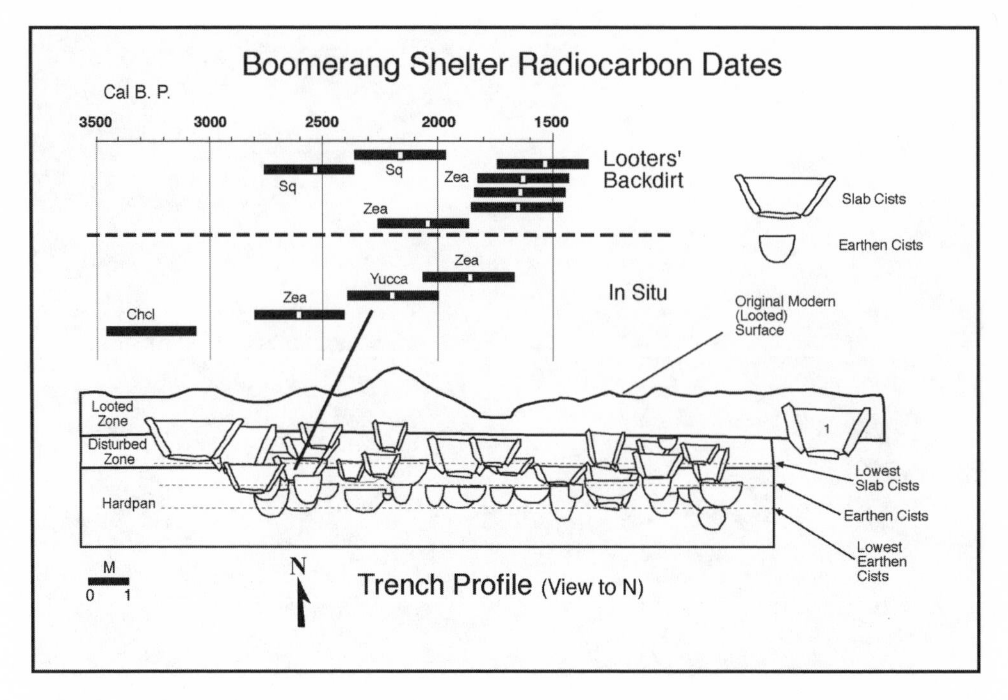

Figure 4. Profile of Boomerang Shelter test trench showing the stratigraphy of storage facilities and construction techniques. The chronometric data obtained from Boomerang Shelter appear in the graph above the profile.

richest biotic zones on the Colorado Plateau (see Hevly 1983). The deep sandy soils in this area support extensive grasslands rich in ricegrass and dropseed that can still be observed today, despite extensive modern-day grazing. Davidson Shelter also lies deep in a riparian canyon, but it is surrounded by mesa-top pigmy conifer forest at an elevation of about 2100 m.

The results of macrobotanical analysis of 12 sediment samples from the two sites appear in Table 1. Grouping the taxa into cool and warm season availability (May-July and August-September, respectively) reveals a stark difference between the two assemblages. Fifty-seven percent of foraged plant food species identified at Boomerang Shelter are cool-season plants, compared to only 7 percent from Davidson Shelter. The dramatic skew toward cool-season resources at Boomerang Shelter is likely of anthropological significance. The occupants of Boomerang Shelter appear to have been far more reliant on ricegrass (*Stipa hymenoides*). We have suggested elsewhere (Smiley and Robins 2003) that access to significant quantities of this plant may have played a pivotal role in enabling foragers to remain settled during the more critical phases of maize farming and also to be less dependent on stored maize during this same period. As we have suggested, early farmers could have survived efficiently on smaller maize crops by depending on ricegrass as a major predictable food crop. Moving through time to a site like Davidson Shelter, in accord with the model of maize evolution proposed by Matson (1991), we might suppose that the Basketmakers were producing more maize per capita. The occupants at Davidson Shelter collected primarily warm-season resources, which we interpret as a result of both reduced logistical foraging mobility (i.e., the foraging range did not include

Table 1. Comparison of foraged seeds and seed parts per 900 ml sample, Boomerang and Davidson Shelters.

Taxon	Boomerang Shelter						Davidson Shelter						Avg Boom.	Avg Davsn.
Cool Season (May–July)														
Descurainia sp.	4	0	68	6	92	0	0	39	6	0	0	0	24	9
S. hymnenoides	835	1125	1067	483	1031	800	232	147	171	7	12	20	796	71
Rhus sp.	140	57	77	20	98	44	0	0	3	0	0	0	62	1
Subtotal													882	81
% cool season	57.2%						7.2%							
Warm Season (Aug–Sept)														
Cheno-Am	850	376	530	317	417	390	68	2559	180	464	30	200	421	687
Cycloloma	33	36	38	54	85	95	0	975	522	7	24	0	49	306
Helianathus sp.	21	34	33	6	21	8	26	33	2	0	0	2	21	7
Corispermum sp.	189	206	122	182	180	133	1	141	18	0	8	9	145	35
Portulaca sp.	25	28	32	13	40	27	0	63	3	0	12	4	24	16
Subtotal													660	1051
% warm season	42.8%						92.8%							

lower elevation grassland and juniper scrub) and a greater dependence on agrestials (plants that grow in association with cultivated field environments) and stored maize. This makes perfect sense in a period of intensified agricultural production. The species that make up 93 percent of the foraged resources recovered from the sediment samples at Davidson Shelter consist primarily of plants that would have grown in and around agricultural fields.

While still tentative, macrobotanical data from these two sites suggest that levels of mobility, and by implication foraging and scheduling strategies, changed through time. While stressing the importance of more data, the emerging pattern from our data favors a model wherein the earliest farmers in the northern Southwest selected areas close to open grassland and juniper scrub with ready access to riparian, arable land. With access to the broad range of forage afforded by these resources, the early groups may have been able to embed an agricultural strategy into the existing hunter-gatherer lifeway with little modification. Mobility may have involved the organization of tasks by age, ability, and gender in order to monitor crops during their most vulnerable cycle (May-July). The necessary plant resources were available within a day's walk of the shelter. Numerous trails cross the precipitous Comb Ridge; at least one bears evidence in the form of rock art of its use by the Basketmakers opening up the rich juniper scrub to the west of the Comb (Till 2001).

CONCLUSIONS

In summary, our research on and evaluations of looted Basketmaker II shelters in the Comb Ridge region of southeastern Utah has revealed long use histories, a developmental sequence for storage, and new chronometric data. In addition, seasonality revealed by wild resource evidence indicates that Boomerang Shelter on the eastern side of Comb Ridge above Butler Wash, with its long occupation sequence and revelation of the nature of the development of Basketmaker II storage, contrasts sharply with Davidson Shelter set deep in a canyon in the Cedar Mesa complex. Macrobotanical evidence at Boomerang suggests multi-season occupation and massive storage—at least massive for early American Neolithic groups living at the edaphic and climatic limits of early farming. Davidson reflects a limited season of occupation, but like Boomerang, Davidson hosted massive storage.

The Basketmakers of southeastern Utah appear to have selected sites carefully, used them intensively, and used them very differently. We think our results indicate that these early Neolithic groups have a great deal to teach us about both the transition to agriculture and the shift to sedentary tribal adaptation in the northern Southwest.

ACKNOWLEDGMENTS

We thank Dale Davidson, BLM archaeologist for his tireless help, advice, and expertise in the archaeology of southeastern Utah and with our projects in the area of the Great Comb Ridge. We thank Nancy Shearin, also of the BLM, for her work at Boomerang Shelter, in particular. We are grateful to numerous students, both graduate and undergraduate, who worked very hard in the heat and the dust and to the people of Bluff, Utah, who were unfailingly hospitable. Thanks also to two anonymous reviewers whose diligence improved this project. Any mistakes or omissions in the work are entirely the responsibility of the authors.

REFERENCES CITED

Ammerman, A., and L. L. Cavalli-Sforza. 1973. A population model for the diffusion of early farming in Europe. In The Explanation of Culture Change: Models in Explanation, edited by C. Renfrew, pp. 343–359. Duckworth Press, London.

Barr, D. 2001. Ancient maize in the northern Southwest: Approaches to the study of variability. Unpublished Master's thesis, Department of Anthropology, Northern Arizona University, Flagstaff.

Berry, M. 1982. Time, Space, and Transition in Anasazi Prehistory. University of Utah Press, Salt Lake City.

Cavalli-Sforza, L. L. 1983. The transition to agriculture and some of its consequences. In How Humans Adapt, edited by D. S. Ortner, pp. 103–127. Smithsonian Press, Washington D.C.

Childe, V. G. 1953. New Light on the Most Ancient East. Routledge & Kegan Paul Ltd Broadway House, 68-74 Carter Lane, London.

Flannery, K. V. 1972. The cultural evolution of civilizations. Annual Review of Ecology and Semantics 3: 399–426.

Geib, P. R., and K. Spurr. 2000. The Basketmaker II–III transition on the Rainbow Plateau. In Foundations of Anasazi Culture: The Basketmaker-Pueblo Transition, edited by P. F. Reed. University of Utah Press, Salt Lake City.

Geib, P. R., N. T. Coulam, V. H. Clark, K. A. Hays-Gilpin, and J. D. Goodman. 1999. Atlatl Rock Cave: Findings from the investigation and remediation of looter damage. Navajo Nation Archaeology Department Report No. 93-121. Manuscript on file, Window Rock AZ.

Guernsey, S. S., and A. V. Kidder. 1921. Basketmaker caves of northeastern Arizona: Report on the explorations, 1916–17. Papers of the Peabody Museum of American Archaeology and Ethnology, Harvard University.

Hevly, R. H. 1983. High-altitude biotic resources, paleoenvironments, and demographic patterns: Southern Colorado plateaus, AD 500–1400. In High-Altitude Adaptations in the Southwest, edited by J. C. Winter, pp. 22–40. USDA Forest Service, Southwest Region, Albuquerque.

Holling, C. S. 1973. Resilience and stability of ecological systems. Annual Review of Ecology and Systematics 4: 1–23.

Huckell, B. 1995. Of Marshes and Maize: Preceramic Agricultural Settlements in the Cienega Valley, Southeastern Arizona. Anthropological Papers of the University of Arizona 59. University of Arizona Press, Tucson.

Hurst, W., and C. Turner II. 1993. Rediscovering the "Great Discovery": Wetherill's first Cave 7 and its record of Basketmaker violence. In Anasazi Basketmakers: Papers from the 1990 Wetherill–Grand Gulch Symposium. Bureau of Land Management.

Kidder, A. V., and S. S. Guernsey. 1919. Archeological Explorations in Northeastern Arizona. Bureau of American Ethnology Bulletin 66. Washington D.C.

Lascaux, A., and I. S. Hesse. 2001. The Early San Pedro Phase Village: Las Capas, AZ AA:12:11 (ASM). SWCA Cultural Resource Report 01-100. SWCA, Inc., Tucson AZ.

Matson, R. G. 1991. The Origins of Southwestern Agriculture. University of Arizona Press, Tucson.

Parry, W., and R. Kelly. 1987. Expedient core technology and sedentism. In The Organization of Core Technology, edited by J. K. Johnson and C. A. Marrow, pp. 285–304. Westview Special Studies in Archaeological Research. Westview Press, Boulder.

Robins, M. R. 1997. Modeling the San Juan Basketmaker socio-economic organization: A preliminary study in rock art and social dynamics. In Early Farmers in the Northern Southwest: Papers on Chronometry, Social Dynamics, and Ecology, edited by F. E. Smiley and M. R. Robins. Animas–La Plata Archaeological Project Research Paper 7. Northern Arizona University, Flagstaff.

Robins, M. R. 2002. Status and Social Power: Rock art as prestige technology among the San Juan Basketmakers of southeast Utah. In Traditions, Transitions, and Technologies: Themes in Southwest Archaeology, edited by S. H. Schlanger, pp. 386–400. University Press of Colorado, Boulder.

Robins, M. R., and K. A. Hays-Gilpin. 2000. The bird in the basket: Gender and social change in Basketmaker iconography. In Foundations of Anasazi Culture: The Basketmaker-Pueblo Transition, edited by P. F. Reed. University of Utah Press, Salt Lake City.

Schaafsma, P. 1980. Indian Rock Art of the Southwest. University of New Mexico Press, Albuquerque.

Service, E. R. 1962. Primitive Social Organization: An Evolutionary Perspective. Random House, New York.

Smiley, F. E. 1984. The Black Mesa Basketmakers: A reevaluation of the chronometry of the Lolomai Phase. Paper presented at the 49th annual meeting of the Society for American Archaeology, Portland OR.

Smiley, F. E. 1985. The chronometrics of early agricultural sites in northeastern Arizona: Approaches to the interpretation of radiocarbon dates. PhD dissertation, Department of Anthropology, University of Michigan, Ann Arbor. University Microfilms, Ann Arbor.

Smiley, F. E. 1994. The agricultural transition in the northern Southwest: Patterns in the current chronometric data. Kiva 60: 165–189.

Smiley, F. E. 1997a. The American Neolithic and the Animas–La Plata Archaeological Project. In Early Farmers in the Northern Southwest: Papers on Chronometry, Social Dynamics, and Ecology, edited by F. E. Smiley and M. R. Robins, pp. 1–12. Animas–La Plata Archaeological Project Research Paper 7. Northern Arizona University, Flagstaff.

Smiley, F. E. 1997b. Regional packing, group dynamics, and the diffusion of innovation. In Early Farmers in the Northern Southwest: Papers on Chronometry, Social Dynamics, and Ecology, edited by F. E. Smiley and M. R. Robins. Animas–La Plata Archaeological Project Research Paper 7. Northern Arizona University, Flagstaff.

Smiley, F. E. 1997c. Toward chronometric resolution for early agriculture in the northern Southwest. In Early Farmers in the Northern Southwest: Papers on Chronometry, Social Dynamics, and Ecology, edited by F. E. Smiley and M. R. Robins. Animas–La Plata Archaeological Project Research Paper 7. Northern Arizona University, Flagstaff.

Smiley, F. E. 1998. Applying radiocarbon models: Lolomai Phase chronometry on Black Mesa. In Archaeological Chronometry: Radiocarbon and Tree-Ring Models and Applications on Black Mesa, Arizona, edited by F. E. Smiley and R. V. N. Ahlstrom, pp. 99–134. Center for Archaeological Investigations Occasional Paper 16. Southern Illinois University, Carbondale.

Smiley, F. E. 2000. Infra-tribal systems and cultural evolutionary studies of early agriculture on the Colorado Plateau. Paper presented at the 65th annual meeting of the Society for American Archaeology, Philadelphia.

Smiley, F. E. 2002. The first Black Mesa farmers. In Prehistoric Culture Change on the Colorado Plateau, edited by S. Powell and F. E. Smiley. University of Arizona Press, Tucson.

Smiley, F. E., and M. R. Robins, editors. 1997. Early Farmers in the Northern Southwest: Papers on Chronometry, Social Dynamics, and Ecology. Animas–La Plata Archaeological Project Research Paper 7. Northern Arizona University, Flagstaff.

Smiley, F. E., and M. R. Robins. 2003. The development of early farming systems in southeastern Utah. Paper presented at the 68th annual meeting of the Society for American Archaeology, Milwaukee WI.

Smiley, F. E., W. J. Parry, and G. J. Gumerman. 1986. Early agriculture in the Black Mesa/Marsh Pass region of Arizona: New chronometric data and recent excavations at Three Fir Shelter. Paper presented at the 51st annual meeting of the Society for American Archaeology, New Orleans.

Smith, B. D. 1995. The Emergence of Agriculture. W. H. Freeman, New York.

Till, J. D. 2001. Chacoan roads and road-associated sites in the lower San Juan region: Assessing the role of Chacoan influences in the northwest periphery. Unpublished Master's thesis, University of Colorado.

Wills, W. H., III. 1988. Early Prehistoric Agriculture in the American Southwest. School of American Research Press, Santa Fe.

Yellen, J. E. 1977. Long term hunter-gatherer adaptation to desert environments. World Archaeology 8 (3): 262–274.

FIRE EFFECTS RESEARCH AND PRESERVATION PLANNING AT WUAPTKI AND WALNUT CANYON NATIONAL MONUMENTS, NORTHEAST ARIZONA

Ian Hough, Jeri DeYoung, and David Barr

Two recent projects at Wupatki and Walnut Canyon National Monuments in northeast Arizona (Figure 1) have focused on the effects of wildfire on archaeological sites. The Wupatki project was designed to study the effects of historically absent grass fires on archaeological sites and to identify potential long-term site preservation issues. The Walnut Canyon project evaluated fuel loads and fire risk at 458 archaeological sites. The primary objectives were to evaluate differences in fuel loading and fire risk potential between vegetation classes and site types, and to identify preservation issues associated with the threat of damage from wildland fires in the monument.

ENVIRONMENTAL SETTING AND CULTURE HISTORY

Wupatki

Wupatki National Monument (35,000 acres) lies 38 miles northeast of Flagstaff at the northern edge of the San Francisco Volcanic Field (Figure 1). Elevation within the monument ranges from just above 5720 feet in the southwest corner to 4300 feet along the Little Colorado River (Anderson 1990:3.1–38). Precipitation at Wupatki ranges from 6.44 inches to 11.40 inches. The four largest vegetation classes are juniper woodland, grassland (with juniper and shrubland patches), mixed grassland-shrubland (with juniper), and mixed herbaceous-shrubland (Hansen et al. 2004). The western section of the monument is covered with juniper woodland and grassland (Figure 2).

Wupatki National Monument protects nearly 2700 archaeological sites, the vast majority of which contain architectural remains. Sites range in size and function from one-room temporary residence fieldhouses to year-round multi-room pueblo habitations. The high density of sites in and around Wupatki represents rapid settlement by three distinct cultural groups—the Sinagua, Kayenta, and Cohonina (Downum and Sullivan 1990). Most sites date to AD 1050–1250. Settlement in what is now the western section of the monument was considerably more concentrated (100 sites per square mile) than in the eastern section (25 sites per square mile), owing to the proximity of productive agricultural land.

Locally available basalt boulders were used to construct one and two-room fieldhouses in this area; they may have been roofed with a permanent structure, or they may have had temporary brush and pole roofs (NPS 2001). Wall construction was typically double stone and dry-laid or dry-laid and mudded.

Walnut Canyon

Walnut Canyon National Monument (3600 acres) is located 6 miles southeast of Flagstaff on the southeast edge of the Little Colorado River drainage (Figure 1). Walnut Creek, a 10-mile section of which forms Walnut Canyon within the monument, drains in an easterly and northeasterly direction and eventually empties into the Little Colorado River. Elevation within the monument

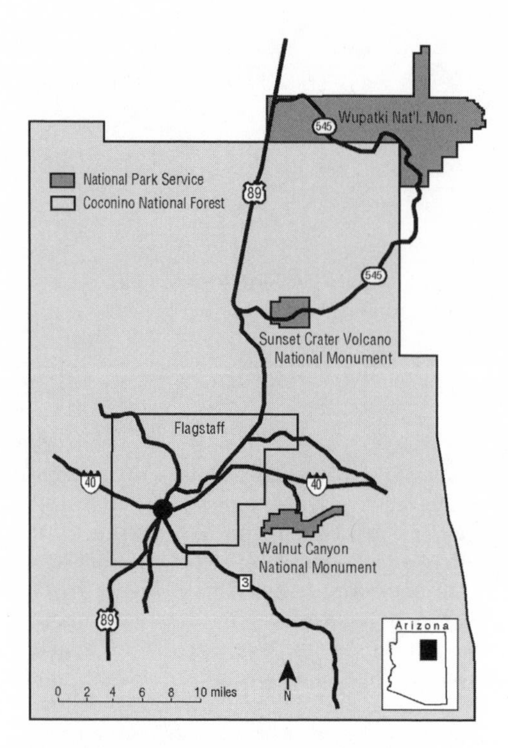

Figure 1. Location of Wupatki and Walnut Canyon National Monuments.

ranges from 6840 feet in the westernmost section to 6220 feet in the canyon bottom in the easternmost section.

Vegetation distribution in the canyon varies closely with elevation, geologic setting, and topography. The main classes are ponderosa-pinyon-juniper-oak woodland, pinyon-juniper–blue grama woodland, Douglas fir–Gambel oak woodland, pinyon-juniper-shrub-succulent woodland, and a riparian corridor along the canyon's narrow bottom (Joyce 1974; Hansen et al. 2004). The ponderosa-pinyon-juniper-oak woodland transitions to pinyon-juniper–blue grama woodland from southwest to northeast, dividing the monument nearly equally. There is a marked difference in vegetation and fuel loads between the Douglas fir–Gambel oak woodland on the north-facing canyon slope and the pinyon-juniper-shrub-succulent woodland on the south-facing canyon slopes (Figure 3).

The oldest evidence of human occupation in Walnut Canyon comes from the Archaic period, dating between 9000 to roughly 2000 BP (Mabry and Faught 1998; Neff and Spurr 2004). A recently completed inventory survey of Walnut Canyon documented 23 Archaic period sites, which are typically scatters of chipped stone waste flakes, chipped stone tools (scrapers, drills, projectile points), and ground stone implements (manos); projectile points serve as the primary diagnostic artifact type.

The period of most rapid settlement and intensive land use around Walnut Canyon began around AD 1150 and continued until AD 1250; the archaeological tradition representing this period is called the Sinagua culture (Neff and Spurr 2004). More than 100 well-preserved masonry cliff dwellings and 150 surface masonry structures are associated with the Northern Sinagua culture at Walnut Canyon.

Cliff dwellings were typically constructed beneath deeply recessed alcoves in the Kaibab Formation. Kaibab limestone blocks at the site were used to build double stone exterior walls and interior dividing walls, and local soils were used to make mortar and wall plaster (NPS 2004). The cool dry conditions in the alcoves provide an excellent environment for the preservation of organic material; basketry, sandals, cordage, cotton textiles, wooden and reed tools, and paleoethnobotanical materials have been found in abundance (Downum 2000). These well-preserved materials are highly vulnerable to damage from fire.

VEGETATION CHANGES AND CURRENT FUEL CONDITIONS

Management of fire has become a primary National Park Service agenda in the past decade, following large devastating fires that resulted in loss of human life and property and severe impacts on regional ecosystems (http://nps.gov/fire/fir_wil_about.html). Of all natural processes, fire has the greatest potential for severe damage as well as lasting benefit. Resource managers are using fire aggressively in the national parks to restore and maintain landscapes and to

Figure 2. Juniper woodland and grassland, southwest corner of Wupatki National Monument.

Figure 3. Four Walnut Canyon vegetation classes. (a) Pinyon-juniper–blue grama woodland; (b) ponderosa-pinyon-juniper-oak woodland; (c) Douglas fir–Gambel oak woodland; (d) pinyon-juniper-shrub-succulent woodland.

reduce the threat of resource loss during wildland fire. As national heritage resources, archaeological sites within many park landscapes are currently at risk from fire damage as well as in need of protection through on-site fire management.

Table 1 presents current and historic fuels conditions on archaeological sites within vegetation classes at each monument. Historic changes in each vegetation class within the past 125 years have been brought about by a combination of grazing, logging, fire suppression, and climate cycles (Davis 1987; Despain and Mosley 1990; Covington and Moore 1994; Menzel 1996; Swetnam and Baisan 1996; Cole 1998; Fulé et al. 2002; Swetnam and Baisan 2003).

At Wupatki, there has been a moderate alteration of the historical natural fire regime in the juniper woodland and grassland (NPS 2005). Wupatki's semi-arid grasslands are fire-dependent systems that historically burned frequently but not intensely (Cole 1998). Since livestock grazing was eliminated at Wupatki after 1989, the grasslands and juniper woodland in the western part of the monument have been allowed to evolve without the pressure from grazing, thus forming a continuous fuel source across the landscape. As demonstrated by the West Fire (1995, 923 acres), the Moon Fire (2000, 550 acres), the State Fire (2002, 180 acres), and the Antelope Fire (2002, 1400 acres), lightning-caused fires can now burn in open expanses across hundreds of acres (NPS 2005). Fires of this size had not been recorded anywhere in the monument prior to 1989.

At Walnut Canyon there has been a moderate or greater level of disruption of the historical natural fire regime, resulting in high fuel loads and low fire frequency in four of the five vegetation classes (NPS 2005). Prior to the late 1800s, the average fire frequency in the ponderosa pine forest at Walnut Canyon was 3 years, with individual stands burning every 14 years (Swetnam et al. 1990; Covington and Moore 1994; Garrett and Soulen 1999; Knox 2004). Between 1876 and 1994, for all tree species in the ponderosa-pinyon-juniper-oak woodland, there was a doubling of tree basal area (sq m/ha) and a fourfold increase in the number of trees per hectare near Walnut Canyon (Menzel 1996). There has been an increase of similar magnitude in tree density and basal area in the pinyon-juniper–blue grama woodland and Douglas fir–Gambel oak woodland (Despain and Mosley 1990; Jenkins et al. 1991; Knox 2004). Fires in the pinyon-juniper-shrub-succulent woodland were probably infrequent due to sparse herbaceous fuels and widely spaced tree crowns, occurring mainly during extreme drought years with high winds.

FIRE'S EFFECTS ON ARCHAEOLOGICAL MATERIALS

For this research we considered the effects of fire on architectural remains, chipped stone artifacts, and ceramic artifacts. These materials are the most common and easily recognized, and are usually seen in abundance on the surface of archaeological sites at Walnut Canyon and Wupatki. Impacts to archaeological materials from wildland fire vary with fuel type, fuel load, fire behavior, proximity of artifacts to fuels, and artifact material type (Buenger 2003).

Architectural Remains

The types of architectural raw material found at Wupatki include vesicular basalt derived from recent lava flows (e.g. Merriam Crater and Woodhouse Mesa) in the southwest corner and surrounding the monument on the south and west, pale reddish brown Moenkopi sandstone found in greatest concentration in the eastern section, and Kaibab limestone found in the central section of the monument (Anderson 1990:3.1). In the southwest corner of Wupatki where the current project is located, basalt is the predominant building material at fieldhouse and small pueblo sites. At Walnut Canyon cliff dwelling structures were constructed mostly from locally available Kaibab limestone. Surface rubble mound sites at Walnut Canyon infrequently contain Coconino sandstone and vesicular basalt.

Architectural remains are found in two primary settings at the two monuments—

Table 1. Vegetation class characteristics, fuel loading conditions, and historic fire trends at Wupatki and Walnut Canyon National Monuments.

Vegetation	Slope/Aspect	Historical Trends (1600–1900)	Current Fuel Condition	Fire Regime
WUPATKI: Juniper woodland	Flat open terraces & ephemeral drainages; noncontiguous in SW and south central; mixed with grassland.	Open canopy with continuous ground fuels between isolated stands.[1] Low-severity ground fires with infrequent crown fires.[1]	Variably dense isolated stands. Historic heavy grazing on grasses; spread of junipers onto the formerly open grasslands.[2]	2-Moderate alteration of historic regime.
Grassland	Flat open terraces and cinder flats; contiguous in SW and south-central monument.	Semi-arid grasslands are fire-dependent systems that historically burned frequently but not intensely.[2]	Historic grazing on grasses drastically reduced finer fuels until recently.[2] Since 1989, grasses forming continuous fuel source; 4 fires (3053 acres) since 1995.[3]	2-Moderate alteration of historic regime.
WALNUT CANYON: Ponderosa-pinyon-juniper-oak woodland	Flat terraces & side drainages. Open local aspect. Western half of monument, both rims; contiguous.	Open canopy, low tree density (25–50 trees/acre) forest with minimal fuel load.[4] Patchy and discontinuous low-severity ground fires; some mixed severity fires every 3–14 years.[5,6]	Dense thickets of young and old trees (50-300 trees/acre);[7] highest fuel load conditions and hazardous fuel risk in monument. 100 yr absence of wild fires. Few recent prescribed fire projects in western section of monument.[3]	2-Moderate alteration of historic regime.[8]
Pinyon-juniper–blue grama woodland	Flat terraces & side drainages. Open local aspect. Eastern half of monument, both canyon rims; contiguous.	Open canopy with continuous ground fuels between isolated stands.[1] Low-severity ground fires with infrequent crown fires.[1]	Increased tree density & fuel continuity, tighter crown spacing.[1] Increased likelihood of crown fires.[1]	2-Moderate alteration of historic regime.[8,3]
Douglas fir-Gambel oak woodland	Cool, moist north-facing canyon slopes; middle/central canyon, non-contiguous transition to pinyon-juniper woodland in eastern section.	Fewer trees, open canopy with continuous ground fuels.[9] Low-severity, patchy fires every 3–4 years; some evidence of small acreage stand-replacing crown fires between ponderosa woodland and Douglas fir stands every 30 years.[6]	Tremendous increase in tree density and canopy bulk in last 100 years.[9] Susceptible to large-scale, high-severity stand-replacing fires that are outside the historical range of variability.[6]	2-Moderate alteration of historic regime.[6]
Pinyon-juniper-shrub-succulent woodland	Dry, hot south-facing canyon slopes; west, central, and eastern sections; continuous throughout.	Sparse herbaceous fuels; widely spaced tree crowns.[6] Fires infrequent, during extreme drought years with high winds.[6]	Moderate increase in tree density, tree height and bulk; finer fuels variably dense.[6]	2-Moderate alterationof historic regime.[8,3]
Riparian corridor	Shady canyon bottom.	Riparian tree and shrub species; variable density and contiguous throughout monument.	Individual trees becoming larger and bushier.	1-Little/no alteration of historic regime.[8,3]

1. Despain and Mosley (1990); 2. Cole (1998); 3. NPS (2005); 4. Covington and Moore (1994), Garrett and Soulen (1999); 5. Davis (1987), Swetnam et al. (1990); 6. Knox (2004); 7. Menzel (1996); 8. Hansen et al. (2004); 9. Jenkins et al. (1991).

open on flat ground and within alcoves. Very few architectural remains are situated in alcoves at Wupatki; most structures were built on open ground, appearing as mounds of rubble, sometimes with still-standing wall sections. At Walnut Canyon, 25 percent (n = 116) of all sites in the monument are well-preserved alcove cliff dwellings. The remaining sites are situated on open ground on the canyon rim tops and adjacent terraces.

The potential for fire damage is low but much more significant in alcove settings. Alcove structures were typically constructed on bare bedrock ledges beneath overhangs where cool dry conditions provide an environment for excellent preservation of organic materials such as wooden entryway lintels, roofing material, wall pegs, arrow shafts, weaving implements, basketry, and post-occupation packrat middens (Romme et al. 1993; Kleidon et al. 2003). In comparison, the potential for fire impacts is typically higher on surface architectural sites but less significant in terms of the fire's impact on perishable material. This is assuming that highly vulnerable organic materials have long since been burned away.

The seven vegetation classes identified at Wupatki and Walnut Canyon provide fuel sources that differ in the amount of heat energy generated during a fire. Experiments with prescribed fire have shown that grass fires are quicker burning, burn at cooler temperatures, and have limited impacts on archaeological materials (Table 2). At the high end of the temperature gradient, fuels in the pinyon-juniper woodland produce the greatest temperatures. Artifact damage associated with heat gradients from each vegetation class range from oxidation on the cool end to fracturing and disintegration on the hot end (Table 3).

Based on the experimental data above and previous post-fire assessments at Wupatki (Brown et al. 1996) and Walnut Canyon (Hough and Johnson 2001), the types of damage expected on surface architectural remains include oxidation, smoke blackening or sooting, fracturing, spalling, and disintegration (see also Eininger 1990; Lentz et al. 1996; Kleidon et al. 2003). Experiments have shown that animal bones, antlers, and teeth are damaged at temperatures lower than 300° C (Ryan and Noste 1985; Buenger 2003). Organic materials and original wall plaster and soil mortar floors at Walnut Canyon alcove sites may be destroyed at about this temperature as well.

Chipped Stone Artifacts

Documented types of damage to chipped stone artifacts include breakage, spalling, potlidding, microfracturing, pitting, bubbling, smudging, discoloration, adhesions (black pine pitch residue), altered hydration, chemically altered residues, and loss of weight and density measures (Eininger 1990:38–40; Romme et al. 1993:28–29; Deal 2003:2; Kleidon et al. 2003:30). Damage to chipped stone artifacts varies considerably with material type, artifact size and weight, surface exposure, duration of heat event, and maximum and minimum temperature. Chipped stone artifacts analyzed in the Wupatki post-fire assessment study were lying on the ground surface, exposed to the heat and direct flames. Chipped stone artifacts included tool production waste flakes and formal tools made from Kaibab chert, Little Colorado River cobble chert (including petrified wood), obsidian (unknown local sources), fine-grained basalt, and chalcedony. Temperatures as low as 200° C can damage chert in the form of oxidation, whereas more severe damage such as fracturing and melting can result from temperatures between 750 and 1000° C (Table 3).

Ceramic Artifacts

Damage from fire to ceramic artifacts can include oxidation, spalling, vitrification, cracking, crazing, warping, adhesions, sooting (smudging), discoloration of sherd surfaces, pigment color distortion, incineration of pigments, incineration of organic carbon, and cracking of a slip when present (Eininger 1990:41–42; Romme et al. 1993:28–29; Brown et al. 1996; Lentz et al. 1996; Kleidon et al. 2003:30). Much of this is detectable with the unaided eye and we expected to find most of these effects to varying degrees

Table 2. Maximum temperature range and artifact damage with vegetation classes.

Vegetation Class	Max Surface Temperature	Heating Duration	Artifact Damage/ Thermal Alteration
Mixed Grass	100–300° C	10–20 sec	Limited
Riparian			
Grass	100–200° C	10–20 sec	Limited
Willow small	100–300° C	1–2 min	Limited
Willow large	300–500° C	2–8 min	Moderate
Mixed Conifer			
Duff/litter	200–400° C	1–2 min	Moderate
Log	400–800° C	5–20 min	Significant
Pinyon-Juniper			
Large litter	700–800° C	2–4 min	Significant

Adapted from Buenger (2003).

Table 3. Temperature range for thermal alteration of archaeological material types.

Artifact Class	Material	Temperature Range	Thermal Alteration
Architecture	Sandstone (Buenger 2003)	300° C 200–700° C 400–700° C	Oxidation Spalling Fracturing
	Limestone (Rice 1987)	650–900° C	Decomposition/ Disintegration
	Basalt (Ryan & Noste 1985)	300–400° C	Fracturing
Chipped Stone	Obsidian (Buenger 2003; Deal 2003)	400–1000° C 600–1000° C 750–1000° C 1000° C	Fracturing Crazing Melting Vesiculation
	Chert (Buenger 2003; Deal 2003)	200–1000° C 350–550° C	Mineral oxidation Fracturing
	Silicified wood (Buenger 2003)	200–1000° C 400–1000° C	Mineral oxidation Fracturing
Ceramic	Fired clay and temper (Buenger 2003)	600–1000° C 600–1000° C	Mineral oxidation Pigment oxidation

during the Wupatki post-fire study. Of particular concern was the effect of the fire on surface pigment, temper, and overall condition of sherds, which would hinder identification of temporal and cultural affiliation. Damage to ceramic artifacts is associated with generally higher temperatures (600–1000° C) than chipped stone or architectural remains (Table 3) because ceramic vessels are originally fired at high temperatures (Buenger 2003).

DATA ANALYSIS

Wupatki Fire Damage Assessment

The area of Wupatki that burned during the 2002 State Fire was surveyed in the early and mid-1980s. In 2002, the entire 180 acres of the burn area was resurveyed to locate previously recorded sites and to document new sites and/or features exposed during the fire. Assessments were completed on 32 previously documented sites, and no new sites were recorded (Figure 4). The following assessments were made at each site: (1) percent of post-fire ground visibility, (2) change in site landscape visibility, (3) presence of fire-related impacts (e.g. sooting, cracking, discoloration), (4) frequency of damage per artifact class, and a narrative statement about preservation issues (Table 4).

All 32 sites in the 2002 burn area are one- to three-room fieldhouses constructed with locally available basalt boulders (Figure 5). Vegetation in the area includes a mix of grasses (*Bouteloua eriopa, Pleuraphis jamesii*), saltbush (*Atriplex canescens*), snakeweed (*Gutierrezia* spp.), Mormon tea (*Ephedra viridis*), and sparse junipers (*Juniperous monosperma*).

At 24 of the 32 sites (67%), the fire burned between 90 and 100 percent of all ground cover. At the remaining 12 sites, all located on bare outcrops or at the edge of the burn, the fire burned between 30 and 85 percent of the ground cover (Table 5). Because this area of the monument is somewhat isolated and within a quarter-mile of open public roads, there was a concern about site vandalism due to the temporary increase in site visibility. However, visibility ratings, given as the distance (in meters) at which a site becomes visible on the landscape, increased only slightly from pre-burn ratings at nearly all sites (Table 5). The few sites that are located on ridge tops experienced the highest visibility rating change, from a pre-burn distance of less than 100 m to being visible from more than 500 m.

Sooting was the most frequent type of fire damage across all artifact classes (Figure 6).

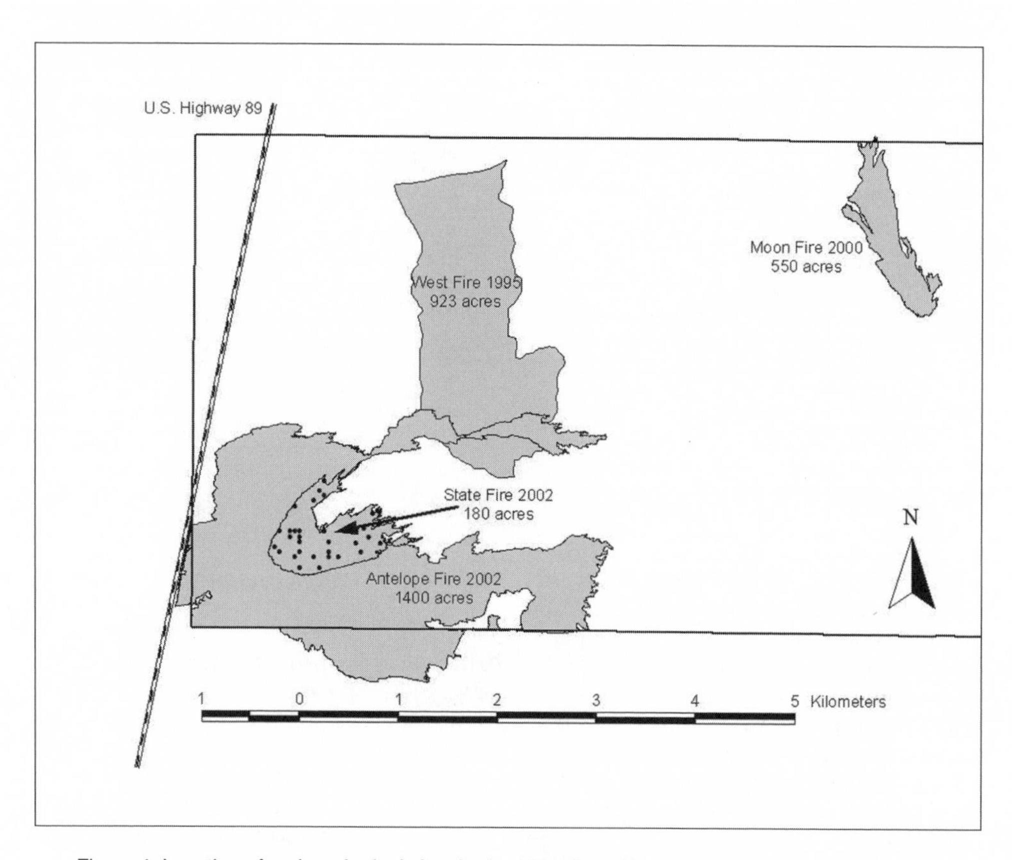

Figure 4. Location of archaeological sites in the 2002 State Fire, Wupatki National Monument.

Table 4. Wupatki post-fire assessment data recording.

Assessment	Post-fire ground visibility	Site landscape visibility	Fire-related impact type	Damage per artifact class	Preservation issues
Expression	Percent ground surface visible	Visibility rating: 1 = < 1 m 2 = 100–500 m 3 = 500–1000m 4 = > 1 km	Number of sites showing impact for each type	Count	Narrative statements about likely outcome

Table 5. Post-fire ground visibility and landscape visibility at Wupatki sites after fire.

Post-fire ground visibility	90–100% n = 24 (67%)	50–89% n = 6 (17%)	0–49% n = 2 (6%)
Landscape visibility change	0 (none – slight change) n = 29 (91%)	+ 1 visibility rating n = 2 (6%)	+ 2 visibility rating n = 1 (3%)

Visibility rating: 1 = < 1 m; 2 = 100–500 m; 3 = 500–1000m; 4 = > 1 km

Figure 5. WS 2464, a fieldhouse, burned during the Wupatki fire.

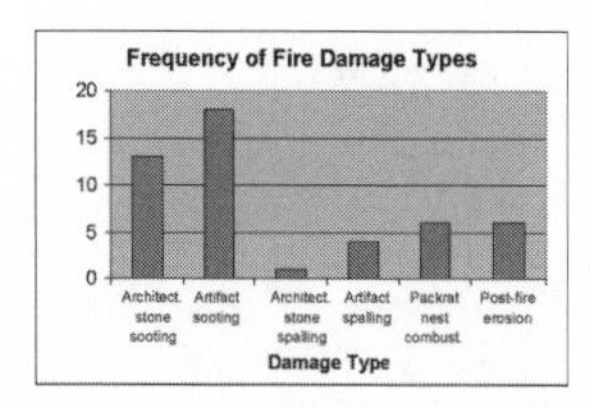

Figure 6. Fire damage types, 2002 Wupatki Fire.

Table 6 presents frequencies of thermal alteration/artifact damage for each artifact class. Ceramic artifacts had the greatest range of damage types at the greatest number of sites. The number of sites where sooting and spalling of architectural remains was documented is significant because this type of damage was intensified by the combustion of packrat nests found to be integrated into subsurface wall bases. The only sites that showed signs of post-fire erosion, and that had the greatest potential for long-term preservation problems, were the five sites where packrat nests burned into subsurface deposits. Combustion of the packrat nests left large voids (1.0 x 0.75 x 0.75 m) and unsupported wall stones (Table 6). Large wall voids and unstable wall stones weaken wall stability and provide an opening for rodents, birds, and reptiles to burrow deeper into the structure, likely causing advanced erosion of the structure from direct precipitation and surface runoff in the future.

The predominance of low-intensity thermal alteration, found at 67 percent of the sites (12 sites showed no signs of artifact damage or post-fire erosion), supports the long-held assumption that grass fires at Wupatki do not generate enough heat to significantly damage surface artifacts. The low percentage of damaged ceramic and chipped stone artifacts within assemblages also supports this assumption (Table 6). The predominance of these low-frequency and low-intensity fire impacts complement experimental data from prescribed fires showing that low maximum temperatures for short duration, such as during grass fires, cause limited impacts (Table 2).

Walnut Canyon Fuel Load Assessment

The Walnut Canyon fuel load assessment project was completed in two phases: the first was during a 2001–02 inventory survey of new lands added to the monument in 1996 and the second was during a fuel load monitoring project in the pre-1996 monument boundary. Both projects provide data for a contiguous block of monument land (Figure 7). The Archaeological Site Hazardous Fuels Assessment Project (ASHFAP) standards followed during the assessment provide a numeric ranking of elements within three sets of data—Fuels Data, Site Topography, and Site Vulnerability (Table 7). The fire risk potential score (1–100) is the sum of percentages from the three subsets of data.

Fuel load assessments were completed at 458 archaeological sites at Walnut Canyon National Monument between 2003 and 2004 (Table 8). We found that nearly half (n = 224) of the sites are at low risk for fire damage, 44 percent (n = 199) are at moderate risk, 6 percent (n = 31) are at high risk, and 1 percent (n = 4) are at very high risk (Table 8). The three most common types of sites at Walnut Canyon, in order of abundance, are artifact scatters (n = 145), fieldhouses (n = 136), and cliff dwellings (n = 116). Sites with intact standing architecture, the alcove cliff dwellings for which the monument was created in 1915, appear to be at a disproportionately high risk from fire damage (Figure 8).

We found significant differences in fire risk potential between artifact scatters, fieldhouses, and cliff dwellings (Table 9). These differences can be attributed to variability in fuel loading, topographic setting, or site vulnerability between site types. There is a significant difference in fuel loading between vegetation classes, but mean fuel loading values in the two classes that contain cliff dwellings—Douglas fir woodland and pinyon-juniper-shrub-succulent woodland—are second and fourth highest among all classes, and probably do not account for the highest fire risk potential rating for cliff dwellings (Table 10). Also, the evaluation of fuel loads at each site did not take into

Table 6. Frequency of sites with thermal alteration/artifact damage by artifact class.

Artifact Class	Thermal Alteration/Damage: Sooting	Spalling	Cracking	Surface Disintegration	Packrat Nest Combustion	Post-fire Erosion
Architectural remains	13 sites	1 site	0	0	5 sites	5 sites
Chipped stone	3 sites	1 site	0	1 site	0	0
Ceramic	17 sites	2 sites	2 sites	3 sites	0	0
Ground surface	0	0	0	0	0	5 sites
Percent of damaged artifacts within assemblage:			Ceramic 0.004–0.17 (only one site, WS 2488, with > 2% (0.02) damaged sherds Chipped stone 0.004 (limited sample, just two sites: WS 2460 and 2477)			

Table 7. Fuel load assessment scoring (using Archaeological Site Hazardous Fuels Assessment Project standards) for fire risk potential. Unless noted, scale for all elements is 1 (low) – 10 (high). Fire Risk Potential Score: 1–34 = Low; 35–49 = Moderate; 50–64 = High; 65–79 = Very High; 80–100 = Extreme.

Fuels Data	Site Topography	Site Vulnerability	Notes
A. Fuel continuity	A. Aspect N = 1; NE, NW = 2; E = 3; W, SE = 4; S = 5; SW = 6	A. Flammability	—
B. Fuel loading	B. Slope	B. Friability	—
C. Ladder fuels	C. Heat trap (1 none – 4 severe)	C. Data potential	—
D. Crown spacing	—	D. Uniqueness	—
E. Max fuel temp (°F) < 90 = 1; 91–100 = 2; 101–110 = 3; > 111 = 4	—	E. Capital outlay	Stabilized or interpreted site, or tribal recognition High score = 10
		F. Visibility	
		G. Post-fire erosion	
Subtotal 1	Subtotal 2 (+Subtotal 1)/max	Subtotal 3/ max	SubT 2 + SubT 3 = Fire Risk Potential Score

Table 8. Archaeological site fire risk potential (see Table 7) in Walnut Canyon National Monument in 2003–04.

Fire Risk Potential	Number of Sites	Artifact Scatters (% all sites)	Fieldhouses (% all sites)	Cliff Dwellings (% all sites)	Forts (% all sites)
Low (11–34)	224 (49%)	120 (26%)	60 (13%)	32 (6.9%)	0
Moderate (35–49)	199 (44%)	25 (5.5%)	72 (15.7%)	64 (13.8%)	5 (1%)
High (50–64)	31 (6%)	0	3 (0.5%)	18 (4.1%)	0
Very high (65–79)	4 (1%)	0	1 (0.5%)	2 (0.5%)	0
Extreme (> 80)	0	0	0	0	0
Total	458 (100%)	145 (31.5%)	136 (29.7%)	116 (25.3%)	5 (1%)

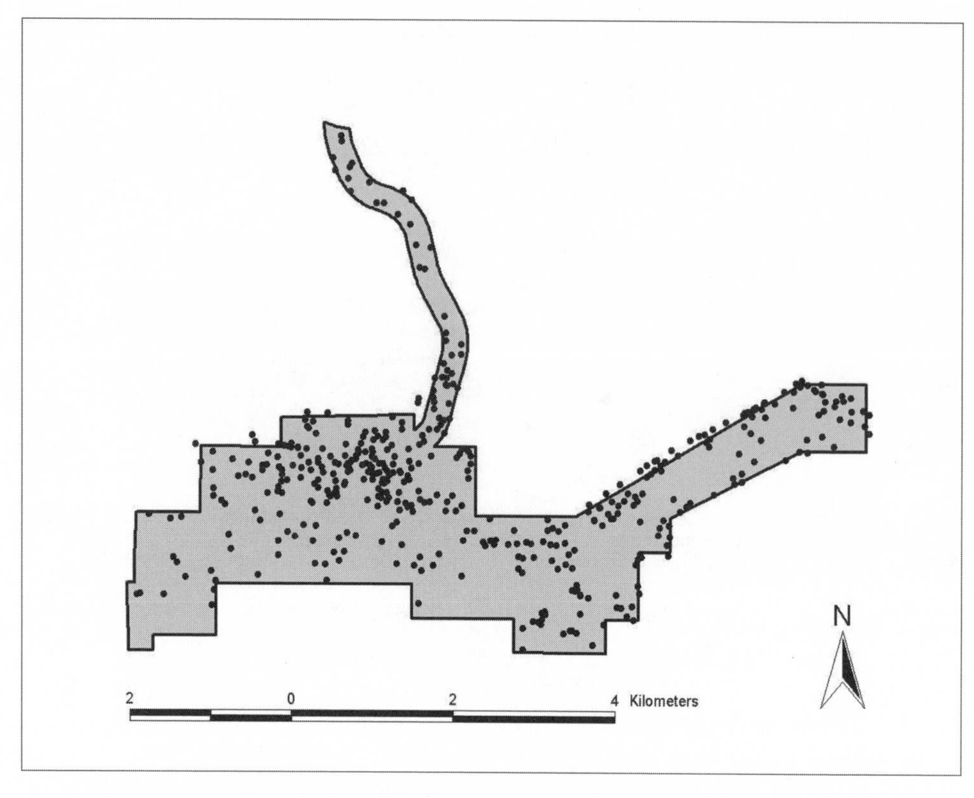

Figure 7. Location of Walnut Canyon National Monument Fuel Load Assessment project.

Figure 8. Walnut Canyon National Monument Sinagua cliff dwelling (Elden Phase, AD 1150–1250).

Table 9. ANOVA results for fire risk potential at three site types in Walnut Canyon.

	Fire Risk Potential			
Site Type	Mean	SD	s^2	n
Artifact scatters	27.70	7.56	57.16	145
Fieldhouses	35.37	6.91	47.74	136
Cliff dwellings	39.90	9.29	86.21	116

$F = 79.82$; $df = 2/394$; $P = 0.0001$.

Table 10. ANOVA results for fuel loading in four vegetation classes in Walnut Canyon.

	Fuel Load			
Vegetation Class	Mean	SD	s^2	n
Ponderosa-pinyon-juniper-oak	37.63	9.56	91.31	149
Douglas fir–Gambel oak	33.88	10.21	104.23	16
Pinyon-juniper–blue grama	33.77	10.07	101.36	177
Pinyon-juniper-shrub-succulent	26.97	8.15	66.37	110

$F = 26.68$; $df = 3/448$; $P = 0.0001$.

account the buildup of flammable packrat nest material or shrubs and saplings growing within room interiors where standing walls with original architecture are the most susceptible to fire damage. Because of their uniqueness in style and short time period of construction, site vulnerability scores are significantly higher for cliff dwellings compared to artifact scatters and fieldhouses. High site vulnerability probably has a greater influence than fuel loading or topography on high fire risk potential for cliff dwellings (Table 11).

DISCUSSION

Results from the Wupatki post-burn assessment and the Walnut Canyon fuel load assessment have shown that resource integrity is at risk from fire-related impacts in the two monuments. The results also show that architectural remains are currently being impacted more severely than artifact assemblages, and that sites with intact, original standing architecture are the most threatened among the three common site types. At Wupatki, although grass fires in the monument do not seem to generate enough heat or burn long enough to cause severe artifact damage, combustion of packrat nests can cause moderately intense impacts to architectural remains. With grass fires burning hundreds of acres every 2 or 3 years (NPS 2005), the chances are greater now than in the past 125 years of fires burning over archaeological sites multiple times per decade. The cumulative effects of multiple burnovers may begin to degrade architectural remains over time, but low-intensity burning may be no more significant than other natural processes. Determining whether the combustion of packrat middens is a significant preservation problem, and whether mitigation is necessary, depends on the distribution of sites with packrat nests across the monument and the study of the long-term effects of fire on architectural remains.

At Walnut Canyon, sites at the highest risk for damage from fire are disproportionately those with standing architecture.

Table 11. ANOVA results for site vulnerability for three site types in Walnut Canyon.

	Site Vulnerability			
Site Type	Mean	SD	s^2	n
Artifact scatters	13.89	4.96	24.64	144
Fieldhouses	20.1	4.30	18.49	123
Cliff dwellings	30.30	8.13	66.15	108

$F = 239.6$; $df = 2/372$ $P = .0001$.

Sinagua cliff dwellings are particularly vulnerable because of historically high fuel loads, an abundance of well-preserved organic remains, and the uniqueness of these alcove structures. Walnut Canyon cliff dwellings are found in such concentration nowhere else in the region. They were built in a relatively short time span in the Northern Sinagua cultural period.

These results point to the need to integrate fire management planning and resource management planning; fire management plans and ruins preservation plans are currently being updated for both Wupatki and Walnut Canyon. There has been significant crossover in defining the goals and objectives for both plans, and in developing projects to meet mutual needs and conditions.

Preservation of cultural resources needs to be a primary objective stated in an FMP. Preservation of archaeological sites is not the same as protection of resources through avoidance by digging fire containment lines around archaeological site boundaries, usually without removing hazardous fuels from within the site. Historically, archaeological site avoidance has been the primary objective during fire suppression and prescribed fire projects. The outcome has been a buildup of hazardous fuel islands centered squarely on sensitive archaeological sites.

We suggest removing hazardous fuels such as tree logs, live saplings, and ladder fuels from archaeological sites *during* large-block fuels treatment projects to allow future prescribed fire and wildland fire to burn over open surface sites when appropriate. This should be done only when the natural processes of fire do not compromise the site integrity (location, design, setting, materials, workmanship, feeling, and association; NPS 1998). At cliff dwelling sites with highly flammable organic materials, we suggest moving hazardous fuels away from standing architectural components and room interiors when adjacent rim tops are treated or prepared for prescribed fire projects. This will serve to reduce the threat of damage from future wildfires, and to mitigate the immediate threat of burning airborne or rolling debris reaching cliff dwelling structures located on slopes directly below rim top burn blocks.

The threat of fire also needs to be brought within the scope of resource management planning, especially ruins preservation planning. Evaluations of ongoing impacts and the threat they pose to architectural remains, and development of treatment plans to mitigate those impacts, are part of preservation planning. Fires need to be recognized as natural occurrences—like water erosion, freeze-thaw cycles, and chemical weathering—that impact archaeological sites. Fire should be included as a factor of deterioration and should lead to a treatment plan to reduce the threat of fire on architectural fabrics (e.g. stone, mortar, wood). Although the immediate impacts of fire may not be readily perceived at sites that have not burned for more than a century, the predicted effects of fires on architectural remains need to be treated alongside active preservation problems. Tree logs should be moved away from walls, and brushy vegetation and surface packrat nest materials should be removed from the area, concurrent with other routine fabric treatments such as repointing, resetting wall stones, and drainage control.

ACKNOWLEDGMENTS

This project would not have been completed without initial planning by Al Remley, and dedication from the field crew. We thank David Barr, Walter Gosart, Jessica Bland, Sara Wendt, Kenny Acord, Mandy Johnson and Bernie Natseway.

LITERATURE CITED

Anderson, B. 1990. The Wupatki Archeological Inventory Survey Project: Final Report. Division of Anthropology, National Park Service, Santa Fe, New Mexico. Professional Paper No. 35.

Brown, G. B., L. Folb, and C. E. Downum. 1996. West Mesa and Walnut Canyon Burn Archeological Survey Report. Anthropology Laboratories report No. 1169. Northern Arizona University, Flagstaff.

Buenger, B. 2003. The impact of wildland and prescribed fire on archaeological resources. Dissertation submitted to the Department of Anthropology, University of Kansas. http: / / www. blm. gov / heritage / docum / Fire / Dissertation_Buenger.htm

Cole, K. L. 1998. Five centuries of environmental change at three southern Colorado Plateau parks. USGS Plan of Work. On file at National Park Service, Flagstaff Area National Monuments.

Covington, W. W., and M. M. Moore. 1994. Southwestern ponderosa forest structure and resource conditions: Changes since Anglo-American settlement. Journal of Forestry 92.

Davis, K. 1987. Fire history of Walnut Canyon National Monument. Unpublished report, National Park Service, Flagstaff Area National Monuments.

Deal, K. 2003. Fire effects on lithic artifacts, for cultural resource protection and fire planning. U.S. Forest Service, Eldorado National Forest, Placerville CA.

Despain, D. W., and J. C. Mosley. 1990. Fire history and stand structure of a pinyon-juniper woodland at Walnut Canyon National Monument, Arizona. Cooperative National Park Resources Studies Unit, Technical Report No. 34. Tucson AZ.

Dowum, C. E. 2000. Archaeological sandal recovery and test excavations at NA 324, Walnut Canyon National Monument. Northern Arizona University, Anthropological Report 1186, Flagstaff.

Downum, C. E., and A. P. Sullivan. 1990. Settlement Patterns. In The Wupatki Archeological Inventory Survey Project: Final Report. Division of Anthropology, National Park Service, Santa Fe, New Mexico. Professional Paper No. 35.

Eininger, S. 1990. Long Mesa Fire: An archeological survey and post-fire assessment. National Park Service, Mesa Verde National Park, Cultural Resources Management.

Fulé, P., W.W. Covington, M. M. Moore, T. A. Heilein, and A. E. M. Waltz. 2002. Natural variability in forests of the Grand Canyon, USA. Journal of Biogeography, Vol. 29.

Garrett, L. D., and M. H. Soulen. 1999. Changes in character and structure of Apache/Sitgreaves Forest Ecology: 1850–1990. Proceedings of the Fourth Biennial Conference of Research on the Colorado Plateau. USGSFRESC/COPL/1999/16; USDI, USGS.

Hansen, M., J. Coles, K. Thomas, D. Cogan, M. Reid, J. Von Loh, and K. Shultz. 2004. USGS-NPS National Vegetation Mapping Program: Wupatki and Walnut Canyon National Monuments, Arizona. Vegetation Classification and Distribution. Southwest Biological Sciences Center, Flagstaff AZ.

Hough, I., and C. Johnson 2001. Archeological site monitoring, Wupatki and Walnut Canyon National Monuments. National Park Service, Flagstaff Area National Monuments. Manuscript on file.

Jenkins, P., F. Reichenbacher, and K. J. A. Gondor. 1991. Vegetation inventory, classification and monitoring at Walnut Canyon National Monument. U.S. Department of Interior, National Park Service, Southern Arizona Group Office, Phoenix.

Joyce, J. F. 1974. A taxonomic and ecological analysis of the flora of Walnut Canyon, Arizona. Masters thesis, Northern Arizona University, Flagstaff.

Kleidon, J., M. Hendrix, D. Long, S. Payne, E. Rezac, B. Shanks, S. Diederichs, J. Karchut, V. MacMillan, and J. Beezley. 2003. Bircher-Pony post-fire assessment project. National Park Service, Mesa Verde National Park Division of Research and Resource Management, Mesa Verde National Park, CO.

Knox, S. 2004. Multi-scale spatial heterogeneity in historical fire regimes of ponderosa pine and Douglas-fir forests in Arizona, USA. Unpublished Masters thesis, Colorado State University, Fort Collins CO.

Lentz, S. C., J. K. Gaunt, and A. J. Willmer. 1996. Ceramic artifact analysis. In Fire Effects on Archaeological Resources, Phase 1: The Henry Fire, Holiday Mesa, Jemez Mountains, NM. Rocky Mountain Forest and Range Experiment Station, General Technical Report RM-GTR-273. USDA Forest Service, Fort Collins CO.

Mabry, J. B., and M. K. Faught. 1998. Chapter 4. In Archaic Complexes of the Early Holocene, edited by J. B. Mabry. Paleoindian and Archaic Sites in Arizona. Center for Desert Archaeology, Tucson AZ.

Menzel, J. P. 1996. Historical changes in forest structure in the ponderosa pine type, Walnut Canyon area, northern Arizona. Unpublished Masters thesis, Northern Arizona University, Flagstaff.

National Park Service. 1998. National Register Bulletin: How to Apply the National Register Criteria for Evaluation. U.S. Department of the Interior, National Park Service, Washington D.C.

National Park Service. 2001. Ruins Preservation Plan and Implementation Guidelines; Wupatki National Monument. Flagstaff Area National Monuments, Division of Resources Management, Flagstaff,AZ.

National Park Service. 2003. Bircher/Pony Fire BAER activity cultural resources assessment accomplishments report. Mesa Verde National Park, Post-Fire Recovery 2000.

National Park Service. 2004. Ruins Preservation Plan and Implementation Guidelines; Walnut Canyon National Monument. Flagstaff Area National Monuments, Division of Resources Management, Flagstaff AZ.

National Park Service. 2005. Wupatki, Sunset Crater Volcano and Walnut Canyon National Monuments Draft Fuels and Fire Management Plan. Flagstaff Area National Monuments, Flagstaff AZ.

Neff, L. T., and K. Spurr. 2004. An Intensive Archaeological Inventory Survey of Walnut Canyon National Monument New Lands. Navajo Nation Archaeology Department Project No. 01-143, Flagstaff AZ.

Rice, P. M. 1987. Pottery Analysis: A Sourcebook. University of Chicago Press, Chicago.

Romme, W. H., L. Floyd-Hanna, and M. Connor. 1993. Effects of fire on cultural resources at Mesa Verde National Park. In Park Science, Summer 1993. National Park Service, Mesa Verde National Park.

Ryan, K., and N. V. Noste. 1985. Evaluating fire effects on cultural resources. USDA Forest Service, Intermountain Research Station, General Technical Report 182. Ogden UT.

Swetnam, T. W., and C. H. Baisan. 1996. Historical fire regime patterns in the southwestern United States since AD 1700. In U.S. Forest Service General Technical Report RM-GTR-286.

Swetnam, T. W., and C. H. Baisan. 2003. Tree-ring reconstruction of fire and climate history in the Sierra Nevada and southwestern United States. In Fire and Climate Changes in Temperate Ecosystems of the Western Americas. Springer, New York.

Swetnam, T. W., W. E. Wright, A. C. Caprio, P. M. Brown, and C. H. Baisan. 1990. Fire scar dates from Walnut Canyon National Monument, Arizona. Tree-Ring Laboratory, University of Arizona. Report to National Park Service, Southern Arizona Group Office, Phoenix AZ.

ARCHAEOLOGICAL SURVEY OF NEW LAND ACQUIRED BY WALNUT CANYON NATIONAL MONUMENT, NORTHERN ARIZONA

Kimberly Spurr and L. Theodore Neff

The Navajo Nation Archaeology Department (NNAD) and the National Park Service (NPS) conducted an archaeological survey in Walnut Canyon National Monument between October 2001 and May 2002, except during a 2-month period when snow prevented fieldwork (Neff and Spurr 2005). The primary purpose of the survey was to document all cultural resources in land acquired by the monument from the Coconino National Forest in 1996, as well as land adjacent to the entrance road corridor, to allow NPS administrators to monitor impacts and implement preservation plans for the resources (Neff 2001). The survey was undertaken to comply with Section 110 (a)(1) of the National Historic Preservation Act (Public Law 89-665; 80 Stat. 915; 16 U.S.C. 470, as amended), which directs federal agencies to "undertake a program to identify historic properties under its jurisdiction or control." A secondary goal was to compare site types and land use/settlement patterns in the new land with those documented by a survey of the monument lands in 1985 (Baldwin and Bremer 1986).

Walnut Canyon National Monument is about 16 km (10 miles) east of Flagstaff, lying south of Interstate 40 and north of Anderson Mesa (Figure 1). The portion of Walnut Canyon encompassed by the monument contains a wealth of archaeological sites, most notably clusters of masonry rooms tucked beneath limestone ledges on the canyon slope. The earliest documentation of these ruins was by James Stevenson, who led an expedition from the Smithsonian Institution through the area in 1883 (Stein 1986:49). The next decade saw numerous visits by scientific researchers, as well as a growing interest in the ruins among local residents. Construction of a wagon road from Flagstaff to the north rim of Walnut Canyon in 1884 facilitated scientific and recreational expeditions and offered more direct access to the well-preserved ruins on the north slopes of the canyon. Growing visitation and extensive artifact collecting, including the use of hand tools and dynamite to remove masonry walls to access the rooms, took a heavy toll on the ruins. Besides damage to the ruins and loss of archaeological materials, the natural beauty of the area was also marred by trash discarded by visitors (Stein 1986:54). By the mid-1890s several local residents and businessmen realized that the ruins in Walnut Canyon were unusual and, if preserved, could be promoted as a tourist and scientific attraction. They began to lobby for establishment of a preserve to protect the canyon and its ruins.

The earliest protection afforded to the ruins was designation of the area as a Forest Reserve around the turn of the twentieth century, which meant that the canyon was patrolled by a ranger charged with monitoring the ruins and visitor activity (Stein 1986:60). The main concentration of ruins on about 960 acres of land was designated a national monument in 1915, but it remained under the jurisdiction of the Coconino National Forest until 1934. Significant improvements came to Walnut Canyon beginning in

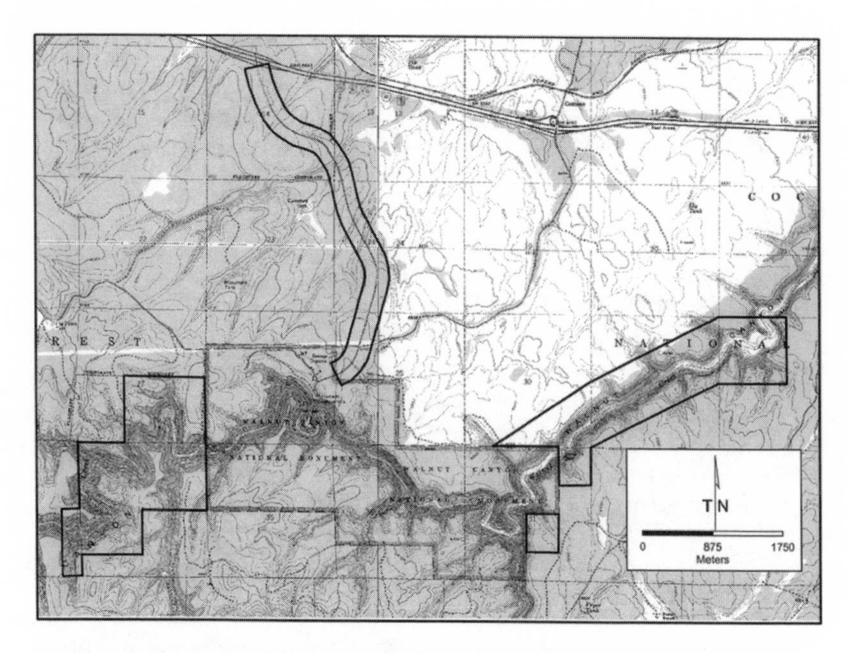

Figure 1. Location of Walnut Canyon National Monument, showing the west and east survey parcels and road corridor in relation to the pre-1996 monument area. Base maps are Flagstaff East, Arizona (photorevised 1983) and Winona, Arizona (photorevised 1974) 7.5' USGS quadrangles.

1938, when the land base of the monument was doubled in size and money was allocated for construction of facilities (Stein 1986:66). Manpower provided by the Civilian Conservation Corps expanded the infrastructure and tourist facilities at the monument, including construction of a visitor center and improvements to the trail around Third Fort (the Island Trail) and to Forest Road 303, then part of the Ocean to Ocean Highway that passed through Flagstaff. In 1996, two blocks of land totaling approximately 1420 acres were added, extending the monument to the west and east. This land was the focus of the present survey.

Walnut Canyon National Monument encompasses the sinuous canyon of Walnut Creek north of Anderson Mesa, as well as portions of numerous tributaries and narrow strips of rim terrain. In addition to spectacular archaeological remains, the monument preserves land that exhibits impressive ecological diversity. Elevation in the monument ranges from about 2090 to 1935 m (6850–6350 ft). The canyon rims support woodland dominated by pinyon and juniper, sometimes mixed with oak, transitioning into ponderosa pine forest in the higher elevations to the west. The topographic variation within Walnut Canyon, however, produces a great variety of microclimates with corresponding biological diversity. Numerous areas along the rim exhibit a transitional woodland environment, with a nearly equal mix of juniper, oak, pinyon and ponderosa pine, mountain mahogany, barberry, serviceberry, and locust, and an understory of grasses and forbs. North-facing slopes in the western end of the monument are covered with ponderosa pine and Douglas fir, with a sparse understory of serviceberry, squawbush, locust, grasses, forbs, and yucca. Farther east the less steep north-facing slopes support pinyon-juniper woodland with more xeric species like prickly pear cactus, ephedra, and abundant grasses. The south-facing slopes of the canyon, particularly toward the eastern end of the monument, exhibit an Upper Sonoran life zone, characterized by fernbush, ephedra, yucca, prickly pear cactus, and various grasses. The canyon bottom supports a riparian zone containing box elder, Arizona walnut, willow, canyon grape, locust, and perennial forbs like bee balm. Walnut Creek once flowed through the canyon on a seasonal basis, but construction of the dam for Lake Mary in 1903–1904 blocked the watershed and stream flow is now rare within the monument, although in February and March of 2005 water could be seen flowing past the visitor center in Walnut Canyon for the first time in more than a decade. Reliable surface water sources are few, and water would have been a concern for occupants of the canyon. Springs and deep potholes in lower Cherry Canyon offer one of the best water sources. Rain and snow captured in small tinajas in the limestone bedrock along the canyon rims and ledges also likely provided a significant amount of water.

Most of Walnut Canyon exposes thick Kaibab limestone bedrock, although the cross-bedded sandstone of the Toroweap Formation lines the canyon bottom westward from the old monument boundary (Chronic 1989:306). Basaltic cinders derived from the eruption of Sunset Crater in the mid-eleventh century form a thin surface deposit on the canyon rim in some areas, but the entire Walnut Canyon area would likely have been available for occupation during, and certainly immediately after, the eruption. The lack of cinder cover may have played a role in the concentration of Sinagua population in this area during post-eruptive times.

SURVEY METHODS

Between October 2001 and May 2002, crews from the Navajo Nation Archaeology Department and the National Park Service surveyed the entire new land acquisition. This included 685 acres west of the pre-1996 monument area and 735 acres on the east end of the monument. In both new areas, land was added to the south and north rims, as well as within the canyon itself, offering a comprehensive view of occupation centered on the canyon. Crews also surveyed a 300 m wide corridor (about 330 acres) of U.S. Forest Service land along the entrance road to the monument. The total survey area was approximately 1750 acres.

Survey crews consisted of 3–5 people, generally walking a series of systematic transects spaced 15 m apart. This approach worked well for most areas on the north and south rims, including the road corridor. Within the canyon itself, where terrain was extremely steep, crews attempted to maintain 15 m spacing (based on GPS readings), but the primary focus was on investigating ledges or flatter areas to ensure complete coverage of all usable ground. The canyon bottom was surveyed by crews of 2–3 people walking parallel transects that extended between the steep canyon slopes. These survey techniques resulted in almost total coverage of the entire new land and road corridor.

Cultural materials were designated as sites or isolated occurrences based on criteria established by the NPS. A site consisted of "a locus of past human activity, at least 50 years old, as represented by 1) one or more humanly constructed or created features (e.g., hearth, wall, check dam, rock art, pithouse, rubble mound, grave); or 2) three or more different kinds of humanly produced items (flakes, bifaces, finished tools, pottery, corncobs, etc.) clustered in a 100 sq m area, where the clustering of materials is clearly not a result of natural landscape processes; or 3) 10 or more humanly produced items of one or two types (e.g., flakes, sherds, cans, bottles) in a 100 sq m area" (National Park Service 2001:2). All sites were recorded on comprehensive forms created by the NPS, supplemented with scale maps and photographs. At each site the crew also completed ASMIS (Archaeological Sites Management Information System), Backcountry Monitoring, and Fuel Assessment forms for inclusion in national NPS databases. For sites with fewer than 100 surface artifacts, the entire assemblage was analyzed in the field. At sites with more than 100 artifacts, or where multiple concentrations were present, crews conducted detailed analysis within a 10 sq m transect that appeared representative of the assemblage, with a more general description of the assemblage as a whole. All diagnostic artifacts or tools were also analyzed at every site. Each site datum was marked with a flat aluminum rebar stake stamped with sequential site numbers (continued from the 1985 survey number sequence). The datum location was plotted with a GPS unit. Artifacts were collected only in extraordinary circumstances, and the four collected items are curated at the NPS Flagstaff Area Parks and Monuments office.

Isolated artifacts or sparse scatters of fewer than 10 artifacts in a 100 sq m area, and some remains less than 50 years old, were recorded as Isolated Occurrences (IOs). These resources were briefly described and plotted with a GPS unit during survey. In addition to a description of the artifacts and their temporal assignment, pertinent information about local topography, vegetation, and geology was also documented. Although some of the IOs are less than 50 years old, documenting them provides information for land managers regarding recent use of the area, which will increase their ability to monitor visitation and protect cultural resources.

Subsequent to survey, all site datum locations and boundaries were incorporated into a GIS database for inclusion in the NPS system. All site forms were entered into a database to facilitate management activities, and photographs were cataloged to NPS standards. All documentation and photographs related to this project are permanently curated at the NPS Flagstaff Area Parks and Monuments Office.

SURVEY RESULTS

Our survey documented 210 archaeological sites in the new land and the road corridor. The sites range in age from the early Archaic period, beginning about 8000 BC, to the present. The most abundantly represented culture is the Sinagua, with most sites occupied between about AD 1050 and 1300. A number of sites have more than one temporal or cultural component, so the survey ultimately documented 263 components, of which 23 (9%) are Archaic, 171 (64%) are Sinagua, 36 (14%) are of unknown Native American affiliation, 4 (2%) are more recent (post AD 1300) Native American, and 29 (11%) are components related primarily to Euro-American use of the region (Figures 2–5).

The most numerous sites are artifact scatters containing lithic and/or ceramic artifacts, of which 121 were documented (57.9% of all sites; Table 1). Twenty-three lithic scatters can be confidently assigned to the Archaic period based on the presence of diagnostic projectile points (see Figure 2); these can be further divided into seven early Archaic (ca. 7000–4500 BC), three middle Archaic (ca. 4500–2000 BC), one late Archaic (ca. 2000 BC–AD 300), and 12 indeterminate Archaic period sites. An additional nine isolated dart points diagnostic of the Archaic period were recorded as IOs. Although 126 whole or fragmentary dart-sized projectile points were recorded, some of these items were clearly curated, reused, and discarded by later Sinagua residents. The entire span of the Archaic period is represented by the 44 dart points that could be classified using established typologies (Table 2), but early Archaic point types dominate the assemblage. Lithic debitage showed no distinct technological differences between the Archaic and Sinagua assemblages. Nearly all sites had a preponderance of biface percussion flakes, followed by direct freehand percussion flakes (Neff and Spurr 2005: Appendix B). Lithic assemblages at the Archaic and presumed Archaic sites did exhibit a much more diverse range of raw material, particularly cherts from the Little Colorado River drainage, suggesting that material was brought into the Walnut Canyon area by mobile foragers whose range extended to the north and east.

Unfortunately, only a few of the lithic scatters exhibit any evidence of intact features. Much of the survey area on the north side of the canyon in the eastern block was

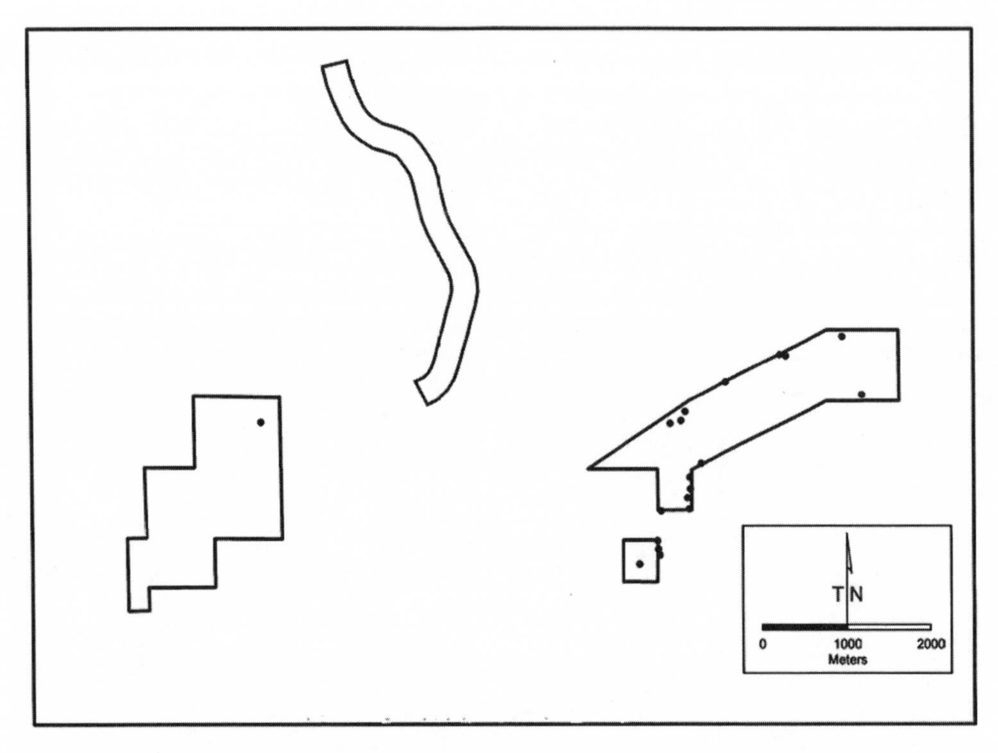

Figure 2. Survey area boundaries, showing the location of all Archaic site components recorded during the survey (each dot represents a site datum).

chained in the 1960s to "improve" range vegetation. This activity displaced artifacts horizontally and resulted in extensive deflation that produced a lag deposit of artifacts and may have destroyed small features such as hearths. Given the severe deflation of the shallow soil deposits overlying bedrock, it seems highly unlikely that many thermal features remain buried. The chaining disturbance also caused interpretive difficulties at a number of sites by smearing together assemblages that appear to represent distinct episodes of activity. Fortunately, a thin margin of adjacent unchained forest contained similar sites, which facilitated interpretation of the disturbed sites to some degree. It became apparent, for example, that the large size of these sites is not just a consequence of chaining and dispersal of artifacts, but reflects reuse of the area over time, often producing expansive sites with overlapping activity areas.

Several artifact scatters in the eastern survey block that lacked diagnostic Archaic dart points did exhibit spatially discrete artifact concentrations with no ceramics, which likely also date to the Archaic period. Some of these sites also contain artifact loci with ceramics, demonstrating use of the rim areas throughout prehistory. Smaller, more discrete scatters of lithic and ceramic artifacts, or ceramics alone, were also recorded in the western survey block and along the road corridor. These likely represent short-term activity areas dating from the Sinagua period, although the exact function of the scatters and their relationship to the larger architectural sites was not entirely clear. In a few cases the depth of soil hinted that buried features, including pit structures, might be

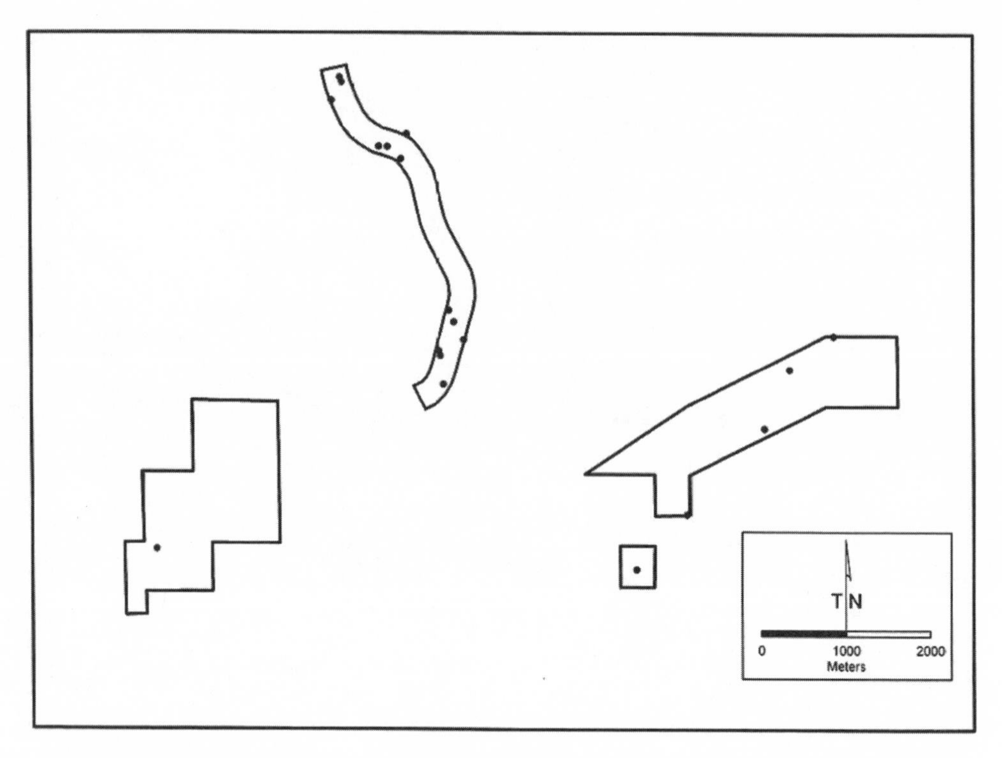

Figure 3. Survey area boundaries, showing the location of all pre-AD 1050 Sinagua site components recorded during the survey (each dot represents a site datum).

present, but most of the small artifact scatters appear to be the remains of brief activity episodes. In general, these small sites showed little evidence of multiple episodes of use, as was often the case in the sprawling Archaic sites at the east end of the project area.

The second most numerous sites are fieldhouses, representing the Sinagua culture and dating primarily after the eruption of Sunset Crater. We documented 30 fieldhouse sites, which typically consist of a one-room masonry structure associated with a small number of ceramics and a few lithic artifacts. The architectural style of these structures is remarkably consistent, containing a single room with double-stone, rubble-core foundations built of unmodified limestone blocks. The amount of rubble suggests that most of these structures were not built entirely of stone. Sinagua fieldhouse sites typically exhibit very small middens or sparse artifact scatters, a pattern evident in the survey area. A few sites were classified as "Two-Room Structures," which are essentially fieldhouse sites with two individual or linked rooms and a somewhat larger artifact assemblage. The fieldhouse sites are most common along the entrance road, some distance from the rim, and are typically situated near arable land.

The quintessential Sinagua sites in Walnut Canyon are the alcove rooms that are so abundant along the canyon slopes, mainly on the north side, facing south. We documented 28 alcove sites, 24 of which are in the western survey area. It was clear that the general lack of alcove sites in the eastern survey blocks is a result of the canyon geology, in that the less steep eastern canyon

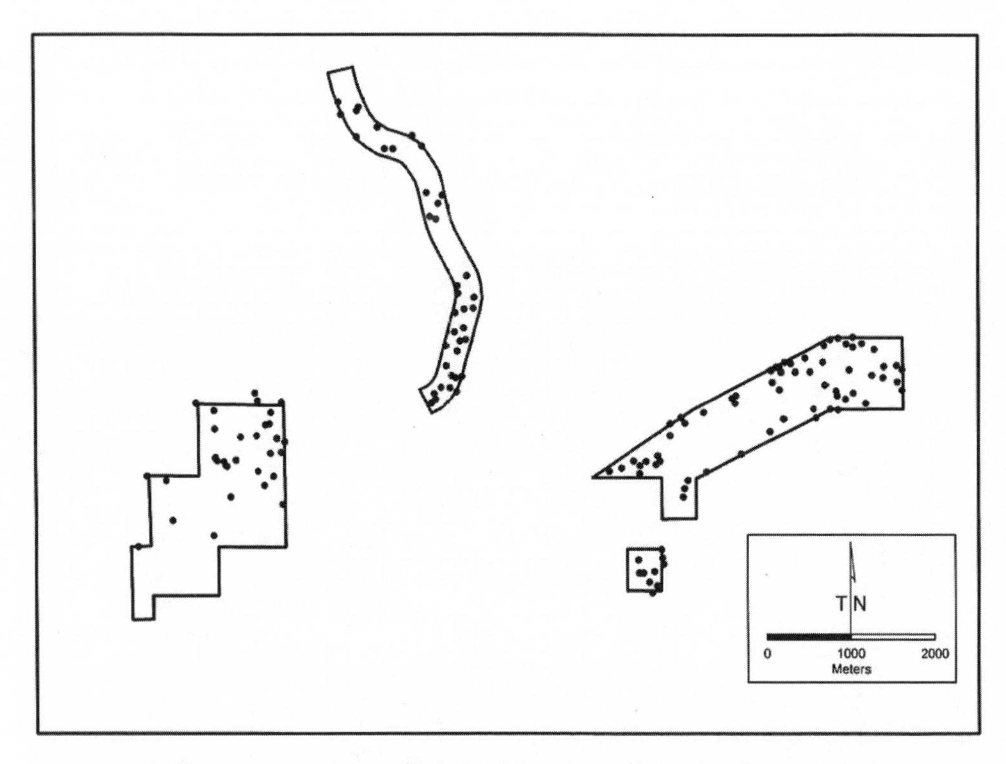

Figure 4. Survey area boundaries, showing the location of all post-AD 1050 Sinagua site components recorded during the survey (each dot represents a site datum).

slopes provide fewer suitable alcoves or shelters. Alcove rooms were built using unshaped or slightly shaped limestone blocks as masonry, and the abundant supply of rock on the canyon slopes minimized the effort involved in construction. These multiroom sites were used for both storage and habitation, the latter clearly indicated by smoke-blackened alcove roofs.

Nearly all of the alcove sites show evidence of vandalism, ranging from removal of the upper courses of stone to near complete destruction of the walls. Modern trash and inscriptions at many of the sites indicate visitation within recent decades. The few sites that are still in good condition offer detailed evidence about construction methods; some even retain intact plaster on their walls. Artifact assemblages at the alcove sites are typically small, and usually lack diagnostic artifacts, probably as a result of years of visitation and collection. This situation frustrated our efforts to confidently date many of the sites and to accurately assess their temporal relationships. Based on small ceramic assemblages and similarities in architectural styles, however, we believe that all of the alcove sites post-date the eruption of Sunset Crater in the eleventh century.

We also recorded a few Sinagua multiroom habitations, but considering the number of fieldhouses and the large quantity of dispersed artifacts reflecting intensive activity in the entire survey area, we were surprised to encounter only 10 two-room structures and 9 multi-room masonry habitations. At least two of these sites were part of larger communities associated with Sinagua forts, and presumably were occupied concurrently with the walled promon-

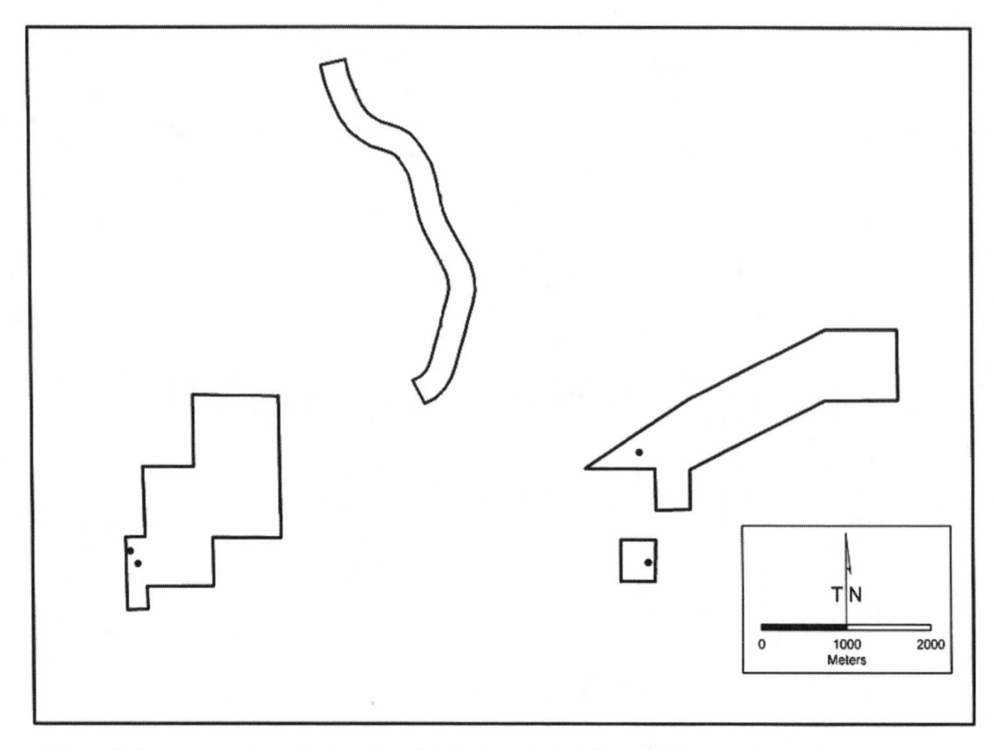

Figure 5. Survey area boundaries, showing the location of all post-AD 1300 Native American site components recorded during the survey (each dot represents a site datum).

tory sites. The amount of rubble at the multi-room sites indicates that the architecture consisted of full-height masonry, in contrast to fieldhouses. The multi-room habitations produced relatively large artifact assemblages, although we found few decorated ceramics or worked lithic artifacts; again this likely reflects collection by visitors.

We documented two Sinagua forts, both of which were first recorded in the 1920s (Colton 1932). These sites are characterized by multiple contiguous masonry rooms built on limestone islands or peninsulas extending into the canyon from the rims. These landforms have constricted necks and steep sides, and masonry walls were usually constructed across the neck and along the perimeter. Most fort sites in Walnut Canyon contain both habitation and storage areas on the island tops and the adjacent canyon slopes. First Fort, the easternmost of the forts in Walnut Canyon, anchors the largest prehistoric community documented in the entire monument. The multi-room fort itself occupies a narrow peninsula with steep scarps and numerous large bedrock tinajas that hold a significant amount of rainwater. The broad canyon bottom below First Fort contains a large masonry "community room," as well as at least 14 pit structures and 21 masonry structures containing 34 rooms. Multiple extramural features and middens are also evident, the latter containing ceramics that date the site to the Elden phase (AD 1150–1225). The multi-room masonry habitations that occupy both canyon rims within a quarter mile of First Fort appear to be contemporaneous.

Fifth Fort, the westernmost fort in Walnut Canyon, occupies a prominent finger ridge

Table 1. Native American site types, frequency of occurrence, and percentage of the total (n = 209).

Site Type	Description[1]	Frequency	%
Alcove (Canyon) site	One or more rooms or other archaeological material located beneath limestone overhangs on canyon slopes.	28	13.4
Ceramic and lithic scatter	A scatter of ceramic and lithic artifacts with ceramic artifacts being more numerous. At times this description was amended with a functional (e.g. camp) or descriptive (e.g. rock alignment) designation.	78	37.3
Ceramic scatter	An artifact scatter consisting solely of ceramic artifacts.	5	2.4
Check dam	A rock wall or berm built perpendicular to a drainage for the purpose of impounding or retaining water and/or soil.	5	2.4
Fieldhouse	Architectural sites exhibiting (1) a single square or rectangular room, also C-, L-, or U-shaped enclosures; (2) use of mostly double-walled masonry as a construction technique; (3) moderate amount of wall fall corresponding to part or all of a masonry wall; (4) diffuse artifact scatter or no artifact scatter; (5) predominance of plain ware ceramic types; (6) interior space of less than 12 sq m; (7) generally distant from habitation sites.	30	14.4
Fort	Fort sites in Walnut Canyon are located on top of ridges that extend into the canyon from both rims; the ridges are along prominent meanders of Walnut Canyon. Forts consist of multiple rooms of various sizes associated with wing walls, compounds/plazas, walls restricting access onto the neck of the island ridge, and a concentration of additional sites or rooms surrounding the fort.	2	1.0
Lithic and ceramic scatter	A scatter of lithic and ceramic artifacts with lithic artifacts being more numerous.	13	6.2
Lithic scatter	A scatter consisting solely of lithic artifacts.	25	12.0
Multiroom masonry structure	Sites of this type, also known as Unit Pueblos, are designated by (1) contiguous multiroom arrangement; (2) abundant trash; (3) full-height, single-story masonry; (4) auxiliary rooms; (5) pit structures; (6) linear arrangement of rooms; and/or (7) a cluster of one and two-room structures.	9	4.3
Rock shelter	A variation of the Alcove (Canyon) site type; typically does not contain masonry.	1	0.5
Quarry	An area where lithic raw material procurement took place.	1	0.5
Trail	A clearly visible pedestrian and/or animal track that is usually associated with artifacts and features.	2	1.0
Two-room structure	A subset of the Fieldhouse site type. One structure with two rooms, or two individual rooms located close together. Construction style and materials are similar to Fieldhouse sites.	10	4.8

[1] Site type definitions from the Walnut Canyon National Monument Archaeological Inventory Survey Site Form Field Manual, Appendix A (National Park Service 2001), which are based on Baldwin and Bremer (1986:101–121).

Table 2. Diagnostic dart points found at Archaic sites.

Projectile Point Type	Temporal Period	Site(s)/Isolated Occurrences (IOs)	Type Total	%
Jay[1]	Early Archaic	Site 362	1	2
Bajada[1]	Early Archaic	Site 314, 333 (2 points), 351, 352, 362; IO27, IO50	8	18
Pinto[2]	Early Archaic	Sites 288, 367B	2	5
Unidentified Early Archaic[3]	Early Archaic	Site 351; IO26, IO34, IO127	4	9
San Rafael[4]	Middle Archaic	Sites 352, 364, 367D; IO142	4	9
Gypsum[5]	Late Archaic	Site 396	1	2
Elko Series[6]	Early to Late Archaic[7]	Sites 273 (2 points), 321, 335, 341, 350 (2 points), 351 (2 points), 367A (2 points), 367C, 367D (3 points), 367E, 373, 396, 399, 401, 421; IO107, IO114, IO148	24	55
Total			44	100

[1] Irwin-Williams (1973).
[2] Amsden (1935); Holmer (1986).
[3] Dart points exhibiting proximal ears or lobes and basal grinding (cf. Berry 1987).
[4] Holmer (1980).
[5] Thomas (1981).
[6] Holmer (1978, 1986).
[7] Elko Series extends to ca. AD 1000 on the northern Colorado Plateau, but is rare in Sinagua assemblages.

extending from the south side of the canyon within a tight meander in Walnut Creek. It consists of two structures on top of the narrow finger ridge, as well as three structures and four artifact scatters on the slopes below the ridge. At least three of the structures are defensible, and access to the upper level is restricted by a stone wall and natural topography. Ceramics at the site suggest occupation during the Elden phase (AD 1150–1225). In contrast to other forts in Walnut Canyon, the "community" of residential sites potentially associated with Fifth Fort is more dispersed, and few are visually linked.

Other documented sites that likely relate to the Sinagua occupation of Walnut Canyon include check dams, a few chert quarries, and two trails. Less than a mile west of the First Fort community we recorded a trail that crosses the north slope of the canyon obliquely and traverses a small side drainage to reduce the steepness of the climb. The distribution of ceramics along the slope, in a sparse scatter that exactly parallels the trail, convinced us that this feature was a prehistoric Native American route; it is currently a well-used game trail. The trail emerges onto the canyon rim in an area with an extremely high site density, supporting a Native American affiliation. A second similar trail was found directly across the canyon from First Fort.

Rock art was recorded at several sites, although few of the elements could be securely dated. We documented geometric pictographs that are probably Sinagua, linear petroglyphs of unknown age, zoomorphic and abstract pictographs associated with probable Apache ceramics, and several panels or elements that are clearly historic in age. Modern graffiti was present at a few sites, but many of the inscribed initials and dates are from the first half of the twentieth century.

The majority of the historic sites in the survey area represent the remains of temporary camps. A few have evidence of informal structures such as tent platforms. Most of these components can be assigned to the period between 1897 and 1940, with a smattering of camps dating from the 1950s to the 1970s. The preponderance of sites from the first four decades of the twentieth century accords well with the results of previous surveys in the monument (Stein 1986:70–75), representing activity associated with logging and ranching in addition to visiting the ruins. Sparse artifact assemblages indicate transitory use of most historic sites.

Two telephone lines, one with an associated trail, were also encountered. According to records at the Coconino National Forest in Flagstaff, the phone line in the western survey block was installed between 1912 and 1915 to connect ranger stations and fire lookouts between Flagstaff and Pine. The line crossed Walnut Canyon and Anderson Mesa and ran past Mormon Lake, roughly along the current route of Lake Mary Road, to Highway 87. It was constructed by stringing telephone wire between ceramic insulators attached to large trees. Many of these insulators still hang from trees, and lengths of wire can be followed between trees in several places. Similar phone lines have been recorded elsewhere on the Coconino National Forest (Hathaway 1999; Herr et al. 2000), demonstrating an important early communication system.

The most visually impressive Euro American site in the monument is the Santa Fe Dam, which was constructed in 1897–1898 by the Santa Fe Railroad to provide water for trains and rail maintenance stations along the main rail line between Winslow and Flagstaff. The dam itself was recorded during the 1985 survey (Stein 1986) and is on the National Register of Historic Places (Rothweiler and Wilson 1979). Our documentation focused on the ancillary components of the site, which include two limestone quarries adjacent to the dam, numerous concrete platforms and equipment foundations, a kitchen area with remains of the cook shack and trash dump, a railroad grade leading to the base of the dam from the main rail line, and portions of a ceramic water pipe installed in the grade after the tracks were removed. The dam was constructed of Kaibab limestone, quarried from ledges on the canyon wall directly adjacent to the site, and capped with sandstone from the Toroweap Formation. Although all quarrying and construction equipment was removed, a number of features and trash scatters remain. The dam never held water on a regular basis due to porous bedrock, although the reservoir behind the dam occasionally fills after heavy winter or spring storms.

Archaeological evidence for Native American use of Walnut Canyon and adjacent lands during the period after AD 1300 is sparse. Numerous accounts by early explorers documented transitory use of the area by indigenous groups, mainly in the context of seasonal resource procurement (Stein 1986:69). Pinyon nut harvesting and collection of medicinal plants continues to draw traditional Navajo and Hopi people to the Walnut Canyon area on occasion (Stein 1986:69–70). Survey of the pre-1996 monument area encountered only one historic Native American artifact, a Jeddito Black-on-yellow sherd (Stein 1986:70), and earlier work documented one sherd of fingernail-incised pottery, probably of Apache origin (Van Valkenburgh 1958). The present survey recorded one campsite in a canyon slope alcove that yielded several pieces of Apache Plain pottery. Isolated sherds of Jeddito Yellow Ware and Hopi Yellow Utility Ware were found at two other sites, and Desert Side-notched projectile points were found at two sites.

DISCUSSION

A comparison of the results of our survey with those of the 1985 survey (Baldwin and Bremer 1986) reveals some strong congruency but also some significant differences. That survey covered 83 percent of the pre-1996 monument area and recorded 241 sites. It emphasized architectural sites and recorded only 14 artifact scatters (5% of total sites). In contrast, artifact scatters comprise more than 50 percent of sites in the present project.

There are several reasons for this discrepancy. Baldwin and Bremer's survey area contained the absolute highest density of alcove sites in the entire canyon, leaving no doubt why this area was designated as a national monument. Our survey demonstrated that the density of alcove sites drops off quickly east of the old monument boundary and more gradually to the west, but with the addition of the 1996 area, all known concentrations of alcove sites are now included in the monument. Although suitable alcoves exist in the western survey block, we found little evidence of their use along the western boundary. This pattern suggests that alcove site distribution was tied to some particular aspect of Sinagua occupation of the canyon, perhaps based on community structure or a perceived need for defense.

Based on previous small surveys (e.g., Gilman 1977) we anticipated that the site density would decrease somewhat away from the canyon, for example moving north along the road corridor, but this did not turn out to be the case. There was no decrease in sites in project parcels near the rim versus along the road corridor. The site density was 14 sites per quarter section in the western survey area, 26 sites per quarter section along the road, and 21 sites in the eastern area, which equates to about 100 sites per section in the most dense areas. The lower density in the western block is probably a function of the rough terrain and less-desirable exposure in the narrower canyon. In total, our survey recorded 62 sites in the western block, 97 in the eastern blocks, and 51 along the road corridor. The density of sites is notably higher on the south-facing slopes and north rim of the canyon and also in the eastern survey area, but even more striking is the greater size of the sites in the eastern area. This size disparity reflects the distribution of site types. Small fieldhouse sites and discrete ceramic and lithic scatters dominate the western block and road corridor. These clearly represent a variety of activities and intensive use of the landscape by Sinagua people during the eleventh and twelfth centuries. Sites in the eastern blocks are predominantly large artifact scatters that represent episodic activity over several thousands of years, evidently related to resource procurement. Small Sinagua fieldhouses are relatively rare in the eastern area.

The most notable divergence between the 1985 survey and ours is in the aceramic, most likely Archaic, site components. More than 95 percent of Baldwin and Bremer's sites dated to the Sinagua period, and no Archaic sites were identified, although several dart points were collected or noted on sites (Baldwin and Bremer 1986: Figures 33–35). Our survey documented 23 Archaic components, as well as nine additional isolated diagnostic dart points. We assigned sites, or components thereof, to the Archaic period based solely on the presence of diagnostic projectile points, although other patterns were noted that probably reflect an Archaic affiliation. Discrete artifact concentrations that completely lack ceramics were noted at several sites, most of which also exhibit lithic assemblages with a more diverse range of raw material than is characteristic of Sinagua sites. Whereas lithics at the later sites are primarily of local Kaibab chert, the aceramic assemblages have higher frequencies of obsidian from local volcanic sources and chert derived from the Little Colorado River drainage. These materials may have been procured and brought to the sites by mobile foragers. Support for this model comes from Archaic camps documented in adjacent areas such as Anderson Mesa (Keller and Dosh 1997), Bellemont (Purcell et al. 1999), the Tuba City–Red Lake area (Neff et al. 2002; Bungart et al. 2004), and the Little Colorado River valley (Wendorf and Thomas 1951; Berry 1987). It seems likely that the Walnut Canyon area would have offered attractive resources, such as riparian plants, pinyon nuts, grasses, and cactus, to foragers moving through a seasonal round. The frequency of probable Archaic sites increases toward the east end of the monument, and all but one of the definitively Archaic sites were found in the eastern survey block (see Figure 2). Peter Pilles, Coconino National Forest Archaeologist (personal communication, 2002) has noted that the majority of Archaic sites on the forest are in the southeastern portion, and Berry (1987) reported assemblages of dart points in the upper Little Colorado River drainage that suggest a regionally distinct Archaic cultural manifestation. Although there is no established Archaic chronology or point-type system for north-central Arizona (cf. Irwin-Williams 1973; Schroedl 1976; Geib 1996), there is little doubt that Archaic hunters and foragers utilized the forests and grassland surrounding the San Francisco Peaks, the plateau country to the north, and the Little Colorado River valley.

Tables 3 and 4 present the temporal assignments of Sinagua sites documented by the present project and all sites from Baldwin and Bremer's (1986) survey of the pre-1996 monument. Of the 242 sites recorded by Baldwin and Bremer, 162 were assigned temporal affiliations that are wholly or partially within the Sinagua era. Five sites (3.1%) were assigned pre-AD 1050 temporal affiliations. Comparison of the two survey areas for the pre-AD 1050 period suggests that the new lands had generally higher relative occupation levels, specifically more Sunset phase (AD 825–1000) occupation. We also noted a proportionally larger number of pre-AD 1050 sites along the entrance road corridor, away from the canyon, hinting that the canyon slopes and rim were less heavily utilized by the Sinagua before the eruption of Sunset Crater. Both surveys suggest that occupation was relatively sparse, but constant, from ca. AD 700, when the first farmers occupied the Flagstaff area and Walnut Canyon, to about AD 1050.

Baldwin and Bremer (1986) assigned the vast majority (93%) of the post-AD 1050 settlement in their survey area to the Elden phase (AD 1150–1225). Based on refined ceramic chronologies developed in the last two decades, and the extensive collection at many sites that has potentially skewed the ceramic database, we assigned three quarters (73%) of our survey assemblages to general post-AD 1050 time spans, rather than a specific phase (Table 3). Despite this, it is clear that the peak in canyon area occupation occurred during the Elden phase. The number and variety of sites representing the post-AD 1050 interval is dramatically higher than for any other period of time, a pattern documented throughout the Sinagua region.

MANAGEMENT ISSUES

The natural ravages of time have taken a toll on sites in Walnut Canyon, mainly due to slumping of architecture and erosion of artifacts down steep slopes. Walls at many of the alcove sites are precarious, and stabilization of some sites will be necessary to prevent collapse. Much of the damage to the architecture came from early tourists and relic hunters who tore down walls to reach the artifacts within. Architectural disturbance is the most obvious damage to the archaeological sites in the monument, but equally destructive was the chaining undertaken across most of the area on the north side of the canyon in the eastern block in the 1960s. This activity universally affected the sites in a negative way, displacing artifacts horizontally and precipitating extensive deflation that left the assemblages without vertical integrity and probably destroyed small features such as hearths. In some cases the tree debris piles were burned, either on purpose or later by natural fire, resulting in additional damage to artifacts.

Overall, however, the main adverse impacts to sites in the monument are due to visitation. Collection of artifacts, unauthorized excavation, and destruction of architecture imposed severe negative impacts on the archaeological resources of the monument. Early accounts document the rich artifact assemblages that were removed from the alcove rooms (Stein 1986), and certainly a wealth of knowledge about temporal and cultural affiliation of the sites throughout the monument has been lost along with these remains. It was clear during our survey that accessibility plays a major part in the impacts to sites. Sites in areas close to roads or trails generally exhibit few surface artifacts, and decorated ceramics or items such as projectile points are extremely rare in the "front country" of the monument. Sites in inaccessible locales generally retain more artifacts and are in better condition.

Table 3. Temporal assignments for site components recorded during survey of new monument lands (after Neff and Spurr 2005: Table 4.11).

Period Designation	Total	% of total (n = 173)	Temporal Designation	Total	% of total (n = 173)
AD 550–825*	1	0.58	AD 1050–1225*	3	1.73
AD 550–1050	1	0.58	AD 1050–1250*	2	1.16
AD 550–1025/Sunset*	1	0.58	AD 1050–1300*	65	35.57
AD 650–1125/Sunset*	1	0.58	AD 1075–1200*	20	11.57
AD 800–1025/Sunset*	2	1.16	AD 1075–1225*	2	1.16
AD 550–1150/Sunset	1	0.58	AD 1075–1225/Padre-Elden*	4	2.31
AD 800–1025/Sunset	2	1.16	AD 1050–1225/Padre-Elden	1	0.58
AD 550–1100*	1	0.58	AD 1075–1225/Padre-Elden	1	0.58
AD 650–1050*	1	0.58	AD 1075–1250*	1	0.58
AD 650–1050	3	1.73	AD 1125–1250/Elden*	2	1.16
AD 700–1050*	1	0.58	AD 1150–1225/Elden*	8	4.62
AD 750–1075/Rio de Flag	1	0.58	AD 1025–1225/Elden	1	0.58
AD 700–1200*	2	1.16	AD 1025–1250/Elden	1	0.58
AD 700–1225*	1	0.58	AD 1150–1225/Elden	2	1.16
AD 800–1150*	1	0.58	AD 1025–1250/predominantly Elden	1	0.58
AD 1000–1100*	1	0.58	AD 1150–1225*	5	2.89
AD 650–1200/Angell-Winona*	1	0.58	AD 1050–1300	1	0.58
AD 1025–1150/Angell-Winona*	6	3.47	AD 1075–1225	1	0.58
AD 1050–1200/Angell-Winona*	1	0.58	AD 1150–1300*	2	1.16
AD 1050–1225/Angell-Winona*	1	0.58	AD 1325–1600	1	0.58
AD 700–1150/Angell-Winona	1	0.58	AD 1540–1900	1	0.58
AD 1025–1150/Angell-Winona	1	0.58	Indeterminate (AD 700–1225)*	1	0.58
AD 1025–1150*	1	0.58	Indeterminate (AD 700–1275)*	1	0.58
AD 1025–1200/Angell-Winona-Padre*	1	0.58	Indeterminate (AD 700–1300)*	2	1.16
AD 1025–1225*	1	0.58	Indeterminate (AD 650–1225)*	1	0.58
AD 1025–1225/Padre*	1	0.58	Indeterminate (AD 650–1225)	1	0.58
AD 1075–1200/Padre*	3	1.73			
AD 1075–1200/Padre	2	1.16			

*Site components exhibiting fewer than 10 decorated ceramics.

Table 4. Frequency of temporal assignments from survey of the pre-1996 Walnut Canyon National Monument lands (after Baldwin and Bremer 1986: 92–100, Table 7).

Temporal Phase	Number of Sites	%
Rio de Flag phase	2	0.83
Rio de Flag / Elden phases	2	0.83
Angell-Winona phase	5	2.07
Angell-Winona / Elden phases	19	7.85
Angell-Winona / Elden / Historic phases(?)	1	0.41
Angell-Winona-Padre / Elden phases	5	2.07
Elden phase	121	50.00
Elden phase / Historic	4	1.65
Elden phase / Historic(?)	1	0.41
Cinder Park and Elden phases	1	0.41
Pueblo II	1	0.41
ca. 1904–1967	1	0.41
Early twentieth century	1	0.41
Historic	5	2.07
Unknown	66	27.27
Unknown / Historic	4	1.65
Unknown / Historic?	3	1.24
Total sites	242	100.00

All the sites recorded during the survey were evaluated for eligibility under the National Historic Preservation Act of 1966, as amended. Sites were evaluated based on their potential to provide data on pertinent research issues, such as temporal and cultural affiliation, the occupational history of Walnut Canyon and its relevance to past environmental conditions (including the eruption of Sunset Crater), and the structure of Sinagua settlement of Walnut Canyon within the greater Flagstaff region (Neff 2001). Sites eligible under Criterion D must retain features or deposits that could provide information on cultural and temporal affiliation, artifact technology and function, subsistence strategies or settlement organization, or function and construction details of features. Although a number of the heavily disturbed sites in the chained area were assessed as not eligible, they do contribute to our understanding of the human history of Walnut Canyon, as do the more numerous sites that are eligible on their own merit. Several sites were also assessed as eligible under Criteria A, B, or C, based on features (e.g., rock art, unique architecture at fort sites) or associations (e.g., significant scientific discoveries, prominent scholars) that reflect important patterns of regional history or development.

The NPS manages sites within national monuments for long-term research, so we also considered the potential of sites with regard to future research methods and questions. The baseline data generated by this survey will allow land managers to understand patterns of past human activity in Walnut Canyon National Monument and what natural or human-based threats must be considered in long-term planning to protect and preserve the remains of the past. One important facet of future preservation would be expansion of the established

Walnut Canyon Archaeological District, which currently encompasses only the pre-1996 monument lands, to include the entire area now within the monument.

ACKNOWLEDGMENTS

NNAD would like to thank the NPS for involving us in this project under the umbrella of the Colorado Plateau Cooperative Ecosystem Study Unit (CPCESU), which facilitates contracts for research, technical assistance, and educational projects among a variety of governmental and academic institutions. We sincerely thank all crew members involved in the survey and recording. We faced some very trying conditions and terrain, and the crews worked hard and were enthusiastic about the project. The core crew of surveyors consisted of Dave Barr, Stewart Deats, Walter Gosart, Ian Hough, Kim Mangum, Ted Neff, Kimberly Spurr, and Cole Wallace. Additional crew members included Kenny Acord, Lyle Balenquah, Roxanne Begay, Jessica Bland, Mark Brehl, Peter Bungart, Jim Collette, Janet Hagopian, Mandy Johnson, Jane Lakeman, Deirdre Morgan, Kerry Thompson, Carissa Tsosie, Neomie Tsosie, Davina Two Bears, and Sara Wendt. Tim Wilcox and Kim Mangum drafted field site maps and produced GIS figures. Laboring diligently behind the scenes to make sure that things ran smoothly were Helen Fairley, Todd Metzger, Al Remley, and Miranda Warburton.

LITERATURE CITED

Amsden, C. A. 1935. The Pinto Basin artifacts. In The Pinto Basin Site, by E. W. Campbell and W. H. Campbell, pp. 35–51. Southwest Museum Paper 9, Los Angeles.

Baldwin, A. R., and M. J. Bremer. 1986. Walnut Canyon National Monument: An archaeological survey. Publications in Anthropology No. 39. National Park Service Western Archaeological and Conservation Center, Tucson. 200 pp.

Berry, C. F. 1987. A reassessment of the Southwestern Archaic. Unpublished Ph.D. dissertation, Department of Anthropology, University of Utah.

Bungart, P. W., J. Collette, and K. Spurr. 2004. A better road ahead: Archaeological excavations along Navajo Route 21 near White Mesa, Arizona. Navajo Nation Archaeology Department Report No. 01-237. Ms. on file, Navajo Nation Historic Preservation Department, Window Rock, AZ.

Chronic, H. 1989. Roadside Geology of Arizona. Mountain Press Publishing, Missoula, MT.

Colton, H. S. 1932. A survey of prehistoric sites in the region of Flagstaff, Arizona. Bureau of American Ethnology Bulletin 104. Smithsonian Institution, Washington, D.C.

Geib, P. R. 1996. Archaic occupancy of the Glen Canyon region. In Glen Canyon Revisited, edited by P. R. Geib, pp. 15–39. Anthropological Papers No. 199. University of Utah, Salt Lake City.

Gilman, P. A. 1977. Archaeological survey for fencing the boundary of Walnut Canyon National Monument. Ms. on file, National Park Service, Flagstaff Area National Monuments, Flagstaff, AZ.

Hathaway, J. B. 1999. Cultural resources surveys of four segments of State Route 87 (between mileposts 226 to 228.7 and mileposts 254.5 to 277.1) in the vicinity of Payson, Pine, and Strawberry, Tonto National Forest (Mesa and Payson Ranger Districts) and Coconino National Forest (Long Valley Ranger District), in Gila and Coconino Counties, Arizona. Project No. N-900-0-903 (ADOT). Ms. on file, Arizona Department of Transportation, Phoenix.

Herr, S. A., P. H. Stein, and P. Cook. 2000. Preliminary report of archaeological data recovery in the Preacher Canyon section and Sharp Creek Campground, State Route 260—Payson to Heber archaeological project, Gila County, Arizona. Project Report No. 00-129. Desert Archaeology, Inc., Tucson.

Holmer, R. N. 1978. A mathematical typology for Archaic projectile points of the eastern Great Basin. Ph.D. dissertation, Department of Anthropology, University of Utah, Salt Lake City. University Microfilms, Ann Arbor.

Holmer, R. N. 1980. Projectile points. In Sudden Shelter, edited by J. D. Jennings, A. R. Schroedl, and R. N. Holmer, pp. 63–83. Anthropological Papers No. 103. University of Utah, Salt Lake City.

Holmer, R. N. 1986. Common point types of the Intermountain West. In Anthropology of the Desert West: Essays in Honor of Jesse D. Jennings, edited by C. J. Condie and D. D. Fowler, pp. 90–115. Anthropological Papers No. 110. University of Utah, Salt Lake City.

Irwin-Williams, C. 1973. The Oshara tradition: Origins of Anasazi culture. Contributions in Anthropology Vol. 5, No. 1. Eastern New Mexico University, Portales.

Keller, D. R., and D. S. Dosh. 1997. Late preceramic utilization of Anderson Mesa, northern Arizona. Ms. on file, Coconino National Forest Supervisor's Office, Flagstaff.

National Park Service. 2001. Scope of work for an intensive archaeological inventory survey of Walnut Canyon National Monument new lands by the Navajo Nation Archaeology Department for the National Park Service, Flagstaff Area National Monuments under Cooperative Agreement 120099009. Ms. on file, National Park Service, Flagstaff Area National Monuments, Flagstaff, AZ.

Neff, L. T. 2001. Research design for an intensive archaeological inventory of Walnut Canyon National Monument new lands. Navajo Nation Archaeology Department Report No. 01-143. Ms. on file, National Park Service, Flagstaff Area National Monuments, Flagstaff, AZ.

Neff, L. T., and K. Spurr. 2005. An intensive archaeological inventory survey of Walnut Canyon National Monument new lands. Navajo Nation Archaeology Department Report No. 01-143. Ms. on file, National Park Service, Flagstaff Area National Monuments, Flagstaff, AZ.

Neff, L. T., K. Anderson, L. Huckell, and M. Robins. 2002. Archaic period occupation, chronostratigraphy, and soil geomorphology on the southern Kaibito Plateau: Results of geoarchaeological investigations along N608–Arizona Boulevard, Tuba City, Arizona. Navajo Nation Archaeology Department Report No. 02-167. Ms. on file, Navajo Nation Historic Preservation Department, Window Rock, AZ.

Purcell, D. E., C. D. North, J. M. Garrotto, and M. J. Boley. 1999. Camp Navajo: A cultural resources survey of 25,000 acres of the western Mogollon Rim at Bellemont, Coconino County, Arizona. Draft ms. on file, SWCA Environmental Consultants, Flagstaff, AZ.

Rothweiler, T. S., and M. H. Wilson. 1979. National Register of Historic Places Inventory–Nomination Form for the Walnut Canyon Dam (Santa Fe Dam). Arizona State Parks, Phoenix.

Schroedl, A. R. 1976. The Archaic of the northern Colorado Plateau. Ph.D. dissertation, University of Utah, Salt Lake City. University Microfilms, Ann Arbor.

Stein, P. H. 1986. Historical resources. In Walnut Canyon National Monument: An Archaeological Survey, edited by A. R. Baldwin and M. J. Bremer, pp. 43–79. Publications in Anthropology No. 39. National Park Service Western Archaeological and Conservation Center, Tucson, AZ. 200 pp.

Thomas, D. H. 1981. How to classify the projectile points from Monitor Valley, Nevada. Journal of California and Great Basin Anthropology 3(1): 7–43.

Van Valkenburgh, S. P. 1958. Archaeological site survey, north rim, Walnut Canyon National Monument. Ms. on file, National Park Service, Flagstaff Area National Monuments, Flagstaff, AZ.

Wendorf, F., and T. H. Thomas. 1951. Early man sites near Concho, Arizona. American Antiquity 23: 107–114.

USING CULTURAL RESOURCES AS PART OF THE PLAN: GRAND CANYON MANAGEMENT AND IMPLICATIONS FOR RESOURCE PRESERVATION

Janet R. Balsom, J. Grace Ellis, Amy Horn, and Lisa M. Leap

The Grand Canyon is a natural wonder, known throughout the world for its scenic beauty and rich geologic history. It is a crown jewel in the National Park system, one of the Seven Wonders of the World, and a World Heritage Site. Yet behind the scenery lies 10,000 years of human history. This history reflects the complex and varied interaction of people with this rugged landscape (Figure 1). An estimated 50,000 archaeological sites are thought to exist in Grand Canyon National Park (GCNP), including Archaic lithic scatters, split-twig figurine caves, Puebloan settlements, and historic period Native American structures. The 870 historic buildings and structures include National Historic Landmarks designed by Mary Elizabeth Jane Colter, rustic miners' cabins, and historic National Park Service (NPS) structures and trails. Historic designed landscapes are located in most of the developed areas of the North and South Rims, and Traditional Culture Properties and Native American sacred sites are present throughout the park.

The grandeur and beauty of the Grand Canyon attracts nearly 5 million visitors per year. The National Park Service endeavors to preserve cultural resources while accommodating visitor enjoyment. Therefore, it is essential that cultural resource concerns and information be a part of the park planning process. Current planning efforts at Grand Canyon National Park are increasing the integration of cultural resource information.

In 1995, Grand Canyon National Park completed and began implementing its General Management Plan (GMP; U.S. Dept. of Interior 1995), which outlines plans for visitor use and infrastructure development in the park. The Fee Demonstration Program allows the park to keep a share of its revenues, providing funding for widespread infrastructure repair and development (Figure 2). Repairs and improvements to Grand Canyon National Park's infrastructure take place within historic districts, in National Historic Landmark buildings, in areas with cultural resources, and in Native American traditional use areas. The GMP provides a broad outline for these activities, with specific resource information helping to guide specific projects. For example, the design of the new Desert View entrance station on the eastern end of the South Rim has been thematically tied to the design of the historic buildings in the area. The Colter-designed Desert View Watchtower, the centerpiece of the area, is a National Historic Landmark building (NPS 2005). In its distinctive yet compatible design, the entrance station will be faced with rubble masonry similar to that found on the Watchtower, but the building's form will be twenty-first century.

Many tools provide an opportunity to use cultural resources as part of the park planning process. Whether it is tribal resource identification, traditional archival materials, archaeological site inventories, or user-discretionary time models, all play a role in helping park management make better decisions for the resources of the park. The varied resources of the Grand Canyon

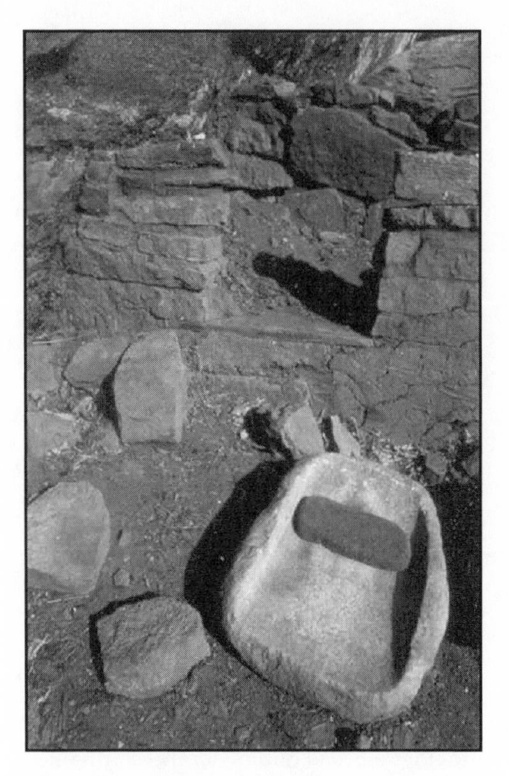

Figure 1. Puebloan habitation site in Grand Canyon National Park (top). Desert View Tower, one of many important historic sites in Grand Canyon National Park (bottom).

require varied approaches to ensure that all resources are considered. We believe that cultural resources are being considered up-front in the planning process, and that we, as stewards of the resources, are being heard.

GENERAL MANAGEMENT PLAN

The integration of cultural resource concerns can be illustrated through various projects undertaken as the park implements its GMP. On the Greenway Trail system, for example, archaeologists have contributed to the design and placement of the trail to minimize impacts to cultural sites. Cultural resources have thus influenced the location of the trail and its appearance. Similarly, trail design in the Grand Canyon Village Historic District (NPS 1997), a National Historic Landmark, is compatible with the surrounding historic cultural landscape. As a result, trails in the district are narrow and trail features such as walls and culvert headwalls are visually compatible within the historic district.

Grand Canyon Village is the focus of much of the GMP's implementation. Rehabilitation of historic buildings, adaptive reuse, preservation maintenance, and many new projects are taking place within this historic cultural landscape. Historic Structure Reports (National Park Service and U.S. Dept. of Interior 1997) document the history of buildings, guide preservation approaches, and help identify appropriate new uses. The Grand Canyon Village Cultural Landscape Report (NPS 2003) is a critical piece of information for including cultural resource concerns into project planning. It describes significant features of the landscape and makes recommendations for planning, orientation, and future developments.

FIRE MANAGEMENT PLANNING

Grand Canyon National Park is currently revising its Fire Management Plan (U.S. Dept. of Interior 2003). Cultural resource considerations are enjoying a prominent role in the development of this revised plan, which includes goals and objectives for the protection and preservation of cultural re-

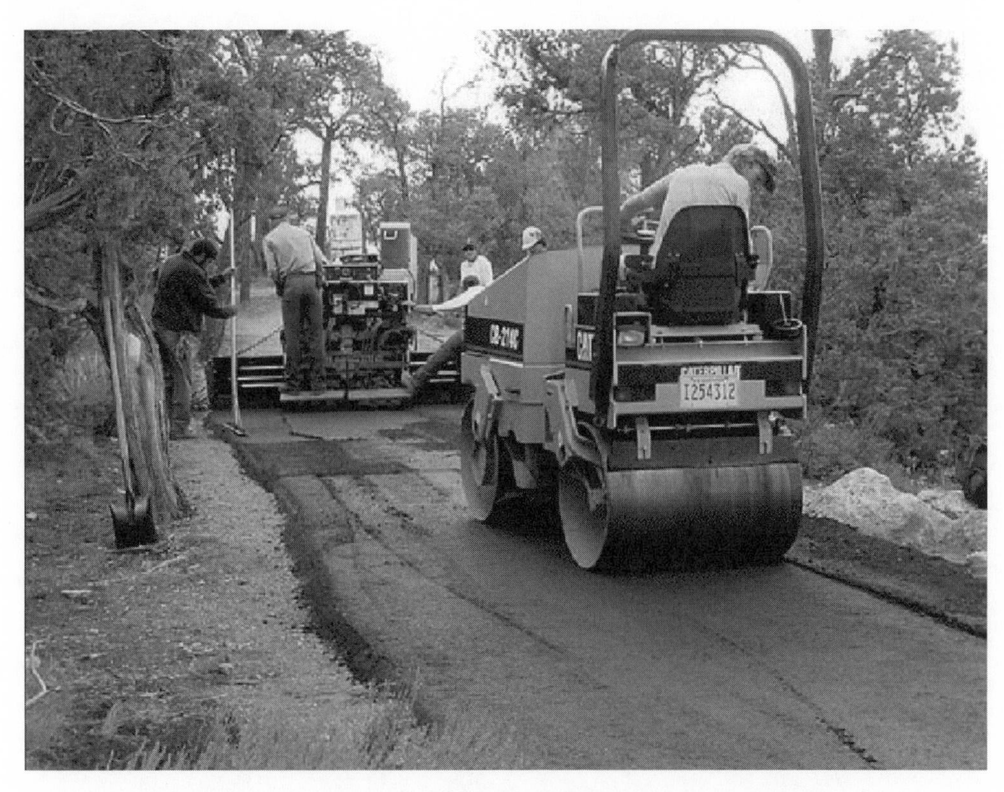

Figure 2. Infrastructure repairs are implemented under the GMP.

sources, and management alternatives and actions required to reach these goals (Figure 3). For example, alternatives include fuel reduction around historic buildings, which not only protects structures from unwanted wildfire, but can also help restore cultural landscapes. Other alternatives include planning for the management of low-intensity wildland fires and prescribed fires, monitored by park archaeologists, as a way to safely reduce fuel loads on open, prehistoric sites. Additionally, the plan requires that management of wildfires will include consultation with park archaeologists to insure that reasonable measures are taken to protect cultural resources from the effects of both fire and firefighting efforts.

HISTORICAL DOCUMENTATION

The Forest Ecosystem Landscape Analysis Project (Crocker-Bedford et al., this volume; Vankat et al., this volume) underscores the utility of historical documents in park management. Review of historical documents, conducted as part of the research for this landscape analysis project, resulted in the discovery of a Park Service-wide vegetation inventory that occurred in 1935. Replication and comparison of the photographs and inventory from this survey has provided a quantifiable body of data that has increased our understanding of how the forest ecosystem and certain cultural landscapes have changed in the last 70 years (Figure 4). These historic contexts will be used in developing an approach for management of forest structure now and into the future.

GLEN CANYON DAM IMPACTS

The Glen Canyon Dam Adaptive Management Program began in 1995 with the completion of the Environmental Impact

Figure 3. Park archaeologists work with fire crews to protect historic sites.

Statement for Glen Canyon Dam Operations (U.S. Dept. of Interior 1996). The purpose of the program is to identify impacts to the physical, cultural, and biological resources as related to the operation of Glen Canyon Dam. Whereas the physical and biological studies are primarily conducted by non-NPS researchers, monitoring and remedial work on historic properties are completed by NPS staff and cooperators. The Grand Canyon River Corridor Monitoring and Treatment Program guides the ongoing process for the identification, monitoring, and protection of historic properties impacted, or potentially impacted, as a result of dam operations. Monitoring and treatment of historic properties are completed under the umbrella of a programmatic agreement for compliance with the 1994 National Historic Preservation Act (U.S. Dept. of Interior et al. 1994). The Bureau of Reclamation is the lead agency in implementation of the agreement, but Glen Canyon National Recreation Area and Grand Canyon National Park are the principal land managers in the study area. Additionally, the Hualapai Tribe and the Navajo Nation, responsible as land managers for the lands affected by dam operations within their respective jurisdictions, and other American Indian tribes with ancestral affiliations to resources in Grand Canyon National Park (i.e. Hopi, Havasupai, Kaibab Band of Paiute Indians, Paiute Indian Tribe of Utah for the Shivwits Paiute, San Juan Southern Paiute, and the Pueblo of Zuni) are regularly consulted as part of the programmatic agreement.

The first step in the Glen Canyon Dam program begun in 1995 was to conduct an archaeological inventory of the Colorado River corridor to comply with the National Environmental Policy Act of 1969 (42 USC 4321 and 4331-4335, as amended) and the National Historic Preservation Act of 1966 (16 USC 470). The area surveyed encompassed a 255-mile stretch of the river corridor (approximately 10,506 acres), extending from Glen Canyon Dam to Separation

Plate 11a. (May 30,1935 10am) Elev. 6,900ft.
300yds. east of the Harvey garage on road, facing south.
Ponderosa pine forest with Pinyon/Juniper bordering to the south.
NPS photo H. Bailey

Figure 4. The Forest Ecosystem Landscape Analysis Project (FELA) match-point photos from 1935 (above) and 1984 (next page) illustrate changes in the forest ecosystem.

Canyon. In total, 475 archaeological sites and 489 isolated occurrences of artifacts or features were located and recorded (Fairley et al. 1994). Of these, 260 sites have the potential to be affected by Glen Canyon Dam operations; however, ongoing monitoring of sites indicates that, currently, 164 sites are actively affected, either directly, indirectly, potentially, or cumulatively (Figure 5; Leap et al. 2003). Of these 164 sites, more than 75 have received some form of preservation treatment such as the building of checkdams to decrease gully cutting, planting vegetation to stabilize dunes, or trail work (obliterating or redirecting trails to avoid historic properties; Figure 6). Protocols for the construction of checkdams, using traditional Zuni methods, were developed as part of ongoing consultations with Native American tribes.

Plate 11b. (July 10,1984 10am) T31N R2E S24
300 yards east of the Fred Harvey public garage in the village.
The road has been widened and the near slope re-graded. Density
of the forest appears the same, ground cover contains more
grasses due to different season and removal of live stock
grazing.
NPS photo J. Armstrong

Figure 4 (continued). The Forest Ecosystem Landscape Analysis Project (FELA) match-point photos from 1935 (previous page) and 1984 (above) illustrate changes in the forest ecosystem.

VISITOR IMPACT PLANNING

Although impacts to cultural sites from the operations of Glen Canyon Dam are outside the direct influence of park management, this is not the case with recreational impacts to cultural sites. Visitors to the Grand Canyon enjoy a wide range of recreational options, all with the potential to impact cultural resources (Figure 7). Visitors backpack on historic trails, raft the Colorado River through the canyon, ride mules into its depths (Figure 8), and enjoy scenic vistas from the rims. The Colorado River Management Plan (CRMP; U.S. Dept. of Interior 2004) and the Backcountry Management Plan (U.S. Dept. of Interior 1988) both focus on visitor use. Integration of cultural resource information into these plans is critical to assure protection of fragile resources in frequently visited areas. As part of the CRMP, a multidisciplinary team of resource and recreation specialists work together to

Figure 5. Coconino Gray Ware recently exposed via alluvial erosion.

identify carrying capacity and a reasonable range of alternatives for recreational use of the Colorado River.

Cultural resource considerations are incorporated into the development of operating requirements for commercial and non-commercial river runners as part of the CRMP. Group size, trip length, time of year, passenger to guide ratios, and the number of trips in the canyon at one time each affect the vulnerability of sites; resource staff members were therefore involved in the development of acceptable ranges for each of these variables.

Several factors that have been extremely valuable in determining carrying capacity include Grand Canyon National Park cultural and natural resource files, the Grand Canyon River Trip Simulator (GCRTS; Roberts and Gimblett 2001; Roberts et al. 2002), public comments, river use statistics, visitor use research, and research on camping beach carrying capacity and visitor impacts. The GCRTS is an integrated statistical and artificial intelligence–based computer simulation that models the complex and dynamic human-environment interactions along the Colorado River in the park. Data on river trip behavior were collected for the GCRTS in the form of trip reports from commercial and non-commercial boaters during the 1998–2000 summer seasons. These data assist the planning team in developing alternatives that are based on predicted levels of use based on various launch schedules.

Georeferenced resource maps that show all known cultural and natural resource areas of concern, as well as recreational

Figure 6. Zuni Conservation Program personnel constructing checkdams near an archaeological site to slow down erosion.

stopping points and their level of use based on the GCRTS, have been developed specifically for the CRMP. These mixed resource maps are important tools for investigating potential visitor versus resource conflicts. When different launch schedules are run in the trip simulator, changes in the intensity of use can be predicted at each of the river stops, and then compared to biophysical impact data (from various Grand Canyon monitoring projects) and the resource map (Figure 9). In this way, we can identify areas of resource vulnerability from visitor impacts based on various launch schedules.

CONCLUSION

Years of previous research conducted through projects in Grand Canyon National Park have provided a baseline of data on cultural and natural resources and visitor use, as well as impacts from visitors, nonnative species, and Glen Canyon Dam. These kinds of data have provided an in-depth understanding of the river corridor environment, both how it has been affected and how it might be affected in the future. These data have also shown the effectiveness and cost of restorative efforts, how visitors affect the environment, and visitor expectations for a river trip. Grand Canyon National Park strives to be a model for the integration of cultural and natural resource data and concerns into planning processes that guide park management. In this way, cultural resource concerns are not merely dependent on the outcome of these plans; they are a part of the plan.

Figure 7. Gullying/social trail near the river's edge leading to and from an archaeological site.

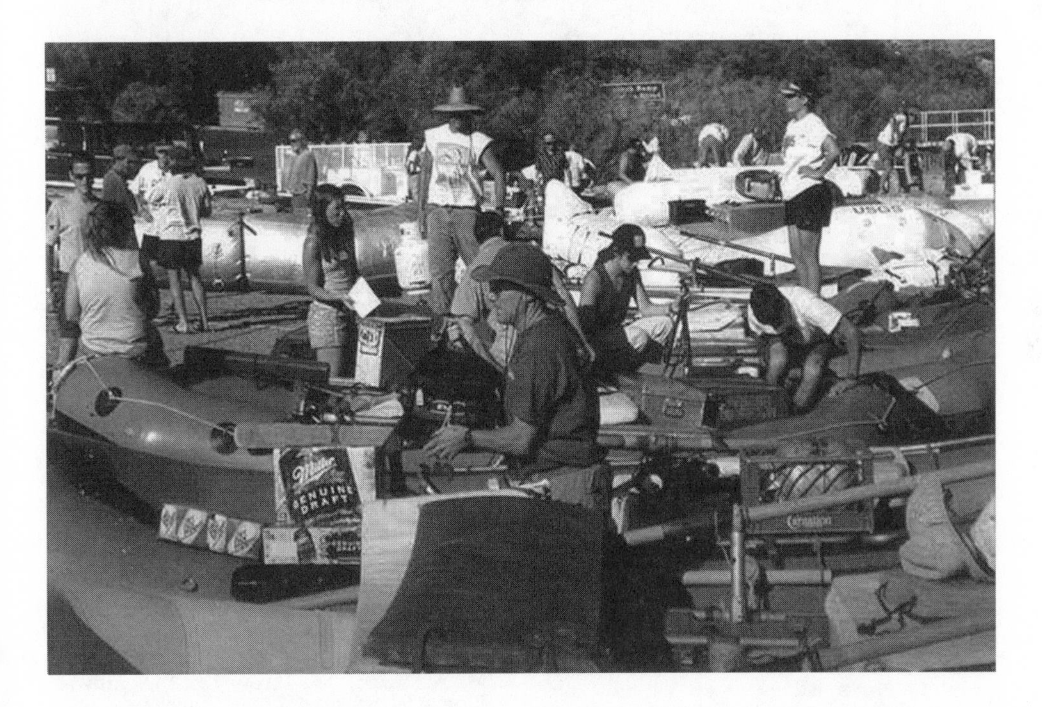

Figure 8. Mule rides (top) offer access to the canyon floor; congestion at Lees Ferry as boaters prepare to begin their river trip (bottom).

Figure 9. Key to the Mixed Resource Map used in CRMP planning.

REFERENCES

Fairley, H. C., P. W. Bungart, C. M. Coder, J. Huffman, T. L. Samples, and J. R. Balsom. 1994. The Grand Canyon River Corridor Survey Project: Archaeological survey along the Colorado River between Glen Canyon Dam and Separation Canyon. Copies available from RCMP #1. Grand Canyon National Park Report Cooperative Agreement 9AA-40-07920, prepared in cooperation with the U.S. Bureau of Reclamation, Salt Lake City, Utah.

Leap, L. M., J. L. Dierker, and N. B.Andrews. 2003. Archaeological site monitoring and management activities along the Colorado River in Grand Canyon National Park. Grand Canyon National Park Report Cooperative Agreement 9AA-40-07920, prepared in cooperation with the U.S. Bureau of Reclamation, Salt Lake City, Utah.

National Park Service. 1997. National Register of Historic Places listings weekly list of actions taken on properties, 9/29/97 thru 10/23/07. Available at http://www.cr.nps.gov/nr/listings/971010.htm.

National Park Service. 2003. Grand Canyon Village national historic landmark district cultural landscape report. Draft Submission. Contract #1443CX2000-98-015. MS on file, Grand Canyon National Park.

National Park Service. 2005. National Historic Landmarks Survey: Listing of national historic landmarks by state. Available online at http://www.cr.nps.gov/nhl/designations/Lists/AZ01.pdf.

National Park Service, and U.S. Department of the Interior. 1997. Cultural Resource Management Guidelines, NPS-28. Washington D.C.

Roberts, C., and R. Gimblett. 2001. Computer simulation for rafting traffic on the Colorado River. In Proceedings of the 5th Biennial Conference of Research on the Colorado Plateau, edited by C. van Ripen III, K. A. Thomas, and M. A. Stuart, pp. 19–30. USGS/Southwest Biological Science Center Colorado Plateau Research Station, Northern Arizona University. Flagstaff.

Roberts, C. A., Stallman, D., and Bieri, J. A. 2002. Modeling complex human-environment interactions. The Grand Canyon River Trip Simulator. Journal of Ecological Modeling 153 (2): 181–196.

U.S. Department of the Interior. 1988. Backcountry Management Plan, Grand Canyon National Park. Available on line at http//www.nps.gov/grca/wilderness/documents/1988_BCMP.pdf (accessed November 20, 2003).

U.S. Department of the Interior. 1995. General Management Plan, Grand Canyon National Park. Denver Service Center, Colorado.

U.S. Department of the Interior. 1996. Operation of Glen Canyon Dam, Colorado River Storage, Arizona—Final Environmental Impact Statement, Bureau of Reclamation, Salt Lake City.

U.S. Department of the Interior. 2003. Fire Management Plan, Grand Canyon National Park. Available on line at hppt//www.nps.gov/grca/fire/plan/index.htm (accessed November 21, 2003).

U.S. Department of the Interior. 2004. Draft Environmental Impact Statement Colorado River Management Plan, vols. 1 & 2, Grand Canyon National Park.

U.S. Department of the Interior, Bureau of Reclamation, National Park Service, Advisory Council on Historic Preservation, Arizona State Historic Preservation Office, Hopi Tribe, Hualapai Tribe, Navajo Nation, San Juan Southern Paiute, Southern Paiute Consortium and Zuni Pueblo. 1994. Programmatic Agreement on Cultural Resources. Bureau of Reclamation, Salt Lake City.

Biophysical Resources
KIM VAN RIPER

FIRE AND SPRINGS: REESTABLISHING THE BALANCE ON THE WHITE MOUNTAIN APACHE RESERVATION

Jonathan Long, Mae Burnette, and Candy Lupe

The Rodeo-Chediski fire of 2002 was the largest wildfire in the history of the Southwest. The fire severely burned large swaths across the northwest quarter of the White Mountain Apache Reservation in Arizona (Figure 1). This part of the Mogollon Rim contains an especially high density of springs (Stevens and Nabhan 2002). Wildfire research and rehabilitation efforts have not emphasized spring-fed ecosystems, despite their ecological and cultural importance. After the fire, we initiated a project to assess and prescribe treatments to rehabilitate wetlands where post-fire flooding threatened important values.

WILDFIRE AS A FORCE OF DESTRUCTION AND REJUVENATION

Many cultures have long recognized the dual nature of fire as a rejuvenating, beneficial element and as a destructive, dangerous force (Pyne 1997), and White Mountain Apache traditions share that fundamental outlook (Long et al. 2003a). Land managers have observed for many decades that wildfire has served to cleanse, rejuvenate, and stabilize ponderosa pine forests of the Apache Reservation (Weaver 1951). However, recent decades have witnessed a profound shift from light surface fires to severe crown fires in ponderosa pine forests across the Southwest (Fulé et al. 1997). Large, severe fires along the Mogollon Rim have become particularly prominent on the reservation in recent years (Table 1). Increased attention to post-wildfire rehabilitation efforts has accompanied the increase in severe wildfires (Robichaud et al. 2000).

Managers weigh the rejuvenating and destructive aspects of wildfire in evaluating the need to treat burned areas (Rieman et al. 1997; Bisson et al. 2003). Much research on wildfires concludes that riparian and aquatic ecosystems recover quickly and even become invigorated following such perturbations (Dwire and Kauffman 2003; Minshall 2003). As revegetation occurs in the first few years following a wildfire, runoff and erosion rates progressively return to normal conditions (Minshall and Brock 1991). Fish and macroinvertebrate populations commonly rebound within a few years after a wildfire (Rieman et al. 1997; Minshall 2003). These findings support the characterization of wildfire as a pulse disturbance that is not expected to alter the long-term equilibrium of an ecosystem.

However, post-wildfire flooding can induce drastic biologic and geomorphic changes that prevent a stream ecosystem from returning to its former structure and function within time frames important to human societies. In Arizona and New Mexico, wildfires have induced ash flows that extirpated local fish populations in several streams, requiring reintroductions of those species (Propst et al. 1992; Rinne 1996). Those severe wildfires induced widening, deepening, and coarsening of stream channels, which in turn limited the regrowth of riparian vegetation for decades (Medina and Martin 1988; Medina and Royalty 2002).

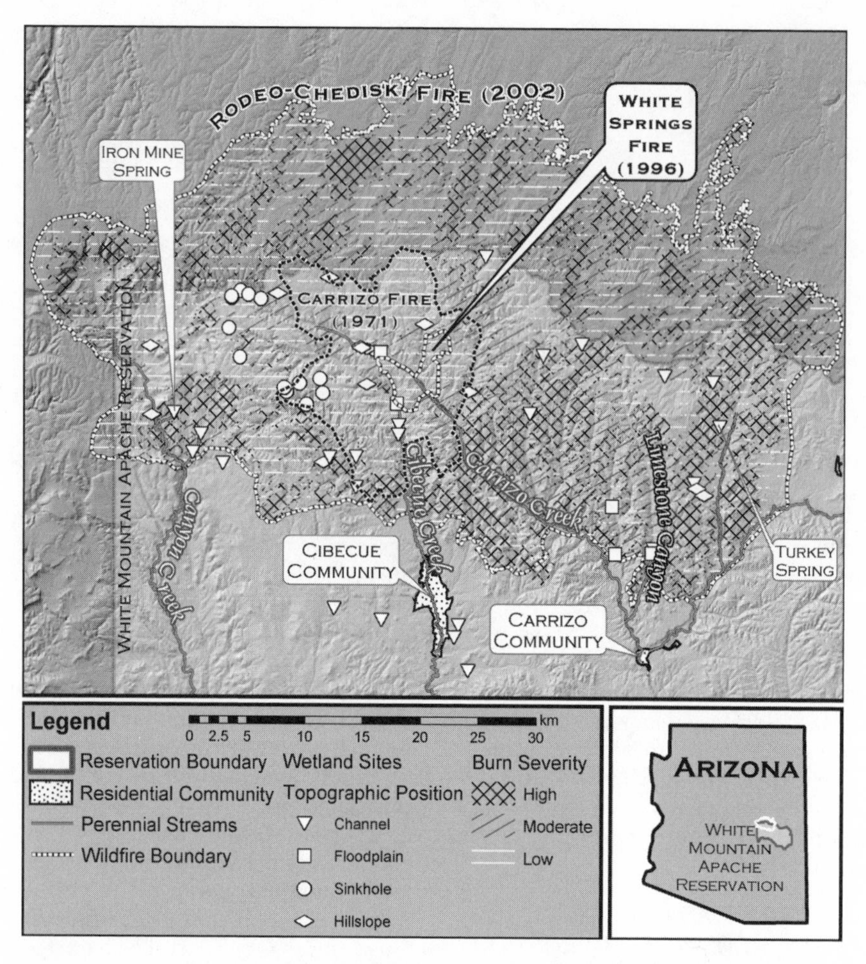

Figure 1. Map of the northwestern corner of the White Mountain Apache Reservation, showing burn severity across the Rodeo-Chediski wildfire, the perimeters of two earlier wildfires, and the location of assessed wetlands with symbols denoting topographic type.

Such degradation often causes extensive loss of riparian soils and lowering of water tables, which greatly reduces the quality of riparian and aquatic habitats (Heede 1986b; Shields et al. 1994).

Debates over post-fire management center on the effectiveness of intervening in stream systems that have been severely disturbed by wildfire (Bisson et al. 2003). Prominent stream ecologists have recently argued that post-fire management should emphasize "natural recovery" processes, and they have discouraged use of in-stream structures on the grounds that they often interfered with those processes (Beschta et al. 2004). On the other hand, Heede (1986a) argued that "corrective actions," including appropriately designed structural treatments, could accelerate natural regenerative processes. In incised channels, active restoration treatments can restore ecological processes such as development of bedforms, deposition of fine sediments, and growth of wetland vegetation (Medina and Long 2004).

Table 1. Year and size of major wildfires on the White Mountain Apache Reservation along the Mogollon Rim, Arizona.

Wildfire Name	Year	Approximate Size (Ha)
Carrizo	1971	23,000
Stago	1990	1,350
White Spring	1996	1,600
Rainbow	1999	2,000
Ridge	2000	3,200
Rodeo-Chediski	2002	187,000*
Kinishba	2003	9,700

*Total including 75,000 ha of National Forest lands.

THE PHILOSOPHY OF REESTABLISHING BALANCE

Heede (1986a) argued that active interventions could be justified by helping stream systems that would otherwise be unstable for long periods to reattain "balance." Fluvial morphologists have long used the term "balance" to describe streams that were neither degrading nor aggrading (Heede 1980). Ecologists have argued that "balance of nature" metaphors can promote imprecise understandings of complex ecological systems, in particular by implying that natural systems are unchanging (Cuddington 2001; Hull et al. 2002). However, Heede (1986a) applied the term "balance" to a dynamic equilibrium, rather than a steady state: "Change is therefore the rule. Why then do we apply balance or equilibrium to an ever-changing world? We do it to contrast an orderly changing condition with severe disturbance or catastrophy [sic]." He did not view disturbance as necessarily destabilizing, since he observed that increases in peak flows and bedload movement due to timber harvest, for example, could accelerate attainment of a new dynamic equilibrium in a system (Heede 1991). That perspective suggests that even severe wildfire could have the potential to restore balance to riparian ecosystems. The use of balance to describe complex interplay within changing ecosystems resonates with one of the authors (Mae Burnette), who applies traditional ecological knowledge to her restoration work. Heede also asserted that stream channels in equilibrium maintained the "health" of their associated riparian areas, and that equilibrium was marked by a "smoothened" longitudinal profile and more "tranquil" flow (Heede 1980, 1986b). These expressions fit well with traditional White Mountain Apache cultural perspectives, which hold that streams have life and agency (Long et al. 2003a). That a male German hydraulic engineer and a self-described traditional White Mountain Apache woman would choose similar terms in English to describe ecological dynamics suggests that these concepts are useful for treating wetlands on the reservation.

IMPORTANCE OF SPRING-FED WETLANDS IN THE REGION

Springs are some of the most important ecosystems on the reservation due to their ecological and cultural value. Springs support diverse and rare wetland communities and provide valuable ecological services such as diminishing downstream flooding (Hendrickson and Minckley 1984; Meyer et al. 2003). Conservation of springs has become a management priority on the Colorado Plateau, particularly on tribal lands (Stevens and Nabhan 2002). The importance of spring-fed ecosystems to members of the White Mountain Apache Tribe is reflected in the numerous place-names that refer specifically to springs and associated wetland plants (Long et al. 2003a). The tribe's Water Quality Code states that cultural uses shall be protected at all springs.

Despite their importance, spring-fed wetlands have not been emphasized in wildfire research. Literature describing the effects of wildfire on springs has focused on changes in discharge (Neary et al. 2003). The Yellowstone fire of 1988 triggered intensive studies of wildfire effects on aquatic ecosystems and channel morphology in the northern Rockies; however, these studies focused on relatively large systems where aggradation and sedimentation predominate (Minshall et al. 1997; Benda et al. 2003; Meyer and Pierce 2003). In the Southwest, wetlands and aquatic habitats are unusually dependent on small, spring-fed reaches (Hendrickson and Minckley 1984). Furthermore, severe stand-replacing fires are an important part of the long-term disturbance regime for the lodgepole pine forests of Yellowstone (Meyer and Pierce 2003), but they are not the norm for the ponderosa pine forests of the Southwest (Fulé et al. 1997). Consequently, severe wildfires along the Mogollon Rim have greater potential to induce changes that are beyond the range of historical variation.

LESSONS LEARNED FROM WHITE SPRING

The tribe's natural resource managers recognized the potential for wildfire to damage springs as a result of the White Spring fire in June of 1996. That fire was named for a vitally important spring at the base of the burned watershed. Despite the fact that the spring had longstanding value as a cultural resource, was one of the main sources of perennial flow to Cibecue Creek, and supported Apache trout, the spring received no direct treatment in the post-fire rehabilitation plan developed through the Burned Area Emergency Rehabilitation (BAER) process. Severe flooding in the wake of the fire led to rapid channel incision below the spring and headcutting toward the spring. A community-led project organized by the tribe's Watershed Program and supported by the local Cibecue Community School, the Rocky Mountain Research Station, the Bureau of Indian Affairs, and the Environmental Protection Agency brought about a variety of treatments that ultimately restored the spring to a healthier condition than residents had witnessed in decades (Long and Endfield 2000). The experience at White Spring led team members writing the BAER plan for the Rodeo-Chediski wildfire to include a specification for assessing threats and designing stabilization treatments for spring-fed ecosystems and sinkhole wetlands affected by the conflagration. The plan provided resources for Mae Burnette to coordinate the project.

METHODS

Site Identification

We consulted with a variety of community members including cultural resource specialists, forestry workers, fence crew workers, and elders to locate sites and obtain information on their pre-fire condition. We obtained photographs of several of the sinkhole lakes, but most of the springs lacked documentation on their historical condition. The specification written into the BAER plan called for assessing 60 springs and 12 sinkhole lakes identified on U.S. Geological Survey topographic maps as lying within the area affected by the fire. The consultations with community members revealed that many springs were inaccurately located or altogether missing on maps. Many sites were difficult to access, because much of the road system was closed intentionally or due to flooding after the fire. Consequently, the results presented here do not constitute all the mapped sites, but rather those that were readily accessible (less than 5 km from a serviceable road) and of greatest concern to community members.

Site Classification

We identified the dominant geologic formations at and above the site using the most detailed geologic maps available (Finnell 1966a, 1966b; McKay 1972). We followed the nomenclature used in those maps, although Blakey (1990) reassigned several members of the Supai Formation to a new Schnebly Hill Formation. We also classified each site by its topographic position: in-channel, in a floodplain, on a hillslope, or in a sinkhole depression (Figures 1 and 2). Finally, we classified the burn severity in the contributing area above each site using maps developed for the BAER plan. The rating classes correspond to visual indicators, with high-severity burn being associated with removal of organic matter and changes in soil structure that increase runoff response (Robichaud et al. 2000).

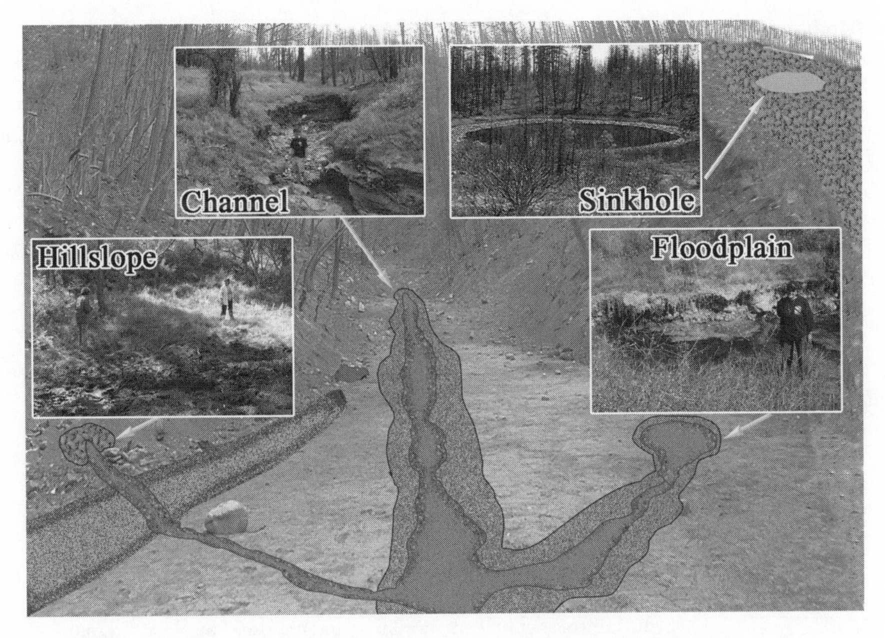

Figure 2. Examples of the four topographic types of wetlands in the study area.

Site Assessment and Evaluation of Stability

We assessed sites through qualitative observation of key vegetative, hydrologic, and geomorphic indicators (Table 2). We identified the presence of core wetland plant species consistent with the list developed by McLaughlin (2003). We evaluated whether ungulate trampling and grazing appeared to compact soils or alter plant structure and composition. We noted indicators of geomorphic instability, such as changes in bar formations, nickpoints, bank erosion, channel incision, changes in substrate size, and gullying (Heede 1980). We took repeat photographs of the sites at least annually for 2 years to evaluate the rates of erosion and vegetative growth. We synthesized the assessments using a checklist of desirable functional processes (Medina et al. 1996); the checklist adds assessment of animal impacts to the Proper Functioning Condition methodology widely used by federal land management agencies (Prichard et al. 1993).

We expected the wildfire to cause short-term changes in vegetation, animal impacts, substrates, and hydrology. To evaluate whether the sites would likely undergo more lasting changes, we focused on geomorphic indicators of degradation, which is typically the most consequential form of channel adjustment to watershed disturbance (Heede 1986b). We rated sites as severely unstable where progressive head-cutting appeared to be rapidly eroding natural grade control features, bed armor, and stream banks (Heede 1991).

Treatment Prescription

We recommended treatments that would reduce constraints on natural recovery, such as excessive grazing pressures and unstable "legacy" roads in riparian areas (Beschta et al. 2004). Specifically, we prescribed fencing of wetland areas, particularly at the heads of springs, where animal impacts appeared to be inhibiting vegetative growth. We recommended rehabilitating old roads and stream crossings that were contributing to channel

Table 2. Indicators used to evaluate condition of wetland sites.

Assessment Criteria	Indicators
Hydrology	Estimated flow: none, low (< 4 l/m), moderate (≥ 4 l/m), high (> 240 l/m)
Vegetation	Presence of core herbaceous wetland species, in particular monkeyflower (*Mimulus guttatus*), watercress (*Rorippa nasturtium-aquaticum*), cardinal flower (*Lobelia cardinalis*), sedges (*Carex* spp.), bulrushes (*Schoenoplectus* spp.), rushes (*Juncus* spp.), cattails (*Typha* spp.), and spikerushes (*Eleocharis* spp.)
Animal Impacts	Soil compaction or erosion due to animal impacts Change in plant structure or composition due to animal impacts
Geomorphology	(see Heede 1980 for explanations of indicators)
Minor instability	Formation, movement, and changes in size of bar formations and log steps Degradation or aggradation of the channel Changes in channel shape Changes in bed particle size due to scour or deposition Bank slumping and erosion due to shearing by water or animals
Severe instability	Formation and migration of channel nickpoints Discontinuous gully formation Scouring that removes armor layer and exposes hardpan bed materials Stream bank failure (collapse of stream banks due to instability)
Road impacts	Gullying due to concentration of flows by roads Potential for culvert to fail or disrupt channel morphology

erosion by dispersing flows and replacing culverts with rock fords.

We also recommended installing grade control features in order to prevent further incision and to restore dynamic equilibrium at rapidly incising sites (Heede 1986b). Specifically, we proposed placing riffle formations composed of large rock, gravel, and transplants of bulrush (*Schoenoplectus pungens*) and sedges (*Carex* spp.; Figure 3). The tribe and the Rocky Mountain Research Station had previously tested this treatment design in incised montane meadow reaches unaffected by wildfire (Medina and Long 2004). We increased the scale of formations from that treatment design so they would withstand the erosive floods occurring in the burned canyons.

We discussed treatment strategies with community members, including representatives of the livestock associations that had grazing privileges in the proposed treatment areas, and representatives of the Tribal Cultural Advisory Board. We conducted a site visit to Swamp Spring with representatives of the Tribal Land Restoration Board to discuss treatment plans. We submitted treatment proposals to the Tribal Land Restoration Fund, the BAER Stabilization and Rehabilitation Program, and the Natural Resource Conservation Service's Environmental Quality Incentives Program. Prior to implementation, the tribe reviewed and permitted the treatments through their plan and project review process.

Results of Assessment

General reconnaissance of the burned area revealed that headwater reaches were affected by scattered debris flows and channel incision, while larger streams such as Carrizo, Cibecue, and Canyon Creeks were altered by aggradation and lateral erosion. These geomorphic changes began with the

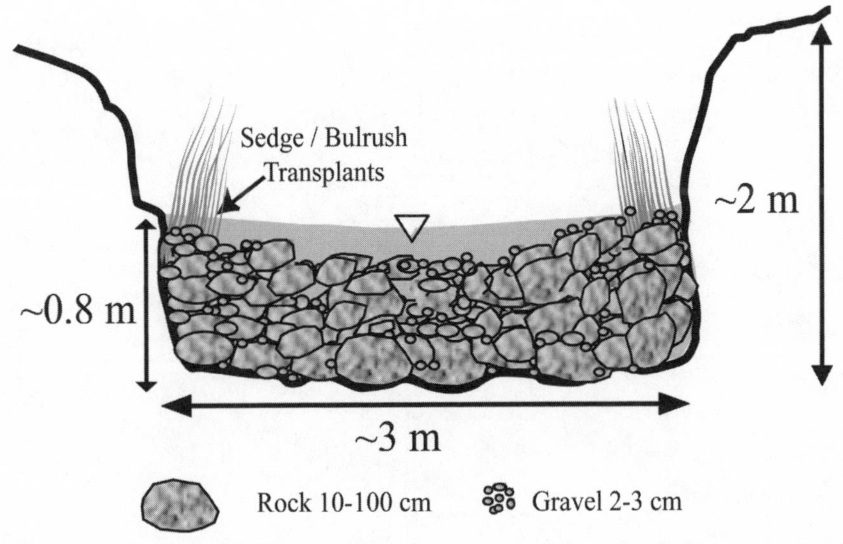

Figure 3. Cross-sectional view of the riffle formation design used to treat rapidly incising channels.

first summer storms after the fire, but they were most pronounced following a late summer storm on September 11, 2002 (Long et al. 2003b). Debris flows were particularly common in steep, short side canyons along the numerous tributaries to Carrizo Creek.

Site Classification

Almost all of the sites examined (51 out of 56) had wetland indicators including surface water and core wetland plant species. Although the remaining five sites had been mapped as springs, they did not have pronounced wetland characteristics. The most common geologic formations at the sites were the upper members of the Supai Formation (25%), the Fort Apache Member of the Supai Formation (18%), Kaibab Limestone (18%), and Coconino Sandstone (16%). Topography corresponded closely to geology at the sites. The red-beds of the Supai Formation were commonly exposed in canyon bottoms and on hillslopes, whereas the gray limestone of the Fort Apache Member formed cliffs in canyon bottoms. Thirteen (23%) of the sites were sinkhole depressions that had formed on ridgetops due to dissolution of the Kaibab Limestone. The relatively resistant Coconino Sandstone, unnamed Cretaceous sandstone, and Mogollon Rim Formation typically surrounded the sinkhole lakes on the ridgetops. The nonsinkhole sites were located within channels (41%), within adjacent floodplains (11%), or on hillslopes (25%). The watershed burn severity ratings across the sites were 22 high, 23 moderate, 7 low, and 4 unburned.

Animal Impacts

The rehabilitation plan deferred livestock grazing from the burned area after the fire. Many areas had already not been grazed by livestock due to the ruggedness of the terrain. The rehabilitation plan also directed the removal of feral horses, which had been problematically common in the area. We attributed reduced plant vigor and undesirable soil impacts to ungulate grazing and trampling at 21 of the 43 lotic sites. We recommended fencing those areas to reduce impacts from the remaining ungulates, chiefly elk. Within the first year after fencing treatments began in the summer of 2003, several sites exhibited rapid growth of wetland vegetation, as shown in repeat photographs (Figure 4).

Figure 4. Two spring-fed wetlands were fenced to exclude ungulates because heavy use after the Rodeo-Chediski wildfire appeared to limit vegetative growth (left). These sites responded to the fencing treatment within a year (right).

Geomorphic Impacts

We observed evidence of geomorphic instability at many of the sites with running water. Fourteen sites showed minor channel instability, and seven sites showed evidence of severe channel instability. Five of the latter seven sites were mapped at or just above contacts of the Fort Apache Member of the Supai Formation. These sites had sandy soils derived from the overlying Coconino Sandstone. Four of the seven highly unstable sites were springs that emerged within a channel, and two others were located in floodplains adjacent to large channels. The remaining highly unstable site was located on a hillslope with a small channel below it. All of the severely unstable sites and 12 of the 14 sites with minor instability were located in areas that burned at moderate to high severity; the two other sites with minor instability were in unburned areas.

We recommended active restoration treatments, including placement of riffle formations and rehabilitation of road impacts, at most of the severely unstable sites. Three of the sites were small enough that riffle formations could be placed by hand using rock materials from adjacent slopes. Four other sites had much more extensive impacts.

Riparian wetlands along a spring-fed creek in Limestone Canyon withstood the initial summer floods, but flooding on September 11, 2002 caused extensive scouring of the main stem and debris flows down side canyons. The main channel deepened by 0.6 m and doubled in width to 17.4 m (Long et al. 2003b). A debris flow from a side canyon

Figure 5. Wetland habitat that existed in Limestone Canyon one year before the Rodeo-Chediski wildfire (inset) was dried out after post-fire debris flows and floods caused the stream channel to divert from this reach and incise.

induced a channel avulsion that left a formerly productive wetland high, dry, and buried with sediment (Figure 5). We did not prescribe in-channel treatments at Limestone Canyon because the large watershed could deliver overwhelming flows. The channel continued to adjust by lateral movement and changes in substrate, but it did not continue to incise.

We separately evaluated conditions in a spring-fed wet meadow that lay in the floodplain of Limestone Canyon. A headcut resulting from the lowered base level in the canyon formed a 1 m tall nickpoint at the edge of the meadow, prompting us to initially consider intervention. After the second year of observation after the fire, however, growth of herbaceous wetland vegetation above and below the nickpoint suggested that the wetland could regain equilibrium without active interventions.

Turkey Spring was another large wet meadow in a severely burned canyon. Headcutting at this site from November 2002 to September 2003 formed a trench approximately 2.0 m wide, 1.5 m deep, and 50 m long (Figure 6), representing an estimated loss of more than 200 metric tons of sediment. Channel incision at the site was discontinuous, as a short (20 m), uneroded, and well-vegetated reach remained 60 m downstream of the uppermost nickpoint. However, the stream plunged off a second, 2.0 m tall nickpoint below the stable reach into another deeply incised reach. Roads paralleling the stream channel appeared to have induced or greatly exacerbated the instability by concentrating runoff and forming new gullies. Headcutting continued through May of 2004, when active treatment began.

The channel at Swamp Spring was deeply incised with steep nickpoints (inset photo in Figure 7) and deposits of slumped materials from the adjacent wet meadow. Erosion of the site appeared to predate the fire, as indi-

Figure 6. The channel at Turkey Spring formed deep, discontinuous gullies as headcuts (insets) eroded through hundreds of meters of wetland soils within the first 2 years after the Rodeo-Chediski wildfire.

cated by relatively continuous downcutting, extensive deposits of slumped materials with some vegetative growth, and an old culvert that showed where a stream crossing had washed out. However, post-fire flooding accelerated upstream migration of nickpoints (Figure 7), exposed fine hardpan bed material in the channel, and increased lateral erosion by undermining the toes of the streambanks.

Due to the persistent instability at Turkey Spring and Swamp Spring, we proposed placing riffle formations along 400 m and 300 m segments at the respective sites. Spacing between formations averaged 20 m; each formation was approximately 0.8 m tall, 6 m long, and 3 m wide; and the median particle size (intermediate diameter) was 0.5 m (see Figure 3). We adjusted the size and spacing of individual formations to the dimensions of each reach, so that formations were larger in more deeply incised reaches and closer together in steeper reaches. We prescribed especially large formations at the bottom of each reach where the channel was less entrenched and had naturally coarse bed materials to control the base level. The volume of rock material and size of individual particles required heavy equipment to deliver the rocks to the designated channel locations, although laborers repositioned the rocks by hand.

Eight of the 13 sinkhole wetlands were located in moderately burned areas, and the remaining five were located in lightly burned areas. None of the sinkhole wetlands showed evidence of severe geomorphic impacts. The deepest of these wetlands, Pumpkin Lake, had been directly treated after the fire through the placement of two concentric circles of straw wattles around the lake. The wattles appeared to prevent most of the ash and sediment from entering the lake. We did not observe obvious changes in wetland vegetation at that lake or others where photographs and accounts from community members provided information on pre-fire conditions. Based on these observations, we

Figure 7. The channel bed at the Swamp Spring site adjusted rapidly following the Rodeo-Chediski wildfire, resulting in migration of nickpoints and extensive bank erosion.

concluded that impacts to the lakes were not consequential enough to warrant additional interventions.

DISCUSSION

Our results showed that, in some circumstances, wildfire can severely impact spring-fed wetlands on the west side on the reservation. Wildfire impacts tend to be greater in small, steep, severely burned watersheds with steep channels, shallow rocky soils, and potential for intense precipitation events (Minshall et al. 1997). Disequilibrium conditions are likely to persist where streams are confined by valley walls, have limited inputs of coarse sediment and large woody debris, have lost bank and floodplain soils, and have exposed rock outcrops (Heede 1985, 1986b). Such qualities typify the steep, narrow, and highly dissected canyons south of the Mogollon Rim. In this region, late summer monsoon storms, fall tropical storms, and winter rain-on-snow events can induce intense runoff in recently burned areas. We identified road impacts as a contributor to instability at four of the highly unstable sites, including the two sites with the most extensive erosion. Extensive road networks tend to reduce the resilience of areas to wildfire impacts (Gresswell 1999; Dwire and Kauffman 2003; Minshall 2003). Consequently, rugged physiography, climatic conditions, and historical conditions explain why this region is particularly vulnerable to severe post-fire impacts.

Our finding that a small percentage of sites became highly unstable demonstrates that the effects of wildfire are highly variable within a particular landscape. By assessing spring-fed wetlands, rather than riparian wetlands more generally, we focused our attention on headwater reaches that tend to respond most rapidly to changes in watershed condition (Minshall et al. 1997). Between periods of watershed disturbance, headwater reaches experience aggradation of sediments, while larger fluvial systems downstream experience higher erosion rates (Benda 1990). However, wildfire reverses these relationships: steep headwater reaches incise while larger valleys with flatter slopes aggrade (Moody and Martin 2001). Wildfire may benefit riparian and aquatic species by depositing sediments needed for rebuilding habitat in the large rivers (Rieman et al. 1997). For instance, the main stem of Carrizo Creek developed dense growth of wetland vegetation within a year after extensive sediment deposition from the Ridge Fire of 2001 (Long et al. 2003b). A spring-fed marsh along East Cedar Creek similarly assimilated fine sediments deposited by floods after the Kinishba fire of 2003 (unpublished data). Lower valley slope, a more finely textured lithology, and reduced burn severity likely contributed to the greater resilience observed at those two streams as compared to Limestone Canyon (Long et al. 2003b).

Due to their ridgetop location, sinkhole wetlands appeared less vulnerable to post-fire impacts than wetlands located in canyons. Ridgetop wetlands naturally had lower slopes and smaller contributing areas, and they were less severely burned. Our results suggest that we could focus future assessments on springs located in channels and on floodplains downstream from moderate to high severity burns. However, because our assessment focused on geomorphic instability, it could have missed less obvious biological changes in the sinkhole wetlands. Although we noted the presence of amphibians at previously uninventoried sites, we did not track amphibian populations. Movement of ash and sediment into sinkhole wetlands has the potential to degrade amphibian habitat through sedimentation or changes in water quality. Pilliod et al. (2003) asserted that such impacts were likely to be inconsequential, but they recommended studies of fire impacts on amphibians to test that assumption. Since the straw wattle treatment did block sediment and ash from entering Pumpkin Lake, it may be appropriate for unusually sensitive waterbodies.

Assessments of responses to wildfire not only need to account for landscape heterogeneity, but they also need to consider temporal variation. We initially assessed two sites as severely unstable, but we later determined that they would probably stabilize without intervention or further loss of wetlands. Due to stochastic post-fire floods and debris flows, streams can shift unpredictably between aggradation and degradation (Heede 1986a; Benda 1990). For example, initial debris flows following the White Spring fire created deposits that inhibited headcutting into White Spring, but those deposits washed out in subsequent floods, triggering rapid headward erosion into the spring. Such dynamics demonstrate the importance of evaluating channel responses for many years after a major fire (Minshall et al. 1997).

Justification for Treatments

Our strategy of treating selected spring-fed wetlands conformed to guidelines established through previous research. These springs were priorities for treatment because they represented small, fragmented habitats in degraded watersheds (Bisson et al. 2003). Research has suggested that fencing the heads of springs and rehabilitating problematic roads are fast and cost-effective conservation measures in disturbed areas (Beever and Brussard 2000; Robichaud et al. 2000). We recommended placing riffle formations to stabilize the wet meadows at Turkey Spring and Swamp Spring based on previous research indicating that reestablishing dynamic equilibrium in such rapidly incising channels would be very slow and costly without interventions (Heede 1986b). Channel degradation and loss of organic wetland soils are not easily reversed in the canyons below the Mogollon Rim, because steep topography and coarse lithology limits the input of the fine sediments needed to rebuild wetlands (Medina and Royalty 2002; Long et al. 2003b). Consequently, on-site treatments that retain fine sediments mobilized by post-fire erosion may be critical to reestablishing dynamic equilibrium. Supporting this idea, we observed that the channel at White Spring did not downcut and wetland vegetation rapidly regrew after the Rodeo-Chediski fire burned through the site. We attribute the site's resilience to the treatments that were applied after the spring's namesake fire 6 years earlier.

The riffle formations complement natural adjustment processes that form wetland habitats, such as landsliding and deposition of alluvial fans (Hendrickson and Minckley 1984; Benda et al. 2003). Under the historic fire regime, patches of high-severity burn could have formed some wetlands through these natural depositional processes while wiping out others. However, the extraordinary size and severity of the Rodeo-Chediski fire likely caused more uniform scouring of headwater reaches and shifted opportunities for wetland formation downstream to larger, flatter rivers. Such a huge shift in resources exacts a toll on plant, animal, and human communities that depend on headwater wetlands. Meanwhile, much of the nutrient-rich sediment exported from the burned area ended up as unwelcome deposits in Roosevelt Reservoir on the Salt River (Ffolliott and Neary 2003). The unnatural severity of the Rodeo-Chediski wildfire and the ensuing loss of high-value wetland habitat seem to justify targeted post-fire interventions in headwater reaches.

Linking Post-Fire Assessment and Treatment to Broader Conservation Efforts

We benefited from having monitored conditions and channel dynamics at several sites for years before the fire. However, our sparse knowledge of previously unvisited sites made it harder to interpret post-fire conditions. Unfortunately, post-fire assessments of tribal lands are often led by resource specialists who are not necessarily familiar with the burned area. Efforts to explicitly link post-fire assessments to broader, longer-term efforts to understand hydrologic conditions, cultural values, and biodiversity would help to conserve these spring-fed wetlands. Furthermore, restoring degraded areas before wildfires strike is likely to be more cost-effective than post-fire rehabilitation (Beschta et al. 2004). In particular, forest management treatments designed to reduce the risk and severity of future wildfires should prioritize rehabilitating roads that concentrate flows in, above, or below spring-fed wetlands.

We sought to ensure that our treatment recommendations would protect tribal cultural values by consulting with cultural advisors and other community members. Fenced exclosures included walk-through gates to allow people to obtain water from the springs. We rejected stabilization measures that would have employed metal gabion baskets in favor of native rock and plant materials that reflect traditional erosion control practices (Long et al. 2003a). The fencing crew posted signs with Apache names of particular wetlands to remind people of their importance. We incorporated site photographs into a database of culturally important sites to guide future conservation efforts. We gave presentations on the restoration effort to help community members and leaders see how their springs have changed after the fire and subsequent treatments.

CONCLUSIONS

Post-fire rehabilitation efforts in the Southwest should include assessment of spring-fed wetlands, because some of these isolated ecosystems are vulnerable to degradation following severe wildfires. Conditions of wetlands following the Rodeo-Chediski wildfire varied with topography, burn severity, and time since the burn. We recommended fencing half of the lotic sites to facilitate recovery of wetland vegetation. Seven (13%) of the assessed sites experienced rapid headcutting after the fire, which signified a loss of dynamic equilibrium that could inhibit long-term productivity. We prescribed road rehabilitation and placement of riffle formations at the four sites that showed continuing degradation 2 years after the fire. Deterioration was most extreme at two wet meadows with deep, finely textured wetland soils and no bedrock to prevent incision. Our finding that most sites did not require active intervention suggests that future assessments can focus on wetlands located in steep, severely burned canyons; however, sinkhole lakes may warrant special attention to evaluate sediment effects on sensitive aquatic life. Integrating post-wildfire assessments of spring-fed wetlands with broader efforts to conserve these important ecosystems will help to ensure that the mixing of water and fire does not sacrifice important community values.

ACKNOWLEDGMENTS

The Burned Area Emergency Rehabilitation (BAER) program, the U.S. Environmental Protection Agency, and the U.S. Forest Service through the National Fire Plan provided funding support for the assessment project. The White Mountain Apache Tribal Land Restoration Fund and the Natural Resources Conservation Service provided additional support to treat degraded sites. We thank three reviewers whose comments substantially improved this manuscript. Finally, we thank Alvin L. Medina and Dr. Burchard H. Heede for their innovative efforts in watershed rehabilitation.

LITERATURE CITED

Beever, E. A., and P. F. Brussard. 2000. Examining ecological consequences of feral horse grazing using exclosures. Western North American Naturalist 60: 236–254.

Benda, L. 1990. The influence of debris flows on channels and valley floors in the Oregon Coast Range, U.S.A. Earth Surface Processes and Landforms 15: 457–466.

Benda, L., D. Miller, P. Bigelow, and K. Andras. 2003. Effects of post-wildfire erosion on channel environments, Boise River, Idaho. Forest Ecology and Management 178: 105–119.

Beschta, R. L., J. J. Rhodes, J. B. Kauffman, R. E. Gresswell, G. W. Minshall, J. A. Karr, D. A. Perry, R. F. Hauer, and C. A. Frissell. 2004. Postfire management on forested public lands of the western United States. Conservation Biology 18: 957–967.

Bisson, P. A., B. E. Rieman, C. Luce, P. F. Hessburg, D. C. Lee, J. L. Kershner, G. H. Reeves, and R. E. Gresswell. 2003. Fire and aquatic ecosystems of the western USA: Current knowledge and key questions. Forest Ecology and Management 178: 213–229.

Blakey, R. C. 1990. Stratigraphy and geologic history of Pennsylvanian and Permian rocks, Mogollon Rim region, central Arizona and vicinity. Geological Society of America Bulletin 102: 1189–1217.

Cuddington, K. 2001. The "balance of nature" metaphor and equilibrium in population ecology. Biology and Philosophy 16: 463–479.

Dwire, K. A., and J. B. Kauffman. 2003. Fire and riparian ecosystems in landscapes of the western USA. Forest Ecology and Management 178: 61–74.

Ffolliott, P. F., and D. G. Neary. 2003. Initial assessment of the Rodeo-Chediski fire impacts on hydrologic processes. Hydrology and Water Resources in Arizona and the Southwest 33: 93–102.

Finnell, T. L. 1966a. Geologic map of the Chediski Peak quadrangle, Navajo County, Arizona. Geologic Quadrangle Map GQ-544. U.S. Geological Survey, Reston VA.

Finnell, T. L. 1966b. Geologic map of the Cibecue quadrangle, Navajo County, Arizona. Geologic Quadrangle Map GQ-545. U.S. Geological Survey, Reston VA.

Fulé, P. Z., W. W. Covington, and M. M. Moore. 1997. Determining reference conditions for ecosystem management in southwestern ponderosa pine forests. Ecological Applications 7: 895–908.

Gresswell, R. E. 1999. Fire and aquatic ecosystems in forested biomes of North America. Transactions of the American Fisheries Society 128: 193–221.

Heede, B. H. 1980. Stream dynamics: An overview for land managers. General Technical Report RM-72. U.S. Department of Agriculture, Forest Service, Rocky Mountain Forest and Range Experiment Station, Fort Collins CO.

Heede, B. H. 1985. Interactions between streamside vegetation and stream dynamics. In Proceedings—Riparian Ecosystems and their Management: Reconciling Conflicting Uses, edited by R. R. Johnson, C. D. Ziebel, D. R. Patton, P. F. Ffolliott, and R. H. Hamre, pp. 54–58. First North American Riparian Conference, Tucson, AZ, April 16–18, 1985. General Technical Report RM-120. U.S. Department of Agriculture, Forest Service, Rocky Mountain Forest and Range and Experiment Station, Fort Collins CO.

Heede, B. H. 1986a. Balance and adjustment processes in stream and riparian systems. In Proceedings—Wyoming Water 1986 and Streamside Zone Conference—Wyoming's Water Doesn't Wait While We Debate, pp. 3–7. Casper, WY, April 28–30, 1986. Wyoming Water Research Center and Agricultural Extension Service, University of Wyoming, Laramie.

Heede, B. H. 1986b. Designing for dynamic equilibrium in streams. Water Resources Bulletin 22.

Heede, B. H. 1991. Response of a stream in disequilibrium to timber harvest. Environmental Management 15: 251–255.

Hendrickson, D. A., and W. L. Minckley. 1984. Ciénegas: Vanishing climax communities of the American Southwest. Desert Plants 6: 130–175.

Hull, R. B., D. P. Robertson, D. Richert, E. Seekamp, and G. J. Buhyoff. 2002. Assumptions about ecological scale and nature knowing best hiding in environmental decisions. Conservation Ecology 6: 12 (http: //www.ecologyand society.org/vol6/iss2/art12/main.html)

Long, J. W., and D. Endfield. 2000. Restoration of White Springs. In Proceedings—Land Stewardship in the 21st Century: The Contributions of Watershed Management, edited by P. F. Ffolliott, M. B. Baker, C. B. Edminster, B. Carleton, M. C. Dillon, and K. C. Mora, pp. 359–360. Tucson, AZ, March 13–16, 2000. RMRS-P-13. U.S. Department of Agriculture, Forest Service, Rocky Mountain Research Station, Fort Collins CO.

Long, J. W., A. Tecle, and B. M. Burnette. 2003a. Cultural foundations for ecological restoration on the White Mountain Apache Reservation. Conservation Ecology 8: 4. (online at http: / / www.consecol.org / vol8 / iss1 / art4)

Long, J. W., A. Tecle, and B. M. Burnette. 2003b. Marsh development at restoration sites on the White Mountain Apache Reservation. Journal of the American Water Resources Association 39: 1345–1359.

McKay, E. J. 1972. Geologic map of the Show Low quadrangle, Navajo County, Arizona. Geological Quadrangle Map GQ-0973. U.S. Geological Survey, Reston VA.

McLaughlin, S. P. 2003. Riparian flora. In Riparian Areas of the Southwestern United States: Hydrology, Ecology, and Management, edited by M. B. Baker Jr., P. F. Ffolliott, L. F. DeBano, and D. G. Neary, pp. 127–167. Lewis Publishers, Boca Raton FL.

Medina, A. L., M. B. Baker Jr., and D. G. Neary. 1996. Desirable functional processes: A conceptual approach for evaluating ecological condition. In Proceedings—Desired Future Conditions for Southwestern Riparian Ecosystems: Bringing Interests and Concerns Together, edited by D. W. Shaw and D. M. Finch, pp. 302–311. Albuquerque, NM, September 18–22, 1995. General Technical Report RM-GTR-272. U.S. Department of Agriculture, Forest Service, Rocky Mountain Forest and Range Experiment Station, Fort Collins CO.

Medina, A. L., and J. W. Long. 2004. Placing riffle formations to restore stream functions in a wet meadow. Ecological Restoration 22: 120–125.

Medina, A. L., and S. C. Martin. 1988. Stream channel and vegetation changes in sections of McKnight Creek, New Mexico. Great Basin Naturalist 48: 373–381.

Medina, A. L., and R. K. Royalty. 2002. A 12-year, post-wildfire geomorphologic evaluation of Ellison Creek, central Arizona. Hydrology and Water Resources in Arizona and the Southwest 32: 77–82.

Meyer, G. A., and J. L. Pierce. 2003. Climatic controls on fire-induced sediment pulses in Yellowstone National Park and central Idaho: A long-term perspective. Forest Ecology and Management 178: 89–104.

Meyer, J. L., L. A. Kaplan, D. Newbold, D. L. Strayer, C. J. Woltemade, J. B. Zedler, R. Beilfuss, Q. Carpenter, R. Semlitsch, M. C. Watzin, and P. H. Zedler. 2003. Where rivers are born: The scientific imperative for defending small streams and wetlands. (On-line at http: / / www.amrivers.org / doc_repository / WhereRiv ersAreBorn1.pdf)

Minshall, G. W. 2003. Responses of stream benthic macroinvertebrates to fire. Forest Ecology and Management 178: 155–161.

Minshall, G. W., and J. T. Brock. 1991. Observed and anticipated effects of forest fire on Yellowstone stream ecosystems. In The Greater Yellowstone Ecosystem: Redefining America's Wilderness Heritage, edited by R. B. Keiter and M. S. Boyce, pp. 123–151. Yale University Press, New Haven CT.

Minshall, G. W., C. T. Robinson, and D. E. Lawrence. 1997. Postfire responses of lotic ecosystems in Yellowstone National Park, U.S.A. Canadian Journal of Fisheries and Aquatic Sciences 54: 2509–2525.

Moody, J. A., and D. A. Martin. 2001. Initial hydrologic and geomorphic response following a wildfire in the Colorado Front Range. Earth Surface Processes and Landforms 26: 1049–1070.

Neary, D. G., G. J. Gottfried, L. F. DeBano, and A. Tecle. 2003. Impacts of fire on watershed resources. Journal of the Arizona-Nevada Academy of Science 35: 23–41.

Pilliod, D. S., R. B. Bury, E. J. Hyde, C. A. Pearl, and P. S. Corn. 2003. Fire and amphibians in North America. Forest Ecology and Management 178: 163–181.

Prichard, D., H. Barrett, J. Cagney, R. Clark, J. Fogg, K. Gebhardt, P. Hansen, B. Mitchell, and D. Tippy. 1993. Riparian area management: Process for assessing Proper Functioning Condition. Technical Report 1737-9. Bureau of Land Management Service Center, CO.

Propst, D. L., J. A. Stefferud, and P. R. Turner. 1992. Conservation status of Gila trout, *Oncorhynchus gilae*. Southwestern Naturalist 37: 117–125.

Pyne, S. J. 1997. World fire: The culture of fire on earth. University of Washington Press, Seattle.

Rieman, B. E., D. C. Lee, G. Chandler, and D. Myers. 1997. Does wildfire threaten extinction for salmonids? Responses of redband trout and bull trout following recent fires in the Boise River Basin, Idaho. Proceedings—Wildfire and Threatened and Endangered Species and Habitats, edited by J. Greenlee, pp. 45–47. Coeur d'Alene, ID, November 13–15, 1995. International Association of Wildland Fire, Fairfield WA.

Rinne, J. N. 1996. Short-term effects of wildfire on fishes and aquatic macroinvertebrates in the southwestern United States. North American Journal of Fisheries Management 16(3): 653–658.

Robichaud, P. R., J. L. Beyers, and D. G. Neary. 2000. Evaluating the effectiveness of postfire rehabilitation treatments. General Technical Report RMRS-GTR-63. U.S. Department of Agriculture, Forest Service, Rocky Mountain Research Station, Fort Collins CO.

Shields Jr., F. J., S. S. Knight, and C. M. Cooper. 1994. Effects of channel incision on base flow stream habitats and fishes. Environmental Management 18: 43–57.

Stevens, L. E., and G. P. Nabhan. 2002. Hydrological diversity: Water's role in shaping natural and cultural diversity on the Colorado Plateau. In Safeguarding the Uniqueness of the Colorado Plateau: An Ecoregional Assessment of Biocultural Diversity, edited by G. P. Nabhan and L. Holter, pp. 33–39. The Center for Sustainable Environments, Terralingua, and the Grand Canyon Wildlands Council, Flagstaff AZ.

Weaver, H. 1951. Fire as an ecological factor in the southwestern ponderosa pine forests. Journal of Forestry 49: 93–98.

A PROTOCOL FOR RAPID ASSESSMENT OF SOUTHWESTERN STREAM-RIPARIAN ECOSYSTEMS

Lawrence E. Stevens, Peter B. Stacey, Allison L. Jones, Don Duff, Chad Gourley, and James C. Catlin

Riparian habitats, including stream margins, springs, wet meadows, and marshes, are areas "inundated or saturated by surface or ground water at a frequency and duration sufficient to support, and which, under normal circumstances do support, a prevalence of vegetation typically adapted for life in saturated soil conditions" (U.S. Department of the Interior 1992). Riparian habitats are inseparably linked to the associated stream's hydrogeologic and geomorphic processes, antecedent and contemporary flows, sedimentology, and local and regional land management practices, as well as to the condition of adjacent upland ecosystems (Hupp 1988; Gregory et al. 1991; Malanson 1993; Mitsch and Gosselink 1993; Stromberg 1993b; Auble et al. 1994; Leopold 1994; McCammon et al. 1998; Chambers and Miller 2004). Stream-riparian ecosystems are among the most biologically diverse, productive, and threatened habitats in the American Southwest (Johnson and Jones 1977; Johnson et al. 1985; Knopf et al. 1988; Ohmart et al. 1988; Johnson 1991; Stromberg 1993a; Minckley and Brown 1994). Southwestern riparian habitats support diverse and unique assemblages of distinctive species not found elsewhere, as well as many facultative species from the surrounding uplands (Johnson 1991; Minckley and Brown 1994; Stacey 1995; Naiman and Decamps 1997; Naiman et al. 2002; Sabo et al. 2005). Stevens and Ayers (2002) reported that natural stream-riparian habitats make up less than 1 percent of the northern Arizona landscape, yet directly support more than 35 percent of the higher plant and bird species, as well as many facultative upland plant and animal species. Desert riparian habitats commonly have ≤ 2 orders of magnitude greater productivity than the surrounding uplands (Perla and Stevens, in press). Southwestern stream-riparian ecosystems also serve as invasion corridors for non-native species (Stevens and Ayers 2002), elevating the need for appropriate management policy and action (Simberloff et al. 2005).

Functioning riparian ecosystems are critical for maintaining surface water quality and limiting erosion and flood impacts (Gregory et al. 1991; Naiman and Decamps 1997; Karr 1999). Widespread flow regulation and introduction of non-native species in stream-riparian ecosystems has resulted in the endangerment and loss of native fish and wildlife and their habitats (Naiman et al. 1995, 2002; Stanford et al. 1996; Poff et al. 1997). Estimates of riparian habitat loss in the Southwest range from 40 to 90 percent (Dahl 1990), and desert riparian habitats are considered to be one of this region's most endangered ecosystems (Minckley and Brown 1994; Noss et al. 1995). Livestock grazing is a common, widespread land management practice in riparian habitats, and reviews of grazing have consistently detected significant impacts (e.g., Kauffman and Krueger 1986; Trimble and Mendel 1995; Jones 2000; Brock and Green 2003). Even low levels of grazing may reduce soil quality, litter cover, and rodent diversity, while moderate to high levels of grazing reduce plant biomass and recovery (Fleischner 1994; Belsky et al. 1999; Dale et al. 2000; Jones 2000; Holechek et al. 2001; Curtin 2002; Starr 2002; Asner et al. 2004). Despite the many, complex impacts on ecosystem integrity, stream-riparian systems are highly resilient, often responding positively and quickly to restoration actions (Naiman et al. 1995; Phillips 1998).

Given the ecological importance and widespread destruction of stream-riparian ecosystems, rapid ecological assessment methods are needed to allow land managers to evaluate existing conditions, detect at-risk components, prioritize management and restoration activities, and monitor changing conditions. Assessment protocols should include consideration of the association of stream, fluvial wetland, and riparian habitats, which we term the stream-riparian habitat complex, as well as adjacent upstream and upslope uplands habitats. A wide array of methods have been proposed and tested in the United States (U. S. Department of the Interior 1991; Karr 1991, 1999; Fore and Karr 1996), in Australia (Ladson et al. 1999; Jansen and Robertson 2001), and elsewhere (e.g., Manel et al. 2000), but these methods need to be reviewed and refined for application to the ecological and sociological context in the American Southwest (National Research Council 2002; Stromberg et al. 2004). Here we review existing methods and describe an improved protocol for rapid ecological assessment of Southwest ecosystems.

Riparian restoration has become a central focus for both public and private sectors in developed nations (National Research Council 2002). Federal agencies in the United States, including the Bureau of Land Management (BLM), the Bureau of Reclamation, the Forest Service (USFS), and the National Park Service, as well as the Department of Defense and Native American tribes, manage more than 90 percent of the land in the Southwest. Federal agencies are required to manage their lands for long-term ecological sustainability (Federal Land Policy and Management Act of 1976). The BLM, which manages most middle- and low-elevation rangelands in the Southwest primarily for grazing, has a stated agency goal to "1) restore and maintain riparian-wetland areas so that > 75% are in proper functioning condition (PFC) by 1997 and 2) to achieve an advanced ecological status, except where resource management objectives, including the PFC, would require an earlier successional stage, thus providing the widest variety of habitat diversity for wildlife, fish, and watershed protection" (USDI 1991:1). Other state and federal agencies, and many private organizations and individuals, have established similar goals for their riparian lands. Standardized, efficient assessment protocols are needed to evaluate ecosystem conditions, monitor the results of management programs, and establish restoration priorities for stream-riparian reaches. If designed well, such protocols should also identify the specific ecosystem components most at risk and help prioritize management sites and activities.

STREAM-RIPARIAN ASSESSMENT

Numerous protocols have been developed to directly or indirectly assess ecosystem health and the condition of stream-riparian and non-riverine wetland ecosystems (e.g., U.S. Fish and Wildlife Service 1980; National Research Council 1992, 1994; Brinson 1993; Davis and Simone 1995; Mageau et al. 1995; Society for Range Management 1995; McCammon et al. 1998; Ladson et al. 1999; Manel et al. 2000; National Research Council 2002; Rapport et al. 2003; Stromberg et al. 2004). Stromberg et al.'s (2004) review revealed four groups of stream-riparian assessment protocols based on (1) the extent of hydrological alteration (Richter et al. 1996; Poff et al. 1997); (2) land cover and water use classification (e.g., Barton and Taylor 1985); (3) biological indicators (e.g., vegetation, avifauna, endangered species; Karr 1991, 1999); and (4) multi-variable approaches, including hydrogeomorphic (HGM) modeling (e.g., Brinson 1993; King et al. 2000). Most assessment protocols focus narrowly on single components or processes (e.g., just riparian vegetation, aquatic habitat, geomorphology; Winward 2000); flow regime patterns (i.e., aquatic habitat quality, particularly for native fish; Duff 1996; Jowett 1997); channel morphology (Rosgen 1996); or water quality and aquatic invertebrate community composition (Merritt and Cummins 1996). In contrast, HGM involves analysis of multiple variables for modeling wetland function. Although many of these protocols have helped guide land management, their usefulness has been compromised by their limited scope, or because they are time-consuming or rely on subjective judgment of ecosystem condition, because of uncertainty over how the focal variables relate to ecosystem health, or because they have not been adapted for use in the Southwest.

One of the most widely used rapid assessment methods for riparian habitat evaluation is the BLM's proper functioning condition assessment (PFCA). Developed by the U.S. National Riparian Service Team for the BLM (USDI 1991), the PFCA emphasizes stream geomorphic function, but qualitatively includes other variables, and provides a method to determine the current status of the study reach (USDI 1992, 1993, 1994, 1998). This methodology has been adopted by the USFS and other federal agencies charged with managing riparian habitats (Winward 2000). It represents an improvement over previous protocols because it is standardized, can be applied to geologically different areas by non-specialists, can be conducted efficiently, and can be used to trigger corrective adaptive management responses when the reach under study is determined to be either nonfunctioning or "functioning at risk with a downward trend" (USDI 1993). However, several features of the protocol have limited its usefulness, both as a scientific tool for evaluation and monitoring and also in making land-management recommendations. Scoring of variables is qualitative ("yes/no" or "not applicable") and open to interpretation. The PFC protocol does not address exotic species, so a riparian habitat that is dominated by non-native species may receive a "properly functioning" rating. The PFCA does not describe what score produces a particular functional rating for the study reach: this determination is left to the survey team. In addition, the assumption of the PFCA is that if geomorphic and vegetative features are properly functioning, succession will lead to the proper function of all other ecosystem components as well (USDI 1993:11). Since the BLM's mandate is to manage for wildlife habitat (USDI 1993:47; USDI 1995), dependence on untested hypotheses about successional trajectories limits the usefulness of this protocol.

A RAPID ASSESSMENT PROTOCOL FOR STREAM-RIPARIAN ECOSYSTEMS

We present an improved method for rapid stream-riparian assessment (RSRA), measuring the functioning condition of these habitats (Table 1; Wild Utah Project 2005: Appendices A–C). We sought to incorporate the best methods from the above protocols with our own research and experience in the Southwest, emphasizing the need for simplicity and efficiency (requiring < 3 hr). We tested the RSRA over two field seasons in the Escalante (Deer Creek, The Gulch, Harris Wash), Colorado, and Paria River basins in southern Utah, and our protocol has been independently reviewed by more than 20 academic scientists, agency personnel, environmental activists, and concerned citizens. The RSRA was designed for use in low to middle elevations in the Southwest, where strong soil moisture, soil texture, and vegetation gradients predominate across riparian-upland boundaries; however, it can be easily modified for most other temperate regions. We recommend several general and specific considerations.

1. Include measurement of all important ecosystem components, including the occurrence of non-native species and human impacts. Assessment protocols involve tradeoffs between detailed measurement and efficiency. Measuring multiple variables in several categories provides a broad depiction of current ecological conditions and indicates priorities for management attention. We include resource or process categories and variables that, in addition to ecological status, involve governmental regulations

Table 1. RSRA checklist categories and variables. (F) refers to measurements or assessments made in the field; (X) refers to indicators that are scored based on data compiled in the office. The field checklist form, scoring criteria, and scoring instructions can be found on the Wild Utah Project Web site at http://www.wildutahproject.org/Templates/submenu%20PFS.dwt.

Category and Variable	Justification, Impacts	Caveats	References
Water Quality (WQ) Qualifier: Interruption of perennial flow	A fundamental ecosystem component: dewatering or flow augmentation affecting ecosystem characteristics.	Historical hydrographic data may not be available.	Stanford et al. 1996; Poff et al. 1997
WQ 1. Algal growth (F)	Dense algal growth may indicate ecological stability, nutrient enrichment, and reduced WQ; may affect invertebrates.	Algal growth can be normal in low-flow SW streams in mid-summer.	Kirk 1983; Stevens et al. 1997b
WQ 2. Base flow turbidity (F)	Fine sediment increases with erosion augmented by livestock; water clarity is reduced by organic and inorganic suspended load; affects aquatic production and bacteria loading.	Southwestern streams often carry natural, remarkably high suspended sediment loads.	Kirk 1983; Graf et al. 1991
WQ 3. Channel shading (F)	Exposure affects stream productivity and temperature. Decreased streambank vegetation cover, increased channel width, and reduced stream depth increase exposure; fisheries impacts.	Dense overstory vegetation may negatively affect autochthonous stream productivity by shading.	Vannote et al. 1980
WQ 4. Water quality (X)	WQ impacts affect aquatic biota and productivity.	Desert streams, particularly calcium-rich systems, may naturally have low WQ; however, detailed WQ data may not be available.	Karr 1999; Asner et al. 2004
Hydrology/Geomorphology (H/G) Qualifier: Flow interruption	Flow affects aquatic and terrestrial riparian habitat amount, integrity and function (??).	See WQ Qualifier.	Richards 1987; Stromberg 1993b; Rosgen 1996; Jowett 1997
H/G 1. Sinuosity (X, F)	Channel configuration is affected by gradient, discharge, and sediment transport interactions, and may become altered by flow regulation, vegetation colonization, and direct human manipulations. Anomalously straight alluvial channels may indicate reduced flow or other anthropogenic modifications.	Sinuosity may be limited in constrained reaches.	Leopold 1994; Rosgen 1996

Table 1 (continued)

Category and Variable	Justification, Impacts	Caveats	References
H/G 2. Hydrograph (X)	Modification of flow timing, duration, frequency, magnitude, and ramping rate by more than about one third may likely influence riparian community structure, either increasing or decreasing riparian vegetation cover and composition in the lower riparian zone.	Flow stabilization or augmentation can greatly increase aquatic production, riparian vegetation, and wildlife habitat.	Stromberg and Patten 1992; Richter et al. 1996; Stevens et al. 1995, 1997a, b; Nilsson and Svedmark 2002
H/G 3. Floodplain inundation frequency (X, F)	Interrupted overbank flooding, or flooding at unusual times of the year, reduces nutrient availability, germination rates of native phreatophytes, and growth and survivorship of vegetation, and can alter plant species composition and wildlife habitat quality.	Flow stabilization or augmentation can greatly increase aquatic production, riparian vegetation, and wildlife habitat.	Johnson 1994; Stevens et al. 1995, 1997a, b
H/G 4. Fine sediment on streambed (F)	Fine sediment cover on the stream floor reduces benthic primary and secondary production and standing mass, and negatively affects fish spawning. It may result from bank erosion, loss of vegetation and erosion of the uplands, and/or upstream channel modification.	Fine sediment is a natural feature of many southwestern streams.	Kirk 1983; Graf et al. 1991; Melis et al. 1996
H/G 5. Bank vertical stability (F)	Oversteepened or vertical cutbanks dominate many southwestern streams, limiting physical dynamics of aquatic ecosystems, preventing flooding, decreasing water tables, in turn leading to the loss of riparian vegetation and invasion by upland species. Also, they may limit wildlife access to water.	Distinguish between climate changes and anthropogenic impacts.	Richards 1987; Beard 2004
H/G 6. Channel lateral stability (F)	Lateral bank stability should vary along the channel, but is augmented by livestock trampling, removal & loss of vegetation, human recreational use, roadways, & other anthropogenic impacts.	Distinguish between climate changes and anthropogenic impacts.	Richards 1987; Rosgen 1996
H/G 7. Hydraulic habitat diversity (F)	Fish diversity & population health is commonly related to habitat diversity: oxbows, side channels, sand bars, gravel/cobble bars, riffles, pools, islands, cut banks, terraces.	Geomorphically constrained streams are unlikely to support high diversity of aquatic habitats.	USDA 1985, 1991

Table 1 (continued)

Category and Variable	Justification, Impacts	Caveats	References
H/G 8. Riparian soil integrity and moisture regime (F)	Riparian and wetland soils reflect existing flow dynamics, management, and vegetation, and affect potential vegetation dynamics and wildlife habitat distribution and quality. Soil moisture regime strongly affects riparian ecosystem productivity and processes, such as germination and herbaceous growth.	Riparian soils vary substantially across the flood frequency gradient, from poorly developed, highly disturbed entisols along stream banks to well-developed mollisols, usually on less-frequently flooded terraces. Constrained bedrock channels may not provide soils or capillary movement of soil moisture.	Mitsch and Gosselink 1993; Stevens et al. 1995; Brock and Green 2003
H/G 9. Beaver dam density and condition (F)	Beavers are keystone species because they modify geomorphology and vegetation, reduce variance in water flows, and provide important fish and plant nursery habitat.	Beaver may not have been present historically in all stream reaches in the Southwest.	Hill 1982
Fish/Aquatic Habitat (F/AH) Qualifier: Interrupted perennial flow	Perennial flow is a fundamental aquatic habitat characteristic, and interruption eliminates many aquatic species.	See WQ Qualifier.	Minckley 1991; Stanford et al. 1996
F/AH 1. Pool distribution (F)	The number, size, distribution, and quality of pools, and pool:riffle ratios are fish habitat quality indicators.	Geomorphically constrained streams may not support high density of pools.	USDA 1985, 1991
F/AH 2a. Underbank cover (F)	Underbank cover is an important component of good fish habitat, used for resting and protection from predators. This cover usually occurs with vigorous vegetative riparian growth, dense root masses, and stable soil conditions.	Geomorphically constrained or ephemeral streams may not support much underbank cover.	USDA 1985; Lloyd 1985
F/AH 2b. Overbank cover (F)	Overhanging terrestrial vegetation is essential for fish production and survival, providing shade, bank protection from high flows, sediment filtering, and organic matter, including insects.	Geomorphically constrained or ephemeral streams may not support much overbank cover.	Lloyd 1985; USDA 1985, 1992
F/AH 3. Channel floor embeddedness (F)	Low levels of embeddedness increase benthic productivity and fish production. The filling of interstitial gravel and boulder spaces with silt, sand and organic material reduces habitat suitability for feeding, cover and spawning (egg to fry survival) by limiting space and macroinvertebrate production.	Many streams in the Southwest have an elevated level of calcium carbonate that precipitates travertine and seals interstitial substrata; therefore, high embeddedness is a natural phenomenon in many streams.	Lloyd 1985; USDA 1985, 1992; Valdez and Ryel 1997; Melis et al. 1996

Table 1 (continued)

Category and Variable	Justification, Impacts	Caveats	References
F/AH 4. Large woody debris (F)	The amount, composition, distribution and condition of large woody debris (LWD) in the channel and on the banks provides important fish habitat for nursery cover, feeding, and cover, and contributes channel stability, nutrients, and food production. Interruption of LWD transport is widely regarded as an ecologically important habitat impact.	Geomorphically constrained or ephemeral streams may not support much vegetation or LWD.	Minckley and Rinne 1985; USDA 1985, 1991, 1992
F/AH 5. Abundance and diversity of aquatic invertebrates (F, X)	The density and composition of aquatic invertebrates are strong indicators of stream functioning. Macroinvertebrate production provides fish food and increases survival, and natural conditions and anthropogenic impacts affect invertebrate habitat quality.	Geomorphically constrained, highly turbid, highly mineralized, or ephemeral streams may naturally support low levels of invertebrate life.	USDA 1985, 1992; Karr 1991; Reice and Wohlenberg 1993; Rosenberg & Resh 1996; Oberlin et al. 1999; Kennedy et al. 2000; Bunn & Arthington 2002
F/AH 6. Terrestrial insect habitat (F)	Terrestrial invertebrate production and input (drop) into the stream are an important food source for fish. The amount of overhead terrestrial vegetation complexity and canopy associated with or in close proximity to the streambank are important factors contributing to high densities and diversity of insect drop.	Geomorphically constrained, highly turbid, or ephemeral streams may not support much riparian vegetation or invertebrate life.	USDA 1985, 1992
F/AH 7a. Native fish and other aquatic faunal populations (X)	The composition and abundance of the native fish community is a strong indicator of a stream's habitat diversity, ecological functionality, and geomorphic consistency.	Geomorphically constrained, highly turbid, and ephemeral streams naturally may not support much native fish or other aquatic life; nor may heavily grazed streams.	Lloyd 1985; USDA 1985; Minckley 1991
F/AH 7b. Non-native fish and other aquatic faunal populations (X)	Non-native fish species are one of the most important threats to native fish assemblages, as predators, competitors, and agents of disease transmission. Management activities may be focused on non-native or sport fishes, but their habitat requirements should not be emphasized to the detriment of the habitat, occupancy, and population needs of the native species.	Geomorphically constrained, highly turbid, and ephemeral streams may not support much non-native fish or other aquatic life. Non-native fish may be a preferred management strategy for the reach.	USDA 1985; Minckley 1991; Stevens et al. 2001

Table 1 (continued)

Category and Variable	Justification, Impacts	Caveats	References
F/AH 8. Habitat availability for aquatic species of special concern (X&F)	Many aquatic species of concern in the western U.S. depend on well-functioning stream ecosystems. Ecosystem management can provide adequate resources and habitat to maintain healthy populations of such species, and to help promote the recovery of extirpated and declining species.	Geomorphically constrained, highly turbid, and ephemeral streams may not support sensitive species or their habitat.	U.S. Fish and Wildlife Service 1998; Stevens et al. 2001
Riparian Vegetation (RV) 1. Composition (F)	Composition of native riparian vegetation provides direct and secondary food, protection, and breeding habitat cover for wildlife and livestock. Riparian plant composition affects avian breeding and foraging patterns.	The riparian zone along geomorphically constrained, highly turbid, and ephemeral streams naturally may not support riparian vegetation.	Carothers et al. 1974; Whitmore 1975; Stevens et al. 1977; Brown & Trossett 1989; Stacey 1995
RV 2. Riparian vegetation structure (F)	A variety of different zones, or heights, of cover, especially ground and shrub cover, may reduce stream energy during floods and promote deposition. Diverse vegetation structure is associated with nesting and migrant bird and other vertebrate associations.	Geologically constrained streams may not naturally support sufficient vegetation to influence stream energy dissipation. Log jams may serve a similar purpose.	Gregory et al. 1991; Auble et al. 1994
RV 3. Dominant native shrub/woody tree species demography (F)	The distribution of size classes of native dominant species indicates recruitment success, ecosystem sustainability, and wildlife habitat availability.	Highly episodic recruitment, particularly in constrained channels at lower eletions, may limit size class representation.	Anderson & Ruffner 1988; Stromberg and Patten 1992; Stevens et al. 2001
RV 4. Non-native plant cover (F)	Strong dominance by non-native plants may eliminate key attributes of wildlife habitat quality, and may limit livestock use. Non-native plant species profoundly influence ecosystem structure, productivity, habitat quality, and processes (e.g., fire frequency, intensity).	Exotic species are ubiquitous in riparian habitats, and low levels of non-native cover may not presently much affect ecological functioning.	Noble 1989; Stevens 1989; Lonsdale 1999; Stevens and Ayers 2002
RV 5. Large woody debris (LWD) production (F)	Large woody debris contributes to riparian soils, and fish and wildlife habitat. Mid-canopy and gallery forest trees contribute to LWD production. Interruption of LWD transport is an important habitat impact on regulated streams.	Constrained bedrock channels may not support tree growth or LWD production.	USDA 1985, 1992; Minckley and Rinne 1985; USDA 1991; see also F/AH 5, above

Table 1 (continued)

Category and Variable	Justification, Impacts	Caveats	References
RV 6a. Mammalian herbivory impacts on ground cover (F)	Ungulate, lagomorph, and rodent (e.g., beaver) herbivores can affect riparian soils, ground cover, and general ecosystem condition. Utilization levels >15% in riparian zones, especially in Rosgen Type C channels, such as meadows, retard vegetation recovery. A moderate level of grazing almost always increases soil compaction.	Some grazing may increase grass and forb production by opening the canopy and increasing site disturbance. Such effects may simulate flood effects in regulated streams.	Fleishner 1994; Belsky et al. 1999; Holechek et al. 2001
RV 6b. Mammalian herbivory impacts on shrub and low canopy cover (F)	Ungulate, lagomorph, and rodent (e.g., beaver) herbivores can affect riparian soils, ground cover, and general ecosystem condition. Utilization levels of 50% or higher generally preclude recovery of shrub and other woody perennial vegetation.	See RV 6a.	See RV 6a.
RV 7. Riparian plant vigor (F)	Vigorous perennial plant growth is typical of alluvial riparian ecosystems where ground water availability is suitable.	Constrained bedrock channels may not provide soils or capillary movement of soil moisture. Plant growth also may be restricted by dewatering, immersion in reaches with augmented flow, and seasonality.	Stevens & Waring 1985; Stromberg and Patten 1992
Wildlife Habitat (WH) 1. Habitat for terrestrial sensitive species (X&F)	Many sensitive terrestrial wildlife species in arid regions depend on riparian ecosystems at some point of their life cycle. A common focus of ecosystem management is to provide food and habitat resources to maintain self-sustaining wildlife populations, including listed species.	Constrained bedrock channels may not provide wildlife habitat.	Endangered Species Act of 1972, as amended; various state legislations; Stevens et al. 1977; U.S. Fish and Wildlife Service 1998; Stevens and Ayers 2002
WH 2a. Riparian shrub patch density (F)	Riparian shrub vegetation often provides food, thermal cover, and nesting or breeding habitat for terrestrial wildlife, including many invertebrates, amphibians, reptiles, birds, and mammals. Such vegetation, both native and exotic, also plays a key role in sediment deposition during periods of over-bank flow.	See WH 1.	See WH 1.
WH 2b. Middle and upper canopy patch density (F)	A high cover of middle canopy trees provides essential habitat for many insects, as well as nesting and foraging sites for birds, bats, and other small mammals.	See WH 1.	See WH 1.

Table 1 (continued)

Category and Variable	Justification, Impacts	Caveats	References
WH 3a. Shrub patch connectivity (F)	Natural alluvial streams in the Southwest often support large, continuous patches of shrub cover. Shrub connectivity provides movement corridors, shade, and improved habitat for invertebrates, herpetofauna, birds, and a various mammals, and limits erosion.	See WH 1.	See WH 1.
WH 3b. Middle-upper canopy patch connectivity (F)	Natural alluvial streams in the Southwest often support large patches of middle and upper canopy overstory. Canopy connectivity provides movement corridors and improved habitat for wildlife, and also shades the understory and channel.	See WH 1.	See WH 1.
WH 4. Fluvial habitat diversity (F)	Natural alluvial streams create a diversity of fluvial landforms for terrestrial wildlife, including terraces, bars, and fluvial marshes (F/AH 1 and 8 above). Conversely, in a highly degraded system with erosion and downcutting, there may be only a single fluvial form—steep and relatively straight banks without vegetation.	Constrained channels are likely to provide a lower frequency and diversity of channel landforms for use as wildlife habitat.	Stacey 1995; Stevens et al. 1997b
WH 5. Wildlife habitat distribution (X, F)	The distribution of fluvial habitats (WH 4) affects species occurrence, density, and gene flow. If critical habitat patches are rare and widely separated, wildlife populations using those habitats may be threatened by isolation.	See WH 4.	Stacey 1995; Stevens and Huber 2004
Human Impacts/Activities (HI) 1. Naturalness of the hydrograph (X, F)	Southwestern streams are subject to many human impacts, including flow alteration, channel geometry, and water quality, grazing, mining, road construction, logging/wood-cutting or burning, chemical treatment, point and non-point source pollution, introduction of non-native species, including competitors, predators and disease organisms, hunting, extirpation of associated species, urbanization, and interactions between these impacts.	The impacts of flow alteration and other human activities are generally poorly known. Basin land use and flow regulation history are often poorly known, making it difficult to distinguish among direct and indirect effects.	Johnson et al. 1985; Naiman et al. 1995; Stanford et al. 1996

Table 1 (continued)

Category and Variable	Justification, Impacts	Caveats	References
HI 2. Condition of uplands (X,F)	Upland conditions may strongly affect stream-riparian hydrology and functionality, particularly vegetative cover and growth. Loss of vegetation may increase erosion, sediment loading in the stream, and the dynamics of flooding.	Uplands in which the parent materials are shale or siltstone are likely to have high erosion rates, high sediment loading, and strong flood impacts, and naturally support little riparian vegetation.	Ellison 1960; USDI 1997; Graf et al. 1991; Stevens et al. 2001
HI 3. Livestock management (X, F)	The presence of livestock in riparian areas can negatively affect ecosystem integrity. Livestock can reduce water quality; trampling reduces bank stability and soil quality; grazing can reduce vegetation complexity, diversity, cover and resilience; and the presence of livestock may influence invertebrate and native wildlife distribution.	Grazing on non-federal lands may not be managed by plan. Stocking rates or site potential may be overestimated or not known.	Fleischner 1994; Belsky et al. 1999; Jones 2000; Holechek 2001
HI4. Development impacts (X, F)	Many human activities affect floodplain integrity, including road, roadworks, building, campground, parking lot, and trail construction; fencing; agricultural development, with fertilizer and pesticide residues; mining; and urbanization. Structures, such as bridges and dams, may block high flows or debris movement and may trigger scouring erosion at the base of the structure.	Adaptive management requires prioritization of values, and stream-riparian ecosystem integrity is often superceded by project construction objectives.	Holling 1978; National Research Council 1986; Gunderson et al. 1995; Naiman et al. 1995; Stanford et al. 1996; Poff et al. 1997; Stromberg et al. 2004
HI 5. Naturalness of channel geomorphology (F)	Water works projects such as impoundment, diversion, channelization, and ground water extraction may alter channel geomorphology. Stream channels adjust to these impacts, in turn affecting other physical and biotic ecosystem properties.	Determination of historic channel geomorphology requires at minimum, historical photographs, and flow and sediment history—information that is often not available for a specific reach. The regional hydrogeological syntheses may provide insight here.	See HI 4.

Table 1 (continued)

Category and Variable	Justification, Impacts	Caveats	References
HI 6a. Extent of road impacts (X, F)	Roads often severely affect stream-riparian ecosystem integrity through impacts on channels, hydrologic processes, riparian vegetation, and nutrient transport. The areal extent of road construction impacts on the floodplain and across the reach, and road use classes (i.e., dirt tracks to fenced state highways or freeways) strongly affects ecosystem functioning.	Stream-riparian ecosystem integrity is often superceded by road construction project objectives.	Forman and Alexander 1998; Burroughs & King 1989
HI 6b. Proximity of road impacts (X, F)	See HI 6a. Proximity of roads of different use classes on floodplains affects stream-riparian ecosystem integrity. Roads also serve as invasion routes for non-native species, and can interrupt or eliminate wildlife movement.	See HI 6b.	Forman and Alexander 1998; Stevens and Ayers 2002
Trend Assessment (X, F)	Stream-riparian ecosystems are characterized by high natural variability, and monitoring is needed to establish the range of system conditions. Trend assessment monitoring requires repeated visits using the same protocols. Trends in the above 6 categories can be established to provide information on variability and triggers to management actions.	Funding availability, changing administrative interests, changing staff and protocols, lack of data management, and other contingencies all work against trend monitoring, and require careful planning.	Holling 1978; National Research Council 1986; Gunderson et al. 1995; Stevens and Gold 2002

(e.g., water quality, native fish and wildlife habitat, listed species; National Rangeland Health Standards, USDI 1995). We also include factors that influence riparian ecosystem function (e.g., non-native plant species distribution, the intensity of mammalian herbivory, human activities). Within each category, we have selected variables that both reflect critical ecosystem processes and inform managers of risk and restoration priorities (e.g., age class structure of dominant trees). Non-native species are included because they threaten riparian ecosystems through impacts on composition, function, habitat availability, productivity, and fire regimes (Mack et al. 2000; Tellman 2002).

2. Base assessment on current rather than predicted future condition. Assessment of ecosystem function should reflect current conditions only, not projected future states or trends (USDI 1993). Riparian succession is characterized as being "perpetual"—frequently reset by unpredictable flood disturbances (Campbell and Green 1968; Nakamura et al. 2000)—limiting the predictability of future conditions. A stream-riparian ecosystem may be considered to be properly functioning with respect to fluvial geomorphology and the ability to dissipate the energy of flood events, but unless sufficient cover and diversity of vegetation exists, that system may not be functional for terrestrial wildlife, including listed species. Such a finding may trigger management action if wildlife habitat is a desired function of that reach.

3. Use quantitative measurements to score variables and rate ecosystem condition. Scoring based on dichotomous categories (functional/non-functional, yes/no) provides insufficient insight into process, and may be difficult to interpret when applied at different times, in different systems, or by different individuals. We used our literature review and experience to create a 5-point scoring system for each variable (Wild Utah Project 2005: Appendix B), with a low score of 1 indicating impaired ecological function and a high score of 5 indicating fully functioning condition, appropriate to its natural geomorphic and fluvial characteristics (see below). A score of N/A is given when the variable is not applicable to the study reach. The 5-point scale provides reasoned value judgments, with sufficient detail to inform managers. A similar ranking is now being used for uplands rangeland assessment (Pellant et al. 2000).

To determine a quantitative measure for each category and for the overall functional condition of the reach, the scores in each category (Water Quality, Hydrology/Geomorphology, Fish/Aquatic Habitat, Riparian Vegetation, and Wildlife Habitat) are first averaged, and those means (excluding the human impacts score) are combined into an average overall score. An overall score of < 2 likely indicates a non-functioning (NF, using the BLM's PFCA terminology) condition, a score of about 2 to < 4 suggests a functioning-at-risk (FAR) condition, and a score of 4 to 5 indicates nearly proper to proper functioning condition (PFC). Meta-analyses of data derived through this protocol may eventually support the use of statistics other than the arithmetic mean to determine the overall score.

Although a single composite site score is desirable for judging site condition and developing regional restoration priorities, we recognize that a grand mean of mean category scores should not constitute the final interpretation of site status. For example, a reach may be functioning well physically but may be biologically degraded, in which case the need for restoration action depends on the management goals for that reach. Alternatively, a reach's hydrology and streamflow patterns may be highly altered but may appear outwardly to be properly functioning, in which case the overall score may be intermediate when it is, in fact, FAR. Such effects are buffered in the RSRA because our system is somewhat weighted toward biological variables. Nonetheless, interpretation of reach conditions should involve an analysis of the overall scores against the mean category scores to improve understanding of ecological function.

Scaling methods are inherently arbitrary, and others may prefer different scoring schemes. Our data protocols permit, and we encourage, future innovative statistical analyses. Initial field tests of the protocol revealed that expert observers from diverse backgrounds generally gave very similar (usually within one point) scores for indicators within a study reach. Regardless of the scoring method used, quantitative and explicitly defined protocols executed by experts reduce the subjectivity of ecosystem assessment. Quantitative methods also allow the RSRA to be used as a research and monitoring tool, permitting comparison of ecological conditions within a reach over time. Scores can be used to evaluate trends, management effectiveness, and interrelationships among variable scores over time, and can provide a foundation for adaptive management.

4. Consider the geomorphic context. Geomorphic characteristics vary dramatically among different reaches and watersheds, particularly in geologically diverse regions. Channels may be sand- or gravel/cobble-floored, geomorphically constrained by talus or bedrock, or alluvial, with wide floodplains and a meandering course. Therefore, scoring of variables must include consideration of the geomorphic context. We use the phrase "geomorphically consistent" and "geomorphically inconsistent" to incorporate the importance of natural channel processes, geometry, antecedent hydrology, and other abiotic, habitat, and biological characteristics. Relationships between channel geometry and riparian vegetation and biota are not well understood and await as-yet-incomplete analyses of a series of reference sites (see consideration 5, below).

5. Use reference sites to understand the natural range of variation. Southwestern riparian ecosystems have a long history of use by Native Americans and Anglo settlers, making it difficult to understand their natural condition. Therefore, we suggest that the analysis of control reference sites be used to scale RSRA scores. Arrays of such reference sites have not been designated by land management agencies, but such reference sites should be located, described in detail, analyzed using the RSRA protocol, and used to scale survey site scores. Reference sites should (1) be selected by an independent scientific panel; (2) display nearly natural conditions and be as free as possible from recent anthropogenic disturbance, especially livestock grazing, water diversion, mining, roads, and ground water pumping; and (3) occur across elevations and a diversity of landform geology, channel gradient, and vegetative community types in the region. Reference site data should be used to develop a regional model of stream-riparian habitat structure and characteristics since there will be inherent variability among reference sites. Reference sites provide essential insight into the range of natural ecological variability, and should be used to train RSRA practitioners. It is important to note that reference sites are likely to reveal shortcomings of the RSRA and biases about cause and effect relationships and human impacts (Brinson and Rheinhardt 1996; Hruby 2001).

Riparian reference sites may be rare (e.g., on BLM land in southern Utah), making their identification all the more important; however, they may exist in state and national parks and wilderness areas. Numerous virtually pristine tributaries of the Colorado River exist in Grand Canyon, Zion, and other nearby national parks, in recreation areas, and in wilderness areas. In New Mexico, relatively unmodified sites exist in the Jemez, San Mateo, and Gila Mountains, and lowland sites occur along the Gila River basin. If no suitable sites exist in a region, the independent science panel may recommend reaches that are functioning reasonably well as surrogate reference sites. Identification and measurement of a network of reference sites will require a new level of cooperation among land management agencies, but will be of great value to stream and riparian research and conservation.

6. Use local and regional background information to help interpret current condition. The RSRA protocol is designed to evaluate current ecosystem functioning, and assessment will be greatly enhanced if the regional cultural, hydrological, and biologi-

cal context is understood. We recommend that the following three regional syntheses of information be compiled in the office before field assessments are conducted.

Land Use History: Antecedent disturbances and site history play a large role in the present condition of any ecosystem and are needed to understand changing land use over time (White 1979; Webb et al. 1991). This regional synthesis should include land-use history, administrative context, legal constraints, economic resources distribution, demographic and economic trends, livestock stocking rates and seasons of use, and road construction history.

Hydrogeological Context: Fluvial geomorphology and flow history are important drivers of riparian ecosystem development: streams with intermittent flow from natural variation or human diversions have vastly different biological characteristics as compared to perennial reaches. Because riparian ecosystem dynamics vary among stream types, and because adequate stream classification systems exist (Rosgen 1996), the existing regional stream type and flow history data should be synthesized. This synthesis should emphasize mean and current annual hydrography, the extent of flow regulation, major flow events, sediment history, and changes in water quality. Ancillary information on ground water supplies, well data, ground and surface water quality, soils, and other physical factors is desirable.

Biological Context: Properly functioning riparian ecosystems are expected to support healthy, self-sustaining populations of regionally appropriate native fish and wildlife species, including endemic, sensitive, threatened or endangered, and key, keystone, and engineering taxa (Gregory et al. 1991; Power et al. 1996; Stanford et al. 1996); however, detection of individual species is unlikely during the brief site visit, and a synthesis is needed to provide necessary background data. This regional biological synthesis will help clarify species distributions and summarize their status. Through the synthesis, information on endangered species surveys and research may be summarized from the U.S. Fish and Wildlife Service, at least on a regional basis. Also, fisheries data may be obtained from state wildlife agencies, interviews with local experts, federal instream habitat surveys (PHABSIM and/or IFIM), the U.S. Forest Service, and BLM general aquatic wildlife surveys (GAWS). These latter two agencies use the biotic condition index (Karr 1991). Lastly, the role of beaver or other ecosystem engineering species may be established from the historical land use, hydrological, and biological syntheses, as beaver impacts may vary substantially between regulated and unregulated streams (Breck et al. 2003).

7. Pay special attention to RSRA team composition. The effectiveness of the RSRA is predicated on the expertise and unbiased scientific attitudes of the assessment team, the members of which should be experts who are familiar with the RSRA techniques, interpretation of geomorphic consistency, site history, and regional background data. We found that a team consisting of a hydrogeomorphologist, a stream/fisheries biologist, and a riparian ecologist was able to conduct an RSRA site visit in less than 3 hours.

8. Establish good data management, archival, and publication protocols. An RSRA program requires sound data management and archival protocols, including rigorous metadata standards. Interpretation of regional patterns and scientific defensibility of management decisions requires ease of access to historical as well as recent data. We recommend that the data collected be electronically compiled, managed for ease of review and comparison with future site visits, and periodically published in peer-reviewed scientific journals. This regimen will help ensure the credibility of the process and feedback into management.

RSRA RESOURCE CATEGORIES AND INDIVIDUAL VARIABLES

We selected a suite of variables from our review of the literature, existing protocols, and our professional experience that are important for accurately determining stream-riparian ecosystem condition and function. We grouped these variables into categories, and describe and justify their use in Table 1 and the Wild Utah Project (2005: Appendices A–C). Several factors strongly modify scoring of these variables. In all cases, the desired condition of the reach is defined by its management criteria: agency mandates, enabling legislation, other legal criteria (e.g., the 1973 Endangered Species Act), management plans, and other administrative decisions that guide management priorities and activities. Geomorphic consistency is a potentially strong modifier for all variables, and its interpretation requires that at least one member of the RSRA team have extensive training in geomorphology. These factors influence interpretation of site conditions and resulting scores.

The human impacts category score is not included in the overall total score because activities such as water works, highway, or campground construction typically involve multiple management entities with diverse mandates. Sound stream-riparian land management in such circumstances is best coordinated through adaptive management (Holling 1978; National Research Council 1986; Gunderson et al. 1995).

One further caveat is that neither this protocol nor others proposed for stream or riparian assessment are appropriate for assessment of springs, lakes, or reservoirs. The geomorphic context, disturbance regime, vegetation structure, and many other ecosystem characteristics of those systems are substantially different from those of stream-riparian ecosystems (Hutchinson 1957–1993; Thornton et al. 1990; van der Kamp 1995; Stevens and Meretsky, in press) and require separate assessment protocols.

TESTING RSRA

We tested the proposed assessment protocol in several drainages in southern Utah during 2000–2001 to evaluate effectiveness, efficiency, and comparability with the BLM's PFCA protocol (Table 2). Our assessment team included several of the authors of this paper: a hydrologist (CG), a stream ecologist (DD), a riparian ecologist (LS), and a wildlife habitat ecologist (PS). Four of the riparian sites we visited were in Grand Staircase–Escalante National Monument—The Gulch, Deer Creek, Harris Wash, and Cottonwood Creek. We also visited Indian Creek, a tributary to the Colorado River just east of Canyonlands National Park. Each visit involved about 3 hours, depending on site access, and team efficiency improved over the course of the assessments. Assessment of two sites per day is likely to be normal for a team, particularly if debriefing and scoring is conducted on-site. Although our team had only limited access to background information on flow, land-use history, and sensitive species (we recommend above that these be compiled prior to field visits), the team members agreed that the RSRA checklist provided adequate detail for reasoned scoring of the variables and the site.

The RSRA documented a much broader array of ecosystem condition issues and more detailed, quantitative information on site conditions than did the BLM's PFCA approach (Table 2). We placed all PFCA questions in RSRA categories for comparison, and found that the BLM approach did not include consideration of water quality, fish or wildlife habitat, or non-native species presence and role. The RSRA provided information on a wide array of human impacts on the site, whereas the primary PFCA human impacts variable was the condition of the uplands. We found striking differences in the final site scores between the two approaches. The PFCA assessments conducted by the BLM concluded that four of the five sites were in proper functioning condition, whereas the RSRA indicated that four of the five sites were rated as functioning at risk and in need of management attention. For example, although not noted in the BLM's PFCA account, The Gulch stream is geomorphically constrained to an old roadway, a condition that strongly affects its channel geometry, bed composition, and vegetation. There, as elsewhere, lower RSRA scores were derived primarily from lower scoring in the hydrogeomorphological and vegetation-related categories, and also from low wildlife habitat scores. Human impacts were not included in the overall site scores, but from this small sample size the human impact scores appear to be positively

Table 2. Comparison of the BLM's PFCA with RSRA results for five study reaches in southern Utah, in summer of 2000 and 2001. TG = The Gulch, DC = Deer Creek, HW = Harris Wash, CW = Cottonwood Creek, and IC = Indian Creek. Some RSRA scores represent the average of composite estimates, and are therefore depicted as decimals to the tenth place.

	RSRA Results				
RSRA Criteria	TG	DC	HW	CW	IC
Water Quality					
WQ Qualifier: Interrupted flow?	no?	part?	no	no	part?
WQ1. Excessive algal growth	4	4	n/a	4	unkn
WQ2. Turbidity	5	5	n/a	4	unkn
WQ3. Channel shading	3	4	2	2	2
WQ4. Water quality	x	x	x	x	x
Section Mean	4	4.3	2.0	3.3	2.0
Hydrology/Geomorphololgy					
HG Qualifier: Interrupted flow?	no?	part?	no	no	part?
HG1. Sinuosity	2	2	2	2	n/a
HG2. Hydrograph resembles the natural flow pattern	n/a	n/a	n/a	n/a	n/a
HG3. Floodplain inundation	3	4	3	3	3
HG4. Fine sediment deposition	4	5	3	3	4
HG5. Vertical bank stability	3	5	4	3	3
HG6. Lateral channel stability	3	5	2	3	n/a
HG7. Hydraulic habitat diversity	2	4	2	3	4
HG8. Riparian soil integrity	4	5	2	5	n/a
HG9. Beaver dam density	1	1	1	n/a	1
Section Mean	2.8	3.9	2.4	3.1	3.0
Fish/Aquatic Habitat					
F/AH Qualifier. Interrupted flow?	no?	part?	no	no	part?
F/AH1. Pool distribution	2	3	2	3	3
F/AH2a. Underbank cover	2	3	1	2	2
F/AH2b. Overbank cover	2	4	2	2	2
F/AH3. Embeddedness	2	4	2	4	3
F/AH54 Large woody debris (LWD) cover	2	3	2	4	1
F/AH5. Aquatic invertebrate distribution	3	4	n/a	3	n/a
F/AH6. Terrestrial insect habitat	3	4	2	4	3
F/AH7a. Native fish distribution	unkn	unkn	unkn	unkn	unkn
F/AH7b. Non-native fish distribution	unkn	unkn	unkn	unkn	unkn
F/AH8. Aquatic sensitive species habitat availability	unkn	unkn	unkn	unkn	unkn
Section Mean	2.3	3.6	1.8	3.1	2.3
Riparian Vegetation					
RV1. Native riparian vegetation composition	2.8	3.9	2.2	4.2	2.2
RV2. Riparian vegetation structure to dissipate energy, support wildlife	2	2.5	2.5	3	2
RV3. Plant demography	2	3	3	3	3
RV4. Non-native plant cover	3	3.5	2.5	3	2.5
RV5. Potential LWD production	2	5	3	3	2
RV6a. Mammal impacts on soil/ground cover	4	4	1	2	2
RV6b. Mammal impacts on native browse cover	5	4	1	2	2
Addressed through HG 8.	x	x	x	x	x
RV7. Riparian plant vigor	4	4.5	2.5	4	x
(Not addressed—difficult to determine in the field)	x	x	x	x	x
Addressed through RV1	x	x	x	x	x
Section Mean	3.1	3.8	2.2	3.0	2.2

Table 2 (continued)

RSRA Criteria	RSRA Results				
	TG	DC	HW	CW	IC
Wildlife Habitat					
WH1. Terrestrial habitat quality for sensitive species	2	3	2	2.5	x
WH2a. Riparian shrub patch density	4	5	3	2	1
WH2b. Mid and upper canopy patch density	3	5	3	2	1
WH3a. Riparian shrub patch connectivity	2	4	2	2	2
WH3b. Middle-upper canopy patch connectivity	2	4	2	2	2
WH4. Fluvial landform diversity	2	4	3	3	4
WH 5. Wildlife habitat distribution	3	4	2	3	x
Section Mean	2.6	4.1	2.4	2.4	2.0
Human Impacts / Activities					
HI1. Naturalness of hydrograph	unkn	unkn	unkn	unkn	unkn
HI2. Upland watershed integrity	2	unkn	2	n/a	x
HI3. Livestock grazing within prescription	unkn	5	unkn	unkn	x
HI4. Extent of development	4	2	4	4	x
HI5. Naturalness of channel geomorphology	2	3	2.4	3	x
HI6a. Extent of road impacts	2	3	2	3	x
HI6b. Proximity of road impacts	2	2	2	3	x
Section Mean	2.4	3.0	2.5	3.3	x
FINAL RSRA RATING (average score for all sections except HI)	2.9	3.9	2.2	3.0	2.3
Trend: Upward, static or downward*					
T1. Upward, static or downward	n/a	n/a	n/a	n/a	n/a
T2. Geomorphic improvement	n/a	n/a	n/a	n/a	n/a
T3. Fish/aquatic habitat improvement	n/a	n/a	n/a	n/a	n/a
T4. Vegetation improvement	n/a	n/a	n/a	n/a	n/a
T5. Wildlife habitat improvement	n/a	n/a	n/a	n/a	n/a
Section Mean	n/a	n/a	n/a	n/a	n/a

* To be answerable only after repeated visits

Table 2 (continued)

Comparison of the BLM's PFCA with RSRA results. TG = The Gulch, DC = Deer Creek, HW = Harris Wash, CW = Cottonwood Creek, and IC = Indian Creek.

BLM Original PFCA Criteria

		PFC Results				
Topic	Checklist Questions	TG	DC	HW	CW	IC
Hydro/Erosion	Sinuosity, w/d ratio, gradient GC	yes	yes	yes	yes	yes
Hydrology	Floodplain inundated relatively frequently	yes	yes	yes	yes	yes
Erosion/Deps'n	No excessive erosion/deposition	yes	yes	yes	yes	yes
Erosion/Deps'n	System is vertically stable	yes	yes	yes	yes	yes
Hydro/Erosion	Lateral stream movement associated with sinuosity	n/a	yes	yes	yes	yes
Hydrology	Active/stable beaver dams	no	no	no	n/a	no
Veg/Erosion	Diverse composition, point bar recolonization	yes	n/a	no	yes	yes
Hydro/Veg/Erosion	Adequate cover to protect banks/dissipate energy	no	yes	no	no	yes
Veg/Erosion	Diverse age structure	yes	yes	yes	yes	yes
Vegetation	Vegetation provides adequate source of LWD	yes	yes	no	n/a	no
Vegetation	Species present indicate riparian soil moisture	yes	yes	yes	yes	yes
Vegetation	Riparian plants exhibit high vigor	yes	yes	yes	yes	yes
Vegetation	Root masses capable of withstanding high flows	no	yes	yes	yes	yes
Erosion/Dep	Point bars are revegetating	n/a	n/a	yes	yes	yes
Hydrology	Upland watershed not contributing to degradation	yes	yes	yes	yes	no
FINAL PFCA RATING						
Trend		unkn	unkn	up	up	unkn

related to RSRA scores. From these analyses, we consider the RSRA to provide an efficient and far more comprehensive and quantitative approach to assessment of stream-riparian ecosystem conditions compared to the existing PRCA approach.

BEYOND ASSESSMENT

Active, scientific land management involves a suite of activities, including inventory, assessment, management planning, implementation, and monitoring, all requiring consistent administration, policy, protocol, and rigorous information management. Here we focused primarily on inventory and assessment because reliable information on stream-riparian ecosystem status is needed by land managers to meet legal and long-term environmental management objectives. The RSRA protocol appears to function well in the Southwest, and may be applicable elsewhere. We recommend quantification of data collection, with defined scaling values for each variable and overall condition, and we encourage the use of reference sites to ground judgments about ecosystem condition. Such information will set the stage for appropriate planning and action. However, it will mean little unless it is brought forth in the context of rigorous, societally motivated adaptive management. Adaptive management programs involve stakeholder-driven definitions of vision, mission, scope, goals, objectives, and implementation triggers, as well as support for monitoring, research, and feedback (Holling 1978; Stevens and Gold 2002). The assessment protocol proposed here will improve the scientific credibility of the assessment process, but we encourage federal and local managers to engage in more rigorous discussion with citizens about societal needs and responsibilities in the adaptive management of these productive, diverse, and threatened ecosystems.

ACKNOWLEDGMENTS

We thank the Southern Utah Landscape Restoration Project for their help with this project. We thank the following for their assistance reviewing and suggesting useful changes to this manuscript: P. Abate, P. Bengeyfield, L. Ellicott, W. Flemming, P. Holden, M. Miller, W. Platts, J. Stromberg, J. Tuhy, R. Valdez, C. van Riper III, and 11 anonymous reviewers. We particularly thank T. Fleischner and various members of the National Riparian Service Team for their thoughtful comments on this research. We also thank J. Gibbon and K. Burke for editorial assistance.

LITERATURE CITED

Anderson, L. S., and G. A. Ruffner. 1988. Effects of the post-Glen Canyon Dam flow regime on the old high water zone plant community along the Colorado River in Grand Canyon. National Technical Information Service PB88-183504/AS.

Asner, G. P., A. J. Elmore, L. P. Olander, R. E. Martin, and A. T. Harris. 2004. Grazing systems, ecosystem responses, and global change. Annual Review of Environmental Resources 29: 11.1–11.39.

Auble, G. T., J. M. Friedman, and M. L. Scott. 1994. Relating riparian vegetation to present and future streamflows. Ecological Applications 4: 544–554.

Barton, D. R., and W. D. Tayor. 1985. Dimensions of riparian buffer strips required to maintain trout habitat in southern Ontario streams. North American Journal of Fisheries Management 5: 364–378.

Beard, R. 2004. Stream channel change in response to cattle exclosures in semi-arid central Arizona. Journal of the Arizona-Nevada Academy of Science 36: 81–87.

Belsky, A. J., A. Matake, and S. Uselman. 1999. Survey of livestock influences on stream and riparian ecosystems in the western United States. Journal of Soil and Water Conservation 54: 419–431.

Breck, S. W., K. R. Wilson, and D. C. Anderson. 2003. Beaver herbivory of willow under two flow regimes: A comparative study on the Green and Yampa Rivers. Western North American Naturalist 63: 463–471.

Brinson, M. M. 1993. A hydrogeomorphic classification for wetlands. U.S. Army Corps of Engineers, Waterways Experiment Station Technical Report WRP-DE-4. Vicksburg MS.

Brinson, M. M., and R. D. Rheinhardt. 1996. The role of reference wetlands in functional assessment and mitigation. Ecological Applications 16: 69–76.

Brock, J. H., and D. M. Green. 2003. Impacts of livestock grazing, mining, recreation, roads, and other land uses on watershed resources. Journal of the Arizona-Nevada Academy of Science 35: 11–22.

Brown, B. T., and M. W. Trossett. 1989. Nesting-habitat relationships of riparian birds along the Colorado River in Grand Canyon, Arizona. The Southwestern Naturalist 34: 260–270.

Bunn, S. E., and A. H. Arthington. 2002. Basic principles and ecological consequences of altered flow regimes for aquatic biodiversity. Environmental Management 30: 492–507.

Burroughs, E. R., Jr., and J. G. King. 1989. Reduction of soil erosion on forest roads. USDA Forest Service Gen. Technical Report INT-264.

Campbell, C. J., and W. Green. 1968. Perpetual succession of stream-channel vegetation in a semiarid region. Journal of the Arizona Academy of Sciences 5: 86–98.

Carothers, S. W., R. R. Johnson, and S. W. Aitchison. 1974. Population structure and social organization of Southwestern riparian birds. American Zoologist 14: 97–108.

Chambers, J. C., and J. R. Miller, editors. 2004. Great Basin riparian ecosystems: Ecology, management, and restoration. Island Press, Washington D.C.

Curtin, C. G. 2002. Livestock grazing, rest, and restoration in arid landscapes. Conservation Biology 16: 840–842.

Dahl, T. E. 1990. Wetland losses in the United States 1780's to 1980's. U.S. Dept. of the Interior Fish and Wildlife Service, Washington D.C.

Dale, V. H., S. Brown, R. A. Haeuber, N. T. Hobbs, N. Huntly, R. J. Naiman, W. E. Riebsame, M. G. Turner, and T. J. Valone. 2000. Ecological principles and guidelines for managing the use of land. Ecological Applications 10: 639–670.

Davis, W., and T. Simone, editors. 1995. Biological assessment and criteria: Tools for water resource planning and decision making. Lewis Publishers, Boca Raton FL.

Duff, D. A. 1996. Conservation assessment for inland cutthroat trout status and distribution. USDA Forest Service, Intermountain Region, Ogden UT.

Ellison, L. 1960. Influence of grazing on plant succession of rangelands. Botanical Review 26: 1–78.

Fleischner, T. L. 1994. Ecological costs of livestock grazing in western North America. Conservation Biology 8: 629–644.

Fore, L. S., and J. R. Karr. 1996. Assessing invertebrate responses to human activities: Evaluating alternative approaches. Journal of the North American Benthological Society 15: 212–231.

Forman, R. T. T., and L. E. Alexander. 1998. Roads and their major ecological effects. Annual Review of Ecology and Systematics 29: 207–231.

Graf, J. B., R. H. Webb, and R. Hereford. 1991. Relation of sediment load and flood-plain formation to climatic variability, Paria River drainage basin, Utah and Arizona. Geological Society of America Bulletin 103: 1405–1415.

Gregory, S. V., F. J. Swanson, W. A. McKee, and K. W. Cummins. 1991. An ecosystem perspective of riparian zones. Bioscience 41: 540–551.

Gunderson, L. H., C. S. Holling, and S. S. Light. 1995. Barriers and bridges to learning in a turbulent human ecology. In Barriers and Bridges to the Renewal of Ecosystems and Institutions, edited by L. H. Gunderson, C. S. Holling, and S. S. Light, pp. 461–488. Columbia University Press, New York.

Hill, E. P. 1982. Beaver. In Wild Mammals of North America: Biology, Management, and Economics, edited by J. A. Chapman and G. A. Feldhamer, pp. 256–281. Johns Hopkins University Press, Baltimore MD.

Holechek, J. L., R. D. Pieper, and C. H. Herbel. 2001. Range Management: Principles and Practices. 4th ed. Prentice Hall, Upper Saddle River NJ.

Holling, C. S., editor. 1978. Adaptive environmental assessment and management. John Wiley & Sons, London.

Hruby, T. 2001. Testing the basic assumption of the hydrogeomorphic approach to assessing wetland functions. Environmental Management 27: 749–761.

Hupp, C. R. 1988. Plant ecological aspects of flood geomorphology and paleoflood history. In Flood Geomorphology, edited by V. R. Baker, pp. 335–357. John Wiley & Sons, New York.

Hutchinson, G. E. 1957–1993. A Treatise on Limnology, vols. I–IV. John Wiley & Sons, New York.

Jansen, A., and A. I. Robertson. 2001. Relationships between livestock management and the ecological condition of riparian habitats along an Australian floodplain river. Journal of Applied Ecology 38: 63–75.

Johnson, R. R. 1991. Historic changes in vegetation along the Colorado River in the Grand Canyon. In Colorado River Ecology and Dam Management, edited by the National Research Council Editorial Board, pp. 178–206. National Academy Press, Washington D.C.

Johnson, W. C. 1994. Woodland expansion in the Platte River, Nebraska: Patterns and causes. Ecological Monographs 64: 45–84.

Johnson, R. R., and D. A. Jones, editors. 1977. Importance, preservation and management of riparian habitat: A symposium. USDA Forest Service General Technical Report RM-43.

Johnson, R. R., C. D. Ziebell, D. R. Patton, P. F. Ffolliott, and R. H. Hamre. 1985. Riparian ecosystems and their management: Reconciling conflicting uses. Proceedings of the First North American Riparian Conference. U.S. Forest Service General Technical Report RM-120. Washington D.C.

Jones, A. J. 2000. Effects of cattle grazing on North American arid ecosystems: A quantitative review. Western North American Naturalist 60: 155–164.

Jowett, I. G. 1997. Instream flow methods: A comparison of approaches. Regulated Rivers: Research & Management 13: 115–127.

Karr, J. R. 1991. Biological integrity: A long-neglected aspect of water resource management. Ecological Applications 1: 66–84.

Karr, J. R. 1999. Defining and measuring river health. Freshwater Biology 41: 221–234.

Kauffman, J. B., and W. C. Krueger. 1986. Livestock impacts on riparian ecosystems and streamside management implications: A review. Journal of Range Management 37: 430–437.

Kennedy, T. B., A. M. Merenlender, and G. L. Vinyard. 2000. A comparison of riparian condition and aquatic invertebrate community indices in central Nevada. Western North American Naturalist 60: 255–272.

King, D. M., L. A. Wainger, C. C. Bartoldus, and J. S. Wakeley. 2000. Expanding wetland assessment procedures: Linking indices of wetland function with services and values. U.S. Army Engineer Research and Development Center Technical Report ERDC/EL TR-00-17. Vicksburg MS.

Kirk, J. T. O. 1983. Light and photosynthesis in aquatic ecosystems. Cambridge University Press, Cambridge.

Knopf, F. L., R. R. Johnson, T. Rich, F. B. Samson, and R. C. Szaro. 1988. Conservation of riparian ecosystems in the United States. Wilson Bulletin 100: 272–284.

Leopold, L. B. 1994. A View of the River. Harvard Press, Cambridge MA.

Ladson, A. R., L. J. White, J. A. Doolan, B. L. Finlayson, B. T. Hart, S. Lake, and J. W. Tilleard. 1999. Development and testing of an index of stream condition for waterway management in Australia. Freshwater Biology 41: 453–468.

Lloyd, J. R. 1985. COWFISH: A guide for estimating the effects of current livestock management on fish production: A FHR habitat capability model. USDA Forest Service, Northern Region, Missoula MT.

Lonsdale, W. M. 1999. Global patterns of plant invasions and the concept of invasibility. Ecology 80: 1522–1536.

Mack, R. N., D. Simberloff, W. M. Lonsdale, H. Evans, M. Clout, and F. Bazzaz. 2000. Biotic invasions: Causes, epidemiology, global consequences and control. Ecological Applications 10: 689–710.

Mageau, M. T., R. Costanza, and R. E. Ulanowicz. 1995. The development, testing and application of a quantitative assessment of ecosystem health. Ecosystem Health 1: 201–213.

Malanson, G. P. 1993. Riparian Landscapes. Cambridge University Press, New York.

Manel, S., S. T. Buckton, and S. J. Ormerod. 2000. Testing large-scale hypotheses using surveys: The effects of land use on the habitats, invertebrates and birds of Himalayan rivers. Journal of Applied Ecology 37: 756–770.

McCammon, B., J. Rector, and K. Gebhardt. 1998. A framework for analyzing the hydrologic condition of watersheds. U.S. Department of the Interior Bureau of Land Management Report BLM/RS/ST-98/004+7210.

Melis, T. S., W. M. Phillips, R. H. Webb, and D. J. Bills. 1996. When the blue-green waters turn red. U.S. Geological Survey Water-Resources Investigations Report 96-4059.

Merritt, R. W., and K. W. Cummins, editors. 1996. Aquatic Insects of North America. 3rd ed. Kendall/Hunt, Dubuque IA.

Minckley, W. C. 1991. Native fishes of the Grand Canyon region: An obituary? Colorado River Ecology and Dam Management, edited by the National Research Council Editorial Board, pp. 124–177. National Academy Press, Washington D.C.

Minckley, W. L., and D. E. Brown. 1994. Wetlands. Biotic Communities: Southwestern United States and Northwestern Mexico, edited by D. E. Brown, pp. 223–287. University of Utah Press, Salt Lake City.

Minckley, W. L., and J. N. Rinne. 1985. Large organic debris in southwestern streams: An historical review. Desert Plants 7: 142–153.

Mitsch, W. J., and J. G. Gosselink. 1993. Wetlands. 2nd ed. Van Nostrand Reinhold, New York.

Nakamura, F., F. J. Swanson, and S. M. Wondzell. 2000. Disturbance regimes of stream and riparian systems—a disturbance-cascade perspective. Hydrological Processes 14: 2849–2860.

Naiman, R. J., and H. Decamps. 1997. The ecology of interfaces: Riparian zones. Annual Review of Ecology and Systematics 28: 621–658.

Naiman, R. J., J. J. Magnuson, D. M. McKnight, and J. A. Stanford. 1995. The freshwater imperative: A research agenda. Island Press, Washington D.C.

Naiman, R. J., S. E. Bunn, C. Nilsson, G. E. Petts, G. Pinay, and L. C. Thompson. 2002. Legitimizing fluvial ecosystems as users of water: An overview. Environmental Management 30: 455–467.

National Research Council. 1986. Ecological Knowledge and Environmental Problem-Solving: Concepts and Case Studies. National Academy Press, Washington D.C.

National Research Council. 1992. Restoration of Aquatic Ecosystems. National Academy Press, Washington D.C.

National Research Council. 1994. Rangeland Health: New Ways to Classify, Inventory and Monitor Rangelands. National Academy Press, Washington D.C.

National Research Council. 2002. Riparian Areas: Functions and Strategies for Management. National Academy Press, Washington D.C.

Nilsson, C., and M. Svedmark. 2002. Basic principles and ecological consequences of changing water regimes: Riparian plant communities. Environmental Management 30: 468–480.

Noble, I. R. 1989. Attributes of invaders and the invading process: Terrestrial vascular plants. In Biological Invasions: A Global Perspective, edited by J. A. Drake et al., pp. 301–313. John Wiley & Sons, New York.

Noss, R. F., E. T. LaRoe III, and S. J. Michael. 1995. Endangered ecosystems of the United States: A preliminary assessment of loss and degradation. Biological Report 28. National Biological Survey, Washington D.C.

Oberlin, G. E., J. P. Shannon, and D. W. Blinn. 1999. Watershed influence on the macroinvertebrate fauna of ten major tributaries of the Colorado River through Grand Canyon, Arizona. The Southwestern Naturalist 44: 17–30.

Ohmart, R. D., B. W. Anderson, and W. C. Hunter. 1988. The ecology of the lower Colorado River from Davis Dam to the Mexico–United States international border: A community profile. USDI Fish and Wildlife Service Biological Report 85 (7.19).

Perla, B. S., and L. E. Stevens. In press. Biodiversity and productivity at an undisturbed spring, in comparison with adjacent grazed riparian and uplands habitats. In Every Last Drop: Ecology and Conservation of North American Aridland Springs Ecosystems, edited by L. E. Stevens and V. J. Meretsky. University of Arizona Press, Tucson.

Pellant, M., D. A. Pyke, P. Shaver, and J. E. Herrick. 2000. Interpreting indicators of rangeland health, version 3. U.S. Bureau of Land Management National Science and Technology Center Technical Reference 1734-6, Denver.

Phillips, F. 1998. The 'Ahakhav Tribal Preserve. Restoration & Management Notes 16: 140–148.

Poff, N. L., K. D. Allan, M. B. Bain, J. R. Karr, K. L. Prestegaard, B. D. Richter, R. E. Sparks, and J. C. Stromberg. 1997. The natural flow regime: A paradigm for river conservation and restoration. BioScience 47: 769–784.

Power, M. E., D. Tilman, J. A. Estes, B. A. Menge, W. J. Bond, L. S. Mills, D. Gretchen, J. C. Castilla, J. Lubchenco, and R. T. Paine. 1996. Challenges in the quest for keystones. BioScience 46: 609–620.

Rapport, D. J., N. O. Nielsen, B. L. Lasley, D. E. Rolston, C. O. Qualset, and A. B. Damania, editors. 2003. Managing for Healthy Ecosystems. Lewis Publishers, Boca Raton FL.

Reice, S. R., and M. Wohlenberg. 1993. Monitoring freshwater benthic macroinvertebrates and benthic processes: Measures for assessment of ecosystem health. In Freshwater Biomonitoring and Benthic Macroinvertebrates, edited by D. M. Rosenberg and V. H. Resh, pp. 287–305. Chapman and Hall, New York.

Richards, K., editor. 1987. River Channel Environment and Process. Basil Blackwell, New York.

Richter, B. D., J. V. Baumgartner, J. Powell, and D. P. Braun. 1996. A method for assessing hydrologic alteration within ecosystems. Conservation Biology 10: 1163–1174.

Rosenberg, D. M., and V. H. Resh. 1996. Use of aquatic insects in biomonitoring. In Aquatic Insects of North America, 3rd ed., edited by R. W. Merritt and K. W. Cummins, pp. 87–97. Kendall/Hunt, Dubuque IA.

Rosgen, D. L. 1996. Applied River Morphology. Wildlands Hydrology, Pagosa Springs CO.

Sabo, J. L., R. Sponseller, M. Dixon, K. Gade, T. Harms, J. Heffernan, A. Jani, G. Katz, C. Soykan, J. Watts, and J. Welter. 2005. Riparian zones increase regional species richness by harboring different, not more, species. Ecology 86: 56–62.

Simberloff, D., I. M. Parker, and P. N. Windle. 2005. Introduced species policy, management, and future research needs. Frontiers in Ecology and the Environment 3: 12–20.

Society for Range Management. 1995. New concepts for assessment of rangeland condition. Journal of Range Management 48: 271–283.

Stacey, P. B. 1995. Biodiversity of rangeland bird populations. In Biodiversity of Rangelands, edited by N. West, pp. 33–41. Utah State University Press, Logan.

Stanford, J. A., J. V. Ward, W. J. Liss, C. A. Frissell, R. N. Williams, J. A. Lichatowich, and C. C. Coutant. 1996. A general protocol for restoration of regulated rivers. Regulated Rivers: Research & Management 12: 391–413.

Starr, D. A. 2002. Livestock exclosure research in the western United States: A critique and some recommendations. Environmental Management 30: 516–526.

Stevens, L. E. 1989. The status of ecological research on tamarisk (Tamaricaceae: *Tamarix ramosissima*) in Arizona. In Tamarisk Control in Southwestern United States, edited by M. R. Kunzman, R. R. Johnson, and P. S. Bennett, pp. 99–105. Cooperative National Park Resources Study Unit Special Report Number 9, Tucson AZ.

Stevens, L. E., and T. J. Ayers. 2002. The biodiversity and distribution of alien vascular plant and animals in the Grand Canyon region. In Invasive Exotic Species in the Sonoran Region, edited by B. Tellman, pp. 241–265. University of Arizona Press, Tucson.

Stevens, L. E., and B. D. Gold. 2002. A long-term monitoring program for adaptive management of the Colorado River ecosystem in Glen and Grand Canyons. In Interdisciplinary Approaches for Evaluating Ecoregional Initiatives, edited by D. Busch and J. Trexler, pp. 101–134. Island Press, Washington D.C.

Stevens, L. E., and R. L. Huber. 2004. Biogeography of tiger beetles (Cicindelidae) in the Grand Canyon Ecoregion, Arizona and Utah. Cicindela 35: 41–64.

Stevens, L. E., and V. J. Meretsky, editors. In press. Every Last Drop: Ecology and Conservation of North American Aridland Springs Ecosystems. University of Arizona Press, Tucson.

Stevens, L. E., and G. L. Waring. 1985. The effects of prolonged flooding on the riparian plant community in Grand Canyon. In Riparian Ecosystems and their Management: Reconciling Conflicting Uses, edited by R. R. Johnson et al., pp. 81–86. USDA Forest Service General Technical Report RM-120, Washington D.C.

Stevens, L.E., B. T. Brown, J. M. Simpson, and R. R. Johnson. 1977. The importance of riparian habitats to migrating birds. Proceedings of a Symposium on the Importance, Preservation and Management of Riparian Habitats, edited by R. R. Johnson and D. Jones, pp. 156–164. USDA National Forest Service General Technical Report RM-43, Ft. Collins CO.

Stevens, L. E., J. S. Schmidt, T. J. Ayers, and B. T. Brown. 1995. Geomorphic influences on fluvial marsh development along the dam-regulated Colorado River in the Grand Canyon, Arizona. Ecological Applications 5: 1035–1039.

Stevens, L. E., K. A. Buck, B. T. Brown, and N. C. Kline. 1997a. Dam and geomorphological influences on Colorado River waterbird distribution, Grand Canyon, Arizona, USA. Regulated Rivers: Research & Management 13: 151–169.

Stevens, L. E., J. P. Shannon, and D. W. Blinn. 1997b. Colorado River benthic ecology in Grand Canyon, Arizona, USA: Dam, tributary and geomorphological influences. Regulated Rivers: Research & Management 13: 129–149.

Stevens, L. E., T. J. Ayers, J. B. Bennett, K. Christensen, M. J. C. Kearsley, V. J. Meretsky, A. M. Phillips III, R. A. Parnell, J. Spence, M. K. Sogge, A. E. Springer, and D. L. Wegner. 2001. Planned flooding and riparian trade-offs along the regulated Colorado River in Glen and Grand Canyons, Arizona. Ecological Applications 11: 701–710.

Stromberg, J. C. 1993a. Fremont cottonwood–Goodding willow riparian forests: A review of their ecology, threats, and recovery potential. Journal of the Arizona-Nevada Academy of Sciences 27: 97–110.

Stromberg, J. C. 1993b. Instream flow models for mixed deciduous riparian vegetation within a semiarid region. Regulated Rivers: Research & Management 8: 225–235.

Stromberg, J. C., and D. T. Patten. 1992. Mortality and age of black cottonwood stands along diverted and undiverted streams in the eastern Sierra Nevada, California. Madroño 39: 205–223.

Stromberg, J., M. Briggs, C. Gourley, M. Scott, and P. Shafroth. 2004. Riparian ecosystem assessments. In Riparian Areas of the Southwestern United States: Hydrology, Ecology, and Management, edited by M. B. Baker, Jr., P. F. Ffolliott, L. DeBano, and D. G. Neary, pp. 315–332. Lewis Publishers, Boca Raton FL.

Tellman, B., editor. 2002. Invasive Exotic Species in the Sonoran Region. University of Arizona Press, Tucson.

Thornton, K. W., B. L. Kimmel, and F. E. Payne. 1990. Reservoir Limnology: Ecological Perspectives. John Wiley & Sons, New York.

Trimble, S. W., and A. C. Mendel. 1995. The cow as a geomorphic agent—a critical review. Geomorphology 13: 233–253.

USDA Forest Service. 1985. Fisheries Surveys Handbook. R4 FSH 2609.23. USDA Forest Service, Intermountain Region, Ogden UT.

USDA Forest Service. 1991. Fisheries and Watershed Field Manual. Nez Perce National Forest, Grangeville ID.

USDA Forest Service. 1992. Integrated Riparian Guide. USDA Forest Service, Intermountain Region, Ogden UT.

U.S. Department of the Interior. 1991. Riparian-Wetland Initiative for the 1990's. Bureau of Land Management, Washington D.C.

U.S. Department of the Interior. 1992. Riparian-Wetland Area Management. U.S. Department of Interior Bureau of Land Management Technical Reference 1737, Washington D.C.

U.S. Department of the Interior. 1993. Riparian Area Management: Process for Assessing Proper Functioning Condition. Bureau of Land Management Technical Reference 1737-9, Denver CO.

U.S. Department of the Interior. 1994. Riparian Area Management: Process for Assessing Proper Functioning Condition for Lentic Riparian-Wetlands Areas. Bureau of Land Management Technical Reference 1737-11, Denver CO.

U.S. Department of the Interior. 1995. Rangeland Health: Standards and Guidelines for Healthy Rangelands. Washington D.C.

U.S. Department of the Interior. 1997. Standards and Guidelines for Healthy Rangelands. Utah State Office of the U.S. Bureau of Land Management, Salt Lake City.

U.S. Department of the Interior. 1998. Riparian Area Management: A User Guide to Assessing Proper Functioning Condition and the Supporting Science for Lotic Areas. Bureau of Land Management Technical Reference 1737-15, Denver CO.

U.S. Fish and Wildlife Service. 1980. Habitat Evaluation Procedures. Ecological Services Manual 102, Washington D.C.

U.S. Fish and Wildlife Service. 1998. Endangered and Threatened Wildlife and Plants: 50 CFR 17.11 and 17.12 (The Red Book). U.S. Fish and Wildlife Service, Washington D.C.

Valdez, R. A., and R. J. Ryel. 1997. Life history and ecology of the humpback chub in the Colorado River in Grand Canyon, Arizona. In Proceedings of the Third Biennial Conference of Research on the Colorado Plateau, edited by C. van Riper III and E. T. Deshler, pp. 3–31. National Park Service Trans. Proc. Serv. PS/NRNAU/NRTP 97/12.

van der Kamp, G. 1995. The hydrogeology of springs in relation to the biodiversity of spring fauna: A review. Journal of the Kansas Entomological Society 68 (supplement): 4–17.

Vannote, R. L., G. L. Minshall, K. W. Cummins, J. R. Sedell, and C. E. Cushing. 1980. The river continuum concept. Canadian Journal of Fisheries and Aquatic Sciences 37: 370–377.

Webb, R. H., S. S. Smith, and V. A. S. McCord. 1991. Historic channel change of Kanab Creek, southern Utah and northern Arizona. Grand Canyon Natural History Association, Grand Canyon AZ.

White, P. S. 1979. Pattern, process, and natural disturbance in vegetation. Botanical Review 45: 229–299.

Whitmore, R. C. 1975. Habitat ordination of passerine birds of the Virgin River Valley, southwestern Utah. Wilson Bulletin 87: 65–74.

Wild Utah Project. 2005. Rapid stream-riparian assessment appendices: Field checklist, scoring criteria, scoring instructions, and illustrations. Available online at http://www.wildutah project.org/Templates/submenu%20PFS.dwt.

Winward, A. H. 2000. Monitoring the vegetation resources in riparian areas. U.S. National Forest Service, Rocky Mountain Research Station General Technical Report RMRS-FTR-47, Ogden UT.

DETERMINING WATERSHED BOUNDARIES AND AREA USING GPS, DEMS, AND TRADITIONAL METHODS: A COMPARISON

Boris Poff, Duncan Leao, Aregai Tecle, and Daniel G. Neary

Defining a watershed's boundary is critical for understanding the movement of water across the landscape. In the past, hydrologists defined watersheds using topographic map features such as contour lines, stream networks, and traditional surveys. However, the more advanced current techniques for defining watershed boundaries use Geographic Information Systems (GIS) in conjunction with Digital Elevation Models (DEMs; Maidment 1999). GIS is an important tool in resource management today. Because of its precision and accuracy, we expect the Global Positioning System (GPS) to be more appropriate for delineating watershed boundaries and calculating area than traditional forestry field methods of using a compass and chain. We also expect GPS surveyed boundaries to be more precise than those determined using DEMs, because DEMs rely on 10 m topographic intervals, which can lead to faulty results, especially in areas with small elevation differences.

After "selective availability" was removed from GPS technology in May of 2000, potential point location error without differential correction dropped from 100 to 3 m. According to Oderwald and Boucher (2003), this indicates that a single point observation on average is within approximately 3 m of its true location. In contrast, on a 1:24,000 topographic map, a 0.5 mm pencil line is 12 m wide (Oderwald and Boucher 2003).

In our study, we chose not to apply post-processing differential correction to our data for three reasons. First, users may not have access to base stations to differentially correct their data or may not have the expertise to do so. Second, low-end, consumer-grade GPS units may not offer the option of differential correction. Third, we assume that for areas greater than 20 hectares (ha) in size cumulative error is insignificant due to the randomness distribution of the horizontal error distance (which averages between 3 and 4 m for each point; Wilson 2000).

DEMs are digital representations of surface elevations laid out over the landscape. A DEM is produced from digitized map contours, spot elevations, and hydrography overlays or from manual scanning of aerial photographs (Elassal and Caruso 1983; Maidment and Djokic 2000). DEMs are valuable because they provide managers with the same information as contour maps, but in a digital format suitable for processing by computer-based systems rather than in an analog format (Cho and Lee 2001). The accuracy of DEMs for use in watershed delineation depends mostly on their cell size and resolution. The smaller the cell size the greater the resolution; therefore, smaller watersheds require a smaller cell size to accurately represent the watershed area. According to Maidment and Djokic (2000) for most hydrologic and geomorphic modeling applications 10 m DEMs are sufficient. Currently, the USGS has DEMs with 30, 100, 500, and 1000 m cell sizes available for the United States. A 30 m DEM means that each cell covers an area of 30 m x 30 m (Maidment 1999). Higher resolution DEMs such as 5 and 10 m are more readily available from state governments and the USGS through its partner ATDI (http://www.atdi-us.com; accessed March 2004).

From October 2002 through January 2004, two large, four medium, and three small watersheds were mapped using a hand-held GPS receiver. The two large watersheds—Watershed 20 (Bar M) and Watershed 19 (Woods Canyon)—as well as three small sub-watersheds (85, 86, and 87) within Watershed 19 are located in the Beaver Creek Experimental Watershed in central Arizona (Figure 1). The four medium-sized watersheds—the East and West Forks of Castle Creek and the North and South Forks of Thomas Creek—are in eastern Arizona (Figure 1). The objective of this study was to compare watershed area and boundary determined using a GPS (without post-processing differential correction of data) with watershed parameters determined using traditional methods of area calculation, and with watersheds determined using DEMs in GIS.

STUDY AREAS

The Forest Service had established experimental watersheds in Arizona to determine the amount and sustainability of water yield produced from various silvicultural practices. Twenty pilot watersheds were set up by the USDA Forest Service between 1957 and 1962 in the Beaver Creek Experimental Watershed, which is located within the Coconino National Forest of northern Arizona (Figure 1). Of the 20 watersheds, 18 range in size from 27 to 824 ha. The remaining two watersheds, Bar M (Watershed 20) and Woods Canyon (Watershed 19), are much larger and are located in the ponderosa pine forest ecosystem, encompassing 6620 and 4893 ha, respectively (Baker and Ffolliott 1999). The four mid-sized watersheds are located 14–25 km south of the town of Alpine in the Apache-Sitgreaves National Forest in eastern Arizona (Figure 1). The West and East Fork experimental watersheds of Castle Creek were established in 1955 and are 354 and 458 ha in size, respectively. The East Fork was subject to preharvest prescribed burning (Gottfried and DeBano 1988). The North and South Forks of Thomas Creek, established in the mid 1960s, are 184 and 221 ha in size, respectively (Dietrich 1980; Ffolliott and Gottfried 1991).

METHODS

The original identification and mapping of the study watersheds involved several steps. The boundary of each watershed was roughly delineated on a topographic map to install a flume at the mouth of each watershed. Following flume installation, the boundary of each watershed was flagged, beginning at the lowest elevation and ending at the highest elevation. Upon marking the boundary, in the Beaver Creek watersheds trees were painted at eye level with yellow paint to make the boundary permanent and visible from a substantial distance in any direction. At Thomas Creek, trees were painted with white paint. In the absence of trees, the paint marks were applied to rocks or stumps. Next, surveyors used traditional forestry surveying techniques to map the boundaries. Beginning at the stream gauging equipment, surveyors measured short segments of the boundary using a 2-chain steel tape while correcting for slope using a clinometer. With a compass, the bearing of each segment was taken and plotted on a Reinhardt Redy Mapper (a hard plastic board with imprinted grid and compass bearings). The Castle Creek watersheds were not delineated on the ground.

Field Methodology

In the Beaver Creek Experimental Watershed we used the paint markings made by surveying crews in the late 1950s as guides to locate ridge points, and we followed these when we delineated the watershed boundaries with our GPS receiver. Originally trees had been marked in close proximity to one another, however not all the trees and markings exist today. Some of the marked trees have since died, burned, or been harvested, or their markings have faded away as a result of exposure to sunlight. Wherever we encountered gaps between marked trees, we depended on the topographic features of the

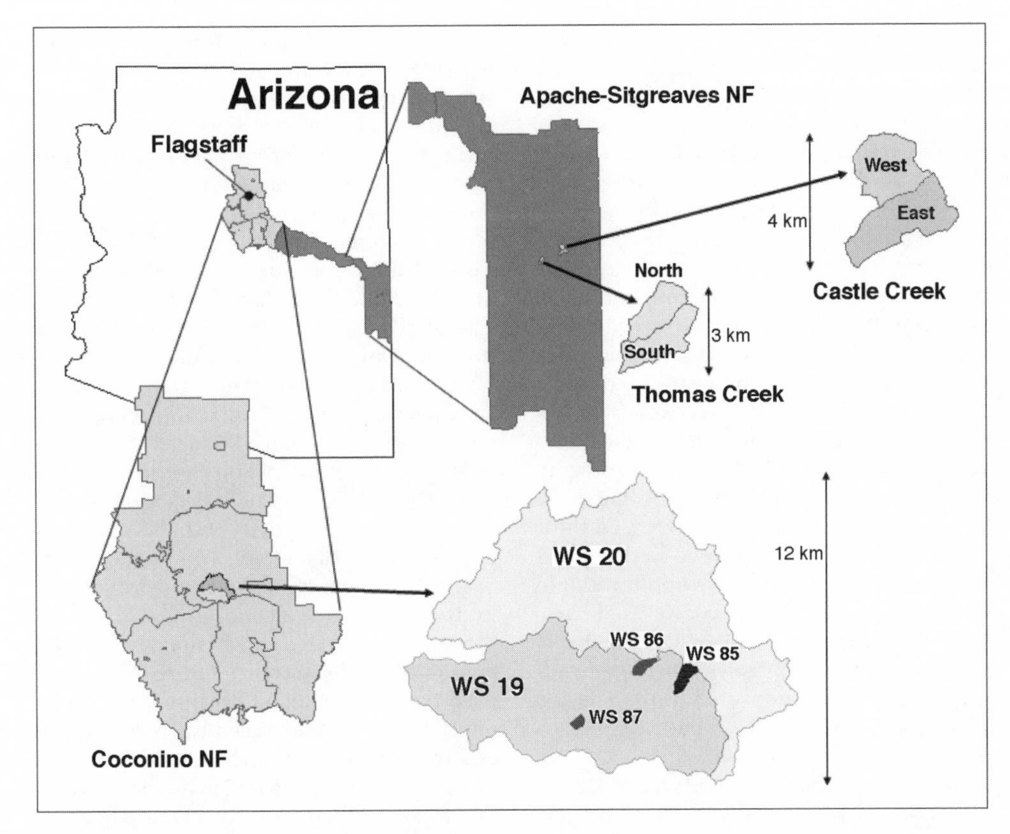

Figure 1. Location of study watersheds within the Coconino and Apache-Sitgreaves National Forests in Arizona.

landscape to delineate the watershed boundary. The longest unmarked boundary line we encountered was approximately 500 m on a well-defined ridgeline. In the Castle Creek watersheds we relied on a distinct ridge as well as a compass and a topographic map with 5-foot contour intervals that circumscribed the watershed boundary. In the Thomas Creek watersheds, we used a combination of topographic features, marked trees, a compass, and a topographic map with 5-foot contours on which the watershed boundary had been marked, to determine the actual watershed boundary on the ground. The beginning and ending location of the survey of each watershed was the respective gauging station of that watershed.

GIS/GPS Methodology

We used a Trimble GeoExplorer3 handheld GPS receiver to record our position along the boundary and to record line data in a kinematic mode. GPS data were recorded in Universal Transverse Mercator (UTM) format and were projected using the North American Datum (NAD) of 1927. The typical distance between each recorded point on the line was about 5 m, but varied up to a maximum of 30 m. The GPS receiver recorded points every few seconds, depending on reception. Reception, in turn, depended on the topographic features and canopy closure (Karsky 2001), which varied from 0 to 100 percent. In rare cases we had to wait several minutes before a point was recorded. The boundaries of small and medium sized watersheds were recorded in a timeframe of one or two days each. The large watershed boundaries, Bar M and Woods Canyon, were recorded over the course of several months. We walked and recorded adjacent sections of the boundary two or three days per week. We then downloaded the line data using Pathfinder Office 2.90 (Trimble 2002) software. We eliminated obvious outliers and artificially created polygons manually in Pathfinder. We viewed the recorded line data files on either the same or the next day to eliminate outlier points. Such outliers would include points that were several meters away from where we walked in a straight line with the GPS. After cleaning up the individual line files, we combined the files for each watershed and exported these data as polyline shapefiles for further processing using a GIS software application. In ArcView GIS 3.3 (ESRI 2003), we used a feature called Xtools to convert the polylines-shapefile into one polygon-shapefile for each watershed. Once imported into ArcMap (ESRI 2004), we determined the area for each respective watershed.

To delineate the watersheds using DEMs, we used 10 m DEMs and the ArcView CRWR Preprocessor (Maidment and Djokic 2000; ESRI 2002). DEMs were obtained from the Arizona Regional Image Archive (ARIA) and projected in UTM NAD 1927. All DEMs were merged using the map calculator to create one complete DEM for analysis. A stream shapefile obtained from the Arizona Land Resource Information Service was burned into the DEM and appropriate steps were followed in the CRWR Preprocessor to fill sinks, compute the flow direction and flow accumulation grids, construct the basic stream network, segment streams into stream links, find link outlets, delineate the watersheds, and vectorize the stream and watershed grids. After the above steps were completed, sub-watersheds were merged to create the specific watersheds of this study.

Determining Discharge Using HEC-HMS

Determining the discharge from all watersheds and their respective areas from each method involved several activities. First, watershed characteristics such as soil type, stream patterns, vegetation type, slope, and land use characteristics were obtained in the form of shapefiles from the Arizona Land Resource Information Service (ALRIS) and the Coconino National Forest. After gathering all watershed data, we used the Army Corp of Engineers surface runoff modeling system, HEC-HMS (Scharffenburg 2001). The surface runoff system is a software program capable of using multiple methods to determine surface runoff from a watershed using various infiltration, baseflow, and runoff calculations. We used the Natural Resource Conservation Service (NRCS; formerly Soil Conservation Service) rainfall-runoff model (McCuen 1982; NRCS 1986; Scharffenburg 2001) to determine the event-based surface runoff under two storm events in average moisture conditions. The NRCS model is categorized as a lumped based and empirical model (Scharffenburg 2001). We used the NRCS method because it is easy to apply, is most widely used by hydrologists, and was developed to evaluate downstream impacts from various management treatments (Woodward et al. 2002). The two events that we used to generate runoff were 25 and 100 yr return period storms covering the entire watershed area. We used the SCS type II 24 hr rainfall distribution pattern for the 25 and 100 yr storms, and the storm depths were determined to be 6.8 and 8.5 cm of precipitation respectively (U.S. Army Corp of Engineers 2000).

RESULTS

The results of this study are threefold. First, we have established the boundaries of the watersheds in a GIS format that has been derived using various methods. Second, we compared the watershed areas determined using our GPS method to the areas determined using traditional cartographic methods as well as to the areas calculated using the computer analysis of DEMs. We also compared the results of computer analysis to the traditional method of determining watershed areas. Third, we used the HEC-HMS surface-runoff software to determine differences between the various methods of delineating a watershed, in terms of modeling output.

Area Discrepancies

Figure 2 shows a section of the Bar M GIS layer created in this study using the GPS data overlaid with a GIS layer currently inuse by the USDA Forest Service. The differences between the two layers are minor in most instances but can easily reach a magnitude of 50 ha or more in some areas. These discrepancies also have a cumulative effect on the entire watershed. Even though DEMs used by government agencies, such as the USDA Forest Service or the USGS, are developed to National Map Accuracy Standards (Longley et al. 2001), the 10 m contour lines in Figure 2 fit our GPS layer arguably better than those currently in use by the Forest Service.

Another example of area discrepancy is given in Figure 3, which shows the difference between a layer we created using a 10 m DEM and our GPS layer. Here the GPS layer is very close to the traditionally determined boundary. However, according to the DEM layer, Woods Canyon should include a section of approximately 200 ha, which it does not. This discrepancy probably results from the DEM's failure to capture the less than 10 m elevation difference between the boundaries.

We encountered another major area difference while creating our DEM layers for sub-watershed 85. Here the DEM created two separate watersheds for the area that had been determined to be one watershed using the traditional methods (Figure 4). However, for the purpose of our area comparisons, we combined the area of these two watersheds into one.

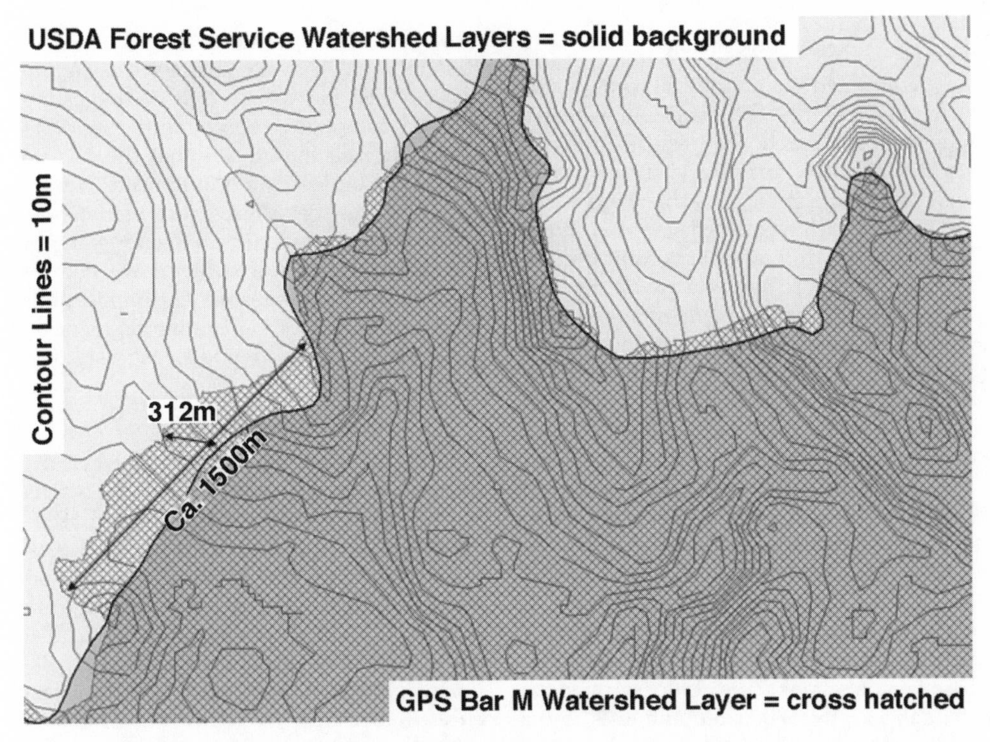

Figure 2. Comparison of a portion of the Bar M watershed with a USDA Forest Service GIS layer; 10 m contour interval derived from USGS digital elevation model.

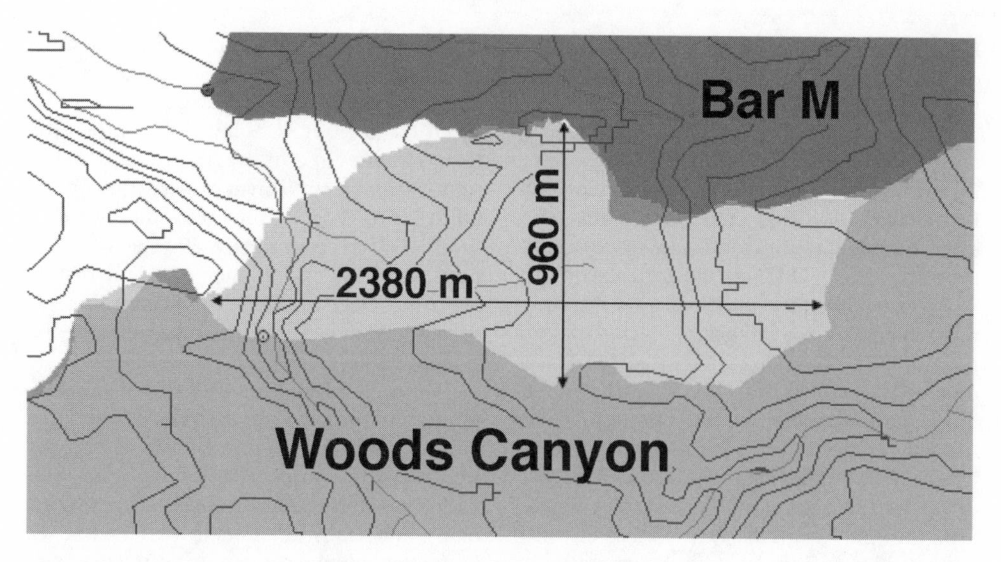

Figure 3. According to our 10 m DEMs, an area of ca. 200 ha should be included in our GPS layer of the Woods Canyon watershed. Even the Woods Canyon gauging station does not fall within the DEM layer.

Area Calculations

We determined the area for all nine watersheds using DEMs, GPS, and GIS technologies and compared the results with the traditional USDA Forest Service approach as well as with each other. Table 1 gives the total area of each watershed surveyed in this study. The net difference in percent is given in Table 2. Because our data are not normally distributed, we ran several nonparametric equivalent tests of a one sample repeated measurement design. Although the subwatersheds are nested within one watershed and some watersheds are adjacent to each other, we believe that our statistics are appropriate because the measurements are separate and watershed areas have a functional independence.

The Friedman procedure tests the null hypothesis that *k* variables come from the same population; in our case these variables are the watershed areas. The test statistic is based on these ranks. We also performed the Kendall's W, the Wilcoxon rank tests, and a multi-response permutation procedure for a block design (MRBP) based on Euclidean distance for unreplicated randomized block design (Table 3). Test results gave us significance values between .2 and .5, suggesting that we do not have enough information to reject the null hypothesis that there is no difference between methods used. Hence, we conclude that there is no statistically significant difference in area among any of the three methods. However, we believe there are several management implications to these differences, depending on management objectives and goals, as discussed below.

Surface Runoff Modeling

Although each method of watershed delineation produced varying areas, the results of the surface runoff modeling using HEC-HMS show a small difference in runoff peak discharges. For the 25 yr return period storm, differences in discharge ranged from 0.06 to 0.17 cubic meters per second (2–6 cubic feet per second) among all methods. For the 100 yr return period storm differences in discharge ranged from 0.23 to 0.53 cms (8–19 cfs) among all methods of area delineation. The difference in discharge was

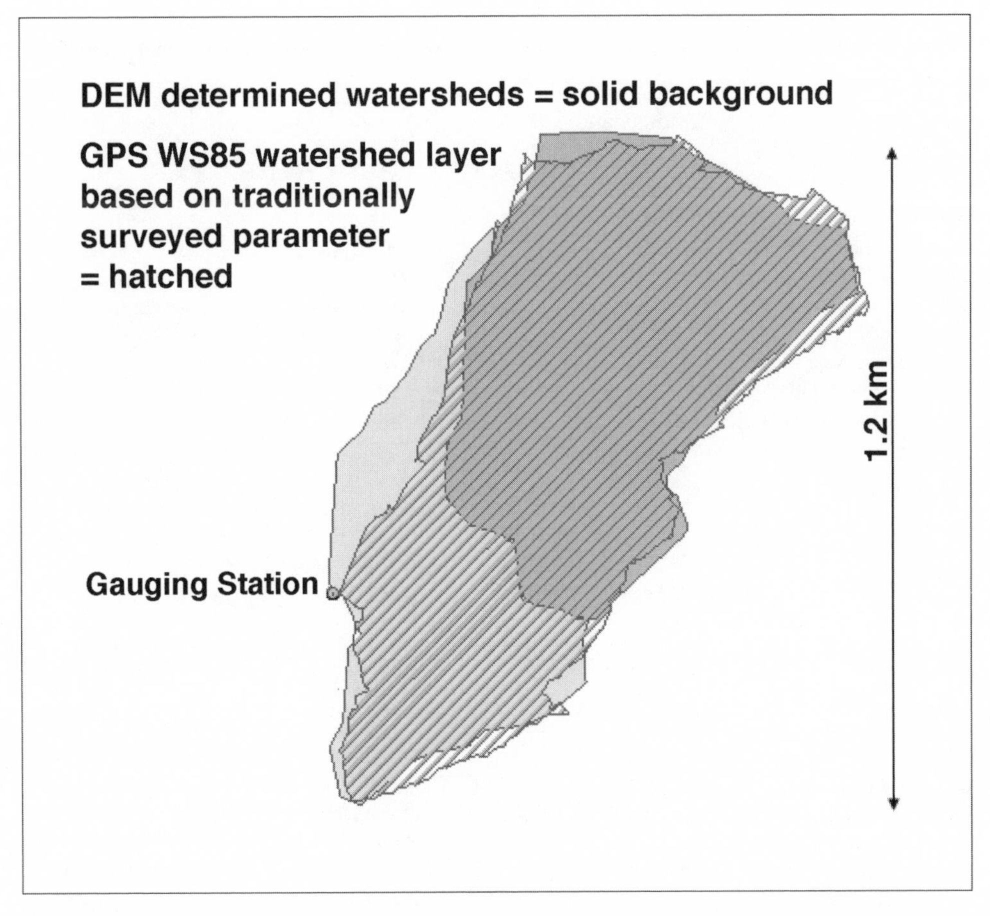

Figure 4. According to our 10 m DEMs, watershed 85 should be two separate watersheds, whereas traditional surveying techniques suggest one watershed.

primarily proportional to the watershed area. For example, watersheds 85, 86, and 87 produced discharge at the lower end of the above discharge ranges, whereas the Bar M, Woods Canyon, Castle Creek, and Thomas Creek watersheds produced discharge at the higher end of the discharge ranges.

DISCUSSION

The results in Tables 1 and 2 show that depending on the size of a watershed it may be advantageous to incorporate GPS technology into the watershed survey procedure. Using a GPS receiver allows for a faster, less complicated estimate when calculating the area of a watershed and determining its boundary. Training field users to collect uncorrected GPS data has a steep learning curve; the equipment is inexpensive and relatively easy to use compared to the traditional forestry field methods of using a compass and chain. Collecting GPS data allows ground truthing of watersheds and avoids adding additional area or leaving out area, which may be the case when using DEMs to determine watersheds. In our study the size of the errors seems relatively constant with respect to watershed size, so

Table 1. Watershed areas (ha) estimated using the traditional Forest Service approach, DEM, and GPS technologies.

Watershed	Traditionally Calculated Area	DEM Area	GPS Area
Bar M	6621	6641	6662
Woods Canyon	4893	5017	4884
Sub-watershed 85	70.4	64.2	61.5
Sub-watershed 86	40.9	40.9	40.5
Sub-watershed 87	23.5	26.1	22.3
Castle Creek East	458	456	467
Castle Creek West	354	398	364
Thomas Creek South Fork	221	236	230
Thomas Creek North Fork	184	185	191

Table 2. Net difference in percent between watershed areas estimated using the traditional Forest Service approach, DEM, and GPS technologies.

Watershed	Traditionally Calculated Area vs. DEM	Traditionally Calculated Area vs. GPS	GPS vs. DEM
Bar M	–0.31	–0.62	–0.31
Woods Canyon	–2.52	0.19	2.65
Sub-watershed 85	8.82	12.64	4.19
Sub-watershed 86	–0.07	0.98	1.05
Sub-watershed 87	–11.21	5.16	14.71
Castle Creek East	0.35	–1.86	–2.22
Castle Creek West	–12.39	–2.94	8.41
Thomas Creek South Fork	–6.58	–4.12	2.31
Thomas Creek North Fork	–0.54	–3.59	–3.03

Table 3. Test statistics indicating similarity between watershed areas estimated using the traditional Forest Service approach, DEM, and GPS technologies.

Test	p-Value	Traditional Method	DEM	GPS
MRBP	.228	6.98	3.49	10.69
Wilcoxon	.495	11.56	15.89	14.56
Friedman	.236	1.67	2.44	1.89
Kendall's W	.236	1.67	2.44	1.89

for large watersheds, these errors seem insignificant. On the other hand, any study using the Woods Canyon gauging station (Figure 3) would need a more precise measurement of watershed area, regardless of monetary significance; otherwise the data become meaningless. The greatest discrepancy we found was in sub-watershed 85. Here the DEM created two separate watersheds for the area that had been determined to be one watershed using the traditional methods. In this instance, which we assume to be a problem primarily with smaller scale watersheds, ground truthing is probably the most reliable technique for determining boundaries. Carrying a GPS receiver while doing so, in turn, is also a cost-effective method of recording and verifying a watershed boundary.

Another concern we have is that certain methods may include or exclude topography or area during delineation. Wu et al. (2003) have pointed out that not only size but also the relief of a watershed can affect the accuracy of its delineation, and they advise caution when delineating watersheds that are small with little or no relief. There may be a negative relationship between the steepness of the terrain (and therefore the watershed divides) and the relative error in measuring watershed boundaries.

The discrepancies in the Woods Canyon watershed delineation using the different methods (see Figure 3) are caused by the flat terrain, and suggest that more precise methods are needed in flat terrain, or in places where watershed divides are low. There may not be a significant correlation between overall watershed relief and the relief of the watershed divides at certain scales.

This study compares the use of the traditional topographic method, DEMs, and GPS to delineate watershed boundaries and derive their areas. However, future comparisons of watershed area should include absolute (rather than net) differences in watershed area size and their effects on watershed water balance. Although each method produced slightly different watershed areas, the difference in runoff modeling using the various watershed areas for model input seemed small. Even so, scientists and land managers need to use special care during the modeling process, and should be aware of how a watershed area was determined and what potential pitfalls exist according to the method of delineation.

ACKNOWLEDGMENTS

We thank Scott Walker and Steve Andariese of Northern Arizona University's Department of Geography and School of Forestry, respectively, for their technical help in the use of GIS; Jonathan Long and Ruihong Huang for their reviews; and Steve Overby and Geoff Seymour for their assistance in the field. Funding for this study was provided by Arizona State Prop. 301, the Bureau of Forestry Research, and the USDA McIntire-Stennis funding as well as the USDA Forest Service Rocky Mountain Research Station in Flagstaff, AZ.

REFERENCES

Baker, M. B., Jr., and P. F. Ffolliott. 1999. Interdisciplinary land use along the Mogollon Rim. In History of Watershed Research in the Central Arizona Highlands, compiled by M. B. Baker, Jr. USDA Forest Service General Technical Report RMRS-GTR-29. Ft. Collins CO.

Cho, S., and M. Lee. 2001. Sensitivity considerations when modeling hydrologic processes with digital elevation models. Journal of the American Water Resources Association 37(4): 931–934.

Dietrich, J. H. 1980. The composite fire interval—A tool for more accurate interpretation of fire history. In Proceedings of the Fire History Workshop, pp. 8–14. USDA Forest Service General Technical Report RM-81. Ft. Collins CO.

Elassal, A. A., and V. M. Caruso. 1983. USGS digital cartographic data standards: Digital elevation models. USGS Geological Survey Circular 895-B.

Environmental Systems Research Institute. 2002. ArcView 3x Computer Software. ESRI, Redlands CA.

Environmental Systems Research Institute. 2003. ArcView GIS. 20 pp. ESRI, Redlands CA.

Environmental Systems Research Institute. 2004. ArcGIS. 9 pp. ESRI, Redlands CA.

Ffolliott, P. F., and G. J. Gottfried. 1991. Mixed conifer and aspen regeneration in small clearcuts within a partially harvested Arizona mixed conifer forest. USDA Forest Service Research Paper RM-294. Ft. Collins CO.

Gottfried, G. J., and L. F. DeBano. 1988. Streamflow and water quality responses to preharvest prescribed burning in an undisturbed ponderosa pine watershed. In Effects of Fire Management of Southwestern Natural Resources, technical coordinator J. S. Krammes, pp. 222–228. USDA Forest Service General Technical Report RM-191. Fort Collins CO.

Karsky, D. 2001. Comparison of GPS receivers under a forest canopy after selective availability has been turned off. SuDoc A 13.137:0171-2809-MTDC. USDA Forest Service, Technology & Development Program.

Longley, P. A., M. F. Goodchild, D. J. Maguire, and D. W. Rhind. 2001. Geographic Information Systems and Science. John Wiley & Sons, New York.

Maidment, D. 1999. Watershed and stream network delineation using digital elevation models. Environmental Systems Research Institute. Available online at http://campus.esri.com/courses/hydrolgy/watshdhy. Last updated in 1999.

Maidment, D., and D. Djokic. 2000. Hydrologic and Hydraulic Modeling Support with Geographic Information Systems. Environmental Systems Research Institute, Redlands CA.

McCuen, R. H. 1982. A Guide to Hydrologic Analysis Using SCS Methods. 145 pp. Prentice-Hall, Englewood Cliffs NJ.

Natural Resource Conservation Service. 1986. National Engineering Manual, Section 4, Hydrology. U.S. Department of Agriculture, Beltsville MD.

Oderwald, R. G., and B. Boucher. 2003. GPS after selective availability: How accurate is accurate enough? Journal of Forestry 101(4): 24–27.

Scharffenburg, W. A. 2001. Hydrologic Modeling System HEC-HMS Users Manual, Version 2.1. U.S. Army Corp of Engineers Hydrologic Engineering Center, Davis CA. 18 pp.

Trimble. 2002. GPS Pathfinder Office: Getting Started Guide. Trimble Navigation Limited, Mapping and GIS Division, Sunnyvale CA.

U.S. Army Corp of Engineers. 2000. Rio de Flag, Flagstaff, Arizona Feasibility Report and Final Environmental Impact Statement. U.S. Army Corp of Engineers, Los Angeles.

Wilson, D. L. 2000. GPS horizontal position accuracy. Available online at http://users.erols.com/dlwilson/gpsacc.htm. Last accessed November 2003.

Woodward D. E., R. H. Hawkins, A. T. Hjelmfelt, J. A. Van Mullem, and Q. D. Quan. 2002. Curve Number Method: Origins, Applications and Limitations. Second Federal Interagency Hydrologic Modeling Conference, Las Vegas NV.

Wu Kansheng, Y. Jun Xu, W. E. Kelso, and D. A. Rutherford. 2003. Application of BASINS for water quality assessment on the Mill Creek Watershed in Louisiana. 1st Interagency Conference on Research in Watersheds, pp. 39–44. ARS.

CONTRIBUTORS

Ethan N. Aumack
Grand Canyon Trust
2601 N. Fort Valley Road
Flagstaff, AZ 86001
eaumack@grandcanyontrust.org

Janet R. Balsom
Grand Canyon National Park
P.O. Box 129
Grand Canyon, AZ 86023

David Barr
National Park Service
Flagstaff Area National Monuments
Flagstaff, AZ 86004

Don R. Bertolette
Science Center
Grand Canyon National Park
P.O. Box 129
Grand Canyon, AZ 86023-0129

Susan R. Boe
Research Branch
Arizona Game and Fish Department
2221 W. Greenway Road
Phoenix, AZ 85023

Mae Burnette
Watershed Program
White Mountain Apache Tribe
P. O. Box 2109
Whiteriver, AZ 85941
watershed@wmat.us

John A. Cannella*
Department of Geography, Planning, and Recreation
Northern Arizona University
* now with National Park Service
Flagstaff Parks and Monuments

James C. Catlin
Wild Utah Project
68 South Main Street
Salt Lake City, UT 84101

Marcelle Coder
Bilby Research Center
Northern Arizona University
Box 6013
Flagstaff, AZ 86011-6013
Marcelle.Coder@nau.edu

Kenneth L. Cole
USGS Colorado Plateau Research Station
Southwest Biological Research Station
P.O. Box 5614
Northern Arizona University
Flagstaff, AZ 96011
ken_cole@usgs.gov

D. Coleman Crocker-Bedford
Science Center
Grand Canyon National Park
P.O. Box 129
Grand Canyon, AZ 86023-0129

Stan C. Cunningham
Research Branch
Arizona Game and Fish Department
2221 W. Greenway Road
Phoenix, AZ 85023
scunningham@gf.state.az.us

Holly Demark
Department of Biology
Carroll College
100 North East Avenue
Waukesha, WI 53186

James C. deVos, Jr.
Arizona Game and Fish Department
Research Branch
2221 W. Greenway Road
Phoenix, AZ 85023

Jeri DeYoung
National Park Service
Flagstaff Area National Monuments
Flagstaff, AZ 86004

Brett G. Dickson
Department of Fishery and Wildlife Biology
Colorado State University
Fort Collins, CO 80523
and USDA Forest Service
Rocky Mountain Research Station
Flagstaff, AZ 86001
dickson@cnr.colostate.edu

Norris L. Dodd
Arizona Game and Fish Department
Research Branch
2221 West Greenway Road
Phoenix, AZ 85023

Charles Drost
U.S. Geological Survey
Southwest Biological Science Center
2255 N. Gemini Drive
Flagstaff, AZ 86001
charles_drost@usgs.gov

Shelli Dubay
Research Branch
Arizona Game and Fish Department
2221 W. Greenway Road
Phoenix, AZ 85023
dubaysa@michigan.gov

Don Duff
USDA Forest Service/Trout Unlimited Program
125 South State Street
Salt Lake City, UT 84138

J. Grace Ellis
Grand Canyon National Park
823 N. San Francisco St., Suite B
Flagstaff Arizona 86001
Grace_Ellis@nps.gov

Sarah Falzarano
Grand Canyon National Park
823 N San Francisco St.
Flagstaff, AZ 86001
Sarah_Falzarano@nps.gov

David Felley
U.S. Geological Survey
Southwest Biological Science Center
Colorado Plateau Research Station
Flagstaff, AZ

Heather L. Germaine
Research Branch
Arizona Game and Fish Department
2221 W. Greenway Road
Phoenix, AZ 85023

Stephen S. Germaine
Research Branch
Arizona Game and Fish Department
2221 W. Greenway Road
Phoenix, AZ 85023

Chad Gourley
Otis Bay Consultants, Inc.
9225 Cordoba Blvd.
Sparks, NV 89436

Timothy B. Graham
USGS-BRD
Canyonlands Field Station
Moab, UT 84532

Haydee M. Hampton
Center for Environmental Sciences and Education
Northern Arizona University
Flagstaff, AZ 86011-5694
haydee.hampton@nau.edu

Walter E. Hecox
Department of Economics and Sustainable Development Workshop
Colorado College
Colorado Springs, CO 80903

F. Patrick Holmes
Department of Economics and
Sustainable Development Workshop
Colorado College
Colorado Springs, CO 80903

Amy Horn
Grand Canyon National Park
P.O. Box 129
Grand Canyon, AZ 86023

Ian Hough
National Park Service
Flagstaff Area National Monuments
Flagstaff, AZ 86004
Ian_Hough@nps.gov

Allison L. Jones
Wild Utah Project
68 South Main Street
Salt Lake City, UT 84101

Zsuzsi Kovacs
Department of Biology
Northern Arizona University
Box 5640
Flagstaff, AZ 86011-5640

Duncan Leao
School of Forestry, NAU
3334 S. Litzler Drive
Flagstaff, AZ
dsl3@dana.ucc.nau.edu

Lisa M. Leap
Grand Canyon National Park
823 N. San Francisco St., Suite B
Flagstaff, AZ 86001

Paul Leatherbury
Science Center
Grand Canyon National Park
P.O. Box 129
Grand Canyon, AZ 86023-0129

Elaine F. Leslie
Canyon de Chelly National Monument
P.O. Box 588
Chinle, AZ 86053
elaine_leslie@nps.gov

Charles H. Lewis
Bureau of Indian Affairs
Branch of Roads
P.O. Box 10
Phoenix, AZ 85001

Jonathan Long
Rocky Mountain Research Station
U.S. Department of Agriculture
Forest Service
2500 S. Pine Knoll Dr.
Flagstaff, AZ 86001
jwlong@fs.fed.us

John Lowry
RSGIS Laboratory, UMC 4275
College of Natural Resources
Utah State University
Logan, UT 84322-5275
jlowry@gis.usu.edu.

Candy Lupe
Watershed Program
White Mountain Apache Tribe
P. O. Box 2109
Whiteriver, AZ 85941
clupe@wmat.us

Jennifer Mathis
Department of Biology
Carroll College
100 North East Avenue
Waukesha, WI 53186

Ted McKinney
Arizona Game and Fish Department
Research Branch
2221 W. Greenway Road
Phoenix, Arizona 85023

Taylor McKinnon
Grand Canyon Trust
2601 N. Fort Valley Road
Flagstaff, AZ 86001

Amber A. Munig
Arizona Game and Fish Department
Game Branch
2221 West Greenway Road
Phoenix, AZ 85023

Gary P. Nabhan
Center for Sustainable Environments
Northern Arizona University
Box 5765
Flagstaff, AZ 86011-5764
Gary.Nabhan@nau.edu

Daniel G. Neary
USDA Forest Service
Rocky Mountain Research Station
2500 S. Pine Knoll Drive
Flagstaf, AZ 86001-6381
dneary@fs.fed.us

L. Theodore Neff
MNA Environmental Solutions
3101 North Fort Valley Road
Flagstaff, AZ 86001
tneff@mna.mus.az.us

Erika M. Nowak
USGS Biological Resources Division
Southwest Biological Science Center
Colorado Plateau Research Station
Northern Arizona University
Flagstaff, AZ 86011
Erika_Nowak@nau.edu

David Ostergren, PhD.
School of Forestry
Center for Environmental Sciences and Education
Box 5694
Northern Arizona University
Flagstaff, AZ 86011
David.Ostergren@nau.edu

Louis L. Pech
Department of Biology
Carroll College
100 North East Avenue
Waukesha, WI 53186
lpech@cc.edu

Boris Poff
School of Forestry, NAU
1555 N. Fort Valley Road #18
Flagstaff, AZ 86001
bp22@dana.ucc.nau.edu

John. W. Prather
Center for Environmental Sciences and Education
Northern Arizona University
Flagstaff, AZ 86011
john.prather@nau.edu

Sarah E. Reed
Department of Environmental Science, Policy & Management
137 Mulford Hall #3110
University of California
Berkeley, CA 94720-3110
sreed@nature.berkeley.edu

Michael R. Robins
Navajo Nation Archaeology Department
Northern Arizona University
Flagstaff, AZ 86011

Tim Rogers
Research Branch
Arizona Game and Fish Department
2221 W. Greenway Road
Phoenix, AZ 85023

Elizabeth J. Ruther
Center for Environmental Sciences and Education
Box 5694
Northern Arizona University
Flagstaff, AZ 86011

Carmen L. Sipe
Science Center
Grand Canyon National Park
P.O. Box 129
Grand Canyon, AZ 86023-0129

Thomas D. Sisk
Center for Environmental Sciences and Education
Northern Arizona University
Flagstaff, AZ 86011-5694
thomas.sisk@nau.edu

Francis E. Smiley
Professor of Anthropology
Northern Arizona University
Flagstaff, AZ 86001

Susan Smith
Bilby Research Center
Northern Arizona University
Box 6013
Flagstaff, AZ 86011-6013
Susan.Smith@nau.edu

Mark K. Sogge
U.S. Geological Survey
Southwest Biological Science Center
Colorado Plateau Research Station
Flagstaff, AZ

Kimberly Spurr
Navajo Nation Archaeology Department
Northern Arizona University Branch Office
P.O. Box 6013
Flagstaff, AZ 86011
kimberly.spurr@nau.edu

Peter B. Stacey
Department of Biology
University of New Mexico
Albuquerque, NM 87131

Lawrence E. Stevens
Grand Canyon Wildlands Council
P.O. Box 1315
Flagstaff, AZ 86002
(928) 774-4923
farvana@aol.com

Aregai Tecle
School of Forestry
Southwest Forest Science Complex
Box 15018
Northern Arizona University
Flagstaff, AZ 86011
Aregai.Tecle@nau.edu

Kathryn Thomas
USGS Southwest Biological Science Center
Colorado Plateau Research Station
P.O. Box 5614
Flagstaff, AZ 86011-5614
Kathryn_A_Thomas@usgs.gov.

John L. Vankat
Science Center
Grand Canyon National Park
and Department of Botany
Miami University
Oxford OH 45056

Brian Wakeling
Game Branch
Arizona Game and Fish Department
2221 W. Greenway Road
Phoenix, AZ 85023
BWakeling@gf.state.az.us

Mark Wotawa
National Park Service
Natural Resource Information Division
Inventory and Monitoring Program
Fort Collins, CO

Yaguang Xu
Center for Environmental Sciences and
Education
Northern Arizona University
Flagstaff, AZ 86011
yaguang.xu@nau.edu

INDEX

C

D

E

S